Springer Proceedings in Mathematics & Statistics

Volume 308

Springer Proceedings in Mathematics & Statistics

This book series features volumes composed of selected contributions from workshops and conferences in all areas of current research in mathematics and statistics, including operation research and optimization. In addition to an overall evaluation of the interest, scientific quality, and timeliness of each proposal at the hands of the publisher, individual contributions are all refereed to the high quality standards of leading journals in the field. Thus, this series provides the research community with well-edited, authoritative reports on developments in the most exciting areas of mathematical and statistical research today.

More information about this series at http://www.springer.com/series/10533

Santanu Manna · Biswa Nath Datta ·
Sk. Safique Ahmad
Editors

Mathematical Modelling and Scientific Computing with Applications

ICMMSC 2018, Indore, India, July 19–21

Springer

Editors
Santanu Manna
Discipline of Mathematics
Indian Institute of Technology Indore
Indore, Madhya Pradesh, India

Biswa Nath Datta
Department of Mathematical Sciences
Northern Illinois University
DeKalb, IL, USA

Sk. Safique Ahmad
Discipline of Mathematics
Indian Institute of Technology Indore
Indore, Madhya Pradesh, India

ISSN 2194-1009 ISSN 2194-1017 (electronic)
Springer Proceedings in Mathematics & Statistics
ISBN 978-981-15-1337-4 ISBN 978-981-15-1338-1 (eBook)
https://doi.org/10.1007/978-981-15-1338-1

Mathematics Subject Classification (2010): 74A10, 74B05, 74F10, 76S05, 65J05, 65N06, 15A69, 35L35, 93B51, 93B20

This Springer imprint is published by the registered company Springer Nature Singapore Pte Ltd.
The registered company address is: 152 Beach Road, #21-01/04 Gateway East, Singapore 189721, Singapore

Conference Organizing Committee

Organizing Chairs

Dr. Santanu Manna, IIT Indore, India
Dr. Sk. Safique Ahmad, IIT Indore, India
Dr. Shanmugam Dhinakaran, IIT Indore, India
Dr. Niraj Kumar Shukla, IIT Indore, India

Program Chairs

Dr. Antony Vijesh, IIT Indore, India
Dr. Swadesh Kumar Sahoo, IIT Indore, India
Dr. Md. Aquil Khan, IIT Indore, India
Dr. M. Tanveer, IIT Indore, India
Dr. Parimal Kar, IIT Indore, India

Publicity Chairs

Dr. Sanjeev Singh, IIT Indore, India
Dr. Vijay Kumar Sohani, IIT Indore, India
Dr. Rajesh Sharma, NIT Hamirpur, India
Dr. Bidyasagar Kumbhakar, NIT Meghalaya, India
Dr. Nilesh Kumar Thakur, NIT Raipur, India
Mr. Somak Das, Auburn University, USA

International Advisory Committee

Prof. Ravi P. Agarwal, Texas A&M University–Kingsville, USA
Prof. S. Chakraborti, University of Alabama, USA
Prof. Biswa N. Datta, Northern Illinois University, USA
Prof. Herbert Huppert, University of Cambridge, UK
Prof. Anuar Ishak, Universiti Kebangsaan Malaysia, Malaysia
Prof. Julius Kaplunov, Keele University, UK
Prof. Volker Mehrmann, Technische Universität Berlin, Germany
Prof. Atma R. Sahu, University System of Maryland, USA
Prof. Kumar Singh, Miami University, USA
Prof. Ivan Slapnicar, University of Split, Croatia
Prof. Shan Zhao, University of Alabama, USA

National Advisory Committee

Prof. Rafikul Alam, IIT Guwahati, India
Prof. T. Amarnath, University of Hyderabad, India
Prof. Shreemayee Bora, IIT Guwahati, India
Prof. Rama Bhargawa, IIT Roorkee, India
Prof. A. Chattopadhyay, IIT Dhanbad, India
Prof. Joydev Chattopadhyay, ISI Kolkata, India
Prof. B. V. Rathish Kumar, IIT Kanpur, India
Prof. J. C. Misra, IIT KGP/IIEST Shibpur, India
Prof. A. K. Nandakumaran, IISc Bangalore, India
Prof. Souymendu Raha, IISc Bangalore, India
Prof. Mythily Ramaswamy, TIFR Bangalore, India
Prof. Govindan Rangarajan, IISc Bangalore, India
Prof. S. Sundar, IIT Madras, India
Prof. R. K. Upadhyay, IIT Dhanbad, India

Local Advisory Committee

Dr. Kapil Ahuja, IIT Indore, India
Prof. Arup Banerjee, RRCAT, Indore, India
Dr. Sandeep Chaudhary, IIT Indore, India
Dr. Biplab Ghosh, BARC/RRCAT, Indore, India
Dr. Hem C. Jha, IIT Indore, India
Dr. Ranjan Kumar Jana, SVNIT, Surat, India
Dr. M. Ashok Kumar, IIT Indore, India

Prof. Krushna R. Mavani, IIT Indore, India
Prof. Suman Mukhopadhyay, IIT Indore, India
Prof. R. B. Pachori, IIT Indore, India
Prof. K. R. Pardasani, MNIT Bhopal, India
Dr. Anand Parkash, IIT Indore, India
Prof. K. N. Rajeshwari, Devi Ahilya University, India
Dr. Shovan K. Majumder, RRCAT, Indore, India
Dr. Mahesh Dumaldar, DAVV, Indore, India

Technical Program Committee

Prof. Kumar Singh, Miami University, USA
Prof. Souymendu Raha, IISc Bangalore
Prof. César Elizondo-González, Universidad Autónoma de Nuevo León
Prof. S. Sundar, IIT Madras
Prof. Le Su, Institute of Computing Technology, Chinese Academy of Sciences, China
Prof. Gauri Shanker Seth, IIT (ISM) Dhanbad
Dr. Ludmila Prikazchikova, Keele University, UK
Dr. Rashmi Agrawal, SR Engineering College, Warangal
Dr. Shailesh Kundalwal, IIT Indore
Dr. Divya Singh, NIT Rourkela
Dr. Nijjwal Karak, University of Jyvaskyla, Finland
Dr. Sabitri Majhi, NIT Jamshedpur
Prof. Fouzi Harrou, Texas A&M University, Qatar
Dr. Jitendra Kumar Singh, VSK University, Bellary
Dr. Sanjeev Singh, IIT Indore
Dr. Vijay Kumar Sohani, IIT Indore
Dr. Niraj K. Shukla, IIT Indore
Dr. Nilesh Kumar Thakur, NIT Raipur
Dr. Antony Vijesh, IIT Indore
Dr. Bidyasagar Kumbhakar, NIT Meghalaya
Dr. Parimal Kar, IIT Indore
Dr. Rajat Tripathi, Government Engineering College, Rewa
Prof. Mahesh N. Dumaldar, Devi Ahilya University
Dr. Sunil Kumar, NIT Jamshedpur
Dr. Sharada Nandan Raw, NIT Raipur
Dr. Sourav Mandal, CSIR-National Institute of Oceanography, Goa
Dr. Subharthi Sarkar, IIT Bhubaneswar
Dr. Sukhendu Ghosh, SRM Institute of Science and Technology, Chennai
Dr. Chirodeep Bakl, IIT Ropar
Dr. Sharad Dwivedi, SRM Institute of Science and Technology, Chennai
Dr. Abhishek Kumar Singh, VIT University, Chennai

Dr. Santhakumar Mohan, IIT Indore
Dr. Manikandan Rangaswamy, Central University of Kerala
Dr. Ranjan Kumar Jana, SVNIT, Surat
Dr. Pradip Roul, VNIT Nagpur
Dr. M. Tanveer, IIT Indore
Dr. Vineet K. Srivastava, ISTRAC/ISRO, Bangalore
Prof. Dilip Maiti, Vidyasagar University
Dr. Santanu Manna, IIT Indore, India
Dr. Sk. Safique Ahmad, IIT Indore, India
Dr. Shanmugam Dhinakaran, IIT Indore, India
Dr. Balasubramaniam Jayaram, IIT Hyderabad
Dr. Sumit Kumar Vishwakarma, BITS Hyderabad
Prof. Rama Bhargawa, IIT Roorkee
Prof. Ivan Slapničar, University of Split
Prof. Kamal Raj Pardasani, MNIT Bhopal
Dr. S. K. Tiwari, BIT Sindri
Dr. G. C. S. Yadav, University of Allahabad
Dr. Shiva Mittal, SPM College, University of Allahabad
Dr. Pooja Singh, Rajkiya Engineering College, Banda
Dr. Aparna Vyas, Chung-Ang University
Prof. S. K. Upadhyay, IIT BHU, Varanasi, India
Dr. Rajesh Pandey, IIT BHU, Varanasi, India
Dr. Mani Mehra, IIT Delhi, India
Dr. Dilip Maiti, Vidyasagar University, West Bengal, India
Dr. Ameeya Nayak, IIT Roorkee, Uttarakhand, India
Dr. Perumal, NIT Surathkal, Karnataka, India
Dr. Krishna Seshagiri, IIT Palakkad, Kerala, India
Dr. Sunil Arolla, IIT Palakkad, Kerala, India
Dr. Manish Mishra, IIT Roorkee, Uttarakhand, India
Dr. P. A. Lakshmi Narayana, IIT Hyderabad, Telangana, India
Dr. S. Kanagaraj, IIT Guwahati, Guwahati, India
Dr. S. K. Sahu, IIT Indore, Madhya Pradesh, India
Dr. Ritunesh Kumar, IIT Indore, Madhya Pradesh, India
Dr. Shyama Prasad Das, IIT Tirupati, Andhra Pradesh, India
Dr. B. S. V. Patanaik, IIT Madras, Tamil Nadu, India
Dr. Supradeepan, BITS Hyderabad, Telangana, India
Dr. Anand Rajkumar, St. Xavier's Catholic College of Engineering, Tamil Nadu, India
Dr. Pratiba Biswal, Shiv Nadar University, Uttar Pradesh, India
Dr. V. M. Rajesh, Shiv Nadar University, Uttar Pradesh, India
Dr. Dhiraj Kumar Garg, Shiv Nadar University, Uttar Pradesh, India
Dr. A. Salih, Indian Institute of Space Science and Technology, Kerala, India
Dr. Ragavendra Gupta, IIT Guwahati, Guwahati, India
Dr. Swati Srivastava, SPM College, University of Allahabad, Allahabad, India

List of Sponsors with Logos

Science and Engineering Research Board (Department of Science and Technology)

National Board of Higher Mathematics (Department of Atomic Energy)

Council of Scientific and Industrial Research

Preface

This book is an outcome of the selected papers presented at the International Conference on Mathematical Modelling and Scientific Computing (ICMMSC 2018) held at IIT Indore during July 19–21, 2018. This highly interdisciplinary conference brought together applied mathematicians and computational scientists with researchers and engineers in several areas of engineering to have an effective exchange of ideas for mutual benefits.

The most important task for attaining a practical solution for any real-life problem arising in science and engineering is to make a good and accurate mathematical model of that problem. Once this model is designed, it is important to numerically solve the mathematical model using appropriate, computationally viable algorithms and the associated high-quality mathematical software, or by developing new suitable algorithms and software, if required. This task is commonly referred to as scientific computing. Thus, this book forms an integral part of research and development in almost all branches of applied sciences, engineering, and technology.

The following types of papers are intended and included in this book: the state-of-the-art review papers or research monograph based on the keynote and plenary talks, given by well-known experts from several areas of science and engineering of interests to the conference; a few highly selected and critically refereed technical papers, based on contributed presentations; and a few short papers, based on ongoing but promising research. In addition to the scientific paper contribution, we also discussed a panel discussion on how to minimize the gap between mathematicians and engineers on the last day (July 21, 2018) of the conference.

The book will be of interests to a wide variety of researchers, students, and the practicing engineers working in diverse areas of science and engineering, ranging from applied and computational mathematics, vibration problem, computer science, and numerical optimization to physics, chemistry, biology, electrical, civil, mechanical, chemical, seismology, aerospace, and medical sciences. Some specific areas of interests whose readership will be highly benefited by this book include numerical linear algebra, control theory, mechanical vibration, signal and image processing, wave propagation, computer vision, computer photography, computational fluid dynamics, heat and mass transfer, biomedical engineering, computational physics, and aeronautics.

The book will serve as a valuable reference book for scientists and engineers working in different areas of science and engineering, and as a textbook for several types of advanced interdisciplinary courses blending science, mathematics, and engineering. The presence of such a book will simulate research and education in science and engineering.

Indore, India
July 2018

Santanu Manna
Biswa Nath Datta
Sk. Safique Ahmad

Acknowledgements

We would like to express our gratitude to all the researchers who provided their constant support and dedication to publish this book as conference proceedings for the International Conference on Mathematical Modelling and Scientific Computing (ICMMSC 2018). We would also like to thank our colleagues who supported and encouraged us in spite of all the time it took us away from them. We would like to thank all the authors for their contributions made to this book. We also wish to thank the Springer team for their continuous assistance and support in the publication of this book.

We express our sincere gratitude to ICMMSC 2018 speakers, reviewers, Technical Program Committee members, International and National Advisory Committee, program and publicity chairs, Local Organizing Committee, and institute administration, without whose support the quality and standard of the conference could not have been maintained.

We thank all participants who had presented their research papers and attended the conference. A special mention of thanks is due to our student volunteers for the spirit and enthusiasm they showed throughout, without which it would have been difficult for us to organize such a successful event.

Talks Given at ICMMSC 2018

Biswa Nath Datta	Computational and Optimization Methods for Quadratic Inverse Eigenvalue Problems Arising in Mechanical Vibration and Structural Dynamics
Volker Mehrmanna	Energy Based Modeling, Simulation, and Control of Coupled Systems
Uday Banerjee	Generalized Finite Element Method: Successes and Challenges
Kumar Singh	Active Control with the Method of Receptances: Recent Progresses and its Application in Active Aeroelastic Control
Ranjit Kumar Upadhyay	Modeling Chaotic Systems and Spatiotemporal Transmission Dynamics of Recent Ebola Spread and Outbreak
Santosh Kapuria	Mixed-Field Multiterm Extended Kantorovich Method for Accurate Prediction of Edge Stresses in Piezothermoelastic Laminates
Rafikul Alam	Fiedler and Generalized Fiedler Pencils for Rational Eigenvalue Problems
Govindan Rangarajan	A Mathematical Model for Storage and Recall of Images in the Human Brain
Mythily Ramaswamya	Control of Some Fluid Models
Soumyendu Rahaa and Saurabh Dixita	Data Assimilation by Directly Projecting the Nominal Dynamics on the Data Manifold
Rama Bhargav	Meshfree Methods: Limitations and Applications
J. C. Misra	Mathematical Modeling of Electro-osmotic Flows in Physiological Systems

J. Prakash	Heat and Mass Transfer Effects on MHD Nanofluid Flow Over a Semi Infinite Flat Plate Embedded in a Porous Medium and Two-phase MHD Flow and Heat Transfer in a Horizontal Channel
Debasis Chakraborty	Numerical Methods in Aeropropulsive Characterization of Flight Vehicles
Mani Mehra	Extension of Wavelets to Topologically Complicated Domains
Muslim Malika	Controllability of Fractional Differential Equations with Non-instantaneous Impulses
Gajendra K. Vishwakarma	Bio-marker Prediction using Bayesian State-Space Modeling in Gene Expression Data Analysis
Punit Sharma and Nicolas Gillis	Computing Nearest Stable Matrices

Abstracts of the Talks Given at ICMMSC 2018

Computational and Optimization Methods for Quadratic Inverse Eigenvalue Problems Arising in Mechanical Vibration and Structural Dynamics

Biswa Nath Datta

The quadratic eigenvalue problem is to find eigenvalues and eigenvectors, a quadratic matrix pencil of the form $P(\lambda) = M\lambda^2 + C\lambda + K$ where the matrices M, C and K are square matrices. Unfortunately, the problem has not been widely studied because of the intrinsic difficulties solving the problem in a numerically effective way. Indeed, the state-of-the-art computational techniques are capable of computing only a few extremal eigenvalues and eigenvectors, especially if the matrices are large and sparse, which is often the case in practical applications. The inverse quadratic eigenvalue problem, on the other hand, refers to constructing the matrices M, C and K given the complete or partial spectrum and the associated eigenvectors. The inverse quadratic eigenvalue problem is equally important and arises in a wide variety of engineering applications, including mechanical vibrations, aerospace engineering, design of space structures, and structural dynamics. Of special practical importance is to construct the coefficient matrices from the knowledge of the only partial spectrum and the associated eigenvectors. The greatest computational challenge is to solve the partial quadratic inverse eigenvalue problem using the small number of eigenvalues and eigenvectors which are all that are computable using the state-of-the-art techniques. Furthermore, computational techniques must be able to take advantage of the exploitable physical properties, such as the symmetry, positive definiteness, and sparsity, which are computational assets for solution of large and sparse problems. This talk will deal with two special quadratic inverse eigenvalue problems that arise in mechanical vibration and structural dynamics. The first one, quadratic partial eigenvalue assignment problem (QPEVAP), arises in controlling dangerous vibrations in mechanical structures. Mathematically, the problem is to find two control feedback matrices such that a small amount of the eigenvalues of the associated quadratic eigenvalue problem, which are responsible for dangerous vibrations, is reassigned to suitably chosen ones while keeping the remaining large number of eigenvalues and eigenvectors unchanged. Additionally, for robust and economic control design, these feedback matrices must be found in such a way that they have the norms as small as possible and the condition number of the modified quadratic inverse problem is minimized. These considerations give rise to two nonlinear unconstrained optimization problems, known, respectively, as robust quadratic partial eigenvalue assignment problem (RQPEVAP) and minimum norm quadratic partial eigenvalue assignment problem (MNQPEVAP). The other one, the finite element model updating problem (FEMUP), arising in the design and analysis of structural dynamics, refers to updating an analytical finite element model so that a set of measured eigenvalues and eigenvectors from a real-life structure are reproduced and the physical and structural properties of the original model are preserved. A properly updated model can be used in confidence for future designs and constructions. Another major application of FEMUP is the damage detections in structures. Solution of FEMUP also gives rise to several constrained nonlinear optimization problems. I will give an overview of the recent developments of computational methods for these difficult nonlinear optimization problems and discuss directions of future research with some open problems for future research. The talk is interdisciplinary in nature and will be of interests to computational and applied mathematicians, and control and vibration engineers and optimization experts.

Energy Based Modeling, Simulation, and Control of Coupled Systems

Volker Mehrmanna

Motivated by modeling modern energy and transport networks, in particular those from different physical domains, the modeling framework of port-Hamiltonian systems is discussed. The classical port-Hamiltonian approach is systematically extended to constrained dynamical systems (partial differential algebraic equations). A new algebraically and geometrically defined system structure is derived, which has many nice mathematical properties. It is shown that this structure is invariant under Galerkin projections and changes of basis, and that a dissipation inequality holds and it is automatically stable and passive. Furthermore, the new representation is very robust to perturbations in the system structure. Using this structure leads to many advantages for tasks like model reduction or large-scale simulation. There exist, however, many open problems associated with port-Hamiltonian systems, which will also be discussed.

Generalized Finite Element Method: Successes and Challenges

Uday Banerjee

The generalized finite element method (GFEM) is used to approximate non-smooth solutions of PDEs associated with many problems of engineering interest, e.g., problems involving voids, inclusions, crack propagation, and interface problems. The GFEM is an extension of the standard finite element method (FEM), where the standard piecewise polynomial trial space of the FEM is augmented by a suitably chosen, problem-dependent, enrichment space that often contains non-polynomial functions. This method (i) allows the use of a simple mesh that does not conform to the features of the underlying problem and (ii) yields optimal order of convergence. However, the GFEM yielding the optimal order of convergence could be badly conditioned for certain choices of the enrichment functions; it could be difficult to solve the underlying linear system.

In this talk, we will present sufficient conditions on the enrichment space that will guarantee that the conditioning of the GFEM is not worse than that of the standard FEM and the conditioning is robust with respect to the position of the mesh (relative to the features of the underlying problem). A stably well-conditioned GFEM is called the stable generalized finite element method (SGFEM). We will present a simple local procedure that changes the GFEM into an SGFEM for a certain class of problems. However, the choice of enrichment functions is non-trivial as this local procedure may not work for other problems. We will illuminate these ideas through two specific applications.

Active Control with the Method of Receptances: Recent Progresses and its Application in Active Aeroelastic Control

Kumar Singh

The design of active control for large and complex engineering structures requires accurate modeling and prediction of their dynamic response and instabilities. The performance of traditional model-based control may be limited due to the errors in model approximation, size of the problem, and/or availability of limited data for realizing active control. To overcome some of these challenges, the method of receptances is developed. This method allows design and computation of controller gains based on modest size of receptance matrices which can be extracted from transfer functions associated with available sensors and actuators. The area of active aeroelastic control deals with developing wing technology for the next-generation aircraft to achieve increased performance by controlling and manipulating the aeroelastic response by active means. In these applications, receptance-based controller design is found to be promising as it eliminates the modeling of complex aeroelastic interaction between elastic structure and surrounding aerodynamics. In this talk, fundamental of the receptance-based control is introduced and recent progress in this area is highlighted. The effectiveness of the controller designed with onboard sensors (embedded) and actuators (control surfaces) for suppressing the flutter instabilities and flutter boundary extension is demonstrated with numerical examples. The performance of the controller such as its ability to control prescribed modes of interest without influencing the other modes including the actuator modes is also presented. Ongoing research in this area and potential for future research are briefly summarized.

Modeling Chaotic Systems and Spatiotemporal Transmission Dynamics of Recent Ebola Spread and Outbreak

Ranjit Kumar Upadhyay

The concept of chaos is one of the major discoveries of the twentieth century. It (deterministic chaos) has been studied extensively in various fields; here, I will discuss a few important areas. In spite of abundant past and current work in this direction, there is still no broadly applicable, convenient, generally accepted definition of the term chaos. In this talk, I will discuss different definitions of (opinion about) chaos, including my own opinion which can be applied very generally to situations that are commonly encountered, including attractors, repellers, periodically and non-periodically forced systems. I will also discuss some related concepts like robust chaos, ecological vs mathematical chaos, crisis, edge of chaos, and wave of chaos and examine the observability of these phenomena in eco-epidemiological systems. I will present the development of a model epidemic and eco-epidemic systems (both spatial and temporal) based on sound ecological principle and discuss why and how the model was developed. What is special about it and what kind of questions of ecological interest can be answered with the help of this model system? Different types of paradoxes in ecology and its resolution. Next, I will discuss my recent work on spatiotemporal transmission dynamics of recent Ebola spread and outbreak in West Africa. In this talk, I will try to explain the modeling and simulation for the virus dynamics to better understand and characterize the transmission trajectories of the Ebola outbreak. I will present the development of a model epidemic and eco-epidemic systems (both spatial and temporal) based on sound ecological/epidemiological principle and discuss why and how the model was developed. What is special about it and what kind of questions of epidemiological interest can be answered with the help of this model system? To understand the Ebola transmission dynamics, we formulate a compartmental epidemic model with exponentially decaying transmission rates and study the impact of control measures on basic public health using an SEIR and SEIRHD models. We compare the simulated results with the most recent reported data of Ebola-infected cases in the three most affected countries Guinea, Liberia, and Sierra Leone. I will discuss my recent work on spatiotemporal transmission dynamics of recent Ebola spread and outbreak in West Africa. The proposed Ebola epidemic model provides an estimate to the potential number of future cases. Two-dimensional simulation experiments show that infectious population and the number of deaths in Sierra Leone will increase up to one and a half year without control, but it will decline after two years.

Mixed-Field Multiterm Extended Kantorovich Method for Accurate Prediction of Edge Stresses in Piezothermoelastic Laminates

Santosh Kapuria

Composite and sandwich laminates are widely used in high-performance lightweight structures in several applications such as aircraft wings, marine propeller blades, wind turbine blades, spacecraft, and robotic arms. Such laminated structures also provide the opportunity to accommodate surface-bonded or embedded smart piezoelectric transducer layers to introduce self-sensing and actuation capabilities. However, the material discontinuity at the interfaces of adjacent dissimilar layers and geometric discontinuity at the free edges of such laminated structures give rise to development of large out-of-plane stresses in the vicinity of free edges (free edge effect) under thermo-electro-mechanical loadings, which are known to be the prime reason for initiation of delamination damage in these laminates. Hence, accurate prediction of free edge stress field, which is truly three-dimensional (3D) in nature, is essential for designing such structures and to ensure its integrity. Since the 3D elasticity-based exact analytical solution for the free edge problem is not known, there has been a continuous effort to develop various numerical and approximate analytical solutions for the problem, starting from the finite difference method-based solution of Pipes and Pagano (1970). The extended Kantorovich method (EKM) is a powerful iterative semi-analytical method for solving partial differential equations (PDEs), which was proposed by Kerr in 1968 based on a solution process given by Kantorovich and Krylov in 1936. The author group has recently generalized the method by developing the mixed-field multiterm extended Kantorovich method (MMEKM) to obtain 3D elasticity solutions of laminated plates. In this method, both displacement and stress variables are considered as field variables and are represented as n-term series of the product of separable functions in the inplane and thickness directions. The iterative solution process alternatively considers the functions in one direction as known (from initial guess or previous iteration) and transforms the PDEs into set of ordinary differential equations (ODEs) for the functions in the other direction, using the Reissner-type mixed variational principle. The ODEs are solved analytically. The iterations over the two directions are continued until the desired convergence is achieved, which generally happens very rapidly. The mixed-field approach ensures satisfaction of all boundary conditions at the free edges and interlaminar continuity conditions exactly and pointwise, and also yields the same order of accuracy for both displacements and stresses. The MMEKM is employed here to obtain accurate 3D piezothermoelasticity solution of the free edge stress field in infinitely long laminated panels under extension, bending, twisting, electric potential, and thermal loadings. The imperfect interlaminar bonding is modeled using the linear spring-layer model, which is incorporated in the variational framework. The numerical results exhibit rapid convergence with respect to the number of terms and iterations and excellent accuracy for a variety of laminate configurations under all loading conditions. The solution is general and is applicable to symmetric, unsymmetric, cross-ply and angle-ply hybrid laminates having surface-bonded or embedded piezoelectric layers. The solution also successfully captures the singular nature of free edge interlaminar stresses in elastic and hybrid laminates under various loadings. The results show significant effect of electromechanical and thermoelectric coupling on the free edge stresses in hybrid laminates. Finally, we show how the free edge stresses due to extension, bending, twisting, and thermal loads can be controlled by applying an appropriate actuation potential.

Fiedler and Generalized Fiedler Pencils for Rational Eigenvalue Problems

Rafikul Alam

Linearization is a standard method for computing eigenvalues, eigenvectors, minimal bases, and minimal indices of matrix polynomials. Frobenius companion pencils are examples of linearizations of matrix polynomials. Fiedler and generalized Fiedler companion pencils have been introduced recently which provide structured and unstructured linearizations of matrix polynomials. Rational eigenvalue problems arise in many applications, and the computation of poles and zeros of rational matrix functions is an important task. For computing eigenvalues, eigenvectors, poles, minimal bases, and minimal indices of rational matrix functions, we introduce Fiedler and generalized Fiedler companion pencils for rational matrix functions and show that these pencils are linearizations of the rational matrix functions in an appropriate sense. We describe the recovery of minimal bases and minimal indices of rational matrix functions from those of the Fiedler and generalized Fiedler pencils.

A Mathematical Model for Storage and Recall of Images in the Human Brain

Govindan Rangarajan

We propose a mathematical model for storage and recall of images in the human brain using coupled maps. We start by theoretically investigating targeted synchronization in coupled map systems, wherein only a desired (partial) subset of the maps is made to synchronize. A simple method is introduced to specify coupling coefficients such that targeted synchronization is ensured. The principle of this method is extended to storage/recall of images using coupled Rulkov maps that are often used to model neuronal networks in the brain. The process of adjusting coupling coefficients between Rulkov maps to store (display) a desired image mimics the process of adjusting synaptic strengths between neurons to store memories. Thus, our method uses both synchronization and synaptic weight modification. The stored image can be recalled by providing an initial random pattern to the dynamical system. The storage and recall of the standard image of Lena are explicitly demonstrated.

Control of Some Fluid Models

Mythily Ramaswamya

After introducing controllability (reaching a desired state in finite time) and stabilizability (reaching a steady state as time tends to infinity), for ODE systems, I will discuss these issues for some PDE systems, arising in fluid models.

Data Assimilation by Directly Projecting the Nominal Dynamics on the Data Manifold

Soumyendu Rahaa and Saurabh Dixita

We describe a differential algebraic equation (DAE) approaches to data assimilation by augmenting the dynamical system model with an algebraic variable and by treating the observed data as a constraint manifold. An additional parameter is analyzed with respect to the state variables and the time variable to study its dynamical and noise characteristics. The augmented dynamics are then used to filter and forecast the state variables. As an application of this approach, the heat equation for a rod is modeled on a graph with the algebraic variable. Together with the measurements, the system is posed as a semi-explicit index-2 DAE. The algebraic variable captures the uncertainty in the measurement data with respect to temperature and time. This is interpreted as the unmodeled dynamics due to the inhomogeneities in the thermal diffusivity of the rod. Further statistical analysis is done to characterize the noisy behavior of the algebraic variable. The nominal dynamics are then modified with the uncertainty quantified with the algebraic variable which is now expressed as a random variable that is used to provide sample paths for the prediction or forecast simulations.

Meshfree Methods: Limitations and Applications

Rama Bhargav

Mathematical modeling provides a tool for conceiving and resolving problems in the real world, ranging from everyday numeracy level to sophisticated equipments or research. They have great potential for multidisciplinary work including designers, manufacturers (CAD and CAM), and academic endeavors. This helps in crossing the boundaries between students of mathematics to the world outside the classroom. These models are usually governed either by optimization approach or by a type of differential equations. For solving these complicated systems, one has to deal with the numerical approach, maybe grid-based or grid tree methods. Various grid-based methods, including finite difference, finite element, and boundary element, sometimes feel themselves limited where the domain is undefined or dealing with phase change or highly irregular domain. In the past two decades, meshfree methods have emerged into a new class of computational methods which successfully deal with such problems. The present lecture will highlight an overview of the development, and type and computational approach of these methods. Given the proper treatment, how these methods can be made more efficient eliminating the existing bottleneck of computational expenses in these methods will also be discussed.

Mathematical Modeling of Electro-osmotic Flows in Physiological Systems

J. C. Misra

Electro-osmotic flows of physiological fluids in micro-vessels will be discussed during the lecture. After explaining the basic concepts and the underlying assumptions, an attempt will be made to develop a mathematical model that would depict the electro-osmotic flow behavior of a non-Newtonian fluid in a microchannel. The model will be analyzed in the sequel. Of particular concern in the investigation will be to examine the effect of heat transfer on electro-osmotic flow in microvessels under the influence of an external magnetic field. The model and its theoretical analysis will be based on the consideration of Cattaneo–Christov heat flux model, because of inherent weaknesses in Fourier's law of heat conduction. Considering that many industrial fluids and most physiological fluids exhibit non-Newtonian behavior, the fluid that flows electro-osmotically in the microvessel will be considered non-Newtonian. The theoretical analysis will be carried out by the use of appropriate analytical and numerical methods. The effects of Hartmann number, surface zeta potential, and Joule heating on electro-osmotic flow velocity and temperature will be discussed. The impact of various material parameters on skin friction and local entropy, as well as on Nusselt number, will also be discussed. Computational results will be presented for the electro-osmotic flow of blood in the microcirculatory system.

Heat and Mass Transfer Effects on MHD Nanofluid Flow Over a Semi Infinite Flat Plate Embedded in a Porous Medium and Two-phase MHD Flow and Heat Transfer in a Horizontal Channel

J. Prakash

There are two problems. In the first, we have investigated the effects of radiation absorption, chemical reaction, and diffusion thermo on free convective heat and mass transfer flow of a nanofluid past a semi-infinite vertical flat plate embedded in a porous medium. The temperature and concentration at the surface are assumed to be oscillatory type. We have considered four types of cubic nanoparticles, namely silver, aluminum, copper, and titanium oxide which are uniform in size with water as a base fluid. In the second, we have investigated steady, mixed convective, laminar flow of incompressible, electrically conducting and heat-absorbing two immiscible viscous fluids in a horizontal channel where the fluid in the region (region I) is saturated with porous medium and the region (region II) is occupied by a clear viscous fluid. A uniform magnetic field is applied in the transverse direction, the fluids rise in the channel driven by uniform pressure, and the heat transfer is influenced by thermal radiation. The equations are modeled using the fully developed flow conditions. An exact solution is obtained for the velocity and temperature distributions. The graphical results are presented, and the physical aspects are discussed in detail to interpret the effect of various significant parameters of the problem. The effect of the skin friction and rate of heat transfer coefficients at the channel walls are tabulated.

Numerical Methods in Aeropropulsive Characterization of Flight Vehicles

Debasis Chakraborty

Over the last few decades, CFD has developed into a rich and diverse subject and emerged as a major component of applied and basic fluid dynamic research along with theoretical and experimental studies. Simultaneous development of new computers, numerical algorithms, physical and chemical models of flow physics is responsible for the greater impact of CFD in both basic and applied scientific/engineering problems. Presently, CFD is an integral part of the design process of airframe and engines for all major aerospace companies in the world. While experimental testing will always remain an integral part of the design, CFD is decreasing the dependence on the more expensive, time-consuming experimental testing or rather using experimental work more effectively and economically. Three-dimensional Euler, Reynolds-averaged Navier–Stokes (RANS), and large eddy simulation (LES) solvers are developed in-house using state-of-the-art numerical techniques and physical models. Systematic validations were carried out through comparisons against reliable experimental results to assess their predictive capabilities and range of applications. The codes are regularly upgraded with new numerical schemes, turbulence models, magneto-hydro-dynamics models, etc. The grid-free CFD solver is integrated with 6-DOF trajectory equations to study the store separation problems. Coupled flow solvers are developed for fluid–structure interaction, hypersonic reentry, and high-speed aerothermal problems. The indigenously developed CFD codes are routinely used in DRDL to predict various aerodynamic parameters pertaining to DRDO missile systems. Aerodynamic characterization of missiles in complete $M - \delta - \phi$ flight regime, control surface deflection studies, heat shield separation of hypersonic air-breathing missiles, store separation from the aircraft, study of plume impingement on jet deflector, low-speed (incompressible limit) characterization of vehicles, etc., are some of the notable applications of the codes.

Core competence is developed in DRDL to simulate turbulent reacting and non-reacting flows for missile propulsion system design. Important user-defined functions (UDFs) are developed to apply commercial CFD solvers in design. Open-source CFD software has been customized to solve many complex propulsion problems. Systematic validations were carried out through comparisons against reliable experimental results before applying these indigenous, open-source, and commercial CFD tools in the design exercises. Extensive non-reacting and reacting simulations were carried out for the development of flightworthy scramjet propulsion system for hypersonic air-breathing cruise vehicle, and CFD simulations guided the experimental testing. Excellent match is obtained between experiment and pretest prediction for various performance and flow parameters. Accurate estimation of heat flux obtained from high-fidelity CFD simulations is used in thermostructural design of the combustor. Aerodynamic and propulsion parameters obtained from end-to-end simulation (comprising of non-reacting flow in external surfaces and reacting flow in the combustor) provide vital input for mission design. CFD-based jet vane correlations are adopted in the flight computer for tactical missiles and form an integral part of missile control and guidance. The problem of high temperature in the base cavity caused due to interaction of free stream and exhaust plume at high altitudes could be analyzed only through CFD methods. Performance prediction of installed air intake of ramjet missiles, solid rocket motor (SRM) port flow field analysis, combustion instability prediction of SRM, plume–canister/plume–launcher interaction, etc., are some of the other notable applications of CFD methods in propulsion system design. Understanding of complex flow phenomena has reduced the developmental cost and time of the system significantly. Development, validation, and application of CFD codes for aeropropulsive design of missiles will be highlighted in the presentation.

Chaos in Mathematical Modelling

Mohammad Sajid

The aim of this presentation is to exhibit chaos in mathematical modeling. Various researches are resolved by using chaos theory in the last few decades. It can easily visualize by using fractals. It is now applicable in almost all disciplines of science, engineering, and technology. Often, chaotic behavior may get in almost all nonlinear systems. Chaos exists everywhere in the world since it seems that it is nonlinear world. It is observed that chaotic behavior came into existence in several mathematical models, like logistic map, delay model, Henon model, Rossler model, Lorenz model, and so on. Some useful ideas of mathematical models associated to chaos with their applications are demonstrated here. It may attract mathematicians and scientists toward chaos in mathematical modeling from different fields of their researches.

Extension of Wavelets to Topologically Complicated Domains

Mani Mehra

Differential equations on topologically complicated domains is a relatively new branch in the theory of differential equations. Some of the examples include differential equations on manifolds or irregularly shaped domains and differential equation on network-like structure. Differential equations on manifolds arises in the areas of mathematical physics, fluid dynamics, image processing, medical imaging, etc. Differential equations on network-like structure also plays a fundamental role in many problems in science and engineering. The aim of this talk is to show how wavelets could be extended to network to solve partial differential equations on network-like structure using spectral graph wavelet.

Controllability of Fractional Differential Equations with Non-instantaneous Impulses

Muslim Malika

We discussed the fractional calculus and controllability concept. Later, we consider a control system governed by an abstract fractional differential equation of order alpha lies between 1 and 2. We are able to prove the exact controllability of the problem by using the cosine family of operators and Banach fixed point theorem. Finally, we have given an example to illustrate the application of these results.

Bio-marker Prediction Using Bayesian State-Space Modeling in Gene Expression Data Analysis

Gajendra K. Vishwakarma

The advancement in computational biology and statistical modeling helps to identify the genes which causes the disease like cancer by comparing its expression levels in diseased and healthy people. Bayesian state-space modeling is a new advancement in statistics which can estimate an unobserved value of a process using the information from an observed outcome and its relationship. Using these two ideas, we are trying to model and estimate the expression values of genes which are significantly different among two groups. The complicated integration of posterior densities is carried out using the Markov chain Monte Carlo (MCMC) simulations. The study sheds light on the use of Bayesian state-space modeling to elucidate the behavior of bio-markers.

Computing Nearest Stable Matrices

Punit Sharma and Nicolas Gillis

The stability of a continuous linear time-invariant (LTI) system $\dot{x} = Ax + Bu$, where $A \in \mathbb{R}^{n,n}$, $B \in \mathbb{R}^{n,m}$ solely depends on the eigenvalues of the stable matrix A. Such a system is stable if all eigenvalues of A are in the closed left half of the complex plane and all eigenvalues on the imaginary axis are semisimple. It is important to know that when an unstable LTI system becomes stable, i.e., when it has all eigenvalues in the stability region, or how much it has to be perturbed to be on this boundary. For control systems, this distance to stability is well understood. This is the converse problem of the distance to instability, where a stable matrix A is given and one looks for the smallest perturbation that moves an eigenvalue outside the stability region. In this talk, I will talk about the distance to the stability problem for LTI control systems. Motivated by the structure of dissipative Hamiltonian systems, we define the DH matrix: A matrix $A \in \mathbb{R}^{n,n}$ is said to be a DH matrix if $A = (J - R)Q$ for some matrices $J, R, Q \in \mathbb{R}^{n,n}$ such that J is skew-symmetric, R is symmetric positive semidefinite, and Q is symmetric positive definite. We will show that a system is stable if and only if its state matrix is a DH matrix. This results in an equivalent optimization problem with a simple convex feasible set. We propose new algorithms to solve this problem. Finally, we show the effectiveness of our method compared to the other approaches and to several state-of-the-art algorithms. These ideas can be generalized to get good approximate solutions to some other nearness problems for control systems like distance to stability for descriptor systems, distance to positive realness, and minimizing the norm of static feedback.

Contents

Scientific Computing and Neural Network

Numerical Mathematics Aspects

A Green's Function Approach to Analyze the Dispersion Characteristics of Love Type Wave Due to an Impulsive Point Source in a Piezoelectric Layered Structure

Anusree Ray and Abhishek K. Singh

Abstract An external source of disturbance in a material, even of point size, give rise to waves propagating away from the concerned region in its interior at a specified time. Such an impulse may be best described with the aid of Dirac delta function. Green's function is primarily utilized in solving these problems of elastodynamics. The present study focuses to investigate the propagation characteristics of Love-type wave influenced by an impulsive point source in a layered structure comprised of a heterogeneous piezoelectric layer lying over a heterogeneous isotropic half-space. Green's function technique is adopted in order to obtain the dispersion equation, which is further reduced to the classical result of Love wave. For sake of computation, numerical data of PZT-5H ceramics for the heterogeneous piezoelectric layer is considered. Influence of heterogeneity, piezoelectricity and dielectric constant associated with the heterogeneous piezoelectric layer; and effect of heterogeneity parameter and corresponding magnification factor concerned with heterogeneity in the isotropic half-space has been reported through graphical delineation.

Keywords Green's function · Point source · Heterogeneity · Piezoelectricity · Love-type wave

1 Introduction

The evolution of wave motions in a material medium may cause due to some external source of disturbances. Green's function provides a visual interpretation of the demeanor associated to the forces concentrated at a point which act as the source of disturbance. These forces are functions of spatial coordinates and time and may originate through a unit impulsive force, defined with the aid of Dirac's delta function.

A. Ray (✉) · A. K. Singh
Department of Applied Mathematics, Indian Institute of Technology (Indian School of Mines), Dhanbad, Jharkhand 826004, India
e-mail: rayanusree07@gmail.com

A. K. Singh
e-mail: abhi.5700@gmail.com

S. Manna et al. (eds.), *Mathematical Modelling and Scientific Computing with Applications*, Springer Proceedings in Mathematics & Statistics 308,
https://doi.org/10.1007/978-981-15-1338-1_1

Green's function is a key to the solution of such problems of Elastodynamics which implicate disturbance due to point source. Green's function is defined depending on the characteristics of the material medium about how it responds mechanically to an impulsive force of disturbance. Chattopadhyay and Singh [1] investigated the propagation behavior of SH-wave propagation due to a point source in a monoclinic layered structure with the aid of Green's function. Some more disquisitions on this ground may be referred from Chattopadhyay et al. [2] and Chattopadhyay and Kar [3].

Piezoelectric structures has been an interest gaining domain the last decade due to their emphatic utilization in many functional devices such as sensors, amplifiers, sonar, air ultrasonic transducers etc. Here, the fact of credit is that this property of piezoelectric material is reversible. Piezoelectric materials have a wide range of applications in efficient designing surface acoustic wave (SAW) devices [4, 5]. The concept of piezoelectricity finds huge applications in the field of mechanical engineering, medical appliances and other modern industrial fields. Piezoelectric materials are generally brittle in nature. Such etiquette of these materials are responsible for generating cavities, impurities, cracks and hence failure of device during handling and bonding procedures. Therefore, the mechanical and electrical performance of piezoelectric materials needs to be improvised so as to achieve a high performance. The most preferred way of doing this involves the consideration of a layer-substrate configuration of piezoelectric structures. Some emphatic works concerning piezoelectric structures/plates are Liu et al. [6], Du et al. [7], Singh et al. [8], Wang [9].

The current study analyses the propagation characteristics of Love-type wave influenced by an impulsive point source in a layered structure constituted of heterogeneous piezoelectric layer lying over a heterogeneous isotropic elastic half-space. Green's function adopted as solution methodology to tackle this of the present study. Closed form of dispersion equation is obtained analytically and reduced to match with the standard result of Love-wave. Moreover, special cases have also been derived from the dispersion relation, and hence compared with some pre-established results. Moreover, the numerical data of PZT-5H ceramics for the heterogeneous piezoelectric layer has been considered to study the effect of heterogeneity, piezoelectricity and dielectric constant associated with heterogeneous piezoelectric layer; and heterogeneity and corresponding magnification factor concerned with heterogeneity of isotropic half-space, on the phase velocity of Love-type wave propagating in such a layered structure.

2 Mathematical Model of the Problem

A heterogeneous piezoelectric layer of thickness H lying over a heterogeneous isotropic half-space is considered. The said model is specified through a Cartesian coordinate system having its origin 'O' on the uppermost boundary surface (free surface), y-axis indicates the direction of wave propagation and x-axis is positive

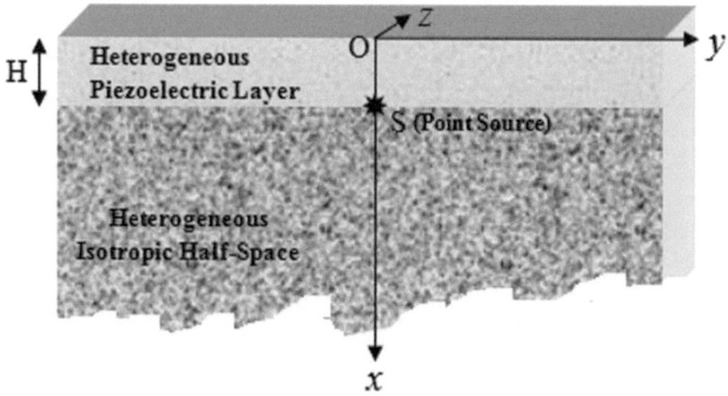

Fig. 1 Geometry of the problem

pointing normally downwards. An impulsive point source of disturbance is located at 'S', interfacial common surface of the layer and half-space. The polling direction of piezoelectric material is parallel to z-axis. Geometrical interpretation of the problem is revealed in Fig. 1. Love type wave propagation is considered through such a layered structure.

3 Governing Equations and Relations

Seeking harmonic solution for propagation of Love-type wave, the displacement (u_1, u_2, u_3) of the constituent particles in the heterogeneous piezoelectric layer and displacements (v_1, v_2, v_3) in heterogeneous isotropic half-space may respectively be taken as

$$u_i(x, y, t) = u_i(x, y)e^{i\omega t}, v_i(x, y, t) = v_i(x, y)e^{i\omega t}, (i = 1, 2, 3). \quad (1)$$

For the characteristic propagation of Love-type wave along y-axis and which causes displacement in z-direction only, it is assumed that

$$\begin{aligned} &u_1 = 0, u_2 = 0, u_3 = u_3(x, y, t), \\ &v_1 = 0, v_2 = 0, v_3 = v_3(x, y, t), \phi = \phi(x, y, t), \end{aligned} \quad (2)$$

where ϕ is the electric potential concerned with heterogeneous piezoelectric layer.

Henceforth, the following notations are used

$$\partial_j \equiv \frac{\partial}{\partial j}, \partial_{jj} \equiv \frac{\partial^2}{\partial j^2}, d_j \equiv \frac{d}{dj}, d_{jj} \equiv \frac{d^2}{dj^2}$$

3.1 Constitutive Relations

The constitutive relations for piezoelectric material are given by the coupled relations [8]

$$\tau_{ij} = C_{ijkl} S_{kl} - e_{kij} E_k,$$
$$D_k = e_{kij} S_{ij} + \varepsilon_{ki} E_i, \tag{3}$$

where the symbols are defined in Table 1.

Indices i, j, k and l span the range (1, 2, 3).

For the sake of clarity, the above relation (3) may be rewritten in expanded matrix form as

$$\begin{bmatrix} \tau_{11} \\ \tau_{22} \\ \tau_{33} \\ \tau_{23} \\ \tau_{31} \\ \tau_{12} \\ D_1 \\ D_2 \\ D_3 \end{bmatrix} = \begin{bmatrix} C_{11} & C_{12} & C_{13} & 0 & 0 & 0 & 0 & 0 & e_{31} \\ C_{12} & C_{11} & C_{13} & 0 & 0 & 0 & 0 & 0 & e_{31} \\ C_{13} & C_{13} & C_{33} & 0 & 0 & 0 & 0 & 0 & e_{33} \\ 0 & 0 & 0 & 2C_{44} & 0 & 0 & 0 & e_{15} & 0 \\ 0 & 0 & 0 & 0 & 2C_{44} & 0 & e_{15} & 0 & 0 \\ 0 & 0 & 0 & 0 & 0 & 2C_{66} & 0 & 0 & 0 \\ 0 & 0 & 0 & 0 & 2e_{15} & 0 & \varepsilon_{11} & 0 & 0 \\ 0 & 0 & 0 & 2e_{15} & 0 & 0 & 0 & \varepsilon_{11} & 0 \\ e_{31} & e_{31} & e_{33} & 0 & 0 & 0 & 0 & 0 & \varepsilon_{33} \end{bmatrix} \begin{bmatrix} S_{11} \\ S_{22} \\ S_{33} \\ S_{23} \\ S_{31} \\ S_{12} \\ E_1 \\ E_2 \\ E_3 \end{bmatrix} \tag{4}$$

Moreover, $C_{66} = (C_{11} - C_{12})/2$.

For an isotropic material the stress components are given by

$$\sigma_{ij} = \lambda \delta_{ij} s_{kk} + 2\mu_2 s_{ij}; \quad i, j, k = 1, 2, 3, \tag{5}$$

where the newly introduced symbols are defined in Table 2.

Table 1 Symbols and their meaning for the heterogeneous piezoelectric layer

Symbols	Meaning	Symbols	Meaning
τ_{ij}	Components of stress for the heterogeneous piezoelectric layer	D_k	Components of electric displacement
ε_{ij}	Dielectric permittivities	e_{kij}	Piezoelectric stress constants
C_{ijkl}	Elastic stiffness constants	$S_{ij} = \partial_j u_i + \partial_i u_j$	Mechanical strains
$E_i (= -\partial_i \phi)$	Intensity of electric field		

Table 2 Symbols and their meaning for the heterogeneous isotropic half-space

Symbols	Meaning	Symbols	Meaning
μ_2	Shear modulus	$s_{ij} = \partial_j v_i + \partial_i v_j$	Components of infinitesimal strain
σ_{ij}	Stress components for an isotropic material	δ_{ij}	Kronecker delta

3.2 *Equations of Motion*

Due to piezoelectricity prevailing in the layer, the governing equations of motion for the heterogeneous piezoelectric layer, in absence of body forces may be expressed as

$$\begin{aligned} \tau_{ij,j} &= \rho_1 \partial_{tt} u_i, \\ D_{i,i} &= 0, \end{aligned} \tag{6}$$

where ρ_1 is its density. The comma followed by the subscript $j\,(or\ i)$ indicates space coordinate differentiation and the repeated subscript index implies summation with respect to that index.

The governing equations of motion for the heterogeneous isotropic half-space in absence of body forces, may be given as

$$\sigma_{ij,j} = \rho_2 \partial_{tt} v_i, \tag{7}$$

where ρ_2 is the density of isotropic material.

4 Dynamics of Heterogeneous Piezoelectric Layer

For Love-type wave propagation in piezoelectric layer, expression for the components of stresses and electric displacements may be obtained with the aid of Eqs. (2) and (4) as

$$\left.\begin{aligned} &\tau_{11} = \tau_{22} = \tau_{33} = \tau_{12} = 0, \\ &\tau_{23} = C_{44}\partial_y u_3 + e_{15}\partial_y \phi,\ \tau_{31} = C_{44}\partial_x u_3 + e_{15}\partial_x \phi, \\ &D_1 = e_{15}\partial_x u_3 - \varepsilon_{11}\partial_x \phi,\ D_2 = e_{15}\partial_y u_3 - \varepsilon_{11}\partial_y \phi,\ D_3 = 0. \end{aligned}\right\} \tag{8}$$

Now, the heterogeneity in the piezoelectric layer is taken in such a way that the elastic constants and density are exponentially varying functions of depth and may be written in the form

$$C_{44} = C_{44}^{(0)} e^{\alpha x},\ e_{15} = e_{15}^{(0)} e^{\alpha x},\ \varepsilon_{11} = \varepsilon_{11}^{(0)} e^{\alpha x},\ \rho_1 = \rho_1^{(0)} e^{\alpha x}, \tag{9}$$

where α is the heterogeneity parameter, and $C_{44}^{(0)}, e_{15}^{(0)}, \varepsilon_{11}^{(0)}, \rho_1^{(0)}$ are the values of $C_{44}, e_{15}, \varepsilon_{11}, \rho_1$ respectively, at the free surface of the layer.

Let us assume that the presence of source of disturbance at the intersection of interfacial surface and x-axis, is responsible for a force density distribution $f_1(r, t)$ in the heterogeneous piezoelectric layer. Thus, with the help of Eqs. (6), (8) and (9), the coupled electromechanical field equations are obtained as

$$C_{44}^{(0)}\left(\nabla^2 u_3 + \alpha \partial_x u_3\right) + e_{15}^{(0)}\left(\nabla^2 \phi + \alpha \partial_x \phi\right) - \rho_1^{(0)} \partial_{tt} u_3 = 4\pi f_1(r, t) e^{-\alpha x}, \quad (10)$$

$$e_{15}^{(0)}\left(\nabla^2 u_3 + \alpha \partial_x u_3\right) - \varepsilon_{11}^{(0)}\left(\nabla^2 \phi + \alpha \partial_x \phi\right) = 0, \quad (11)$$

where $\nabla^2 = \partial_{xx} + \partial_{yy}$.

Now, Eq. (11) may be rewritten as

$$\nabla^2 \phi + \alpha \partial_x \phi = \frac{e_{15}^{(0)}}{\varepsilon_{11}^{(0)}}\left(\nabla^2 u_3 + \alpha \partial_x u_3\right), \quad (12)$$

which when substituted in Eq. (10), yields

$$\nabla^2 u_3 + \alpha \partial_x u_3 - \frac{\rho_1^{(0)}}{\mu_1} \partial_{tt} u_3 = \frac{4\pi f_1(r, t) e^{-\alpha x}}{\mu_1}, \quad (13)$$

where $\mu_1 = C_{44}^{(0)} + \frac{\left(e_{15}^{(0)}\right)^2}{\varepsilon_{11}^{(0)}}$.

In view of Eq. (1), and the substitutions $f_1(r, t) = f_1(r) e^{i\omega t}$ and $u_3(x, y) = U_3(x, y) e^{-\alpha x/2}$, Eq. (13) leads to

$$\nabla^2 U_3 + k_1^2 U_3 = \frac{4\pi f_1(r) e^{-\alpha x/2}}{\mu_1}, \quad (14)$$

where $\omega (= kc)$, k and c is the angular frequency, wave number and phase velocity respectively. Moreover, $k_1^2 = \frac{\rho_1^{(0)}}{\mu_1}\omega^2 - \frac{\alpha^2}{4}$.

Now, we define the Fourier transform $\overline{U}_3(x, \xi)$ of $U_3(x, y)$ as

$$\overline{U}_3(x, \xi) = \frac{1}{2\pi} \int_{-\infty}^{\infty} U_3(x, y) e^{i\xi y} dy, \quad (15)$$

and the consequent inverse Fourier transform as

$$U_3(x, y) = \int_{-\infty}^{\infty} \overline{U}_3(x, \xi) e^{-i\xi y} d\xi. \quad (16)$$

Therefore, in light of Eq. (15), application of Fourier transform on Eq. (14), lead to

$$d_{xx}\overline{U}_3 - \eta_1^2\overline{U}_3 = 4\pi f_1(x)e^{-\alpha x/2}, \tag{17}$$

where $f_1(r) = \delta(y)\delta(x - H)$, $\eta_1^2 = \xi^2 - k_1^2$ and $f_1(x) = \frac{\delta(x-H)}{2\pi\mu_1}$.

5 Dynamics of Heterogeneous Isotropic Half-Space

The only non-vanishing equation of motion for the heterogeneous isotropic semi-infinite medium derived from Eq. (7) with the aid of Eqs. (2) and (5) may be written as

$$\partial_x\sigma_{13} + \partial_y\sigma_{23} = \rho_2\partial_{tt}v_3, \tag{18}$$

where

$$\sigma_{13} = \mu_2\partial_x v_3, \quad \sigma_{23} = \mu_2\partial_y v_3. \tag{19}$$

The vertical heterogeneity in elastic constants and density of the heterogeneous isotropic semi-infinite medium (half-space) is considered as follows:

$$\begin{aligned} \mu_2 &= \mu_2^{(0)} + \varepsilon \sinh b(x - H), \\ \rho_2 &= \rho_2^{(0)} + \varepsilon^* \sinh b(x - H), \end{aligned} \tag{20}$$

where b is the heterogeneity parameter associated with the hyperbolic type of heterogeneity in the isotropic half-space; and ε, ε^* are the magnification factors of hyperbolic type of heterogeneity in shear modulus (μ_2) and density (ρ_2) respectively, of heterogeneous isotropic half-space. $\mu_2^{(0)}$, $\rho_2^{(0)}$ are the values of μ_2, ρ_2 respectively, at the interfacial surface i.e. at $x = H$. Using Eq. (1) and substituting the relations from Eqs. (19) and (20) into Eq. (18) we get

$$\begin{aligned} \mu_2^{(0)}\big(\partial_{xx}v_3 + \partial_{yy}v_3\big) &= -\varepsilon b \cosh b(x - H)\partial_x v_3 - \Big(\rho_2^{(0)} + \varepsilon^* \sinh b(x - H)\Big)\omega^2 v_3 \\ &\quad - \varepsilon \sinh b(x - H)\big(\partial_{xx}v_3 + \partial_{yy}v_3\big). \end{aligned} \tag{21}$$

Now, we define the Fourier transform $\overline{V}_3(x, \xi)$ of $v_3(x, y)$ as

$$\overline{V}_3(x, \xi) = \frac{1}{2\pi}\int_{-\infty}^{\infty} v_3(x, y)e^{i\xi y}dy, \tag{22}$$

and the consequent inverse Fourier transform may be defined as

$$v_3(x, y) = \int_{-\infty}^{\infty} \overline{V}_3(x, \xi) e^{-i\xi y} d\xi, \tag{23}$$

so that when it is employed to Eq. (21), yields an ordinary differential equation given by

$$d_{xx}\overline{V}_3 - \eta_2^2 \overline{V}_3 = 4\pi f_2(x), \tag{24}$$

where

$$4\pi f_2(x) = \frac{-\varepsilon}{\mu_2^{(0)}} \Big[\sinh b(x - H) d_{xx}\overline{V}_3 + b\cosh b(x - H) d_x \overline{V}_3 \\ - \left(\xi^2 \sinh b(x - H) - \frac{\varepsilon^*}{\varepsilon}\omega^2 \sinh b(x - H)\right)\overline{V}_3,\Big] \tag{25}$$

$$\eta_2^2 = \xi^2 - \frac{\rho_2^{(0)}\omega^2}{\mu_2^{(0)}}. \tag{26}$$

6 Boundary Conditions and Dispersion Relation

In this section we aim to explore the solution of Eqs. (17) and (24) employing Green's function, The admissible boundary conditions may be of the following type:

(i) The upper boundary surface of the heterogeneous layer is free from mechanical traction and electric potential. Therefore, the boundary conditions at the traction free boundary surface may be written in mathematical form as follows:
Mechanical traction free condition at $x = 0$, i.e.

$$\tau_{31} = 0, \tag{27}$$

Electrical boundary condition at $x = 0$, i.e.

$$\phi(x) = 0. \tag{28}$$

(ii) At the interfacial surface of layer and half-space, the admissible boundary conditions are
Continuity of stresses at $x = H$, i.e.

$$\tau_{31} = \sigma_{31}, \tag{29}$$

Continuity of displacements at $x = H$, i.e.

$$\overline{U}_3(x) = \overline{V}_3(x), \tag{30}$$

Electrical boundary condition at $x = H$, i.e.

$$\phi(x) = 0. \tag{31}$$

With the aid of Eqs. (28) and (31), the boundary condition in Eqs. (27) and (29) respectively may be rewritten as

$$d_x\overline{U}_3 - \frac{\alpha}{2}\overline{U}_3 = 0, \quad \text{at } x = 0 \tag{32}$$

and

$$d_x\overline{V}_3 = \frac{C_{44}^{(0)}}{P}\left[d_x\overline{U}_3 - \frac{\alpha}{2}\overline{U}_3\right]e^{\alpha H/2}, \quad \text{at } x = H. \tag{33}$$

It is worthy to mention here that Dirac-delta function ($\delta(x)$) is used to represent a source of disturbance.

Keeping this into our account, let us take an arbitrary point x_0 in the upper layer, and hence define the Green's function, $G_1(x/x_0)$ for the layer which satisfy the boundary conditions given by

$$d_x G_1 - \frac{\alpha}{2}G_1 = 0 \ \text{ both } \ \text{at } x = 0 \ \text{ and } \ x = H. \tag{34}$$

Therefore, $G_1(x, x_0)$ will satisfy the equation

$$d_{xx}G_1(x, x_0) - \eta_1^2 G_1(x, x_0) = \delta(x - x_0). \tag{35}$$

Now, multiplying Eqs. (17) and (35) by $G_1(x, x_0)$ and $\overline{U}_3(x)$ respectively, then subtracting and further integrating with respect to x from $x = 0$ to H, it is obtained that

$$\int_0^H \left[G_1(x, x_0)d_{xx}\overline{U}_3(x) - \overline{U}_3(x)d_{xx}G_1(x, x_0)\right]dx$$

$$= \int_0^H 4\pi f_1(x)e^{-\alpha x/2}G_1(x, x_0)dx - \int_0^H \delta(x - x_0)\overline{U}_3(x)dx, \tag{36}$$

which on further simplification, and making use of the results of Eqs. (34) and (32), result in

$$G_1(H, x_0)\Big[d_x\overline{U}_3 - \frac{\alpha}{2}\overline{U}_3\Big]_{x=H} = \frac{2}{\mu_1}G_1(H, x_0)e^{-\alpha H/2} - \overline{U}_3(x_0). \quad (37)$$

Interchanging x_0 by x in Eq. (37) and using $G_1(H, x) = G_1(x, H)$,$\overline{U}_3(x)$ may be expressed as

$$\overline{U}_3(x) = (2/\mu_1)G_1(x, H)e^{-\alpha H/2} - G_1(x, H)\big[d_x\overline{U}_3 - \alpha/2\overline{U}_3\big]_{x=H}. \quad (38)$$

With a similar approach, for any arbitrary point x_0 in the half-space, let us assume $G_2(x, x_0)$ be the Green's function for the half-space satisfying the conditions:

$$d_x G_2 = 0 \text{ at } x = H \text{ and tends to zero as } x \to \infty. \quad (39)$$

Therefore, $G_2(x, x_0)$ is the solution of the equation

$$d_{xx}G_2(x, x_0) - \eta_2^2 G_2(x, x_0) = \delta(x - x_0). \quad (40)$$

Multiplying Eqs. (26) and (40) by $G_2(x, x_0)$ and $\overline{V}_3(x)$ respectively, then subtracting and further integrating with respect to x from $x = H$ to ∞, yields

$$\int_H^\infty \big[G_2(x, x_0)d_{xx}\overline{V}_3(x) - \overline{V}_3(x)d_{xx}G_2(x, x_0)\big]dx$$
$$= \int_H^\infty 4\pi f_2(x)G_2(x, x_0)dx - \int_H^\infty \overline{V}_3(x)\delta(x - x_0)dx, \quad (41)$$

which after simplification results in

$$G_2(H, x_0)\big[d_x\overline{V}_3(x)\big]_{x=H} = \overline{V}_3(x_0) - \int_H^\infty 4\pi f_2(x)G_2(x, x_0)dx. \quad (42)$$

Now, interchanging x_0 by x in Eq. (42), the expression for $\overline{V}_3$ at any point x in the half space may be written as

$$\overline{V}_3(x) = G_2(x, H)\big[d_x\overline{V}_3(x)\big]_{x=H} + 4\pi\int_H^\infty f_2(x_0)e^{\gamma x_0/2}G_2(x, x_0)dx_0, \quad (43)$$

where $G_2(H, x) = G_2(x, H)$ and $G_2(x, x_0) = G_2(x_0, x)$.

Applying the boundary condition as provided in Eqs. (30), (38) and (43) collectively yield

$$\frac{2}{\mu_1}G_1(H,H)e^{-\alpha H/2}-G_1(H,H)\Big[d_x\overline{U}_3-\frac{\alpha}{2}\overline{U}_3\Big]_{x=H}$$
$$=G_2(H,H)\big[d_x\overline{V}_3\big]_{x=H}+4\pi\int_H^\infty f_2(x_0)G_2(H,x_0)dx_0. \tag{44}$$

In reference to Eqs. (33), (44) leads to

$$\Big[d_x\overline{U}_3-\frac{\alpha}{2}\overline{U}_3\Big]_{x=H}=\frac{2}{\mu_1}\frac{\mu_2^{(0)}G_1(H,H)e^{-\alpha H/2}}{\mu_2^{(0)}G_1(H,H)+C_{44}^{(0)}G_2(H,H)e^{\alpha H/2}}$$
$$-\frac{4\pi\mu_2^{(0)}\int_H^\infty f_2(x_0)G_2(H,x_0)dx_0}{\mu_2^{(0)}G_1(H,H)+C_{44}^{(0)}G_2(H,H)e^{\alpha H/2}}. \tag{45}$$

With the aid of the Eqs. (25) and (45), Eq. (38) results in

$$\overline{U}_3(x)=\frac{G_1(x,H)}{\mu_2^{(0)}G_1(H,H)+C_{44}^{(0)}G_2(H,H)e^{\alpha H/2}}\left(\frac{2C_{44}^{(0)}}{\mu_1}G_2(H,H)-\varepsilon\int_H^\infty t_1G_2(H,x_0)dx_0\right), \tag{46}$$

where t_1 is provided in Appendix.

Again, in view of the Eqs. (33) and (43), the value of $\overline{V}_3(x)$ may be found as

$$\overline{V}_3(x)=\frac{2C_{44}^{(0)}}{\mu_1}\frac{G_1(H,H)G_2(x,H)}{\mu_2^{(0)}G_1(H,H)+C_{44}^{(0)}G_2(H,H)e^{\alpha H/2}}$$
$$+\frac{\varepsilon C_{44}^{(0)}}{\mu_2^{(0)}}\frac{G_1(x,H)}{\mu_2^{(0)}e^{-\alpha H/2}G_1(H,H)+C_{44}^{(0)}G_2(H,H)}$$
$$\times\int_H^\infty t_1G_2(H,x_0)dx_0-\frac{\varepsilon}{\mu_2^{(0)}}\int_H^\infty t_1G_2(x,x_0)dx_0. \tag{47}$$

The value of $\overline{V}_3(x)$ may be derived from Eq. (47) by the methodology of successive approximations. Neglecting the higher powers of ε, taking the first approximation of $\overline{V}_3(x)$, it may be established that

$$V_3(x)=\left(\frac{2C_{44}^{(0)}}{\mu_1}\right)\frac{G_1(H,H)G_2(x,H)}{\mu_2^{(0)}G_1(H,H)+C_{44}^{(0)}G_2(H,H)e^{\alpha H/2}}. \tag{48}$$

With the aid of Eqs. (48), (46) leads to

$$\overline{U}_3(x)=\frac{2C_{44}^{(0)}}{\mu_1}\frac{G_1(x,H)G_2(H,H)}{\mu_2^{(0)}G_1(H,H)+C_{44}^{(0)}G_2(H,H)e^{\alpha H\alpha H2-2}}$$

$$
-\frac{2\varepsilon C_{44}^{(0)}}{\mu_1}\frac{G_1(H,H)G_1(H,H)}{\left[\mu_2^{(0)}G_1(H,H)+C_{44}^{(0)}G_2(H,H)e^{\alpha H/\alpha H2-2}\right]^2}\int\limits_H^\infty\{\sinh b(x_0-H)
$$
$$
\times d_{xx}G_2(x_0,H)+b\cosh b(x_0-H)d_xG_2(x_0,H)-\Big(\xi^2\sinh b(x_0-H)
$$
$$
-\frac{\varepsilon^*}{\varepsilon}\omega^2\sinh b(x_0-H)\Big)G_2(x_0,H)\Big\}G_2(H,x_0)dx_0. \tag{49}
$$

To evaluate $\overline{U}_3(x)$, the values of $G_1(x, H)$ and $G_2(x, H)$ must be determined.

Now, we aim to obtain a solution of Eq. (35). For this let us assume two independent solutions of the equation

$$
d_{xx}G_1(x,x_0)-\eta_1^2G_1(x,x_0)=0, \tag{50}
$$

which vanishes at $x = -\infty$ and $x = \infty$, as

$P_1(x) = e^{\eta_1 x}$ and $P_2(x) = e^{-\eta_1 x}$.

Therefore, the solution of Eq. (50) for an infinite medium is

$$
\frac{P_1(x)P_2(x_0)}{W} \quad \text{for} \quad x < x_0,
$$

$$
\frac{P_1(x_0)P_2(x)}{W} \quad \text{for} \quad x > x_0,
$$

where $W = P_1(x)P_2'(x) - P_2(x)P_{12}'(x) = -2\eta_1$.

Therefore, for an unbounded medium the solution of Eq. (35) may be expressed as $-\frac{e^{-\eta_1|x-x_0|}}{2\eta_1}$.

Since $G_1(x, x_0)$ should satisfy conditions indicated in Eq. (33), therefore, it may be assumed that

$$
G_1(x,x_0)=-\frac{e^{-\eta_1|x-x_0|}}{2\eta_1}+Ae^{\eta_1 x}+Be^{-\eta_1 x}, \tag{51}
$$

where A and B are arbitrary constants.

After, evaluating the values of A and B by employing condition (34), $G_1(x, x_0)$ may be expressed as

$$
G_1(x,x_0)=-\frac{T_0}{M}, \tag{52}
$$

where T_0 and M are provided in Appendix.

In view of Eq. (52), the expressions for $G_1(x, H)$ and $G_1(H, H)$ may be obtained as

$$
G_1(x,H)=-\frac{T_1}{M} \quad \text{and} \quad G_1(H,H)=-\frac{T_2}{M} \tag{53}
$$

respectively, where T_1 and T_2 are given in Appendix.

Adopting a similar approach, the expression for $G_2(x, x_0)$ may be deduced from Eq. (40) as

$$G_2(x, x_0) = -\frac{1}{2\eta_2}\left[e^{-\eta_2|x-x_0|} + e^{-\eta_2(x+x_0-2H)}\right]. \tag{54}$$

Now, the followings may be easily computed from Eq. (54):

$$G_2(H, x_0) = -\frac{1}{\eta_2}e^{-\eta_2(x_0-H)} \quad \text{and} \quad G_2(H, H) = -\frac{1}{\eta_2}. \tag{55}$$

Substitution of the values obtained in Eqs. (53) and (55) into Eq. (49), it may be deduced that

$$\overline{U}_3(x) = -\left(\frac{2C_{44}^{(0)}}{\mu_1}\right)\frac{T_1}{M_0 + N_0}\left[1 + \frac{T_2 b}{M_0 + N_0}\left\{\frac{\varepsilon\eta_2^2 - \varepsilon^*\omega^2 + \varepsilon\xi^2}{4\eta_2^2 - b^2}\right\}\right], \tag{56}$$

where $M_0 = MC_{44}^{(0)}e^{\alpha H/2}$ and $N_0 = \mu_2^{(0)}\eta_2 T_2$.

Eliminating higher powers of ε and in reference to Eq. (16), the displacement at any point in layer may be obtained as

$$u_3(x, y) = -\left(\frac{2C_{44}^{(0)}}{\mu_1}\right)e^{-\alpha x/2}\int_{-\infty}^{\infty}\frac{\left(4\eta_2^2 - b^2\right)T_1 e^{-i\xi y}}{\left(4\eta_2^2 - b^2\right)(M_0 + N_0) - T_2 b\left(\varepsilon\eta_2^2 - \varepsilon^*\omega^2 + \varepsilon\xi^2\right)}d\xi. \tag{57}$$

Now, the integral value in Eq. (57) depends entirely on the poles of the integrand which are located at the roots of equation

$$(4\eta_2^2 - b^2)(M_0 + N_0) - T_2 b\left(\varepsilon\eta_2^2 - \varepsilon^*\omega^2 + \varepsilon\xi^2\right) = 0. \tag{58}$$

Replacing η_1 by $i\eta_1$ and ξ by k, Eq. (58) may be reduced to

$$\tan\eta_1 H = \frac{4\eta_1\left\{\mu_2^{(0)}\eta_2\left(4\eta_2^2 - b^2\right) - b\left(\varepsilon\eta_2^2 - \varepsilon^*k^2c^2 + \varepsilon k^2\right)\right\}}{\left\{c_{44}^{(0)}e^{\alpha H}\left(4\eta_1^2 + \alpha^2\right) - 2\alpha P^{(2)}\eta_2\right\}\left(4\eta_2^2 - b^2\right) + 2\alpha b\left(\varepsilon\eta_2^2 - \varepsilon^*k^2c^2 + \varepsilon k^2\right)}, \tag{59}$$

which represents the dispersion relation of Love-type wave propagation under the influence of a point source in a heterogeneous piezoelectric layer lying over a heterogeneous half-space.

7 Cases and Validation

Following are some special cases discussed in order to compare and validate the obtained result with some previously established results.

Case I When isotropic half-space is not heterogeneous i.e. $\varepsilon = 0$, $\varepsilon^* = 0$ and $b = 0$, the dispersion Eq. (59) results in

$$\tan \eta_1 H = \frac{4\eta_1 P^{(2)} \bar{\eta}_2}{C_{44}^{(0)} e^{\alpha H} \left(4\eta_1^2 + \alpha^2\right) - 2\alpha P^{(2)} \bar{\eta}_2}, \tag{60}$$

where $\bar{\eta}_2^2 = k^2\left(\frac{\rho_2^{(0)} c^2}{\mu_2} - 1\right)$.

Equation (60) is the dispersion relation [8] for the propagation of Love wave in a heterogeneous piezoelectric layer lying over a homogeneous isotropic half-space.

Case II When both layer and half-space are not heterogeneous i.e. $\alpha = 0$, $\varepsilon = 0$, $\varepsilon^* = 0$, $b = 0$ the dispersion Eq. (59) takes the form

$$\tan \bar{\eta}_1 H = \frac{\mu_2 \bar{\eta}_2}{C_{44}^{(0)} \bar{\eta}_1}, \tag{61}$$

where $\bar{\eta}_1^2 = k^2\left(1 - \frac{\rho_1^{(0)} c^2}{\mu_1}\right)$.

Equation (61) is the dispersion relation [6] for the propagation of Love wave in a (homogeneous) piezoelectric layer lying over a homogeneous isotropic half-space.

Case III In absence of heterogeneity in both layer and half space; and in absence of piezoelectric effect in the layer i.e. $\alpha = 0$, $b = 0$, $\varepsilon = 0$, $\varepsilon^* = 0$, $e_{15} = 0$, the dispersion Eq. (59) takes the form

$$\tan\left(kH\sqrt{c^2/\beta_1^2 - 1}\right) = \frac{\mu_2\sqrt{1 - c^2/\beta_2^2}}{\mu_1\sqrt{c^2/\beta_1^2 - 1}}, \tag{62}$$

where $c_{44}^{(0)}$ and $P^{(2)}$ are replaced by μ_1 and μ_2 respectively and β_1, β_2 are provided in Appendix.

Equation (62) shows that the dispersion relation in Eq. (59) reduces to the standard Love-wave equation [10] which authenticates the present problem.

8 Computational Results and Discussion

The derived dispersion equation establishes a dependency relationship of the phase velocity of Love-type wave with the wave number and other parameters associated to the geometrical configuration concerned with the present study. In order to unravel the effect of various influential parameters on the phase velocity of Love-type wave, numerical computation is performed taking the following data into account:

For the heterogeneous piezoelectric layer [6]:

$$C_{44}^{(0)} = 2.30 \times 10^9 \mathrm{N/m^2},\ \rho_1^{(0)} = 7500\,\mathrm{Kg/m^3} (\mathrm{PZT} - 5\mathrm{H}).$$

For the heterogeneous isotropic half-space [11]:

$$\mu_2^{(0)} = 7.8 \times 10^9 \mathrm{N/m^2},\ \rho_2^{(0)} = 3535\,\mathrm{Kg/m^3}.$$

Unless otherwise stated:

$$\alpha H = 0.1, 0.5, 0.9;\ bH = 0.1, 0.3, 0.5;\ e_{15} = 0, 7, 17, 27(C/m^2);$$
$$\varepsilon_{11} = 217, 277, 347(*10^{-9}C^2/Nm^2);\ \varepsilon = 0.1, 0.3, 0.5(*10^{-9}C^2/Nm^2);$$
$$\varepsilon^* = 0, 50, 100(kg/m^3).$$

Figures 2, 3 and 4 graphically illustrates first mode of Love-type wave propagation portraying the variation of dimensionless phase velocity against dimensionless wave number for different values of the influencing parameters. It is observed that the phase velocity decreases with increase in wave number.

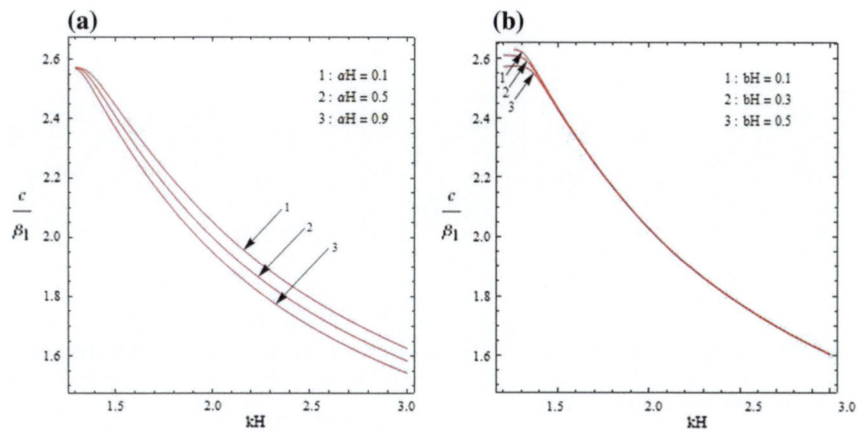

Fig. 2 Phase velocity profile (c/β_1 vs. kH) of Love wave for different values of **(a)** heterogeneity parameter (αH) associated with the heterogeneous piezoelectric layer and **(b)** heterogeneity parameter (bH) associated with the heterogeneous isotropic half-space

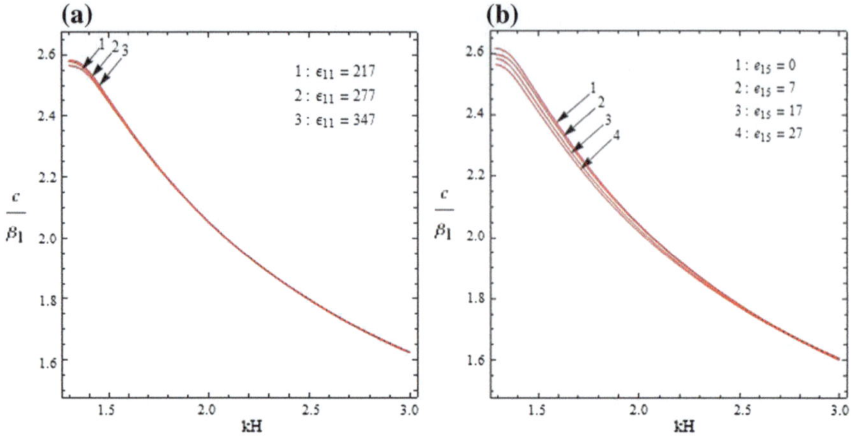

Fig. 3 Phase velocity profile of Love wave for different values of **(a)** dielectric constant (ε_{11}) and **(b)** piezoelectric constant (e_{15}), associated with the heterogeneous piezoelectric layer

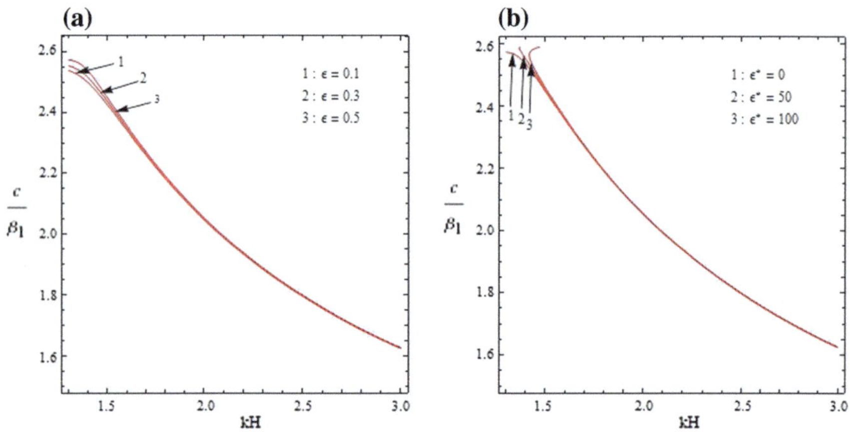

Fig. 4 Phase velocity profile (c/β_1 vs. kH) of Love wave for different values of **(a)** magnification factor (ε) associated with the rigidity and **(b)** of magnification factor (ε^*) associated with the density of the heterogeneous isotropic half-space

9 Concluding Remarks

The numerically computed results show the efficacy of heterogeneities, piezoelectricity and dielectric constant on the propagation behaviour of Love-type wave influenced due to a point source in a heterogeneous piezoelectric layer lying over a heterogeneous isotropic half-space. The outcomes of the current study may be concluded as follows:

- Wave number disfavours the phase velocity of Love-type wave in both the cases when the heterogeneous half-space is with or without heterogeneity.
- Vertical heterogeneity (αH) of exponential type causing heterogeneity in piezoelectric layer suppresses the phase velocity of Love-type wave.
- Heterogeneity parameter (bH) responsible for hyperbolic type of variation in shear modulus as well as density associated with heterogeneous isotropic half-space has a significant disfavouring effect on phase velocity of Love-type wave.
- Piezoelectric parameter $\left(e_{15}/\mu_{11}^2\right)$ has a significant effect on the phase velocity of Love-type wave. With increase in magnitude of piezoelectricity in the velocity decreases considerably.
- The magnification factor (ε and ε^*) associated to the hyperbolic type of heterogeneity in the half-space has a favouring effect on the phase velocity of Love-type wave. Meticulous examination conclude that magnification factor of heterogeneity (ε) concerned with shear modulus have prominent effect on phase velocity of Love-type wave as compare to that of magnification factor of heterogeneity (ε^*) concerned with the density.
- The deduced dispersion equation matches with the well established previous results and the classical one, which authenticates the present study.

The well known smart materials, piezoelectric materials, are inherently capable of actuation and sensing. Therefore, it is of great interest to the researchers and engineers to study the behaviour of a composite structure comprised of such materials when interacted with waves and vibrations influenced due to some source of disturbance, holding a great deal of potential for commercial and industrial applications.

Acknowledgements The authors convey their sincere thanks to Indian Institute of Technology (Indian School of Mines), Dhanbad, for providing all the necessary facilities to carry out the research work.

Appendix

$$t_1 = \sinh b(x_0 - H) d_{xx} V_3(x_0) + b \cosh b(x_0 - H) d_x V_3(x_0) - \left(\xi^2 \sinh b(x_0 - H) - \frac{\varepsilon^*}{\varepsilon}\omega^2 \sinh b(x_0 - H)\right) V_3(x_0)$$

$$\begin{aligned} T_0 = {} & \left(\eta_1^2 - \alpha^2/4\right)\left(e^{\eta_1 H} - e^{-\eta_1 H}\right)e^{-\eta_1 |x - x_0|} \\ & + (\eta_1 + \alpha/2)e^{\eta_1 x}\left[(\eta_1 - \alpha/2)e^{-\eta_1 (H + x_0)} + (\eta_1 + \alpha/2)e^{-\eta_1 (H - x_0)}\right] \\ & + (\eta_1 - \alpha/2)e^{-\eta_1 x}\left[(\eta_1 - \alpha/2)e^{\eta_1 (H - x_0)} + (\eta_1 + \alpha/2)e^{-\eta_1 (H - x_0)}\right], \end{aligned}$$

$$\begin{aligned} T_1 = {} & \left(\eta_1^2 - \alpha^2/4\right)\left(e^{\eta_1 H} - e^{-\eta_1 H}\right)e^{-\eta_1 |x - H|} \\ & + (\eta_1 + \alpha/2)e^{\eta_1 x}\left[(\eta_1 - \alpha/2)e^{-2\eta_1 H} + (\eta_1 + \alpha/2)\right] + 2\eta_1(\eta_1 - \alpha/2)e^{-\eta_1 x}, \end{aligned}$$

$$T_2 = \left(\eta_1^2 - \alpha^2/4\right)\left(e^{\eta_1 H} - e^{-\eta_1 H}\right) + (\eta_1 + \alpha/2)e^{\eta_1 H}\left[(\eta_1 - \alpha/2)e^{-2\eta_1 H} + (\eta_1 + \alpha/2)\right] + 2\eta_1(\eta_1 - \alpha/2)e^{-\eta_1 H},$$

$$M = 2\eta_1\left(\eta_1^2 - \alpha^2/4\right)\left(e^{\eta_1 H} - e^{-\eta_1 H}\right), \beta_1 = \sqrt{\mu_1/\rho_1}, \beta_2 = \sqrt{\mu_2/\rho_2}.$$

References

1. Chattopadhyay, A., Singh, A.K.: Effect of point source and heterogeneity on the propagation of magnetoelastic shear wave in a monoclinic medium. Int. J. Eng. Sci. Technol. **3**(2) (2011)
2. Chattopadhyay, A., Gupta, S., Kumari, P., Sharma, V.: Effect of point source and heterogeneity on the propagation of SH-Waves in a viscoelastic layer over a viscoelastic half space. Acta Geophys. **60**(1), 119–139 (2012)
3. Chattopadhyay, A., Kar, B.K.: Love waves due to a point source in an isotropic elastic medium under initial stress. Int. J. Non Linear Mech. **16**(3), 247–258 (1981)
4. Vives, A.A.: Piezoelectric Transducer and Applications. Springer, Berlin (2008)
5. Wu, T.T., Chen, Y.Y.: Surface acoustic waves in layered piezoelectric media and its applications to the analyses of SAW devices. Chin. J. Mech. Eng-En. **19**, 207–214 (2003)
6. Liu, H., Wang, Z.K., Wang, T.J.: Effect of initial stress on the propagation behavior of Love waves in a layered piezoelectric structure. Int. J. Solids Struct. **38**(1), 37–51 (2001)
7. Du, J., Jin, X., Wang, J., Xian, K.: Love wave propagation in functionally graded piezoelectric material layer. Ultrasonics **46**(1), 13–22 (2007)
8. Singh, A.K., Kumar, S., Chattopadhyay, A.: Love-type wave propagation in a piezoelectric structure with irregularity. Int. J. Eng. Sci. **89**, 35–60 (2015)
9. Wang, Q.: SH wave propagation in piezoelectric coupled plates. IEEE T. Ultrason. Ferr. **49**(5), 596–603 (2002)
10. Ewing, W.M., Jardetzky, W.S., Press, F.: Elastic Waves in Layered Media. McGraw-Hill, New York (1957)
11. Gubbins, D.: Seismology and Plate Tectonics. Cambridge University Press, Cambridge (1990)

The Role of Adaptation in Plankton System with Beddington-DeAngelis Type Functional Response

Nilesh Kumar Thakur, Archana Ojha and S. K. Tiwari

Abstract In this paper two interacting species in presence of adaptation (dormancy of the predators such as resting eggs) has been discussed. The dormant stage is an equipment to survive in harsh environment. We have discussed the stability analysis of system without diffusion and in presence of diffusion. Our numerical investigation reveals that above the critical value of interference among the zooplankton the system become stable. Spatiotemporal pattern shows a transient complex spatiotemporal pattern by increasing the time and space.

Keywords Adaptation · Dormancy · Plankton · Spatiotemporal pattern

1 Introduction

In ecological system, many organisms try to find the way to avoid the harsh condition which impacts their survival. Dispersal of the organisms may be one of the ways to avoid harsh condition but it has been observed that certain adaptations such as change in habitat, diet or dormancy enables organism to survive in harsh environmental condition [1–5]. Many planktonic species produce dormant eggs that can survive for a long period [6]. For example, the freshwater zooplankton Daphnia can reproduce both asexually and sexually [7, 8]. Not only dormancy as adaptation but also pools of dormant eggs in sediment plays an important role for survival of zooplankton dynamics in both freshwater and marine environments [9]. There is very few information regarding the dormancy and emerging patterns of plankton dynamics.

N. K. Thakur (✉) · A. Ojha
Department of Mathematics, National Institute of Technology Raipur, Raipur, India
e-mail: nkthakur.maths@nitrr.ac.in

A. Ojha
e-mail: archanaojha1991@gmail.com

S. K. Tiwari
Department of Mathematics, BIT Sindri, Dhanbad 828123, India
e-mail: sktiwari.ism@gmail.com

S. Manna et al. (eds.), *Mathematical Modelling and Scientific Computing with Applications*, Springer Proceedings in Mathematics & Statistics 308,
https://doi.org/10.1007/978-981-15-1338-1_2

Recently, dormancy of the predators reflects much attention due to avoid the harsh environmental condition. A number of research papers available on the dormancy of the predators because it helps to understand the different biological study [10–14]. But very few study on the spatiotemporal of the system in presence of dormancy. In this purposed paper we have considered a minimal plankton model with Beddington-DeAngelis type functional response, and introduced the predator's dormancy. The characteristic feature of Beddington-DeAngelis type functional response $\left((P, Z) = \frac{P}{a+bZ+cP}\right)$ is that if $b = 0$, then it changes into a Holling type-II functional response. If $b < 0$, predators benefit from co-feeding [15]. In this paper our objective to understand the role of adaptation (resting eggs, dormancy state of predator) and emergence of spatiotemporal patterns in enriched environment through a simple mathematical model.

2 Model System

Consider a reaction-diffusion model where phytoplankton and zooplankton populations satisfy the following

$$\begin{aligned}\frac{\partial P}{\partial T} &= rP\left(1 - \frac{P}{K}\right) - \frac{\nu PZ}{\alpha + \beta Z + \gamma P} + D_P \nabla^2 P,\\ \frac{\partial Z}{\partial T} &= \frac{e\nu PZ}{\alpha + \beta Z + \gamma P} - mZ + D_Z \nabla^2 Z.\end{aligned} \tag{2.1}$$

The meaning of parameters and their units are defined in [15].
We take the following notations for non-dimensional form of the model system

$$\begin{aligned}&u = \frac{P}{K}, v = \frac{\nu Z}{K}, x = \frac{X}{L}, y = \frac{Y}{L}, t = rT, \theta = \frac{K}{\alpha r}, \theta' = e\nu\theta,\\ &m = cr, \frac{\beta K}{\alpha \nu} = \xi, \frac{\gamma K}{\alpha} = \eta, d_u = \frac{D_P}{L^2 r}, d_v = \frac{D_Z}{L^2 r}.\end{aligned}$$

Dimensionless form of the model is as follows:

$$\begin{aligned}\frac{\partial u}{\partial t} &= u(1 - u) - \frac{\theta uv}{1 + \xi v + \eta u} + d_u \nabla^2 u,\\ \frac{\partial v}{\partial t} &= \frac{\theta' uv}{1 + \xi v + \eta u} - cv + d_v \nabla^2 v.\end{aligned} \tag{2.2}$$

Now, we introduce the predator dormancy in the model system (2.2), we have

$$
\begin{aligned}
\frac{\partial u}{\partial t} &= u(1-u) - \frac{\theta uv}{1+\xi v+\eta u} + d_u \nabla^2 u, \\
\frac{\partial v}{\partial t} &= \mu(u)\frac{\theta' uv}{1+\xi v+\eta u} + \delta w - cv + d_v \nabla^2 v, \\
\frac{\partial w}{\partial t} &= \tau(u)\frac{\theta' uv}{1+\xi v+\eta u} - \delta w - \rho w + d_w \nabla^2 w,
\end{aligned} \tag{2.3}
$$

with initial condition

$$u(x,0) > 0,\ v(x,0) > 0,\ w(x,0) > 0 \quad \text{for} \quad x \in [0, L]. \tag{2.4}$$

and zero flux boundary conditions

$$\left.\frac{\partial u}{\partial x}\right|_{x=0,L} = \left.\frac{\partial v}{\partial x}\right|_{x=0,L} = \left.\frac{\partial w}{\partial x}\right|_{x=0,L} = 0. \tag{2.5}$$

The zero flux boundary conditions are used for spatially bounded aquatic ecosystems [16].
Here the sigmoid function defined as

$$\mu(u) = \frac{1}{1+\exp(-2\kappa(u-\sigma))} = \frac{1+\tanh(\kappa(u-\sigma))}{2}, \tag{2.6}$$

and $\tau(u) = 1 - \mu(u)$; κ and σ denote the sharpness and the level of the switching effect; δ denotes the hatching of dormant predator and ρ denotes the mortality rate of dormant predator respectively.

3 Stability Analysis of the Model System

Now we analytically study the model system (2.3) without diffusion as well as with diffusion. Without diffusion we take only interaction part of the model system.

3.1 Local Stability of Non-spatial Model

First of all we study the non-spatial model by dropping the diffusion terms from the model system (2.3). In this case, the model system reduces to the form:

$$\begin{aligned}\frac{du}{dt} &= u(1-u) - \frac{\theta uv}{1+\xi v+\eta u},\\ \frac{dv}{dt} &= \mu(u)\frac{\theta' uv}{1+\xi v+\eta u} + \delta w - cv,\\ \frac{dw}{dt} &= \tau(u)\frac{\theta' uv}{1+\xi v+\eta u} - \delta w - \rho w,\end{aligned} \tag{3.1}$$

with Initial conditions

$$u(0) > 0, v(0) > 0, w(0) > 0.$$

Lemma 1 $\Omega = \left\{(u(t), v(t), w(t)) : 0 \le \frac{u(t)}{\theta} + \frac{v(t)+w(t)}{\theta'} \le \frac{(1+\zeta)}{\theta\zeta}\right\}$ *is a region of attraction for all solutions initiating in the interior of the positive octant, where* $\zeta = \min(c, \rho)$.

Proof We define a function

$$\omega = \frac{u(t)}{\theta} + \frac{v(t)+w(t)}{\theta'} \tag{3.2}$$

Differentiating with respect to t, we get

$$\begin{aligned}\frac{d\omega}{dt} &= \frac{1}{\theta}\frac{du}{dt} + \frac{1}{\theta'}\frac{dv+dw}{dt},\\ &= \frac{1}{\theta}u(1-u) - \frac{1}{\theta'}(cv+\rho w),\\ &\le \frac{1}{\theta}u - \frac{1}{\theta'}(cv+\rho w),\\ &\le \frac{u}{\theta} - \min(c,\rho)\frac{(v+w)}{\theta'}.\end{aligned}$$

We choose $\zeta = \min(c, \rho)$, then

$$\frac{d\omega}{dt} + \zeta\omega \le \frac{(1+\zeta)}{\theta}.$$

By comparison lemma for $t \ge \tilde{T} \ge 0$, we obtain

$$\omega(t) \le \frac{(1+\zeta)}{\theta\zeta} - \left[\frac{(1+\zeta)}{\theta\zeta} - \omega(\tilde{T})\right]e^{-\zeta(t-\tilde{T})}.$$

If $\tilde{T} = 0$, then

$$\omega(t) \le \frac{(1+\zeta)}{\theta\zeta} - \left[\frac{(1+\zeta)}{\theta\zeta} - \omega(0)\right]e^{-\zeta t}.$$

For large value of t, we have

$$\omega(t) \leq \frac{(1+\zeta)}{\theta\zeta}.$$

Thus $\omega(t) = \frac{u(t)}{\theta} + \frac{v(t)+w(t)}{\theta'} \leq \frac{(1+\zeta)}{\theta\zeta}$, then all species are uniformly bounded for initial value in R^3_+. This completes the proof of Lemma 1.

For a biologically realistic model, system (3.1) has to be dissipative.
Existence of the interior equilibrium point $E^*(u^*, v^*, w^*)$ has been established below. Here u^*, v^* and w^* are defined as

$$\frac{\theta' u^*}{(1+\xi v^* + \eta u^*)}\left(\mu(u^*)\frac{\tau(u^*)\delta}{(\delta+\rho)}\right) = c, \tag{3.3}$$

$$v^* = \frac{(1-u)(1+\eta u^*)}{(\theta - \xi(1-u^*))}, \tag{3.4}$$

$$w^* = \frac{\tau(u^*)\theta' u^* v^*}{(\delta+\rho)(1+\xi v^* + \eta u^*)}. \tag{3.5}$$

Variational matrix at $E^*(u^*, v^*, w^*)$ is as follows:

$$V = \begin{pmatrix} a_{11} & a_{12} & a_{13} \\ a_{21} & a_{22} & a_{23} \\ a_{31} & a_{32} & a_{33} \end{pmatrix}, \tag{3.6}$$

where

$$a_{11} = 1 - 2u^* - \frac{\theta v^*(1+\xi v^*)}{(1+\xi v^* + \eta u^*)^2}, \quad a_{12} = -\frac{\theta u^*(1+\eta u^*)}{(1+\xi v^* + \eta u^*)^2}, \quad a_{13} = 0,$$
$$a_{21} = \frac{\theta' v^*\big((1+\xi v^* + \eta u^*)u^*\mu'(u^*) + \mu(u^*)(1+\xi v^*)\big)}{(1+\xi v^* + \eta u^*)^2},$$
$$a_{22} = \frac{\theta' u^*\mu(u^*)(1+\eta u^*)}{(1+\xi v^* + \eta u^*)^2} - c, \, a_{23} = \delta,$$
$$a_{31} = \frac{\theta' v^*\big((1+\xi v^* + \eta u^*)u^*\tau'(u^*) + \tau(u^*)(1+\xi v^*)\big)}{(1+\xi v^* + \eta u^*)^2},$$
$$a_{32} = \frac{\theta' u^*\tau(u^*)(1+\eta u^*)}{(1+\xi v^* + \eta u^*)^2}, \, a_{33} = -(\delta+\rho).$$

Now the characteristic equation of matrix V is given by

$$\lambda^3 + A_1\lambda^2 + A_2\lambda + A_3 = 0, \tag{3.7}$$

where

$$\begin{aligned}
A_1 &= -(a_{11} + a_{22} + a_{33}),\\
A_2 &= a_{11}a_{33} + a_{11}a_{22} - a_{12}a_{21} + a_{22}a_{33} - a_{23}a_{32},\\
A_3 &= a_{11}(a_{23}a_{32} - a_{22}a_{33}) + a_{12}(a_{21}a_{33} - a_{31}a_{23}),\\
A_1A_2 - A_3 &= a_{12}(a_{21}a_{22} + a_{23}a_{31}) - a_{11}^2(a_{22} + a_{33}) - (a_{22} + a_{33})\\
&\quad \times (a_{22}a_{33} - a_{23}a_{32}) + a_{11}(a_{12}a_{21} - (a_{22} + a_{33})^2).
\end{aligned}$$

In the following theorem, we are able to find conditions for the positive equilibrium point $E^*(u^*, v^*, w^*)$ to be locally asymptotically stable.

Theorem 1 *If the positive equilibrium point $E^*(u^*, v^*, w^*)$ of the model system (3.1) exists. The equilibrium point $E^*(u^*, v^*, w^*)$ is locally asymptotically stable if the following conditions hold:*

(i) $1 < 2u^* + \frac{\theta v^*(1+\xi v^*)}{(1+\xi v^*+\eta u^*)^2}$,
(ii) $\frac{\theta' u^* \mu(u^*)(1+\eta u^*)}{(1+\xi v^*+\eta u^*)^2} < c$,
(iii) $\frac{\delta\theta' u^* \tau(u^*)(1+\eta u^*)}{(1+\xi v^*+\eta u^*)^2} < (\delta + \rho)\left(\frac{\theta' u^* \mu(u^*)(1+\eta u^*)}{(1+\xi v^*+\eta u^*)^2} - c\right)$.

3.2 Local Stability of Spatial Model

To understand the spatial dynamics of the model system (2.3) we take the form of the system about $E^*(u^*, v^*, w^*)$ as follows:

$$\begin{aligned}
\frac{\partial \bar{u}}{dt} &= a_{11}\bar{u} + a_{12}\bar{v} + a_{13}\bar{w},\\
\frac{\partial \bar{v}}{dt} &= a_{21}\bar{u} + a_{22}\bar{v} + a_{23}\bar{w},\\
\frac{\partial \bar{w}}{dt} &= a_{31}\bar{u} + a_{32}\bar{v} + a_{33}\bar{w}.
\end{aligned} \tag{3.8}$$

where we introduce small perturbations $\bar{u} = u - u^*$, $\bar{v} = v - v^*$ and $\bar{w} = w - w^*$. Let us assume the solution of (3.8) is of the form

$$\begin{pmatrix} \bar{u} \\ \bar{v} \\ \bar{w} \end{pmatrix} = \begin{pmatrix} a_1 \\ a_2 \\ a_3 \end{pmatrix} \exp(\beta t)\cos(kx),$$

where a_1, a_2 and a_3 are sufficiently small constants. k is the wave number and β is the wavelength. The variational matrix of linearized model (3.8) is given by

$$J = \begin{pmatrix} a_{11} - d_u k^2 & a_{12} & 0 \\ a_{21} & a_{22} - d_v k^2 & a_{23} \\ a_{31} & a_{32} & a_{33} - d_w k^2 \end{pmatrix}.$$

The characteristic equation of matrix J is given by

$$\beta^3 + B_1\beta^2 + B_2\beta + B_3 = 0, \tag{3.9}$$

where

$$\begin{aligned} B_1 &= (d_u + d_v + d_w)k^2 + A_1, \\ B_2 &= (d_u d_v + d_v d_w + d_w d_u)k^4 - (d_u(a_{22} + a_{33}) + d_v(a_{33} + a_{11}) \\ &\quad + d_w(a_{11} + a_{22}))k^2 + A_2, \\ B_3 &= d_u d_v d_w k^6 - (a_{11} d_v d_w + a_{22} d_u d_w + a_{33} d_u d_v)k^4 + (d_u(a_{22}a_{33} - a_{23}a_{32}) \\ &\quad + d_v a_{11} a_{33} + d_w(a_{11}a_{22} - a_{12}a_{21}))k^2 + A_3. \end{aligned}$$

Theorem 2 *The equilibrium point $E^*(u^*, v^*, w^*)$ is locally asymptotically stable in the presence of diffusion if and only if:*

$$B_1 > 0, \tag{3.10}$$

$$B_3 > 0, \tag{3.11}$$

$$B_1 B_2 - B_3 > 0, \tag{3.12}$$

From Eq. (3.9) and using the Routh–Hurwitz criterion, the above theorem follows immediately.

4 Numerical Investigation

In this section, present the numerical investigation of the model system (2.3). To understand the intensity of interference between individuals of zooplankton we have plotted the bifurcation diagram, time series and phase portrait of the model system (3.1) for a particular set of parameters value [15].

$$\theta = 6.5, \theta' = 6.05, \xi = 0.02, \eta = 2.5, c = 0.4, \delta = 0.14, \rho = 0.0,$$
$$\kappa = 1.0, \sigma = 1.4. \tag{4.1}$$

All other parameters are same as in Eq. (4.1). Oscillatory dynamics are presented by isolated "." marks in the range $\xi < \xi_c = 3.11$.

Figure 1 describes the bifurcation diagram of the model system (3.1). In Fig. 1 the successive variation of phytoplankton in range $0 \leq u \leq 0.9$ as a function of intensity of interference between individuals of zooplankton (i.e. ξ) which is in the range $0 \leq \xi \leq 5$. Under the bifurcation analysis of model system (3.1), we observe that at the lower value of ξ, the dynamics shows the limit cycle oscillations and higher value of ξ is responsible for the stable dynamics. Oscillatory dynamics are presented by the isolated "dot" marks in the range $\xi < \xi_c$, where $\xi_c = 3.11$ is the critical point of bifurcation and it shows that after the critical value the dynamics become stable (c.f. Fig. 1). In Fig. 2a–d we have observed the time series and phase plane diagram of the model system (3.1). The system dynamics is in oscillatory behaviour at $\xi = 0.02$. As we increase the value of $\xi = 4.0$ the dynamics of model system becomes stable (c.f. Fig. 2c, d). Hence we can say that the model system attains its stability at higher value of ξ which can also be seen from bifurcation diagram. To understand the spatiotemporal dynamics of the model system (2.3) we have taken same set of parameters value as in (4.1) and the diffusion coefficients are $d_u = d_w = 0.0001, d_v = 0.001$. For the above set of values of parameters (4.1), we note that the positive interior equilibrium point E^* exists, and is given by$(u^*, v^*, w^*) =$

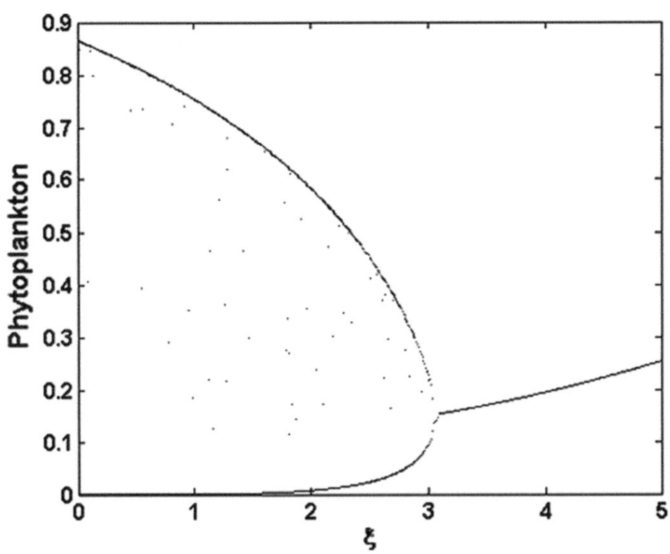

Fig. 1 Bifurcation diagram of phytoplankton for the model system (3.1) with respect to ξ

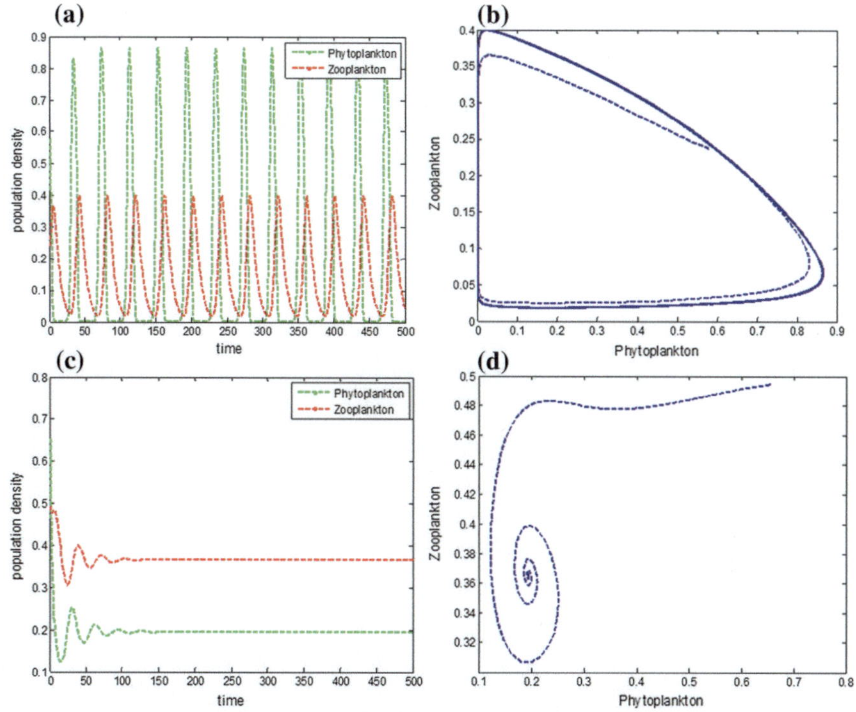

Fig. 2 Time series and phase plane diagram of (3.1) at **a**, **b** $\xi = 0.02$, and **c**, **d** $\xi = 4.0$

(0.0795, 0.1702, 0.4540). For spatiotemporal pattern we use the semi-implicit (in time) finite-difference method [17]. We consider the following initial condition.

$$u(x,0) = u^* + \varepsilon_1 \sin\left(\frac{2\pi(x - x_0)}{0.2}\right), v(x,0) = v^* + \varepsilon_1 \sin\left(\frac{2\pi(x - x_0)}{0.2}\right)$$
$$w(x,0) = w^* + \varepsilon_1 \sin\left(\frac{2\pi(x - x_0)}{0.2}\right) \tag{4.2}$$

where $\varepsilon_1 = 5 \times 10^{-4}$, $x_0 = 0.1$, $(u^*, v^*, w^*) = (0.0795, 0.1702, 0.4540)$.
To obtain the time series of the model system (2.3) we take $(u_0, v_0, w_0) = (0.8, 0.6, 0.001)$. With the above initial condition, we observed that the system shows the oscillatory behaviour (c.f., Fig. 3) and dormancy reduces the fluctuation in population densities. Figure 4 shows the complex spatiotemporal pattern (mixture of standing and travelling waves) in the presence as well as in the absence of dormancy of the predator.

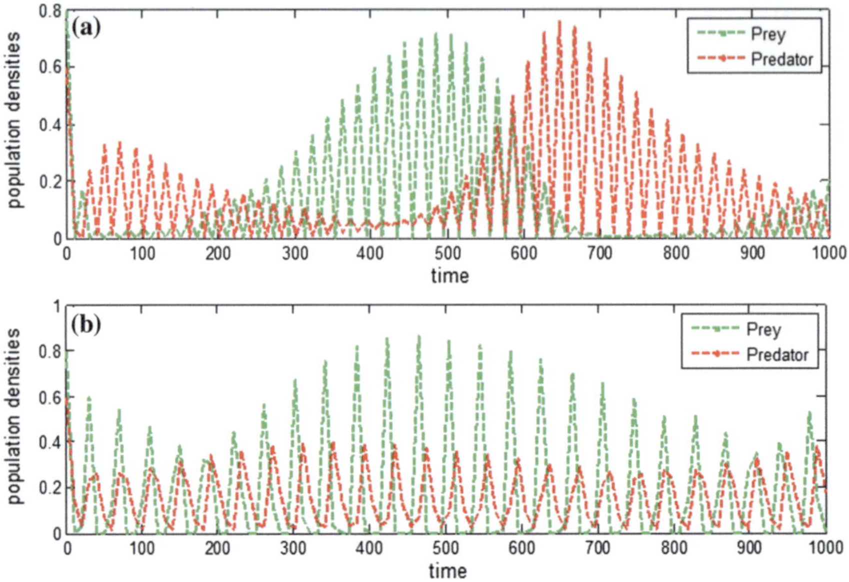

Fig. 3 Time series of the model system (2.3), **a** without dormancy, **b** with dormancy

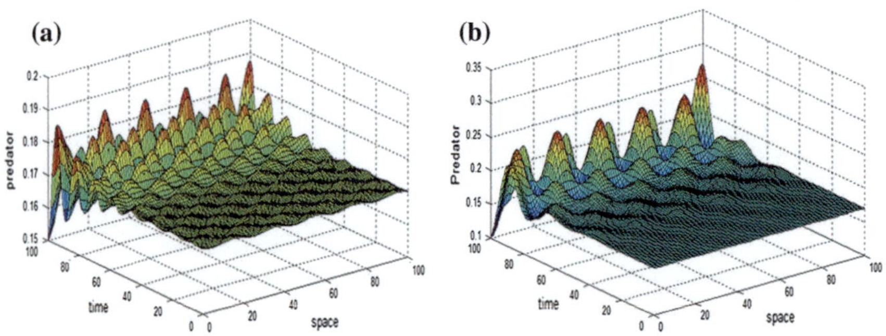

Fig. 4 Complex spatiotemporal patterns of predator density of the model system (2.3), **a** without dormancy, **b** with dormancy

Now, to understand the effect of space and time we have plotted the spatiotemporal patterns of the model system (2.3). First of all we have increased the value of time from $t = 5$ to $t = 70$, and observed that for $t = 5$ a spatially periodic pattern appears, and on increasing the value $t = 70$ the system exhibits periodic pattern in space and time both (c.f., Fig. 5). Similarly, as we increase the value of space $x = 1$ to $x = 19$ the system exhibits transient complex spatiotemporal complex patterns (c.f., Fig. 6).

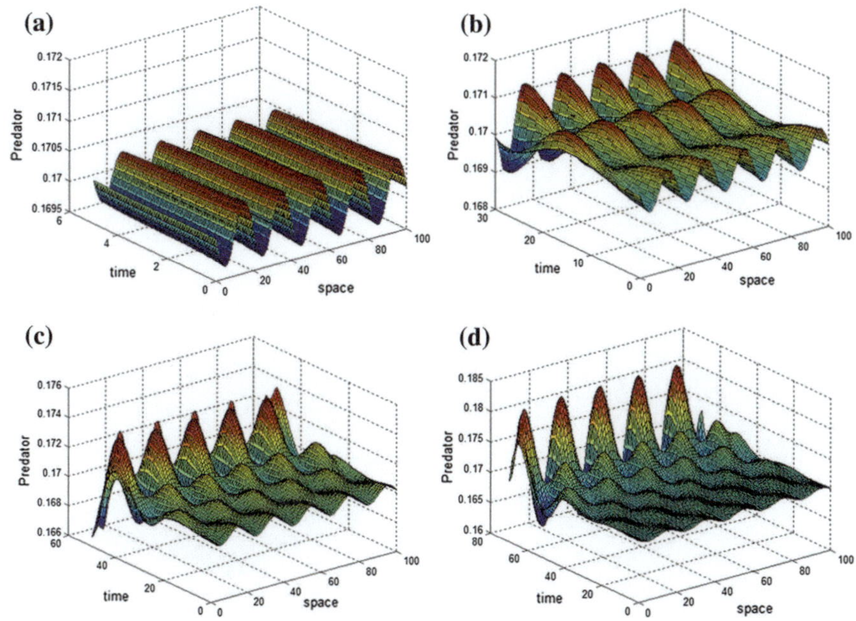

Fig. 5 Complex spatiotemporal patterns of predator density of the model system (2.3) with dormancy for **a** $t = 5$, **b** $t = 30$, **c** $t = 50$ and **d** $t = 70$

5 Discussions and Conclusions

In this paper we have analysed the dynamical behaviour of plankton system (phytoplankton-zooplankton) in the presence of predator's dormancy. The dormancy enables many animals and plants to survive harsh environmental conditions. Therefore, the theoretical as well as numerical study presented in this paper is essential for understanding the ecological implications of dormancy. Our theoretical study shows that the system is uniformly bounded and locally asymptotic stable in the presence of predator's dormancy under some conditions. With the help of numerical simulation we have plotted bifurcation diagram, time series and spatiotemporal pattern of the model system (2.3) without diffusion as well as with diffusion. Our numerical investigation reveals that for $\xi < \xi_c$, where $\xi_c = 3.11$ the dynamics shows the limit cycle oscillations and for $\xi > \xi_c$ the system become stable dynamics (c.f. Figs. 1 and 2). Time series and spatiotemporal patterns of the model system (2.3) show that the dormancy suppresses the fluctuation in population densities (c.f., Figs. 3 and 4). From Fig. 5 we observed that at $t = 5$ spatially periodic pattern appears and as we increase the value of t system shows periodic pattern in space and time both. Similarly, as we increase the value of space from $x = 1$ to $x = 19$ the system exhibits transient complex patterns which is a mixture of spatially periodic steady states and travelling waves (c.f., Fig. 6).

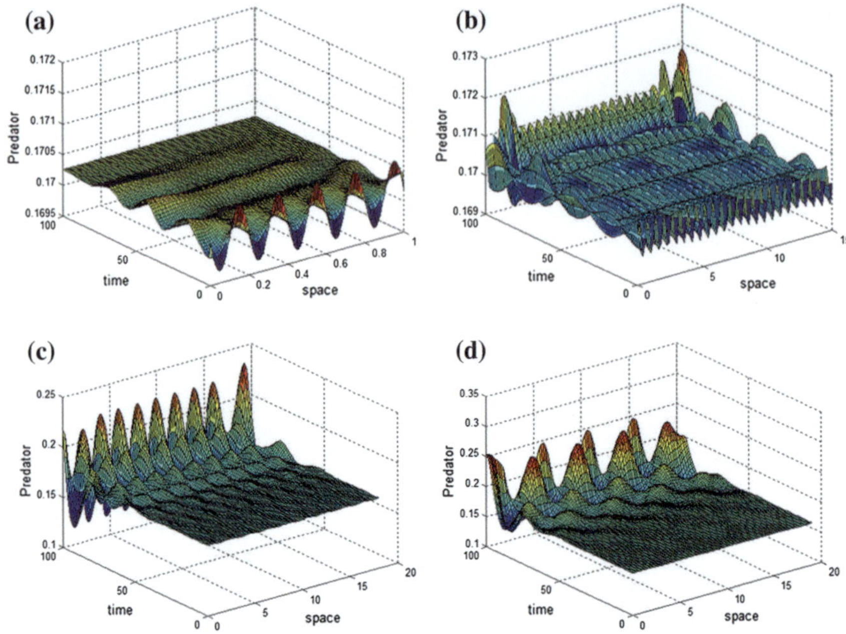

Fig. 6 Complex spatiotemporal patterns of predator density of the model system (2.3) with dormancy with different values of space **a** $x = 1$, **b** $x = 15$, **c** $x = 18$ and **d** $x = 19$

Acknowledgements This research work is supported by Chhattisgarh Council of Science and Technology, India under grant no. 2238/CCOST/MRP/2015 to the corresponding author (Nilesh Kumar Thakur).

References

1. Harper, J.L.: Population Biology of Plants. Academic Press, London (1977)
2. Wiggins, G.B., Mackay, R.J., Smith, I.M.: Evolutionary and ecological strategies of animals in annual temporary pools. Arch. Hydrobiol. Suppl. **58**, 97–206 (1980)
3. Henis, Y.: Survival and Dormancy of Microorganisms. Wiley, Hoboken (1987)
4. Piltz, S.H., Porter, M.A., Maini, P.K.: Prey switching with a linear performance trade-off. SIAM J. Appl. Dyn. Syst. **13**, 658–682 (2014)
5. De Stasio Jr., B.T.: The role of dormancy and emergence patterns in the dynamics of a freshwater zooplankton community. Limnol. Oceanogr. **35**, 1079–1090 (1990)
6. Hutchinson, G.E.: A Treatise on Limnology. Wiley, Hoboken (1967)
7. Slusarczyk, M.: Food threshold for diapause in Daphnia under the threat of fish predation. Ecology **82**, 1089–1096 (2001)
8. Slusarczyk, M.: Predator-induced diapauses in Daphnia. Ecology **76**, 1008–1013 (1995)
9. De Stasio, B.T.: The role of dormancy and emergence patterns in the dynamics of a freshwater zooplankton community. Limnol. Oceanogr. **35**, 1079–1090 (1990)
10. Kuwamura, M., Nakazawa, T., Ogawa, T.: A minimum model of prey-predator system with dormancy of predators and the paradox of enrichment. J. Math. Biol. **58**, 459–479 (2009)

11. Kuwamura, M., Chiba, H.: Mixed-mode oscillations and chaos in a prey-predator system with dormancy of predators. Chaos **19**, 043121 (2009)
12. Nakazawa, T., Kuwamura, M., Yamamura, N.: Implications of resting eggs of zooplankton for the paradox of enrichment. Popul. Ecol. **53**, 341–350 (2011)
13. Wang, J., Jiang, W.: Bifurcation and chaos of a delayed predator-prey model with dormancy of predators. Nonlinear Dyn. **69**, 1541–1558 (2012)
14. Kuwamura, M.: Turing instabilities in prey-predator systems with dormancy of predators. J. Math. Biol. **71**, 125–149 (2015)
15. Upadhyay, R.K., Thakur, N.K., Rai, V.: Diffusion driven instabilities and spatio-temporal patterns in an aquatic predator-prey system with Beddington-DeAngelis type functional response. Int. J. Bifur. Chaos **21**, 663–684 (2011)
16. Scheffer, M.: Ecology of Shallow Lakes. Chapman and Hall, London (1998)
17. Garvie, M.R.: Finite difference schemes for reaction-diffusion equations modeling predator-prey interactions in MATLAB. Bull. Math. Biol. **69**, 931–956 (2007)

Exact Traveling Wave Solutions and Bifurcation Analysis for Time Fractional Dual Power Zakharov-Kuznetsov-Burgers Equation

Amiya Das

Abstract In this paper, we introduce the time fractional dual power Zakharov-Kuznetsov-Burgers equation in the sense of modified Riemann-Liouville derivative. We briefly describe one direct ansatz method namely (G'/G)-expansion method in adherence of fractional complex transformation and applying this method exploit miscellaneous exact traveling wave solutions including solitary wave, kink-type wave, breaking wave and periodic wave solutions of the equation. Next we investigate the dynamical behavior, bifurcations and phase portrait analysis of the exact traveling wave solutions of the system in presence and absence of damping effect. Moreover, we demonstrate the exceptional features of the traveling wave solutions and phase portraits of planar dynamical system via interesting figures.

Keywords Fractional differential equation · Time fractional dual power ZK-Burgers equation · Traveling wave solution · (G'/G)-expansion method · Bifurcation analysis

1 Introduction

Fractional differential equations (FDEs) has extensive applications in almost all fields of science and engineering [1–5]. FDEs can be applied in execution of the modeling with long-range time memory and spatial interaction far better than integer order partial differential equations. The existence and uniqueness of the solution of Cauchy type problems, the boundary value problems using fixed-points methods, different methods to obtain exact and numerical solutions and their stability [6–9] can be easily handled using FDEs. In literature a number of definitions of fractional derivatives are available. Some of the popular fractional derivatives are due to Riemann, Liouville, Grünwald, Weyl [2] etc. Very recently a modified form of the Riemann-Liouville fractional derivative has been introduced by Jumarie [10]. In the field of nonlinear

A. Das (✉)
Department of Mathematics, Kazi Nazrul University, Asansol 713340, India
e-mail: amiya620@gmail.com

S. Manna et al. (eds.), *Mathematical Modelling and Scientific Computing with Applications*, Springer Proceedings in Mathematics & Statistics 308,
https://doi.org/10.1007/978-981-15-1338-1_3

evolution equations, a large number of powerful direct methods are available such as (G'/G)-expansion method, Kudryashov method, exponential function method, first integral method, F-expansion method [11–18] etc. to exploit exact analytical solutions of FDEs.

Nonlinear partial differential equations (NLPDEs) has been remained at the center of immense practical importance due to its presence in diversified physical systems such as water wave theory, plasma physics, condensed matters physics, lattice dynamics, nonlinear optics etc. [19–24]. The pioneer NLPDEs are KdV equation [25], sine-Gordon equation [26], nonlinear Schrödinger equation [27] etc. The exact solutions helps us to understand the mechanism of the complicated physical phenomena and dynamical processes modelled by these nonlinear evolution equations [28] in a simplified way. Nowadays another aspect of studying NLPDEs become investigation of dynamical behavior of the exact traveling wave solutions and their phase portrait analysis in view of dynamical system theory. Jiang et al. [29] investigated the dynamical behavior of equilibrium points and the bifurcations of phase portraits of the traveling wave solutions for the CH-γ equation. Das et al. [30, 31] studied the existence and stability analysis of dispersive solutions of the KP-BBM equation and KP equation in presence of dispersion effect. Scientific research has been initiated along the generalization of physically important NLPDEs in the context of fractional derivatives and discuss the existence of exact traveling wave solutions of these equations [32, 33]. We plan to enquire the dynamical behavior of these traveling wave solutions and their stability analysis using dynamical system theory. A study in this direction may enrich the understanding of dynamical behavior of nonlinear fractional partial differential equations.

We consider the dimensionless form of the Zakharov-Kuznetsov (ZK) equation with dual-power law nonlinearity [34–36]

$$q_t + (aq^n + bq^{2n})q_x + c(q_{xx} + q_{yy})_x = 0\,. \tag{1}$$

Here the exponents n and $2n$ depicts dual-power laws with $n > 0$ and a, b and $c \neq 0$ are real constants. In case of $n = 1$, Eq. (1) reduces to the dual power ZK equation

$$q_t + (aq + bq^2)q_x + c(q_{xx} + q_{yy})_x = 0\,. \tag{2}$$

In Yan et al. [37] demonstrated the symmetry reductions and obtained some polynomial solutions, elliptic periodic solutions and triangular function solutions of Eq. (2).

The 1-soliton solution of Eq. (1) with dual-power law nonlinearity has been derived by Biswas et al. [36] using the solitary wave ansatz method.

Now we consider the time fractional Burgers equation

$$q_t + aqq_x = bq_{xx}\,, \tag{3}$$

where x and t indicates the space variable and time variable respectively and q denote electrostatic potential. Equation (3) describes dust-acoustic shock wave [38], nonlinear ion-acoustic shock wave [39], dust-ion acoustic shock wave [40] and positron-acoustic shock wave [41] in plasmas.

In this article we introduce the time fractional dual power ZK-Burgers equation as

$$D_t^\alpha q + (aq + bq^2)q_x + c(q_{xx} + q_{yy})_x + \mu(q_{xx} + q_{yy}) = 0\,, \tag{4}$$

where a, b, c are real parameters and μ is a damping parameter.

2 Overview on Fractional Calculus

A few well known fractional order derivatives are due to Riemann, Liouville, Riemann-Liouville, Grünwald, Weyl, Grünwald-Letnikov, Caputo and Riesz fractional derivatives [2]. In this paper, we apply the Jumarie's modified Riemann-Liouville form [10]. The fractional derivative of a function $f : \mathbb{R} \to \mathbb{R}$ of order α in the sense of Jumarie's modified Riemann-Liouville form is defined as

$$D_x^\alpha = \begin{cases} \frac{1}{\Gamma(-\alpha)} \frac{\mathrm{d}}{\mathrm{d}x} \int_0^x (x-\eta)^{-\alpha-1}[f(\eta) - f(0)]\,\mathrm{d}\eta\,, \alpha < 0\,, \\ \frac{1}{\Gamma(1-\alpha)} \frac{\mathrm{d}}{\mathrm{d}x} \int_0^x (x-\eta)^{-\alpha}[f(\eta) - f(0)]\,\mathrm{d}\eta\,, 0 < \alpha < 1\,, \\ (f^{(n)}(x))^{(\alpha-n)}\,, n \le \alpha < n+1\,, n \ge 1\,. \end{cases}$$

Moreover, some interesting properties of the modified Riemann-Liouville derivative can be listed as

$$D_x^\alpha x^s = \frac{\Gamma(1+s)}{\Gamma(1+s-\alpha)} x^{s-\alpha}\,, \quad s > 0\,, \tag{5}$$

$$D_x^\alpha [f(x)g(x)] = g(x) D_x^\alpha f(x) + D_x^\alpha g(x)\,, \tag{6}$$

$$D_x^\alpha f[u(x)] = f_u'[u(x)] D_x^\alpha u(x)\,. \tag{7}$$

The complex fractional transformation [42] is defined as

$$q(x, y, t) = q(\xi)\,, \qquad \xi = \frac{kt^\alpha}{\Gamma(\alpha+1)} + \frac{lx^\beta}{\Gamma(\beta+1)} + \frac{my^\gamma}{\Gamma(\gamma+1)}\,, \tag{8}$$

where k, l, m are arbitrary constants which is frequently used to convert a fractional differential equation into an ordinary differential equation.

3 (G'/G)-Expansion Method for Fractional Differential Equations

Consider a general nonlinear FDE in an unknown function $q = q(x, y, t)$

$$P(q, D_t^{\alpha} q, D_x^{\beta} q, D_y^{\gamma} q, D_t^{2\alpha} q, D_x^{2\beta} q, D_y^{2\gamma} q, D_t^{\alpha} D_x^{\beta} D_y^{\gamma} q, \ldots) = 0, \quad 0 < \alpha, \beta, \gamma < 1, \tag{9}$$

where P is a polynomial of q and its fractional partial derivatives.

The traveling wave variable can be considered as the fractional complex transformation

$$q(x, y, t) = q(\xi), \qquad \xi = \frac{Kt^{\alpha}}{\Gamma(\alpha + 1)} + \frac{Lx^{\beta}}{\Gamma(\beta + 1)} + \frac{My^{\gamma}}{\Gamma(\gamma + 1)}, \tag{10}$$

which transforms Eq. (9) into a nonlinear ordinary differential equation of integer order

$$P(q, q', q'', q''', \ldots) = 0, \tag{11}$$

where $'$ depicts derivative with respect to ξ. If possible, Eq. (11) is to be integrated term by term once or more.

The (G'/G)-expansion method [11, 12] can be briefed in the following steps.
Step 1: Assume that the solution of Eq. (11) can be expressed through a polynomial in (G'/G) as

$$q(\xi) = \sum_{i=1}^{m} (G'/G)^i, \quad a_m \neq 0, \tag{12}$$

where $a_i (i = 0, 1, 2, \ldots, m)$ are constants and $G = G(\xi)$ satisfies the following auxiliary equation

$$G'' + \lambda G' + \mu G = 0, \tag{13}$$

where λ and μ are constants.
Step 2: The homogeneous balance between the highest order derivative and the nonlinear term appearing in Eq. (11) provides the value of positive integer m.
Step 3: After substituting Eq. (12) along with (13) into Eq. (11) and collecting all terms having same power of $(G'/G)^i, i = 0, 1, 2, \ldots$, equate each coefficient of the polynomial to zero. This generates a set of algebraic equations in $a_i (i = 0, 1, 2, \ldots, m)$, λ, μ, K, L and M.
Step 4: The solution of the algebraic equation along with the values of $a_i (i = 0, 1, 2, \ldots, m)$, λ, μ, K, L, M and the general solutions of Eqs. (13) into (12) yields a variety of exact solutions of (9).

4 Application of the (G'/G)-Expansion Method on Time Fractional Dual Power ZK-Burgers Equation

In this section we apply the (G'/G)-expansion method on the time fractional dual power ZK-Burgers equation (4) and obtain exact analytical solution. Consider Eq. (25)

$$-vq + \frac{al}{2}q^2 + \frac{bl}{3}q^3 + clq'' + \mu q' = 0\,, \tag{14}$$

where "q'" $= \frac{\mathrm{d}q}{\mathrm{d}\xi}$.

The homogeneous balance between q'' and q^3 in (14) yields

$$2 + m = 3m \quad \Rightarrow m = 1\,. \tag{15}$$

Assume that the solution of (14) can be expressed by a polynomial in (G'/G) as

$$q(\xi) = a_0 + a_1(G'/G)\,, \quad a_1 \neq 0\,, \tag{16}$$

where $G = G(\xi)$ satisfies the second order ordinary differential equation

$$G'' + \lambda G' + \mu G = 0\,. \tag{17}$$

The values of a_0, a_1 are unknowns and to be determined later. Substituting (16) in (14) with the help of (17), we find a polynomial equation in (G'/G). Collecting the coefficients of $(G'/G)^i$, $i = 0, 1, 2, 3$ and equating each coefficients to zero we find a set of algebraic equations. Solving the set of algebraic equations using Mapple we get

$$\begin{aligned} a_0 &= -\frac{1}{2}\frac{al\sqrt{-\frac{6c}{b}} + 6cl\lambda - 2m}{bl\sqrt{-\frac{6c}{b}}}\,, a_1 = \sqrt{-\frac{6c}{b}}, \\ v &= -\frac{6bl^2c^2\lambda^2 + 2bm^2 + 3ca^2l^2 - 24bl^2c^2\mu}{12bcl}, \end{aligned} \tag{18}$$

where λ, μ, l and m are arbitrary constants.

Using (18), expression (16) becomes

$$q(\xi) = -\frac{1}{2}\frac{al\sqrt{-\frac{6c}{b}} + 6cl\lambda - 2m}{bl\sqrt{-\frac{6c}{b}}} + \sqrt{-\frac{6c}{b}}(G'/G)\,, \tag{19}$$

where $\xi = lx + my + \frac{6bl^2c^2\lambda^2 + 2bm^2 + 3ca^2l^2 - 24bl^2c^2\mu}{12bcl}\frac{t^\alpha}{\Gamma(\alpha+1)}$.

We substitute the general solution of the auxiliary equation (17) in (19) and obtain following traveling wave solutions of (4) as follows:

When $\lambda^2 - 4\mu > 0$,

$$q_1(\xi) = -\frac{1}{2}\frac{al\sqrt{-\frac{6c}{b}} + 6cl\lambda - 2m}{bl\sqrt{-\frac{6c}{b}}} + \frac{1}{2}\sqrt{\lambda^2 - 4\mu}\sqrt{-\frac{6c}{b}}\left(\frac{C_1 \sinh\frac{1}{2}\sqrt{\lambda^2 - 4\mu}\xi + C_2 \cosh\frac{1}{2}\sqrt{\lambda^2 - 4\mu}\xi}{C_1 \cosh\frac{1}{2}\sqrt{\lambda^2 - 4\mu}\xi + C_2 \sinh\frac{1}{2}\sqrt{\lambda^2 - 4\mu}\xi}\right), \tag{20}$$

where $\xi = lx + my + \frac{6bl^2c^2\lambda^2 + 2bm^2 + 3ca^2l^2 - 24bl^2c^2\mu}{12bcl}\frac{t^\alpha}{\Gamma(\alpha+1)}$, C_1 and C_2 are arbitrary constants.

For particular values of C_1 and C_2 for instance, $C_1 > 0$, $C_1^2 > C_2^2$, $q_1(\xi)$ can be expressed as

$$q_1(\xi) = -\frac{1}{2}\frac{al\sqrt{-\frac{6c}{b}} + 6cl\lambda - 2m}{bl\sqrt{-\frac{6c}{b}}} + \frac{1}{2}\sqrt{\lambda^2 - 4\mu}\sqrt{-\frac{6c}{b}}\tanh\left(\frac{1}{2}\sqrt{\lambda^2 - 4\mu}\xi + \xi_0\right), \tag{21}$$

where

$$\xi_0 = \tanh^{-1}\frac{C_2}{C_1}, \xi = lx + my + \frac{6bl^2c^2\lambda^2 + 2bm^2 + 3ca^2l^2 - 24bl^2c^2\mu}{12bcl}\frac{t^\alpha}{\Gamma(\alpha+1)}.$$

It is a kink-type solution (see Figs. 1 and 2).

When $\lambda^2 - 4\mu < 0$,

$$q_2(\xi) = -\frac{1}{2}\frac{al\sqrt{-\frac{6c}{b}} + 6cl\lambda - 2m}{bl\sqrt{-\frac{6c}{b}}} + \frac{1}{2}\sqrt{4\mu - \lambda^2}\sqrt{-\frac{6c}{b}}\left(\frac{-C_1 \sin\frac{1}{2}\sqrt{4\mu - \lambda^2}\xi + C_2 \cos\frac{1}{2}\sqrt{4\mu - \lambda^2}\xi}{C_1 \cos\frac{1}{2}\sqrt{4\mu - \lambda^2}\xi + C_2 \sin\frac{1}{2}\sqrt{4\mu - \lambda^2}\xi}\right), \tag{22}$$

where $\xi = lx + my + \frac{6bl^2c^2\lambda^2 + 2bm^2 + 3ca^2l^2 - 24bl^2c^2\mu}{12bcl}\frac{t^\alpha}{\Gamma(\alpha+1)}$, C_1 and C_2 are arbitrary constants.

When $\lambda^2 - 4\mu = 0$,

$$q_3(\xi) = -\frac{1}{2}\frac{al\sqrt{-\frac{6c}{b}} + 6cl\lambda - 2m}{bl\sqrt{-\frac{6c}{b}}} + \sqrt{-\frac{6c}{b}}\frac{C_2}{C_1 + C_2\xi}, \tag{23}$$

where $\xi = lx + my + \frac{2bm^2 + 3ca^2l^2}{12bcl}\frac{t^\alpha}{\Gamma(\alpha+1)}$, C_1 and C_2 are arbitrary constants.

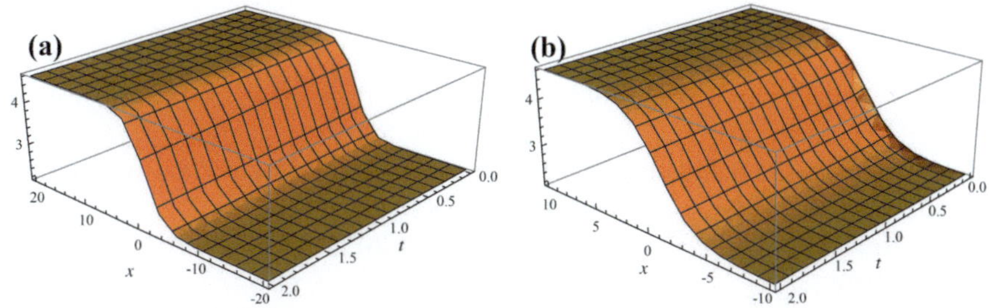

Fig. 1 Kink-type solution (21) for **a** $\alpha = 1$, **b** $\alpha = 1/2$ with $a = b = 1, c = -1, l = m = \frac{1}{\sqrt{2}}, \lambda = 1, \mu = 0, \xi_0 = 0.5$

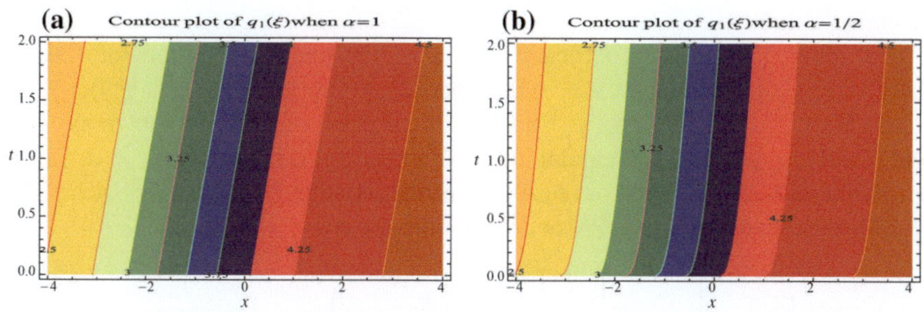

Fig. 2 Contour plot of the kink-type solution (21) for **a** $\alpha = 1$, **b** $\alpha = 1/2$ with the same parameters

5 Bifurcation Analysis and Phase Portraits of Time Fractional Dual Power ZK-Burgers Equation

In this section we discuss all possible periodic motions of Eq. (4) by considering the traveling wave transformation $\xi = lx + my - \frac{vt^\alpha}{\Gamma(\alpha+1)}$, where l and m are direction cosines of the line of propagation of the traveling wave with velocity v in the xy-plane such that $l^2 + m^2 = 1$. Then Eq. (4) becomes

$$-vq_\xi + l(aq + bq^2)q_\xi + lcq_{\xi\xi\xi} + \mu q_{\xi\xi} = 0. \tag{24}$$

Integrating Eq. (24) we obtain

$$-vq + \frac{al}{2}q^2 + \frac{bl}{3}q^3 + clq_{\xi\xi} + \mu q_\xi = 0, \tag{25}$$

where we choose the value of constant of integration as zero. Clearly Eq. (25) is equivalent to the planar dynamical system

$$q_\xi = z\,,$$
$$z_\xi = \frac{v}{cl}q - \frac{a}{2c}q^2 - \frac{b}{3c}q^3 - \frac{\mu}{cl}z\,. \tag{26}$$

Here we discuss the qualitative behavior of the traveling waves of the system (26) by using bifurcation theory of planar dynamical system [43].

The critical points for the dynamical system (26) are given by $(0,0)$, $(q_1,0)$, $(q_2,0)$, where

$$q_{1,2} = \frac{-3al \pm \sqrt{9a^2l^2 + 48blv}}{4bl}\,. \tag{27}$$

Note that the number of critical points depends on the value of v. The details are given in Table 1.

Let $M(\phi_e,\psi_e)$ be the coefficient matrix of the linearized system of (26) at fixed point (ϕ_e,ψ_e) and $J=\det(M(\phi_e,\psi_e))$, $T_1=$trace$(M(\phi_e,\psi_e))$, $T_2=(\text{trace}(M(\phi_e,\psi_e)))^2$. Let (ϕ_e,ψ_e) be an critical point of a planar integrable system. Then (ϕ_e,ψ_e) is a saddle point if $J<0$, a centre point if $J>0$, $T_1=0$, a cusp if $J=0$ and Poincaré index of (ϕ_e,ψ_e) is zero, a node if $J>0$ and $T_2-4J>0$.

The linearized system of (26) at critical point q_0 is

$$\begin{bmatrix} q' \\ z' \end{bmatrix} = \begin{bmatrix} 0 & 1 \\ \frac{v}{cl} - \frac{a}{c}q_0 - \frac{b}{c}q_0^2 & -\frac{\mu}{cl} \end{bmatrix} \begin{bmatrix} q-q_0 \\ z \end{bmatrix} \equiv M \begin{bmatrix} q-q_0 \\ z \end{bmatrix}, \tag{28}$$

where M denotes the coefficient matrix.

Now we discuss the bifurcations of traveling waves via phase portraits of Eq. (26) in the (q,z) phase plane different parametric values as given in Table 1.

Case 1: $v>0>-\frac{3a^2l}{16b}$. At $(0,0)$, $J(0,0)=-\frac{v}{cl}<0$ and hence $(0,0)$ is a saddle point. At $(q_1,0)$, $J(q_1,0)>0$, $T_1(q_1,0)=-\frac{\mu}{cl}$ and $T_2(q_1,0)-4J(q_1,0)<0$ with $3a^2l+16bv-\sqrt{3a^2l(3a^2l+16bv)}>2b\mu^2$. Thus $(q_1,0)$ is a stable spiral. Again at $(q_2,0)$, $J(q_2,0)>0$, $T_1(q_2,0)=-\frac{\mu}{cl}$ and $T_2(q_2,0)-4J(q_2,0)<0$ with $3a^2l+16bv+\sqrt{3a^2l(3a^2l+16bv)}>2b\mu^2$. Thus $(q_2,0)$ is also a stable spiral (see Fig. 3a).

Table 1 Critical points of (26) for different choices of v

Value of v	Number of critical points	Critical point
$v>-\frac{3a^2l}{16b}$, $v\neq 0$ *either* $v>0>-\frac{3a^2l}{16b}$ *or* $-\frac{3a^2l}{16b}<v<0$	3	$(0,0)$, $(q_1,0)$, $(q_2,0)$
$v=-\frac{3a^2l}{16b}$	2	$(0,0)$, $(-\frac{3a}{4b},0)$
$v<-\frac{3a^2l}{16b}$	1	$(0,0)$
$v=0$	2	$(0,0)$, $(-\frac{3a}{2b},0)$

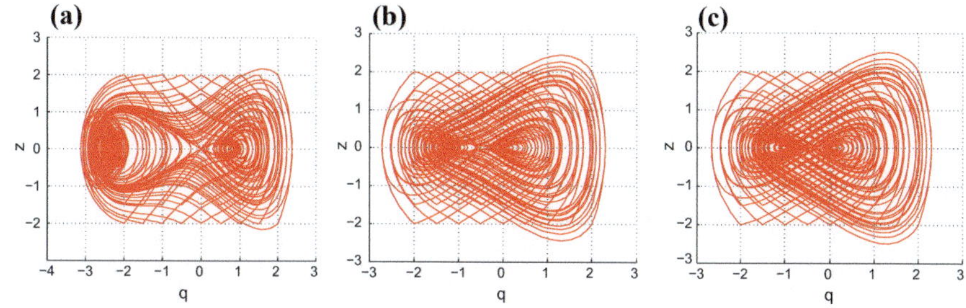

Fig. 3 Phase portrait of (26) for **a** $v > 0 > -\frac{3a^2l}{16b}$, **b** $-\frac{3a^2l}{16b} < v < 0$, **c** $v = \frac{3a^2l}{16b}$ with $a = b = c = 1$ and $l = m = \frac{1}{\sqrt{2}}$

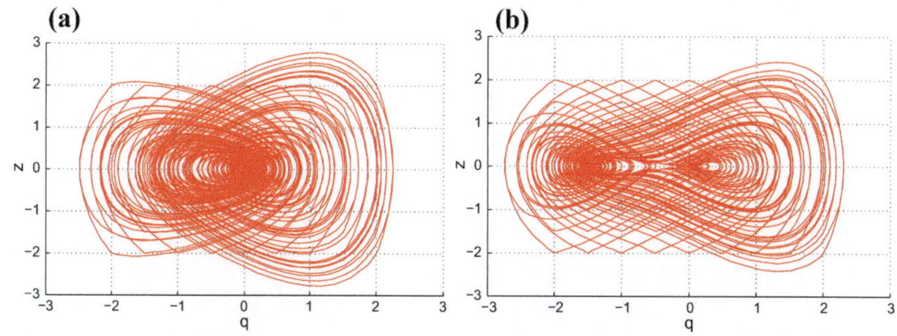

Fig. 4 Phase portrait of (26) for **a** $v < -\frac{3a^2l}{16b}$, **b** $v = 0$ with $a = b = c = 1$ and $l = m = \frac{1}{\sqrt{2}}$

Case 2: $-\frac{3a^2l}{16b} < v < 0$. At $(0, 0)$, $J(0, 0) = -\frac{v}{cl} > 0$, $T_1(0, 0) = -\frac{\mu}{cl}$ and $T_2(0, 0) - 4J(0, 0) < 0$ with $-4vcl > \mu^2$. Hence $(0, 0)$ is a stable spiral. At $(q_1, 0)$, $J(q_1, 0) < 0$ and $(q_1, 0)$ is a saddle point. At $(q_2, 0)$, $J(q_2, 0) > 0$, $T_1(q_2, 0) = -\frac{\mu}{cl}$ and $T_2(q_2, 0) - 4J(q_2, 0) < 0$ with $3a^2l + 16bv + \sqrt{3a^2l(3a^2l + 16bv)} > 2b\mu^2$. Thus $(q_2, 0)$ is also a stable spiral (see Fig. 3b).
Case 3: $v = \frac{3a^2l}{16b}$. At $(0, 0)$, $J(0, 0) = \frac{3a^2}{16bc} > 0$, $T_2(0, 0) - 4J(0, 0) < 0$ with $-4vcl > \mu^2$. Hence $(0, 0)$ is a stable spiral. At $\left(-\frac{3a}{4b}, 0\right)$, $J\left(-\frac{3a}{4b}, 0\right) = 0$ and $\left(-\frac{3a}{4b}, 0\right)$ is a high order equilibrium point (see Fig. 3c).
Case 4: $v < -\frac{3a^2l}{16b}$. At $(0, 0)$, $J(0, 0) = -\frac{v}{cl} > 0$, $T_2(0, 0) - 4J(0, 0) < 0$ and $(0, 0)$ is a stable spiral (see Fig. 4a).
Case 5: $v = 0$. At $(0, 0)$, $J(0, 0) = -\frac{v}{cl} = 0$ and $(0, 0)$ is a high order equilibrium point. At $\left(-\frac{3a}{2b}, 0\right)$, $J\left(-\frac{3a}{2b}, 0\right) = \frac{3a^2}{4bc} > 0$, $T_2\left(-\frac{3a}{2b}, 0\right) - 4J\left(-\frac{3a}{2b}, 0\right) < 0$ with $-4vcl > \mu^2$. Thus $\left(-\frac{3a}{2b}, 0\right)$ is a stable spiral (see Fig. 4b).

In case of no damping $\mu = 0$ and the dynamical system (26) takes the form

$$q_\xi = z\,,$$
$$z_\xi = \frac{v}{cl}q - \frac{a}{2c}q^2 - \frac{b}{3c}q^3\,. \tag{29}$$

This is a planar Hamiltonian system with Hamiltonian function

$$\mathcal{H}(q,z) = z^2 - \frac{v}{cl}q^2 - \frac{a}{3c}q^3 - \frac{b}{6c}q^4. \tag{30}$$

The linearized system of (29) at critical point q_0 is

$$\begin{bmatrix} q' \\ z' \end{bmatrix} = \begin{bmatrix} 0 & 1 \\ \frac{v}{cl} - \frac{a}{c}q_0 - \frac{b}{c}q_0^2 & 0 \end{bmatrix} \begin{bmatrix} q - q_0 \\ z \end{bmatrix} \equiv N \begin{bmatrix} q - q_0 \\ z \end{bmatrix}, \tag{31}$$

where N denotes the coefficient matrix.

We discuss the bifurcations of traveling waves via phase portraits of Eq. (29) in the (q, z) phase plane for different parametric values as given in Table 1.

Case 1: $v > 0 > -\frac{3a^2l}{16b}$. At $(0, 0)$, $J(0, 0) < 0$ and hence $(0, 0)$ is a saddle point. At $(q_{1,2}, 0)$, $J(q_{1,2}, 0) > 0$ and $T_1(q_{1,2}, 0) = 0$. Thus $(q_{1,2}, 0)$ are two centers and we obtain a pair of homoclinic orbits at $(0, 0)$ and two families of periodic orbits about $(q_{1,2}, 0)$ (see Fig. 5a).

Case 2: $-\frac{3a^2l}{16b} < v < 0$. At $(q_1, 0)$, $J(q_1, 0) < 0$ and $(q_1, 0)$ is a saddle point. At $(0, 0)$ and $(q_2, 0)$, $J(0, 0)$, $J(q_2, 0) > 0$ and $T_1(0, 0) = T_1(q_2, 0) = 0$. Thus $(0, 0)$ and $(q_2, 0)$ are two centers and a pair of homoclinic orbits are obtained at $(q_1, 0)$ and two families of periodic orbits about $(0, 0)$ and $(q_2, 0)$ (see Fig. 5b).

Case 3: $v = \frac{3a^2l}{16b}$. At $(0, 0)$, $J(0, 0) > 0$ and $T_1(0, 0) = 0$. Hence $(0, 0)$ is a center point. At $\left(-\frac{3a}{4b}, 0\right)$, $J\left(-\frac{3a}{4b}, 0\right) = 0$ and $\left(-\frac{3a}{4b}, 0\right)$ is a high order equilibrium point and a family of periodic orbits about $(0, 0)$ and a homoclinic orbit at $\left(-\frac{3a}{4b}, 0\right)$ are found. Moreover, system (29) has infinite periodic orbits outside the homoclinic orbit (see Fig. 5c).

Case 4: $v < -\frac{3a^2l}{16b}$. At $(0, 0)$, $J(0, 0) > 0$ and $T_1(0, 0) = 0$ and hence $(0, 0)$ is a center point. Thus a family of periodic orbits about $(0, 0)$ are found (see Fig. 6a and 7).

Case 5: $v = 0$. At $(0, 0)$, $J(0, 0) = 0$. Thus $(0, 0)$ is a high order equilibrium point. At $\left(-\frac{3a}{2b}, 0\right)$, $J\left(-\frac{3a}{4b}, 0\right) > 0$, $T_1\left(-\frac{3a}{4b}, 0\right) = 0$ and $\left(-\frac{3a}{4b}, 0\right)$ is a center point. We obtain a homoclinic orbit at $(0, 0)$ and a family of periodic orbits about $\left(-\frac{3a}{4b}, 0\right)$ (see Fig. 6b).

The first equation of system (29) yields

$$\begin{aligned} \xi = \int \frac{\mathrm{d}q}{z} &= \pm \int \frac{\mathrm{d}q}{q\sqrt{\frac{v}{cl} + \frac{a}{3c}q + \frac{b}{6c}q^2}} \\ &= \mp\sqrt{\frac{cl}{v}} \ln\left|\frac{1}{q}\left\{\left(\frac{v}{cl} + \frac{a}{6c}q\right) + \sqrt{\frac{v}{cl}\left(\frac{v}{cl} + \frac{a}{3c}q + \frac{b}{6c}q^2\right)}\right\}\right|. \end{aligned} \tag{32}$$

We take exponential on both sides of (32) and simplifying obtain

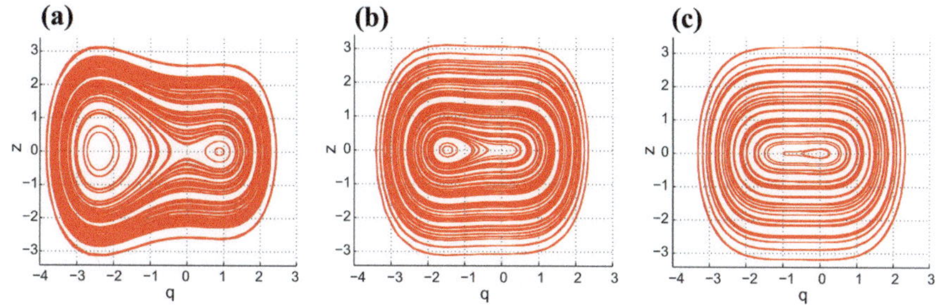

Fig. 5 Phase portrait of (26) for **a** $v > 0 > -\frac{3a^2l}{16b}$, **b** $-\frac{3a^2l}{16b} < v < 0$, **c** $v = \frac{3a^2l}{16b}$ with $a = b = c = 1$ and $l = m = \frac{1}{\sqrt{2}}$

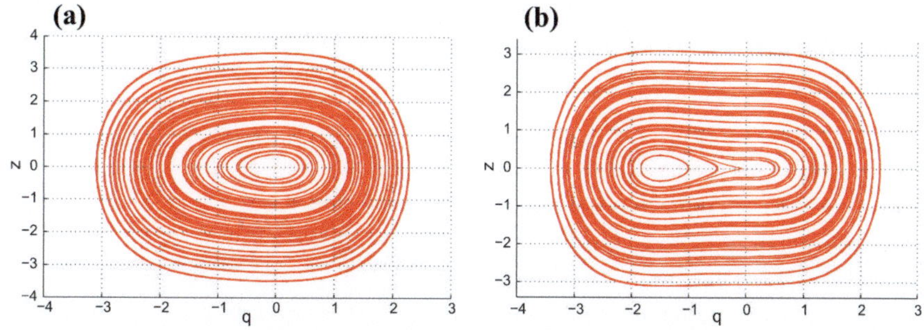

Fig. 6 Phase portrait of (26) for **a** $v < -\frac{3a^2l}{16b}$, **b** $v = 0$ with $a = b = c = 1$ and $l = m = \frac{1}{\sqrt{2}}$

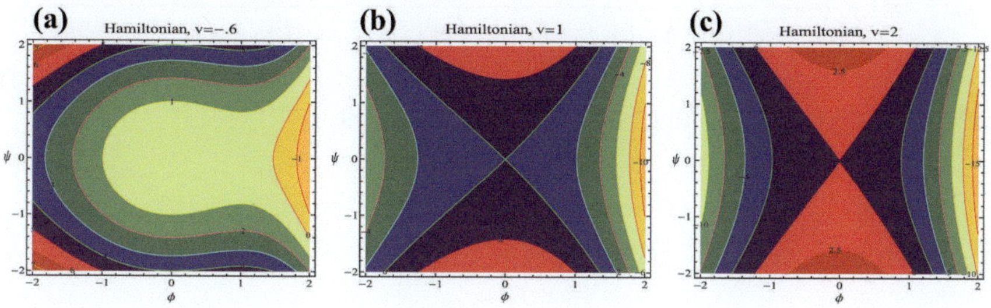

Fig. 7 Contour plot of $\mathscr{H}(q, z) = \text{const}$ for **a** $v = -0.6$, **b** $v = 1$, **c** $v = 2$ with $a = b = c = 1$ and $l = m = \frac{1}{\sqrt{2}}$

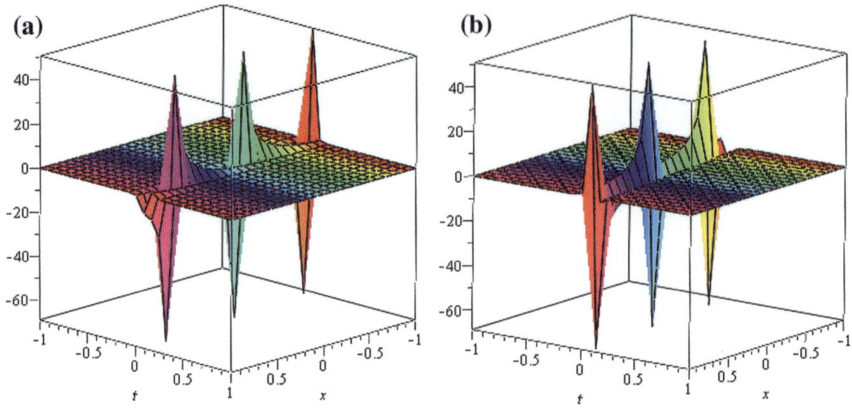

Fig. 8 Profile of (33) **a** positive case, **b** negative case with $l = 1$, $v = 10$ and other parameters are same as Fig. 6

$$q(\xi) = \frac{2\,\mathrm{e}^{\pm\sqrt{\frac{v}{cl}}\xi}}{\frac{cl}{v}\left(1 - \frac{a}{6c}\,\mathrm{e}^{\pm\sqrt{\frac{v}{cl}}\xi}\right)^2 - \frac{b}{6c}\,\mathrm{e}^{\pm 2\sqrt{\frac{v}{cl}}\xi}}. \tag{33}$$

Note that it is a breaking wave solution (see Fig. 8). We simplify Eq. (32) and obtain the explicit form

$$q(\xi) = \frac{2}{\left(\frac{a^2 l}{36cv} - \frac{b}{6c} + \frac{cl}{v}\right)\cosh\left(\sqrt{\frac{v}{cl}}\xi\right) \pm \left(\frac{a^2 l}{36cv} - \frac{b}{6c} - \frac{cl}{v}\right)\sinh\left(\sqrt{\frac{v}{cl}}\xi\right) - \frac{al}{3c}}, \tag{34}$$

which also depicts a breaking wave solution (see Fig. 9).

6 Conclusions

In this paper, the time fractional dual power ZK-Burgers equation with fractional temporal evolution (4) is introduced for the first time. The successful application of (G'/G)-expansion method along with Jumarie's modified Riemann-Liouville derivative produces miscellaneous exact traveling wave solutions in terms of hyperbolic, trigonometric and rational functions. These traveling wave solutions turns in to kink wave and periodic wave solutions for particular choice of arbitrary constants. Furthermore, we discuss the dynamical behavior, bifurcation analysis and phase portraits of the dynamical system in presence and absence of damping effect. We hope that this approach can be executed to a variety of NLPDEs which are generalized in context of fractional temporal evolution.

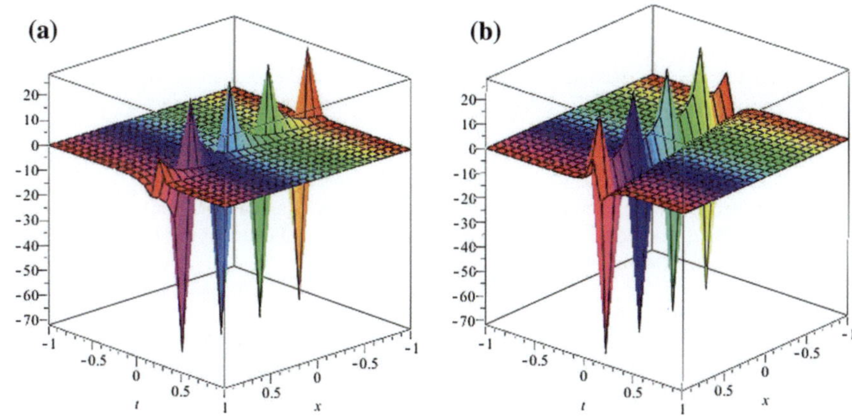

Fig. 9 Profile of (34) **a** positive case, **b** negative case with $l = 1$, $v = 6$ and other parameters are same as Fig. 6

References

1. Oldham, K., Spanier, J.: The Fractional Calculus. Academic Press, New York (1974)
2. Podlubny, I.: Fractional Differential Equations. Academic Press, San Diego (1999)
3. Kilbas, A., Srivastava, H.M., Trujillo, J.J.: Theory and Applications of Fractional Differential Equations. Elsevier, Amsterdam (2006)
4. Baillie, R.T.: Long memory processes and fractional integration in econometrics. J. Econometrics **73**, 5–59 (1996)
5. Metzler, R., Klafter, J.: The restaurant at the end of the random walk: recent developments in the description of anomalous transport by fractional dynamics. J. Phys. A **37**, 161–208 (2004)
6. Saadatmandi, A., Dehghan, M.: A new operational matrix for solving fractional-order differential equations. Comput. Math. Appl. **59**, 1326–1336 (2010)
7. Zhou, Y., Jiao, F., Li, J.: Existence and uniqueness for p-type fractional neutral differential equations. Nonlinear Anal. **71**, 2724–2733 (2009)
8. Galeone, L., Garrappa, R.: Explicit methods for fractional differential equations and their stability properties. J. Comput. Appl. Math. **228**, 548–560 (2009)
9. Trigeassou, J.C., Maamri, N., Sabatier, J., Oustaloup, A.: A Lyapunov approach to the stability of fractional differential equations. Signal Process. **91**, 437–445 (2011)
10. Jumarie, G.: Modified Riemann-Liouville derivative and fractional Taylor series of nondifferentiable functions further results. Comput. Math. Appl. **51**, 1367–1376 (2006)
11. Wang, M.L., Li, X.Z., Zheng, J.L.: The (G'/G)-expansion method and travelling wave solutions of nonlinear evolution equations in mathematical physics. Phys. Lett. A **372**, 417–423 (2008)
12. Zheng, B.: (G'/G)-Expansion method for solving fractional partial differential equations in the theory of mathematical physics. Commun. Theor. Phys. (Beijing, China) **58**, 623–630 (2012)
13. Hosseini, K., Ayati, Z.: Exact solutions of space-time fractional EW and modified EW equations using Kudryashov method. Nonlinear Sci. Lett. A **7**(2), 58–66 (2016)
14. Zheng, B.: Exp-function method for solving fractional partial differential equations. Sci. World J. **2013**, 465723 (2013)
15. Bekir, A., Guner, Ö., Ünsal, Ö.: The first integral method for exact solutions of nonlinear fractional differential equations. J. Comput. Nonlinear Dyn. **10**(2), 021020 (2015)
16. Zhou, Y., Wang, M., Wang, Y.: Periodic wave solutions to a coupled KdV equations with variable coefficients. Phys. Lett. A **308**, 31–36 (2003)

17. Wang, M., Zhou, Y.: The periodic wave solutions for the Klein-Gordon-Schrödinger equations. Phys. Lett. A **318**, 84–92 (2003)
18. Rizvi, S.T.R., Ali, K.: Jacobian elliptic periodic traveling wave solutions in the negative-index materials. Nonlinear Dyn. **87**, 1967–1972 (2017)
19. Hasegawa, A.: Plasma Instabilities and Nonlinear Effects. Springer, Berlin (1975)
20. Gedalin, M., Scott, T.C., Band, Y.B.: Optical solitary waves in the higher order nonlinear Schrödinger equation. Phys. Rev. Lett **78**, 448–451 (1997)
21. Gray, P., Scott, S.: Chemical Oscillations and Instabilities. Clarendon, Oxford (1990)
22. Ablowitz, M.J., Clarkson, P.A.: Solitons, Nonlinear Evolution Equations and Inverse Scattering Transform. Cambridge University Press, Cambridge (1991)
23. Adomian, G.: Solving Frontier Problems of Physics: The Decomposition Method. Kluwer Academic Publishers, Dordrecht (1994)
24. Agrawal, G.P.: Nonlinear Fiber Optics. Academic Press, New York (1995)
25. Gardner, C.S., Greene, J.M., Kruskal, M.D., Miura, R.M.: Method for solving the Korteweg-deVries equation. Phys. Rev. Lett. **19**, 1095–1097 (1967)
26. Yu, J., Lou, S.Y.: Deformation and $(3+1)$-dimensional integrable model. Sci. China Ser. A **43**, 655–660 (2000)
27. Lou, S.Y.: Searching for higher dimensional integrable models from lower ones via Painlevé analysis. Phys. Rev. Lett. **80**, 5027–5031 (1998)
28. El-Wakil, S.A., Abdou, M.A., Elhanbaly, A.: New solitons and periodic wave solutions for nonlinear evolution equations. Phys. Lett. A **353**, 40–7 (2006)
29. Jiang, B., Liu, Y., Zhang, J., et al.: Bifurcations and some new traveling wave solutions for the CH-γ equation. Appl. Math. Comput. **228**(1), 220–233 (2014)
30. Ganguly, A., Das, A.: Explicit solutions and stability analysis of the $(2+1)$ dimensional KP-BBM equation with dispersion effect. Commun. Nonlin. Sci. Numer. Simulat **25**, 102–117 (2015)
31. Das, A., Ganguly, A.: Existence and stability of dispersive solutions to the Kadomtsev-Petviashvili equation in the presence of dispersion effect. Commun. Nonlin. Sci. Numer. Simulat **48**, 326–339 (2017)
32. Zhou, Y., Peng, L.: Weak solutions of the time-fractional Navier-Stokes equations and optimal control. Comput. Math. Appl. **73**, 1016–1027 (2017)
33. Unsal, O., Guner, O., Bekir, A.: Analytical approach for space-time fractional Klein-Gordon equation. Optik **135**, 337–345 (2017)
34. Hongsit, N., Allen, M.A., Rowlands, G.: Growth rate of transverse instabilities of solitary pulse solutions to a family of modified Zakharov-Kuznetsov equations. Phys. Lett. A **372**(14), 2420 (2008)
35. Wazwaz, A.M.: The extended tanh method for the Zakharov-Kuznetsov (ZK) equation, the modified ZK equation, and its generalized forms. Commun. Nonlinear Sci. Numer. Simul. **13**(6), 1039–47 (2008)
36. Biswas, A., Zerrad, E.: 1-soliton solution of the Zakharov-Kuznetsov equation with dual-power law nonlinearity. Commun. Nonlinear Sci. Numer. Simul. **14**, 3574–3577 (2009)
37. Yan, Z.L., Liu, X.Q.: Symmetry reductions and explicit solutions for a generalized Zakharov-Kuznetsov equation. Commun. Theor. Phys. (Beijing, China) **45**, 29–32 (2006)
38. Ferdousi, M., Miah, M.R., Sultana, S., Mamun, A.A.: Dust-acoustic shock waves in an electron depleted nonextensive dusty plasmas. Astrophys. Space Sci. **360**, 43 (2015)
39. Jannat, N., Ferdousi, M., Mamun, A.A.: Ion-acoustic shock waves in nonextensive multi-ion plasmas. Commun. Theor. Phys. **64**, 479–484 (2015)
40. Ema, S.A., Ferdousi, M., Sultana, S., Mamun, A.A.: Dust-ion-acoustic shock waves in nonextensive dusty multi-ion plasmas. Eur. Phys. J. Plus **130**, 46 (2015)
41. Uddin, M.J., Alam, M.S., Mamun, A.A.: Positron-acoustic shock waves associated with cold viscous positron fluid in superthermal electron-positron-ion plasmas. Phys. Plasmas **22**, 062111 (2015)

42. Li, J., Chen, G.: Bifurcations of travelling wave solutions for four classes of nonlinear wave equations. Int. J. Bifurcation Chaos **15**, 3973 (2005)
43. Guckenheimer, J., Holmes, P.J.: Nonlinear Oscillations, Dynamical Systems and Bifurcations of Vector Fields. Springer, New York (1983)

Stability Analysis for an SEIQR Epidemic Model with Saturated Incidence Rate

Deepti Mokati, Nirmala Gupta and V. H. Badshah

Abstract Mathematics plays an important role in study of biological systems through mathematical models. In the present paper, we extended the work of Nirwani et al. (Nonlinear Anal Differ Equ 4:43–50, 2016) [5] by introducing the transmission rate η from the exposed class E to infectious class I and converted the model into an Susceptible-Exposed-Infectious-Quarantine-Recovered epidemic model with saturated incidence rate. Determine the equilibrium points of the model and basic reproduction number R_q is obtained. Stability analysis have been discussed of both equilibrium points by Routh-Hurwitz criteria and Lyapunov function criteria. Also, Numerical simulations are carried out for the model.

Keywords Epidemic model · Compartmental model · Equilibrium points · Quarantine · Basic reproduction number

Mathematics Subject Classification 92D30 · 92D25 · 34D20

1 Introduction

The mathematical analysis of epidemiological models has highly developed in the last twenty years. Mathematical modeling of infectious diseases has a long history in mathematical biology. Hethcote [3] developed various models, analyzed mathematically and applied to infectious disease for many aspects such as passive immunity, gradual loss of vaccine, stages of infection, disease vectors, vaccination, quarantine. The first Susceptible-Infected-Recovered model was proposed by Kermack and McKendrick in the year 1927. Pathak et al. studied an SIR epidemic model with an asymptotically homogeneous transmission function in [6]. Zang et al. [8] discussed an epidemic model with dependent vaccination. Many Authors studied epidemic

D. Mokati (✉) · V. H. Badshah
School of Studies in Mathematics, Vikram University, Ujjain, Madhya Pradesh, India
e-mail: mahe.deepti@gmail.com

N. Gupta
Govt. Girls P.G. College, Ujjain, Madhya Pradesh, India

S. Manna et al. (eds.), *Mathematical Modelling and Scientific Computing with Applications*, Springer Proceedings in Mathematics & Statistics 308,
https://doi.org/10.1007/978-981-15-1338-1_4

models with saturated incidence rate. Adebimpe et al. [1] investigated an epidemic model including population of susceptibility, exposing, infected and recovered with non-linear incidence rate and discussed stability of disease-free and endemic equilibria if $R_0 < 1$ and $R_0 > 1$, respectively with numerical simulations to illustrate the analytical results. Feng et al. [2] have been formulated a quarantine model for childhood diseases. Quarantine plays an important role in infectious disease epidemiology. It is used to separate and restrict the movement of persons. It helps to control the spread of infectious diseases. Hethcote et al. [4] described the dynamical behavior of six endemic models: three for quarantine to susceptible (SIQS) and three for quarantine to recovered (SIQR) for infectious diseases with three forms of incidence. Wa et al. [7] described the global stability of a vaccinated epidemic model with quarantine strategy. Nirwani et al. [5] analyzed the stability of a quarantine endemic model with saturated incidence rate. Here we have proposed a more generalized SEIQR epidemic model with saturated incidence rate. Numerical results are also provided.

2 The Mathematical Model

In the model, consider a population into five classes: susceptible (S), exposed (E), infectious (I), quarantine (Q) and recovered (R) individuals at time t and the total number of population at time t is $S + E + I + Q + R = N$.

The flow of individual depicted in the following transfer diagram (Fig. 1).

The symbol are used here stand for

A = Recruitment rate,
β = Average number of adequate contact rate,
d = natural mortality rate,
α = constant rate,

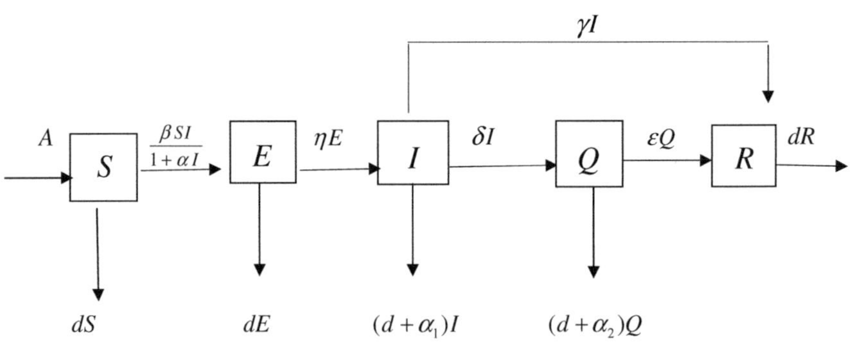

Fig. 1 Transfer diagram for SEIQR epidemic model

η = progression rate to symptoms development (the rate at which an Infected individual becomes infections per unit time),
ε, δ = removal rate constant,
α_1, α_2 = disease-related death rates,
γ = Recovery rate from the disease.

All parameters are assumed nonnegative.

3 Formulation of the Model

The differential equation corresponding to the transfer diagram are

$$\left.\begin{array}{l} \frac{dS}{dt} = A - \frac{\beta SI}{1+\alpha I} - dS \\ \frac{dE}{dt} = \frac{\beta SI}{1+\alpha I} - (\eta + d)E \\ \frac{dI}{dt} = \eta E - (\gamma + \delta + d + \alpha_1)I \\ \frac{dQ}{dt} = \delta I - (\varepsilon + d + \alpha_2)Q \\ \frac{dR}{dt} = \gamma I + \varepsilon Q - dR \end{array}\right\} \tag{1}$$

The feasible region of human population corresponding to the system (1) will be

$$\Omega = \left\{(S, E, I, Q, R){:}S \geq 0, E \geq 0, I \geq 0, Q \geq 0, R \geq 0, S + E + I + Q + R \leq \frac{A}{d}\right\}$$

Thus, the proposed model is mathematically well posed and is epidemiologically reasonable.

4 Equilibrium Points

4.1 Disease-Free Equilibrium

All the equations of the system (1) equate to zero. Then the system of equation becomes

$$\left.\begin{array}{l} A - \frac{\beta SI}{1+\alpha I} - dS = 0 \\ \frac{\beta SI}{1+\alpha I} - (\eta + d)E = 0 \\ \eta E - (\gamma + \delta + d + \alpha_1)I = 0 \\ \delta I - (\varepsilon + d + \alpha_2)Q = 0 \\ \gamma I + \varepsilon Q - dR = 0. \end{array}\right\} \tag{2}$$

Assume that if the disease is not occur, then $I = 0$.

Thus, the disease-free equilibrium is $E^0 = (S, E, I, Q, R)$, i.e., $E^0 = (\frac{A}{d}, 0, 0, 0, 0)$.

4.2 *Endemic Equilibrium* (E**)

The system (1) can be written as

$$\left.\begin{array}{l} A - \frac{\beta S^* I^*}{1+\alpha I^*} - dS^* = 0 \\ \frac{\beta S^* I^*}{1+\alpha I^*} - (\eta + d)E^* = 0 \\ \eta E^* - (\gamma + \delta + d + \alpha_1)I^* = 0 \\ \delta I^* - (\varepsilon + d + \alpha_2)Q^* = 0 \\ \gamma I^* + \varepsilon Q^* - dR^* = 0. \end{array}\right\} \tag{3}$$

On solving the equations, finally, we have endemic equilibrium $E^{**} = (S^*, E^*, I^*, Q^*, R^*)$, where $S^* = \frac{(\eta+d)(\gamma+d+\alpha_1+\delta)}{\eta\beta}\left[1 + \frac{\alpha d(R_q-1)}{\beta+d\alpha}\right]$, $E^* = \frac{(\gamma+d+\alpha_1+\delta)}{\eta}\left[\frac{d(R_q-1)}{\beta+d\alpha}\right]$, $I^* = \frac{d(R_q-1)}{\beta+d\alpha}$, $Q^* = \frac{\delta}{(\varepsilon+d+\alpha_2)}\left[\frac{d(R_q-1)}{\beta+d\alpha}\right]$, $R^* = \frac{(R_q-1)}{(\beta+d\alpha)}\left[\gamma + \frac{\varepsilon\delta}{(\varepsilon+d+\alpha_2)}\right]$.

The quarantine reproduction number is given by $R_q = \frac{\eta\beta A}{d(\eta+d)(\gamma+\delta+d+\alpha_1)}$.

5 Stability Analysis

5.1 *Local Stability of Disease Free Equilibrium*

Theorem 1 *If* $R_q < 1$, *then the disease-free equilibrium is locally asymptotically stable.*

Proof The variational matrix of the system (1) at $E_0 = \left(\frac{A}{d}, 0, 0, 0, 0\right)$ is given by

$$J = \begin{bmatrix} -d & 0 & -\frac{\beta A}{d} & 0 & 0 \\ 0 & -(\eta + d) & \frac{\beta A}{d} & 0 & 0 \\ 0 & \eta & -(\gamma + \delta + d + \alpha_1) & 0 & 0 \\ 0 & 0 & \delta & -(\varepsilon + d + \alpha_2) & 0 \\ 0 & 0 & \gamma & \varepsilon & -d \end{bmatrix}.$$

Now, its characteristic equation will be

$$|J - \lambda I| = 0.$$

$$\begin{vmatrix} -(d+\lambda) & 0 & -\frac{\beta A}{d} & 0 & 0 \\ 0 & -(\eta+d+\lambda) & \frac{\beta A}{d} & 0 & 0 \\ 0 & \eta & -(\gamma+\delta+d+\alpha_1+\lambda) & 0 & 0 \\ 0 & 0 & \delta & -(\varepsilon+d+\alpha_2+\lambda) & 0 \\ 0 & 0 & \gamma & \varepsilon & -(d+\lambda) \end{vmatrix} = 0$$

On solving, we have

$$-(d+\lambda)(d+\lambda)(\varepsilon+d+\alpha_2+\lambda)\left[(\eta+d+\lambda)(\gamma+\delta+d+\alpha_1+\lambda) - \frac{\eta\beta A}{d}\right] = 0.$$

From this, we conclude that three eigen values
$\lambda = -d, \lambda = -d, \lambda = -(\varepsilon + d + \alpha_2)$ are negative.
For obtaining other two eigenvalues, we have

$$\lambda^2 + \lambda\left[(\eta+\gamma+\delta+\alpha_1+2d\right] + (\eta+d)(\gamma+\delta+d+\alpha_1)\left[1-R_q\right] = 0.$$

Or $\lambda^2 + a\lambda + b = 0$.
Here,

$$a = \eta + \gamma + \delta + \alpha_1 + 2d > 0.$$

$$b = (\eta+d)(\gamma+\delta+d+\alpha_1)(1-R_q) > 0.$$

This is possible if $R_q < 1$.
Hence, it is clearly seen that *a > 0, b > 0* and *ab > 0.*
Thus, by Routh-Hurwitz criteria the theorem is proved.

5.2 Global Stability of Disease-Free Equilibrium

Theorem 2 *If* $R_q < 1$, *the disease-free equilibrium is global asymptotically stable and the disease dies out. But if* $R_q > 1$, *then it is unstable.*

Proof Consider, the Lyapunov function

$$L = \eta E + (\eta + d)I$$

then $\frac{dL}{dt} = \eta\frac{dE}{dt} + (\eta+d)\frac{dI}{dt}$
or $\frac{dL}{dt} = \left[\eta(\eta+d)(\gamma+\delta+d+\alpha_1)\right]\left(\frac{R_q}{\eta(1+\alpha I)} - 1\right)I \leq 0$ if $R_q < 1$.

Hence, the maximal compact invariant set in $\{(S, E, I, Q, R) \in D : \frac{dL}{dt} = 0\}$ is the singleton $\{E^0\}$. Using Lasalle's invariance principle we have proved the theorem.

5.3 Local Stability of Endemic Equilibrium

Theorem 3 *If $R_q > 1$, the endemic equilibrium E^{**} is locally asymptotically stable.*

Proof At the endemic equilibrium points $E^{**} = (S^*, E^*, I^*, Q^*, R^*)$, the variational matrix will be

$$J^* = \begin{bmatrix} -J_1 - d & 0 & -J_2 & 0 & 0 \\ J_1 & -(\eta + d) & J_2 & 0 & 0 \\ 0 & \eta & -(\gamma + \delta + d + \alpha_1) & 0 & 0 \\ 0 & 0 & \delta & -(\varepsilon + d + \alpha_2) & 0 \\ 0 & 0 & \gamma & \varepsilon & -d \end{bmatrix}$$

where, $\frac{\beta I^*}{1+\alpha I^*} = J_1$ and $\frac{\beta S^*}{(1+\alpha I^*)^2} = J_2$.

Then, its characteristic equation will be

$$\left|J^* - \lambda I\right| = 0.$$

$$\begin{vmatrix} -(J_1 + d + \lambda) & 0 & -J_2 & 0 & 0 \\ J_1 & -(\eta + d + \lambda) & J_2 & 0 & 0 \\ 0 & \eta & -(\gamma + \delta + d + \alpha_1 + \lambda) & 0 & 0 \\ 0 & 0 & \delta & -(\varepsilon + d + \alpha_2 + \lambda) & 0 \\ 0 & 0 & \gamma & \varepsilon & -(d + \lambda) \end{vmatrix} = 0.$$

On solving, we have

$$(d + \lambda)(\varepsilon + d + \alpha_2 + \lambda)[(J_1 + d + \lambda)\{(\eta + d + \lambda)(\gamma + \delta + d + \alpha_1 + \lambda) - J_2\eta\} + \eta J_1 J_2] = 0$$

Clearly seen that two eigen values $\lambda = -d$ and $\lambda = -(\varepsilon + d + \alpha_2)$ are negative and other eigen values are obtained on solving the equation

$$[(J_1 + d + \lambda)\{(\eta + d + \lambda)(\gamma + \delta + d + \alpha_1 + \lambda) - J_2\eta\} + \eta J_1 J_2] = 0.$$

$$\lambda^3 + [\eta + 3d + \gamma + \delta + \alpha_1 + J_1]\lambda^2 + [(d + \gamma + \delta + \alpha_1)(\eta + 2d + J_1) + (J_1 + d)(\eta + d) - \eta J_2]\lambda + [(J_1 + d)(\eta + d)(d + \gamma + \delta + \alpha_1) - \eta J_2 d] = 0$$

or $\lambda^3 + a\lambda^2 + b\lambda + c = 0$,

where

$$a = \eta + 3d + \gamma + \delta + \alpha_1 + J_1,$$

$$b = (d + \gamma + \delta + \alpha_1)(\eta + 2d + J_1) + (J_1 + d)(\eta + d) - \eta J_2,$$

$$c = (J_1 + d)(\eta + d)(d + \gamma + \delta + \alpha_1) - \eta J_2 d.$$

It is clearly seen that $a > 0, b > 0, c > 0$ and $ab - c > 0$.

Hence, by Routh-Hurwitz criteria, the endemic equilibrium is locally asymptotically stable.

6 Numerical Analysis and Graphical Representation

In this section, we have analyzed the model numerically and graphically by considering the set of parameters values. From practical point of view, numerical solutions are very important beside analytical system.

Case I

$$S(0) = 10, E(0) = 8, I(0) = 5, Q(0) = 5, R(0) = 5,$$
$$A = 3, d = 2.29, \gamma = 1.5, \delta = 0.5, \alpha_1 = 0.01, \beta = 0.5,$$
$$\alpha = 3.1, \eta = 0.05, \varepsilon = 0.2, \alpha_2 = 0.03, R_q = 0.00325 < 1.$$

Figure 2 shows that $S(t)$ approaches to its steady state value while $E(t)$, $I(t)$, $Q(t)$ and $R(t)$ approaches to zero as time progresses , the disease dies out.

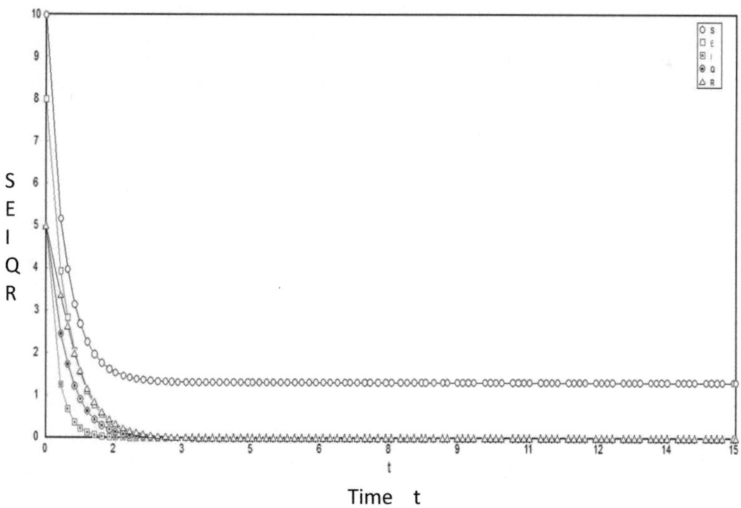

Fig. 2 Graph of SEIQR epidemic model when $R_q < 1$

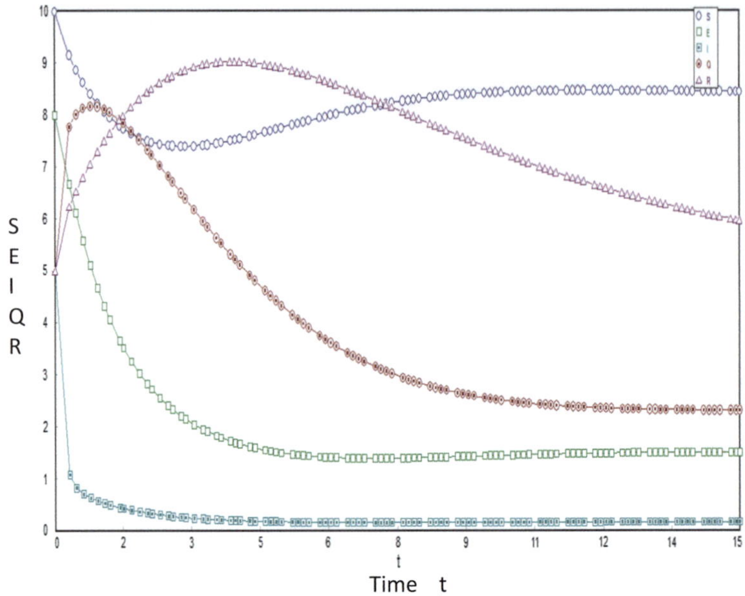

Fig. 3 Graph of SEIQR epidemic model when $R_q > 1$

Case II

$$S(0) = 10, E(0) = 8, I(0) = 5, Q(0) = 5, R(0) = 5,$$
$$A = 3, d = 0.15, \gamma = 1.5, \delta = 5, \alpha_1 = 2, \beta = 2,$$
$$\alpha = 4, \eta = 1, \varepsilon = 0.2, \alpha_2 = 0.03, R_q = 4.021 > 1.$$

Figure 3 shows that $S(t)$, $E(t)$, $I(t)$, $Q(t)$ and $R(t)$ approaches to its steady state value as time progresses, the disease becomes endemic.

7 Conclusion

The mathematical analysis of epidemiological models has highly developed in the last twenty years. In this paper, we have considered an SIQR epidemic model of Nirwani et al. [5] with saturated incidence rate and converted into SEIQR epidemic model which is a modified form. The SEIQR model is used in the modeling of infectious diseases by computing the amount of people in a closed population that are susceptible, exposed, infected, quarantined or recovered at a given period of time. The model can also be used by researchers and health officials to explain the increase and decrease in people needing medical care for a certain disease during an epidemic.

We have found disease-free and endemic equilibria for the model and analyzed the stability criteria for the both equilibria. We have seen that the disease-free equilibria and endemic equilibria are locally asymptotically stable by Routh-Hurwitz criteria if $R_q < 1$ and $R_q > 1$ respectively. We have also discussed globally asymptotically stability for disease-free equilibrium by Lyapunov function criteria. Also, numerical simulations are carried out for the model with graphical representation for ordinary differential equation and numerically found that if $R_q < 1$, then the disease dies out and if $R_q > 1$, the disease becomes endemic. This can be more clearly seen in the graphs.

In future, we will study of various types of epidemiological models and also analyze how we can cure the diseases with the help of vaccinated epidemiological models and various techniques theoretically as well as numerically with graphical representation.

References

1. Adebimpe, O., Waheed, A.A., Gbadamosi, B.: Modeling and analysis of an SEIRS epidemic model with saturated incidence. Int. J. Eng. Res. Appl. **3**(5), 1111–1116 (2013)
2. Feng, Z., Thieme, H.: Recurrent outbreaks of childhood diseases revisited: the impact of isolation. Math. Biosci. **128**, 93–130 (1995)
3. Hethcote, H.W.: The mathematics of infectious disease. SIAM Rev. **42**, 599–653 (2000)
4. Hethcote, H.W., Zhien, M., Shengbing, L.: Effects of quarantine in six endemic models for infectious diseases. Math. Biosci. **180**, 141–160 (2002)
5. Nirwani, N., Badshah, V.H., Khandelwal, R.: Dynamical study of an SIQR model with saturated incidence rate. Nonlinear Anal. Differ. Equ. **4**(1), 43–50 (2016)
6. Pathak, S., Maiti, A., Samanta, G.P.: Rich dynamics of an SIR epidemic model. Nonlinear Anal. Model. Control **15**(1), 71–81 (2010)
7. Wang, F., Yang, Y., Zhang, Y., Ma, J.: Global analysis of a SEIQV epidemic model for scanning worms with quarantine strategy. Int. J. Netw. Secur. **17**(4), 423–430 (2015)
8. Zhang, X., Jia, J.: Stability of an SIR epidemic model with information variable and limited medical resources. Int. J. Res. Rev. Appl. Sci. (IJRRAS) **16**(1), 91–103 (2013)

Impact of Rectangular/Parabolic Shaped Irregularity on the Propagation of Shear Horizontal Wave in a Slightly Compressible Layered Structure

M. S. Chaki, S. Guha and A. K. Singh

Abstract The present paper enunciates the propagation behaviour of Shear Horizontal (SH) wave in a layered structure comprised of a slightly compressible layer overlying a slightly compressible half-space with rectangular shaped irregularity and parabolic shaped irregularity at the common interface in two distinct cases. Dispersion relations have been deduced analytically in closed form for both the cases. As a particular case, the obtained dispersion equations have been found to be in agreement with the classical Love wave equation in isotropic case. The impact of various parameters, viz. wave number, rectangular shaped irregularity and parabolic shaped irregularity associated at the common interface, on the phase velocity of SH wave has been studied for the layered structure. Numerical computation and graphical illustration has been done in order to analyse the impact of irregularity parameters (rectangular and parabolic) more profoundly for slightly compressible varga material which serves as one of the major highlight of the present study.

Keywords Slightly compressible · SH wave · Irregularity · Perturbation method

1 Introduction

Since material technology is capable of producing materials which can bear massive pressure and large deformations in aspect of industrial applications, slightly compressible (finitely deformed) and incompressible materials are being widely used now-a-days, viz. engine mounts, off-shore structure flex-joints, vibration insulators etc. The stress analysis of rubber-like materials offers two rather unique features;

M. S. Chaki · S. Guha (✉) · A. K. Singh
Department of Applied Mathematics, Indian Institute of Technology (Indian School of Mines), Dhanbad, Jharkhand 826004, India
e-mail: sayantanguha.maths@gmail.com

M. S. Chaki
e-mail: mriganka.chaki@gmail.com

A. K. Singh
e-mail: abhi.5700@gmail.com

S. Manna et al. (eds.), *Mathematical Modelling and Scientific Computing with Applications*, Springer Proceedings in Mathematics & Statistics 308,
https://doi.org/10.1007/978-981-15-1338-1_5

firstly, these materials are nearly compressible (i.e. their bulk modulii are much larger than their shear modulii) and secondly, they are capable of experiencing large deformations. Motivated by the desire to explicate such response, there have been a number of recent papers aiming at explaining the effect of pre-stress on wave propagation in elastic plates, laminates, half-spaces, and other layered structure. Since the slightly compressibility seems to be a more accurate presumption as compared to ideal incompressibility, some significant studies pertaining to this topic have been conducted by Dowaikh and Ogden [1], Rogerson and Fu [2], Rogerson and Murphy [3], Sandiford and Rogerson [4], Chattopadhyay and Sahu [5], Rogerson et al. [6], Chatterjee and Chattopadhyay [7], Scott [8] and so on. A good example of such type of material is varga material which has drawn attention of many authors due to its mathematical feasibility.

It is not always feasible to find parallel boundaries in problems of wave propagation in elastic media and hence, mathematical presumption of irregular boundaries due to its closeness to the real world scenarios have harnessed abundant attention of geophysicists and seismologists to understand the seismic activity at mountain basins and continental margins. As a result, many authors have studied the wave propagation in different types of irregularities (rectangular, parabolic, corrugated, triangular etc.) at the interfaces.

Bhattacharya [9] considered the irregularity in the transversely isotropic crustal layer. Chattopadhyay [10] studied Love wave propagation in the crystal layer with irregularity and non-homogeneity. Singh et al. [11] analytically discussed Love-type wave propagation in a piezoelectric structure with irregularity. Recently, Singh et al. [12] discussed about the impact of irregularity on SH-type wave propagation in micropolar elastic composite structure.

Till today, no study has been done on horizontally polarized shear wave propagation in slightly compressible finitely deformed layer/half-space configuration having rectangular and parabolic shaped irregular interfaces.

2 The Governing Equations

Let B_0 and B_e be a natural configuration and a finitely deformed equilibrium configuration of an elastic solid respectively where X_A and $\chi_i(X_A)$ are position vectors in B_0 and B_e. Now, upon the finite deformation $B_0 \to B_e$, if an infinitesimal time-dependent motion is super-imposed in the current configuration B_t with position vector $\bar{\chi}_i(X_A, t)$. The deformation gradients associated with the deformations $B_0 \to B_t$ and $B_0 \to B_e$ are defined through the component relation [13]

$$F_{iA} = (\delta_{ij} + u_{i,j})\bar{F}_{jA}, \quad (\pi_{iA}\bar{F}_{pA})_{,p} = \rho\ddot{u}_i, \quad \pi_{iA} = \frac{\partial W}{\partial F_{iA}} \tag{1}$$

where $u_i(\mathbf{X}, t)$ is a small time-dependent displacement in $B_e \to B_t, \delta_{ij}$ the Kronecker delta, π_{iA} denotes Piola-Kirchhoff stress tensor, ρ is the material density and $W(\mathbf{F})$ is the strain energy function per unit volume.

A system of linearized equations of motion can be obtained with the help of Taylor series expansion of the stress tensor about B_0 which yields by using Eq. (1) as

$$B_{jilk}u_{k,lj} = \rho\ddot{u}_i, \quad B_{jilk} = \bar{J}^{-1}\bar{F}_{jA}\bar{F}_{lC}\frac{\partial^2 W}{\partial F_{iA}\partial F_{kC}}\bigg|_{\mathbf{F}=\bar{\mathbf{F}}}, \tag{2}$$

where $J = \det \mathbf{F}$.

Now, Eq. (2) may be rewritten for small order parameter $(J - 1)$ [14] as

$$B_{jilk} = \bar{J}^{-1}\bar{F}_{jA}\bar{F}_{lC}\frac{\partial^2 W}{\partial F_{iA}\partial F_{kC}}\bigg|_{\mathbf{F}=\bar{\mathbf{F}}} + \kappa\bar{J}\delta_{ij}\delta_{kl} + \kappa\big(\bar{J} - 1\big)(\delta_{ij}\delta_{kl} - \delta_{jk}\delta_{il}) \tag{3}$$

where $\kappa = \frac{\partial^2 W}{\partial J^2}$ is the bulk modulus and $\kappa \sim O(J - 1)^{-1}$.

Hence the equation of small-amplitude motions in a slightly incompressible material may be obtained as

$$B^0_{jilk}u_{k,lj} + \kappa\bar{J}u_{k,ki} = \rho\ddot{u}_i, B^0_{jilk} = \bar{J}^{-1}\bar{F}_{jA}\bar{F}_{lC}\frac{\partial^2 W}{\partial F_{iA}\partial F_{kC}}\bigg|_{\mathbf{F}=\bar{\mathbf{F}}}, \tag{4}$$

where $\mathbf{B}^0$ is associated with the leading-order term $W_0(\mathbf{F}, 1)$ of the strain energy function.

Surface traction at a point along with outward normal $\mathbf{n}$ in B_e may be obtained by expanding π_{iA} in the form of Taylor series about B_e as

$$\tau_i = B^0_{jilk}u_{k,l}\, n_j + \big[\kappa(\bar{J} - 1) + \kappa\bar{J}\big]u_{k,k}\, n_i - \kappa(\bar{J} - 1)u_{j,i}\, n_j\,. \tag{5}$$

3 Constitutive Relations and Formulation of the Problem

From strain energy function associated with incompressible elastic material, we obtain

$$W_0(\mathbf{F}, 1) = W_0(I_1, I_2), I_1 = \mathbf{tr}\, C, I_2 = \frac{1}{2}\big(I_1^2 - \big(\mathbf{tr}\, C^2\big)\big), \tag{6}$$

where I_1 and I_2 represent the first two principal invariants of the modified right Cauchy-Green strain tensor $C = \mathbf{F}^T\mathbf{F}$ and $\mathbf{tr}$ denotes its trace.

Here we have considered propagation of SH wave in a slightly compressible finitely deformed layer of finite thickness H lying over a slightly compressible finitely

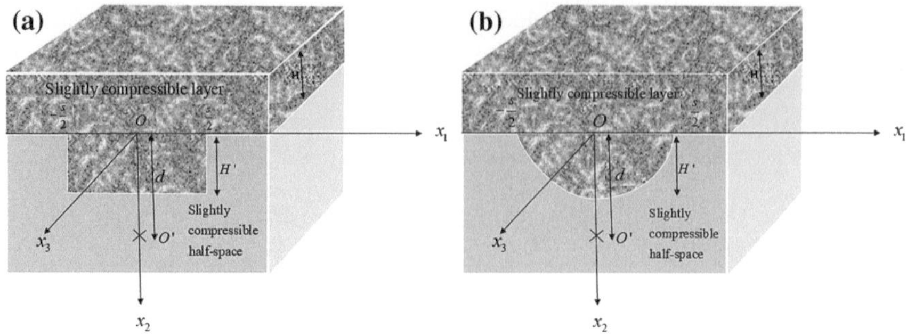

Fig. 1 Geometry of the problem for the case of **a** rectangular shaped irregularity and **b** parabolic shaped irregularity

deformed half-space as shown in Figs. 1a and b. Rectangular and parabolic shaped irregularities os span s have been considered at the interface.

A rectangular Cartesian coordinate system with origin O at the middle point of the span of the irregularities has been considered, where x_2-axis is taken vertically downwards and x_1-axis lies along the propagation of SH wave. The depth of both irregularities H' is considered below $x_2 = 0$ along with a source of disturbance at the point O' on the x_2-axis having depth d $\left(d > H'\right)$.

The equations of irregularity is given by

$$x_2 = \varepsilon h(x_1) \tag{7}$$

where for rectangular irregularity

$$h(x_1) = \begin{cases} 0 \ for \ |x| > \frac{s}{2}; \\ s \ \ for \ |x| \leq \frac{s}{2}. \end{cases}$$

and for parabolic irregularity

$$h(x_1) = \begin{cases} 0 & for \ |x| > \frac{s}{2}; \\ s\left(1 - \frac{4x_1^2}{s^2}\right) & for \ |x| \leq \frac{s}{2}. \end{cases}$$

where $\varepsilon = \frac{H'}{s} \ll 1$ is small perturbation parameter.

Due to SH wave propagation, the displacement components are assumed as

$$u_1 = 0, \quad u_2 = 0, \quad u_3 = u_3(x_1, x_2, t), \tag{8}$$

Equation (8) may now be used in Eq. (4) which upon use of the fact that all non-zero components of the elasticity tensor take one of the three forms B^0_{iijj}, B^0_{ijji} or $B^0_{ijij}(i, j = 1, 2, 3)$, enables the two non-trivial equations of motion for the layer and half-space to be expressed in the form

$$B_{1313}^{(1)} u_{3,11}^{(1)} + B_{2323}^{(1)} u_{3,22}^{(1)} = \rho_1 \ddot{u}_3^{(1)} \tag{9}$$

$$B_{1313}^{(2)} u_{3,11}^{(2)} + B_{2323}^{(2)} u_{3,22}^{(2)} = \rho_2 \ddot{u}_3^{(2)} \tag{10}$$

where $B_{1313}^{(1)}$ and $B_{2323}^{(1)}$ are components of elasticity tensor with respect to slightly compressible layer and $B_{1313}^{(2)}$ and $B_{2323}^{(2)}$ are components of elasticity tensor with respect to slightly compressible half-space and $u_3^{(i)}$ and ρ_i are displacement, initial stress and density of the slightly compressible layer and half-space for $i = 1, 2$ respectively.

The two non-zero incremental tractions are obtained from (5) for the layer and half-space as

$$\tau_3^{(1)} = \frac{1}{\sqrt{\varepsilon^2 (h'(x))^2 + 1}} \left[-\varepsilon h'(x_1) B_{1313}^{(1)} u_{3,1}^{(1)} + B_{2323}^{(1)} u_{3,2}^{(1)} \right] \tag{11}$$

$$\tau_3^{(2)} = \frac{1}{\sqrt{\varepsilon^2 (h'(x))^2 + 1}} \left[-\varepsilon h'(x_1) B_{1313}^{(2)} u_{3,1}^{(2)} + B_{2323}^{(2)} u_{3,2}^{(2)} \right]. \tag{12}$$

4 Boundary Conditions

1. The upper surface of the slightly compressible layer is stress free:

$$\tau_3^{(1)} = 0 \text{ at } x_2 = -H \tag{13}$$

2. The stresses are continuous at the interface $x_2 = \varepsilon h(x_1)$:

$$\tau_3^{(1)} = \tau_3^{(2)} \tag{14}$$

3. The displacements are continuous at the interface $x_2 = \varepsilon h(x_1)$:

$$u_3^{(1)} = u_3^{(2)} \tag{15}$$

5 Solution of the Problem

The solution of the equations of motion (9) and (10) can be considered as

$$u_3^{(j)} = U_3^{(j)}(x_1, x_2) e^{i\omega t} \quad \text{for } j = 1, 2. \tag{16}$$

where ω is the circular frequency. Using (16), the Eqs. (9) and (10) reduce to

$$(B_{1313}^{(1)})\frac{\partial^2 U_3^{(1)}}{\partial x_1^2} + (B_{2323}^{(1)})\frac{\partial^2 U_3^{(1)}}{\partial x_2^2} + \rho_1\omega^2 U_3^{(1)} = 0 \tag{17}$$

$$(B_{1313}^{(2)})\frac{\partial^2 U_3^{(2)}}{\partial x_1^2} + (B_{2323}^{(2)})\frac{\partial^2 U_3^{(2)}}{\partial x_2^2} + \rho_2\omega^2 U_3^{(2)} = 0 \tag{18}$$

Defining the Fourier transform $\bar{U}_3^{(r)}(\eta, x_2)$, we obtain

$$\frac{\partial^2 \bar{U}_3^{(1)}}{\partial x_2^2} + q_1^2 \bar{U}_3^{(1)} = 0\,, \tag{19}$$

$$\frac{\partial^2 \bar{U}_3^{(2)}}{\partial x_2^2} - q_2^2 \bar{U}_3^{(2)} = 0 \tag{20}$$

where $q_1^2 = \frac{\rho_1\omega^2 - \eta^2(B_{1313}^{(1)})}{B_{2323}^{(1)}} = \sqrt{\frac{\omega^2}{\beta_1^2} - \frac{B_{1313}^{(1)}}{B_{2323}^{(1)}}}$, $q_2^2 = \frac{-\rho_2\omega^2 + \eta^2(B_{1313}^{(2)})}{B_{2323}^{(2)}} = \sqrt{-\frac{\omega^2}{\beta_2^2} + \frac{B_{1313}^{(2)}}{B_{2323}^{(2)}}}$; $\beta_1 = \sqrt{\frac{B_{2323}^{(1)}}{\rho_1}}$ represents the shear wave velocity for the upper slightly compressible finitely deformed layer and $\beta_2 = \sqrt{\frac{B_{2323}^{(2)}}{\rho_2}}$ represents the shear wave velocity for the lower slightly compressible finitely deformed half-space.

The solutions of Eqs. (19) and (20) are

$$\bar{U}_3^{(1)} = A\cos(q_1 x_2) + B\sin(q_1 x_2) \tag{21}$$

$$\bar{U}_3^{(2)} = De^{-q_2 x_2} \tag{22}$$

where A, B and D are functions of η.

The displacement components of the layer and half-space are

$$U_3^{(1)}(x_1, x_2) = \frac{1}{2\pi}\int_{-\infty}^{\infty} (A\cos(q_1 x_2) + B\sin(q_1 x_2))e^{-i\eta x_1}\,d\eta \tag{23}$$

$$U_3^{(2)}(x_1, x_2) = \frac{1}{2\pi}\int_{-\infty}^{\infty} (De^{-q_2 x_2} + \frac{2}{q_2}e^{q_2 x_2}e^{-q_2 d})e^{-i\eta x_1}\,d\eta. \tag{24}$$

In Eq. (24), the second term in the integrand of $U_3^{(2)}$ physically interprets the presence of the source in the half-space.

The following approximations can be considered due to small value of ε

$$A \cong A_0 + A_1\varepsilon, \quad B \cong B_0 + B_1\varepsilon,$$
$$D \cong D_0 + D_1\varepsilon, \quad e^{\pm\nu\varepsilon h} \cong 1 \pm \nu\varepsilon h, \cos\nu\varepsilon h \cong 1, \sin\nu\varepsilon h \cong \nu\varepsilon h \tag{25}$$

where ν is any quantity.

Using the boundary condition (1), we get

$$(A_0 \sin(q_1 H) + B_0 \cos(q_1 H)) + \varepsilon(A_1 \sin(q_1 H) + B_1 \cos(q_1 H)) = 0 \tag{26}$$

Defining the Fourier transform of $h(x_1)$ and putting $\eta + \lambda = k$, the boundary condition (2) and (3) gives

$$\left(A_0 - D_0 - \frac{2}{q_2}e^{-q_2 d}\right) + \varepsilon(A_1 - D_1) = \varepsilon R_1(k) \tag{27}$$

$$B_{2323}^{(2)}[-q_2 D_0 - D_1 q_2\varepsilon + 2e^{-q_2 d}] - B_{2323}^{(1)}[B_0 q_1 + B_1 q_1 \varepsilon] = \varepsilon R_2(k) \tag{28}$$

where

$$R_1(k) = \frac{1}{2\pi}\int_{-\infty}^{\infty}[2e^{-q_2 d} - B_0 q_1 - D_0 q_2]\bar{h}(\lambda)d\lambda. \tag{29}$$

$$R_2(k) = \frac{1}{2\pi}\int_{-\infty}^{\infty}[k\lambda B_{1313}^{(1)} A_0 - B_{2323}^{(1)} A_0 q_1^2 - D_0 k\lambda B_{1313}^{(2)}$$
$$- \frac{2}{q_2}k\lambda B_{1313} e^{-q_2 d} - D_0 B_{2323}^{(2)} q_2^2 - 2B_{2323}^{(2)} q_2 e^{-q_2 d}]\bar{h}(\lambda)d\lambda \tag{30}$$

Taking into consideration the terms not containing ϵ and the coefficient of ϵ from Eqs. (26), (27) and (28), we acquire

$$A_0 \sin(q_1 H) + B_0 \cos(q_1 H) = 0,$$

$$(A_0 - D_0) - \frac{2}{q_2}e^{-q_2 d} = 0,$$

$$B_{2323}^{(2)}\left[-q_2 D_0 + 2e^{-q_2 d}\right] - B_{2323}^{(1)} B_0 q_1 = 0,$$

$$A_1 \sin(q_1 H) + B_1 \cos(q_1 H) = 0,$$

$$A_1 - D_1 = R_1(k),$$

$$-B_{2323}^{(2)} D_1 q_2 - B_{2323}^{(1)} B_1 q_1 = R_2(k).$$

Solving the above equations, the following may be deduced

$$A_0 = \frac{4e^{-q_2 d} B_{2323}^{(2)}}{q_2 B_{2323}^{(2)} - q_1 B_{2323}^{(1)} \tan(q_1 H)},$$

$$B_0 = \frac{-4e^{-q_2 d} B_{2323}^{(2)} \tan(q_1 H)}{q_2 B_{2323}^{(2)} - q_1 B_{2323}^{(1)} \tan(q_1 H)},$$

$$D_0 = \frac{2e^{-q_2 d}[q_1 B_{2323}^{(1)} \tan(q_1 H) + q_2 B_{2323}^{(2)}]}{q_2[q_2 B_{2323}^{(2)} - q_1 B_{2323}^{(1)} \tan(q_1 H)]},$$

$$A_1 = \frac{R_1(k) q_2 B_{2323}^{(2)} - R_2(k)}{q_2 B_{2323}^{(2)} - q_1 B_{2323}^{(1)} \tan(q_1 H)},$$

$$B_1 = \frac{-[R_1(k) q_2 B_{2323}^{(2)} - R_2(k)] \tan(q_1 H)}{q_2 B_{2323}^{(2)} - q_1 B_{2323}^{(1)} \tan(q_1 H)},$$

$$D_1 = \frac{R_1(k) q_1 B_{2323}^{(1)} \tan(q_1 H) - R_2(k)}{q_2 B_{2323}^{(2)} - q_1 B_{2323}^{(1)} \tan(q_1 H)}.$$

The displacement in the slightly compressible layer is then given by

$$U_3^{(1)} = \frac{1}{2\pi} \int_{-\infty}^{\infty} \frac{[4e^{-q_2 d} B_{2323}^{(2)} + \varepsilon[R_1(k) q_2 B_{2323}^{(2)} - R_2(k)]]}{q_2 B_{2323}^{(2)} - q_1 B_{2323}^{(1)} \tan(q_1 H)} \times [\cos(q_1 x_2) - \tan(q_1 H) \sin(q_1 x_2)] e^{-ikx_1} dk \tag{31}$$

Case 1 (*For Rectangular Shaped Irregularity*) From Eq. (7) for rectangular irregularity, we have

$$\bar{h}(\lambda) = \frac{2s}{\lambda} \sin\left(\frac{\lambda s}{2}\right) \tag{32}$$

Using (32) and (31), we get

$$R_1 q_2 B_{2323}^{(2)} - R_2 = \frac{2s B_{2323}^{(2)}}{\pi} \int_{-\infty}^{\infty} [\phi(k-\lambda) + \phi(k+\lambda)] \frac{1}{\lambda} \sin \frac{\lambda s}{2} d\lambda \tag{33}$$

where

$$\phi(k-\lambda) = \frac{e^{-q_2 d}}{q_2[q_2 B_{2323}^{(2)} - q_1 B_{2323}^{(1)} \tan(q_1 H)]} [2B_{2323}^{(2)} q_1^3 - 2B_{2323}^{(1)} q_2^2 q_1 \tan(q_1 H) + 2B_{2323}^{(2)} q_1 q_2^2 \tan(q_1 H) + k\lambda q_2 B_{1313}^{(2)} + 2q_2 q_1^2 B_{2323}^{(1)} - 2q_2 B_{1313}^{(1)} k\lambda + k\lambda B_{1313}^{(2)} q_2]^{\eta=k-\lambda}.$$

with $\eta + \lambda = k$.

Asymptotic formulae presented by Willis [15] and Tranter [16] and neglection of powers of $2/s$ for large s results in

$$\int_{-\infty}^{\infty} [\phi(k-\lambda) + \phi(k+\lambda)]\frac{1}{\lambda}\sin\frac{\lambda s}{2}d\lambda \cong \frac{\pi}{2}2\phi(k) = \pi\phi(k) \tag{34}$$

Then, using (34) in (33), we obtain

$$R_1 q_2 B_{2323}^{(2)} - R_2 = 2B_{2323}^{(2)}\frac{H'}{\varepsilon}\phi(k) \tag{35}$$

Therefore, the displacement in the slightly compressible layer is

$$U_3^{(1)} = \frac{1}{2\pi}\int_{-\infty}^{\infty} \frac{4e^{-q_2 d}B_{2323}^{(2)}}{\left[q_2 B_{2323}^{(2)} - q_1 B_{2323}^{(1)}\tan(q_1 H)\right]} \frac{[\cos(q_1 x_2) - (\tan(q_1 H)\sin(q_1 x_2)]}{(1 - \frac{H'}{2}\phi(k)e^{q_2 d})} e^{-ikx_1}dk \tag{36}$$

Since the integration in Eq. (36) depends entirely on the poles, the following equation may be considered, in order to find the poles, as the dispersion relation

$$\tan(q_1^* kH) = \frac{H' B_{2323}^{(2)} k(q_1^*)^3 + H' k q_2^*(q_1^*)^2 B_{2323}^{(1)} - (q_2^*)^2 B_{2323}^{(2)}}{H' k B_{2323}^{(1)} q_1^*(q_2^*)^2 - H' k q_1^*(q_2^*)^2 B_{2323}^{(2)} - q_1^* q_2^* B_{2323}^{(1)}} \tag{37}$$

where $\omega = ck$ (where ω is the circular frequency and k is the wave number), $q_1 = kq_1^*$ and $q_2 = kq_2^*$ where $q_1^* = \sqrt{\frac{c^2}{\beta_1^2} - \frac{B_{1313}^{(1)}}{B_{2323}^{(1)}}}$, $q_2^* = \sqrt{-\frac{c^2}{\beta_2^2} + \frac{B_{1313}^{(2)}}{B_{2323}^{(2)}}}$ and β_1 and β_2 represent the shear wave velocities for the upper slightly compressible finitely deformed layer and for the lower slightly compressible finitely deformed half-space respectively.

Case 2 (*For Parabolic Shaped Irregularity*) From Eq. (11) for parabolic irregularity, we have

$$\bar{h}(\lambda) = \frac{8H's}{\varepsilon}\left(\frac{2\sin(\frac{\lambda s}{2}) - (\lambda s)\cos(\frac{\lambda s}{2})}{(\lambda s)^3}\right) \tag{38}$$

Hence,

$$R_1 q_2 B_{2323}^{(2)} - R_2 = \frac{B_{2323}^{(2)}}{2\pi}\frac{8H's}{\varepsilon}\int_0^{\infty}[\phi(k-\lambda) + \phi(k+\lambda)]\sqrt{\frac{\pi}{4}}\frac{J_{3/2}\left(\frac{\lambda s}{2}\right)}{(\lambda s)^{3/2}}d\lambda, \tag{39}$$

where $J_{3/2}\left(\frac{\lambda s}{2}\right)$ is a Bessel function of the first kind of order $\frac{3}{2}$.

Using asymptotic formulae [15, 16], and proceeding similarly as before, we obtain

$$U_3^{(1)} = \frac{1}{2\pi}\int_{-\infty}^{\infty} \frac{4e^{-q_2 d} B_{2323}^{(2)}}{[\, q_2 B_{2323}^{(2)} - q_1 B_{2323}^{(1)} \tan(q_1 H)]} \times \frac{[\cos(q_1 x_2) - (\tan(q_1 H)\sin(q_1 x_2)]}{(1 - \frac{H'}{6\pi}\phi(k) e^{q_2 d})} e^{-ikx_1} dk \tag{40}$$

Since the integration in Eq. (40) depends entirely on the poles, the following equation may be considered, in order to find the poles, as the dispersion relation

$$\tan(q_1^* kH) = \frac{H' B_{2323}^{(2)} k (q_1^*)^3 + H' k q_2^* (q_1^*)^2 B_{2323}^{(1)} - 3\pi (q_2^*)^2 B_{2323}^{(2)}}{H' k B_{2323}^{(1)} q_1^* (q_2^*)^2 - H' k q_1^* (q_2^*)^2 B_{2323}^{(2)} - 3\pi q_1^* q_2^* B_{2323}^{(1)}} \tag{41}$$

where the notations are already described after Eq. (37) in **Case 1**.

6 Particular Cases

6.1 *Case 1*

When slightly compressible finitely deformed layer overlies on a slightly compressible finitely deformed half-space having a perfect bonding (i.e. $H' = 0$) at the interface, the dispersion equations for both cases of irregularities (rectangular and parabolic) i.e. Equations (37) and (41) transform to Eq. (25) of [7]

$$\tan(q_1^* kH) = \frac{q_2^* B_{2323}^{(2)}}{q_1^* B_{2323}^{(1)}} \tag{42}$$

6.2 *Case 2*

When the upper layer as well as the lower half-space are considered to be isotropic (i.e.$B_{1313}^1 = B_{2323}^1 = \mu_1$ and $B_{1313}^2 = B_{2323}^2 = \mu_2$) and have a perfect bonding (i.e. $H' = 0$) at the interface, the dispersion equations i.e. Equations (37) and (41) reduce to the dispersion equation for the classical isotropic elasticity for Love wave propagation

$$\tan\left(kH\sqrt{\frac{c^2}{\beta_1^2}-1}\right)=\frac{\mu_2\sqrt{1-\frac{c^2}{\beta_2^2}}}{\mu_1\sqrt{\frac{c^2}{\beta_1^2}-1}} \tag{43}$$

7 Numerical Results and Discussion

In this section, numerical calculation and graphical interpretation have been done in order to reveal the effect of irregularity parameter (H'/H, i.e. the ratio of the irregularity span with the thickness of the layer) on SH wave propagation in a slightly compressible finitely deformed layer overlying a half-space in both cases of rectangular and parabolic shaped irregularities at the interface.

We shall now specify the strain-energy function $W_0(\bar{F}, 1)$ to the Neo-Hookean strain-energy function

$$W=\frac{\mu}{2}(\lambda_1^2+\lambda_2^2+\lambda_3^2-3-2ln\,J)+\frac{k'}{2}(J-1)^2 \tag{44}$$

within which $\lambda_1,\ \lambda_2,\ \lambda_3$ are principal stretches, $k'=k-2/3\mu$, where μ and k' (often denoted by λ) are the Lame' moduli and k is the bulk modulus of the material in the unstressed configuration [13]. In this case for the strain-energy function, we have [6]

$$B^0_{ijij}=\mu J^{-1}\lambda_i^2 \tag{45}$$

For numerical purpose, the following data are taken into consideration [6]

$$\begin{cases}\lambda_1=1.7, \lambda_2=2.0, \lambda_3=1.6, \mu=1.0\\ \lambda_1=1.1, \lambda_2=0.91, \lambda_3=1.0, \mu=0.7\end{cases}$$

In Fig. 2, the dimensionless phase velocity c/β_1 has been plotted against the dimensionless wave number kH for different values of the ratios of the irregularity span with the thickness of the layer for the case of rectangular shaped irregular interface. Curve 1 represents the case of no irregularity condition i.e. $H'=0$ whereas curves 2 and 3 depict the case of rectangular shaped irregular interface. It is evident from the figure that phase velocity of SH wave decreases with increase in rectangular shaped irregularity at the interface.

Similarly, Fig. 3 manifests variation of the parabolic shaped irregularity at the interface. Curve 1 represents the case of no irregularity condition i.e. $H'=0$ whereas curves 2 and 3 depict the case of parabolic shaped irregular interface. It has also been found that parabolic shaped irregularity at the interface decreases the phase velocity of SH wave.

Fig. 2 Variation of dimensionless phase velocity with dimensionless wave number for different values of irregularity parameter for the case of rectangular shaped irregularity

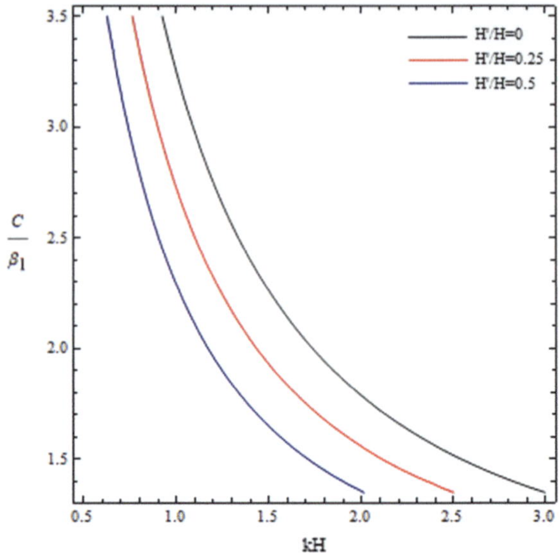

Fig. 3 Variation of dimensionless phase velocity with dimensionless wave number for different values of irregularity parameter for the case of parabolic shaped irregularity

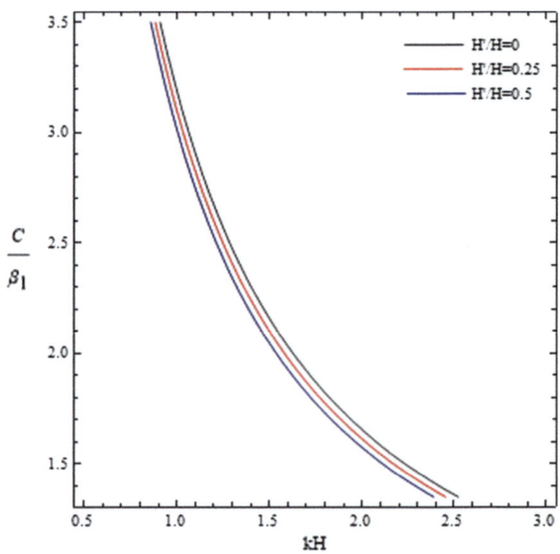

In order to reveal the influence of the irregularity parameter more explicitly, surface plots showing variation of dimensionless phase velocity against dimensionless wave number and irregularity parameter for the case of rectangular shaped irregularity and parabolic shaped irregularity have been drawn in Fig. 4a and b respectively.

A meticulous observation of the Figs. 2, 3 and 4 specifies that the presence of distinctly shaped irregularity (rectangular/parabolic) decreases the phase velocity

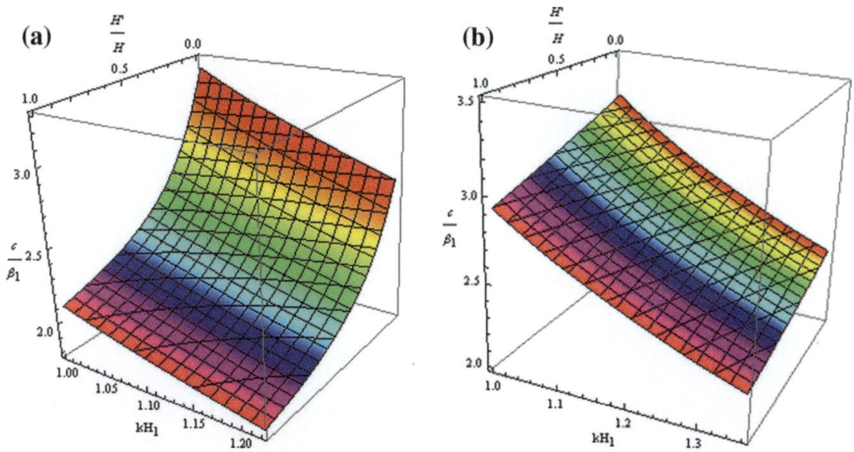

Fig. 4 Surface plots showing variation of dimensionless phase velocity with dimensionless wave number and irregularity parameter for the case of **a** rectangular shaped irregularity and **b** parabolic shaped irregularity

of SH wave. An overview and comparative study of the figures concludes that the rectangular shaped irregularity decreases the phase velocity of SH wave more as compared to the parabolic shaped irregularity.

8 Conclusion

In the present study, a layer/half-space configuration comprised of slightly compressible finitely deformed material has been considered. At the interface, two different shapes of irregularities i.e. rectangular and parabolic, have been studied in two distinct cases for which the dispersion equations for both the cases have been obtained analytically in closed form. As a particular case, the obtained dispersion equations have been found to be in agreement with the classical Love wave equation in isotropic case. The influence of two types of irregularities i.e. rectangular and parabolic, on the propagation of shear horizontal wave has been studied through numerical computations and graphical illustrations. The outcomes of the present study may be encapsulated as follows:

(i) Phase velocity of SH wave gets decreased with the increase in wave number.
(ii) The presence of rectangular as well as parabolic shaped irregularity at the interface decreases the phase velocity of SH wave.
(iii) The rectangular shaped irregularity decreases the phase velocity of SH wave more as compared to the parabolic shaped irregularity.

The idealization of incompressibility in modelling the mechanical behavior of rubber like materials using the theory of finite elasticity is found to be extremely effective in predicting the response of rubber when both the original geometry and boundary conditions are relatively simple. Hence, the findings of the present work can be employed in designing vibration insulators in bridges and tall buildings which have direct relevance to modern methods of earthquake protection [17].

Acknowledgements The Authors sincerely thank Indian Institute of Technology (Indian School of Mines), Dhanbad, India for facilitating us with best environment for research.

References

1. Dowaikh, M.A., Ogden, R.W.: On surface waves and deformations in a pre-stressed incompressible elastic solid. IMA J. Appl. Math. **44**(3), 261–284 (1990)
2. Rogerson, G.A., Fu, Y.B.: An asymptotic analysis of the dispersion relation of a pre-stressed incompressible elastic plate. Acta Mech. **111**, 59–77 (1995)
3. Rogerson, G.A., Murphy, J.G.: Surface waves in slightly compressible, finitely deformed elastic media. Mech. Res. Commun. **25**(5), 493–502 (1998)
4. Sandiford, K.J., Rogerson, G.A.: Some dynamic properties of a pre-stressed, nearly incompressible (rubber-like) elastic layer. Int. J. Non-Linear Mech. **35**, 849–868 (2000)
5. Chattopadhyay, A., Sahu, S.A.: Stresses produced in slightly compressible, finitely deformed elastic media due to normal moving load. Arch. Appl. Mech. **82**, 699–708 (2012)
6. Rogerson, G.A., Sandiford, K.J., Prikazchikova, L.A.: Abnormal long wave dispersion phenomena in a slightly compressible elastic plate with non-classical boundary conditions. Int. J. Non-Linear Mech. **42**, 298–309 (2007)
7. Chatterjee, M., Chattopadhyay, A.: Propagation, reflection, and transmission of SH-waves in slightly compressible, finitely deformed elastic media. Appl. Math. Mech. (English Edition) **36**(8), 1045–1056 (2015)
8. Scott, N.H.: The slowness surface of incompressible and nearly incompressible elastic materials. J. Elast. **16**, 239–250 (1986)
9. Bhattacharya, J.: On the dispersion curve for Love waves due to irregularity in the thickness of the transversely isotropic crustal layer. Beitr. Geophys. **71**, 324–333 (1962)
10. Chattopadhyay, A.: On the dispersion equation for Love wave due to irregularity in the thickness of the non-homogeneous crustal layer. Acta Geophys. Pol. **23**, 307–317 (1975)
11. Singh, A.K., Kumar, S., Chattopadhyay, A.: Love-type wave propagation in a piezoelectric structure with irregularity. Int. J. Eng. Sci. **89**, 35–60 (2015)
12. A.K. Singh, M.S. Chaki, A. Chattopadhyay: Remarks on impact of irregularity on SH-type wave propagation in micropolar elastic composite structure. Int. J. Mech. Sci. (2017)
13. Ogden, R.W.: Non-linear Elastic Deformations. Ellis Horwood, New York (1984)
14. Cole, J., Huth, J.: Stresses produced ill a half-plane by moving loads. J. Appl. Mech. **25**, 433–443 (1958)
15. Willis, H.F.: A formula for expanding an integral as a series. Phil. Mag. **39**, 455–459 (1948)
16. Tranter, C.J.: Intergral Transform in Mathematical Physics, pp. 63–67. Methuen and Co., Ltd., London (1966)
17. Sheridan, P.M., James, F.O., Miller, T.S.: Design of components. In: Engineering with Rubber, pp. 209–235. Hanser, Munich (1992)

Effect of Source Geometry on Interdependent Calcium and Inositol 1; 4; 5-Trisphosphate Dynamics in a Cardiac Myocyte Cell

Nisha Singh and Neeru Adlakha

Abstract Intracellular calcium governs the most versatile and universal signalling mechanism in living systems which includes contraction of the cardiac tissues, information processing in the brain, release of digestive enzymes by the liver etc. Various investigations have been made on study of calcium signalling in cardiac myocyte to understand its mechanisms. But most of existing investigations have mainly focused on study of calcium signalling in cardiac myocyte cell without paying attention on interdependence of calcium signalling and inositol 1; 4; 5-trisphosphate signalling. In this paper, we propose a mathematical model to understand the impact of the source geometry of calcium on these coupled signalling processes. This study suggests that the source geometry plays a vital role in these signalling processes. Also, calcium and inositol 1; 4; 5-trisphosphate shows a beautiful coordination with each other, which explains the role of inositol 1; 4; 5-trisphosphate in calcium signalling in cardiac myocyte cell. Such studies will provide the better understanding of various factors involved in calcium signalling in cardiac myocytes, which as a result will be of great use to biomedical scientists for making protocols for various heart diseases and their cure.

Keywords Calcium signalling · Interdependent dynamics · Inositol 1; 4; 5-trisphosphate signalling · Source geometry

1 Introduction

A wealth of experimental data is available detailing elementary Ca^{2+} release events and functional properties of the IP_3 receptors [3]. Luo and Rudy [14] proposed a model of electrical signalling of the cardiac ventricular. They studied concentration changes due to simulations of ionic currents excluding the impact of IP_3. Michailova et al. [18] proposed a model to study spatiotemporal features of Ca^{2+} buffering and

N. Singh (✉) · N. Adlakha
Applied Mathematics and Humanities Department, SVNIT, Surat, Gujarat 395007, India
e-mail: nishasingh.maths@gmail.com

S. Manna et al. (eds.), *Mathematical Modelling and Scientific Computing with Applications*, Springer Proceedings in Mathematics & Statistics 308,
https://doi.org/10.1007/978-981-15-1338-1_6

diffusion in atrial cardiac myocytes with inhibited *ER* without considering role of IP_3 dynamics.

Also, there are many attempt to study these complex signalling in various cells like neuron [9, 31, 32], astrocyte [10], fibroblast [12, 13], oocyte [19, 20, 22] and pancreatic acinar cell [16, 17] etc. are reported in the literature and some attempts are reported in the literature for the study of calcium dynamics in myocytes [14, 15, 18, 26, 27]. Most of the studies reported on calcium diffusion in myocytes are experimental [8]. Pathak et al. [23–25] have developed mathematical models to understand Ca^{2+} signalling dynamics in myocyte cell. But they have not considered the role of IP_3 dynamics in their model. Wagner et al. [33] have worked on a wave of IP_3 production accompanies the fertilization Ca^{2+} wave in the egg of the frog, Xenopus laevis. The interdependent study of calcium and IP_3 dynamics is reported in most of the above mentioned cells without taking impact of IP_3 on calcium dynamics and vice-versa [1, 4, 29].

Experiments are limited by protocol and therefore computational studies, across different spatial and temporal scales, are a significant and indispensable tool [3, 5, 28, 30, 34]. In this study, a systematic effort has been made to develop a mathematical model to understand the dependence of Ca^{2+} and IP_3 dynamics on each other and also, the effect of source of IP_3 and Ca^{2+} on cell signalling process [11]. The proposed model is used to explain intracellular calcium signalling in a myocyte cell in the presence of IP_3 receptors. A successful model should be able to predict and explain functional properties involved in the process of calcium signalling [33].

2 Mathematical Model

2.1 Calcium and IP_3 Dynamics Model

Here, we have proposed a coupled mathematical model of calcium and IP_3 dynamics in a myocyte cell. The reaction-diffusion equations for Ca^{2+} and IP_3 are given by [33],

$$\frac{\partial[Ca^{2+}]}{\partial t} = D_{cy}\nabla^2[Ca^{2+}] + \frac{J_{ipr} - J_{serca} + J_{leak}}{F_{cy}} \quad (1)$$

where $[Ca^{2+}]$ shows concentrations of Ca^{2+} in the cytosol, D_{cy} represents the diffusion coefficient, and various flux terms involved are expressed as,

$$J_{ipr} = V_{ipr} m^3 h^3 ([Ca^{2+}]_{er} - [Ca^{2+}]) \quad (2)$$

$$J_{leak} = V_{leak}([Ca^{2+}]_{er} - [Ca^{2+}]) \quad (3)$$

$$J_{serca} = V_{serca} \frac{[\mathrm{Ca}^{2+}]^2}{K_{serca}^2 + [\mathrm{Ca}^{2+}]^2} \tag{4}$$

In this proposed mathematical model, F_{cy} and F_{er} are the volume fractions, relative to total cell volume, of the cytosol and ER, respectively (with $F_{cy} + F_{er} = 1$).

According to the Li-Rinzel model, however, utilizes time scale arguments to remove two of the three binding reactions, replacing them with the equilibrium equation [33],

$$m = \frac{[IP_3]}{[IP_3] + K_{IP_3}} \frac{[\mathrm{Ca}^{2+}]}{[\mathrm{Ca}^{2+}] + K_{ac}} \tag{5}$$

Here, h variable shows the fraction of subunits not yet inactivated by Ca^{2+} which is expresses as follows,

$$h = \frac{K_{inh}}{K_{inh} + [\mathrm{Ca}^{2+}]}, \tag{6}$$

The reaction-diffusion equation for the cytosolic IP_3 concentration in a myocyte cell can be expressed as [33],

$$\frac{\partial [IP_3]}{\partial t} = D_{ip} \nabla^2 [IP_3] + \frac{J_{production} - \lambda (J_{kinase} + J_{phosphatase})}{F_{cy}} \tag{7}$$

where D_{ip} represent the diffusion coefficient.

$$J_{production} = V_{production} \frac{[\mathrm{Ca}^{2+}]^2}{[\mathrm{Ca}^{2+}]^2 + K_{production}^2} \tag{8}$$

$$J_{kinase} = (1 - \theta) V_1 \frac{[IP_3]}{[IP_3] + 2.5} + \theta V_2 \frac{[IP_3]}{[IP_3] + 0.5} \tag{9}$$

$$J_{phosphatase} = V_3 \frac{[IP_3]}{[IP_3] + 30} \tag{10}$$

Also, $J_{production}$, J_{kinase} and $J_{phosphatase}$ are in terms of total cell volume.

The Ca^{2+} dependence of the 3-kinase degradation pathway is described by a Hill function,

$$\theta = \frac{[\mathrm{Ca}^{2+}]}{[\mathrm{Ca}^{2+}] + 0.39} \tag{11}$$

Now, this model is formulated using finite difference method to study Ca^{2+} and IP_3 dynamics and the role of source influx on these dynamics.

2.2 *Mathematical Model to Study* Ca^{2+} *and* IP_3 *dynamics*

The calcium diffusion in a myocyte cell in the presence of inositol 1; 4; 5-trisphosphate receptors (IP_3R'*s*) for unsteady state case is represented as [33],

$$\frac{\partial[\mathrm{Ca}^{2+}]}{\partial t} = D_{cy}\nabla^2[\mathrm{Ca}^{2+}] + \frac{J_{ipr} - J_{serca} + J_{leak}}{F_{cy}} \tag{12}$$

$$\frac{\partial[IP_3]}{\partial t} = D_{ip}\nabla^2[IP_3] + \frac{J_{production} - \lambda(J_{kinase} + J_{phosphatase})}{F_{cy}} \tag{13}$$

Here, $\nabla = (\frac{\partial}{\partial r}, \frac{\partial}{\partial \theta}, \frac{\partial}{\partial z})$ and $p = (r, \theta, z)$ represents the position coordinates where r = 0, 1, 2, 3 and 4, $0 \leq \theta \leq 2\pi$ and z = 0, 1, 2. The calcium source is located at $(r, \theta, z) = (4, \pi, 1)$ and IP_3 source is located at $(r, \theta, z) = (4, 0, 1)$. The initial and boundary conditions governing the Ca^{2+} and IP_3 diffusion process are given by [7, 25],

(i) Initial conditions

$$[\mathrm{Ca}^{2+}]_{t=0} = 0.1\ \mu\mathrm{M} \tag{14}$$

$$[IP_3]_{t=0} = 0.16\ \mu\mathrm{M} \tag{15}$$

(ii) Boundary conditions

It is assumed that near the source the calcium concentration (at position a), calcium concentration which can be written as follows:

$$\lim_{p\to a}\left(-D_{cy}\frac{\partial[\mathrm{Ca}^{2+}]}{\partial r}\right) = \sigma \tag{16}$$

And at the boundary far away from the source the calcium concentration (at position b) is maintained at background calcium concentration which can be written as follows:

$$\lim_{p\to b}[\mathrm{Ca}^{2+}] = [\mathrm{Ca}^{2+}]_\infty = 0.1\ \mu\mathrm{M} \tag{17}$$

Also, it is assumed that near the source of IP_3 (at position c), the $[IP_3]$ can be written as follows:

$$\begin{aligned}\lim_{p\to c}[IP_3] = {} & 0.1882(t)^6 + 1.3121(t)^5 + 3.5391(t)^4 + 4.5312(t)^3 \\ & + 2.5893(t)^2 + 0.3648(t) + 0.1691 \leq 3\end{aligned} \tag{18}$$

And at the boundary far away from the source of IP_3 (at position d), the $[IP_3]$ is maintained at background concentration which can be written as follows:

$$\lim_{p \to d} [IP_3] = 3\ \mu\text{M}; \quad t > 0 \tag{19}$$

Moreover, at rest of the boundary points there is no flux. The mathematical model equations from (12) to (19) are solved numerically using Crank Nicholson Method. Then the resulting system provides simultaneous algebraic equations for unknown nodal concentrations. To solve the resulting equations for each time step Gaussian elimination method has been employed.

3 Results and Discussion

The proposed mathematical model is solved using numerical technique in which various standard parameters are used for the computation of numerical results as shown in Table 1 [33]. To facilitate our discussion, all calculations utilize the standard model parameters in Table 1, except where noted otherwise.

The calcium concentration profile at $t = 0.03\ s$, $z = 1$ ($i.e.$ 50 μm) in myocyte cell for (a) Point source, (b) Line source, (c) Surface source is shown in Fig. 1. It is observed from the figure that maximum calcium concentration is at $r = 4\ \mu m$, $\theta = \pi$ i.e. source of Ca^{2+}. The calcium concentration falls sharply between $r = 3$ μm and $r = 4$ μm and then falls gradually and achieves background concentration of 0.1 μM away from source of Ca^{2+}.

The peak calcium concentration is highest in Fig. 1c as compared to Fig. 1a and b i.e. 0.042 μM. Also, the area of spread of calcium distribution in Fig. 1c is greater than that in Fig. 1a and b for the same points of time interval i.e. $t = 0.03\ s$. This

Table 1 Parameters for Ca^{2+} and IP_3 signalling

Parameters	Values [33]	Parameters	Values [33]
V_{ipr}	8.5 s^{-1}	V_{leak}	0.01 s^{-1}
K_{ip}	0.15 μM	K_{ac}	0.8 μM
V_{serca}	0.65 μM/s	K_{serca}	0.4 μM
K_{inh}	1.9 μM	τ	2 s
$V_{production}$	0.075 μM/s	$K_{production}$	0.4 μM
V_1	0.001 μM/s	V_2	0.005 μM/s
V_3	0.02 μM/s	λ	30
D_{cy}	16 $\mu m^2/s$	D_{er}	16 $\mu m^2/s$
D_{ip}	283 $\mu m^2/s$	$[Ca^{2+}]_T$	1.7 μM
F_{cy}	0.83	F_{er}	0.17

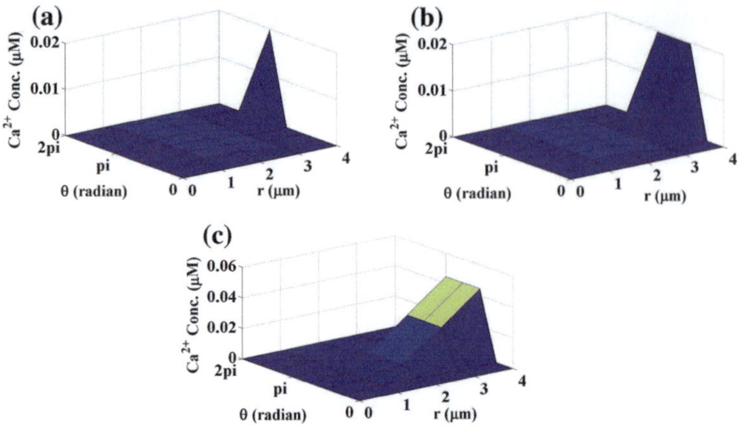

Fig. 1 Ca^{2+} concentration profile in a cardiac myocyte cell at time $t = 0.03$ s in the cytosol at $z = 1$ (*i.e.* 50 μm) position for different source geometry **a** point source, **b** line source, **c** surface source

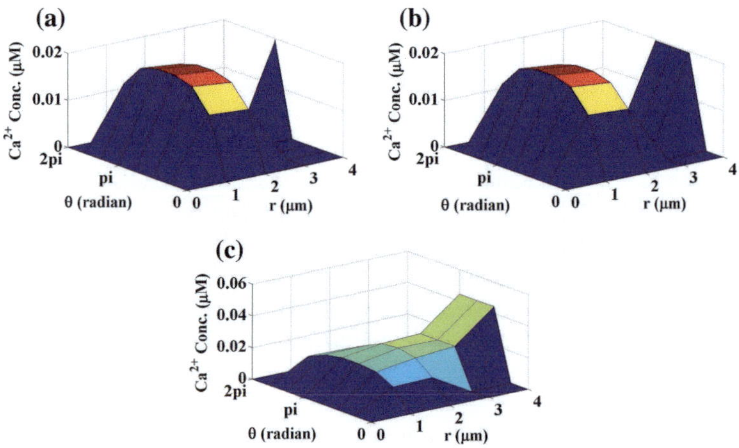

Fig. 2 Ca^{2+} concentration profile in a cardiac myocyte cell at time $t = 0.06$ s in the cytosol at $z = 1$ (*i.e.* 50 μm) position for different source geometry **a** point source, **b** line source, **c** surface source

is because the calcium influx through plasma membrane channel in case of surface source is higher than that of line and point source.

At time $t = 0.06$ s, $z = 1$ (*i.e.* 50 μm) the calcium concentration profile in myocyte cell for (a) Point source, (b) Line source, (c) Surface source is represented in Fig. 2. It is observed from the figure that maximum calcium concentration is at source of Ca^{2+} ($r = 4$ μm, $\theta = \pi$). The calcium concentration falls sharply between $r = 3$ μm to $r = 4$ μm and then rises gradually till $r = 1$ μm, after that it falls to achieves background concentration of 0.1 μM.

The peak of calcium concentration in Fig. 2c is 0.05 μM, which is higher than that in Fig. 2a and b. Also, the area of spread of calcium distribution in Fig. 2c is greater than that in Fig. 2a and b for the same points of time interval i.e. $t = 0.06$ s. This is due to the calcium influx through plasma membrane channel in case of surface source is higher than that of line and point source.

4 Conclusion

Mathematical modelling and numerical simulations of calcium and IP_3 dynamics in a myocyte cell predict interesting interdependence on each other and shows impact of source geometry on these signalling processes. Above results conclude that the source plays a vital role in calcium and IP_3 signalling. In various models given in literature [6, 21, 25] they have excluded the role of inositol 1; 4; 5-trisphosphate (IP_3), but many experimental works [1, 2, 4] have shown the importance of IP_3 dynamics in calcium dynamics of cell. So, a mathematical model is proposed to understand effect of inositol 1; 4; 5-trisphosphate (IP_3) dynamics on intracellular Ca^{2+} dynamics and vice versa in a cardiac myocyte cell, which describes the properties of elemental calcium release events in the presence of inositol 1; 4; 5-trisphosphate (IP_3).

This model shows the effect of source geometry on calcium profile is quite significant. Obtained results suggests that the surface source is more effective as well as efficient to regulate calcium signalling in a myocyte cell as compared to line and point source. Here, the finite difference method was successfully employed to solve the problem of calcium diffusion in cylindrical shaped myocyte cell for three dimensional case. Such studies can generate new information and knowledge which can be useful to understand the mechanisms involved in cardiac cell.

Acknowledgements The authors are thankful to the Department of Biotechnology, New Delhi, India for providing support in the form of Bioinformatics Infrastructure Facility for carrying out this work.

References

1. Adkins, C.E., Taylor, C.W.: Lateral inhibition of inositol 1, 4, 5-trisphosphate receptors by cytosolic Ca^{2+}. Current Biol. **9**(19), 1115–1118 (1999)
2. Allbritton, N.L., Meyer, T., Stryer, L.: Range of messenger action of calcium ion and inositol 1, 4, 5-trisphosphate. Science New York then Washington **258**, 1812–1812 (1992)
3. Ciapa, B., Pesando, D., Wilding, M., Whitaker, M.: Cell-cycle calcium transients driven by cyclic changes in inositol trisphosphate levels. Nature **368**(6474), 875–878 (1994)
4. De Young, G.W., Keizer, J.: A single-pool inositol 1, 4, 5-trisphosphate-receptor-based model for agonist-stimulated oscillations in Ca^{2+} concentration. Proc. Natl. Acad. Sci. **89**(20), 9895–9899 (1992)
5. Dupont, G., Goldbeter, A.: One-pool model for Ca^{2+} oscillations involving Ca^{2+} and inositol 1, 4, 5-trisphosphate as co-agonists for Ca^{2+} release. Cell Calcium **14**(4), 311–322 (1993)

6. Falcke, M.: Buffers and oscillations in intracellular Ca^{2+} dynamics. Biophys. J. **84**(1), 28–41 (2003)
7. Fink, C.C., Slepchenko, B., Moraru, I.I., Watras, J., Schaff, J.C., Loew, L.M.: An image-based model of calcium waves in differentiated neuroblastoma cells. Biophys. J. **79**(1), 163–183 (2000)
8. Goonasekera, S.A., Hammer, K., Auger-Messier, M., Bodi, I., Chen, X., Zhang, H., Reiken, S., Elrod, J.W., Correll, R.N., York, A.J., et al.: Decreased cardiac l-type Ca^{2+} channel activity induces hypertrophy and heart failure in mice. J. Clin. Investig. **122**(1), 280 (2012)
9. Jha, A., Adlakha, N.: Finite element model to study the effect of exogenous buffer on calcium dynamics in dendritic spines. Int. J. Model., Simul., Sci. Comput. **5**(02), 1350027 (2014)
10. Jha, B.K., Adlakha, N., Mehta, M.: Two-dimensional finite element model to study calcium distribution in astrocytes in presence of excess buffer. Int. J. Biomath. **7**(03), 1450031 (2014)
11. Klipp, E., Liebermeister, W.: Mathematical modeling of intracellular signaling pathways. BMC Neurosci. **7**(1), S10 (2006)
12. Kotwani, M., Adlakha, N., Mehta, M.: Numerical model to study calcium diffusion in fibroblasts cell for one dimensional unsteady state case. Appl. Math. Sci. **6**(102), 5063–5072 (2012)
13. Kotwani, M., Adlakha, N., Mehta, M.: Finite element model to study the effect of buffers, source amplitude and source geometry on spatio-temporal calcium distribution in fibroblast cell. J. Med. Imaging Health Inform. **4**(6), 840–847 (2014)
14. Luo, C.h., Rudy, Y.: A dynamic model of the cardiac ventricular action potential. i. Simulations of ionic currents and concentration changes. Circ. Res. **74**(6), 1071–1096 (1994)
15. Luo, C.h., Rudy, Y.: A dynamic model of the cardiac ventricular action potential. ii. After depolarizations, triggered activity, and potentiation. Circ. Res. **74**(6), 1097–1113 (1994)
16. Manhas, N., Pardasani, K.: Modelling mechanism of calcium oscillations in pancreatic acinar cells. J. Bioenerg. Biomembr. **46**(5), 403–420 (2014)
17. Manhas, N., Sneyd, J., Pardasani, K.: Modelling the transition from simple to complex Ca^{2+} oscillations in pancreatic acinar cells. J. Biosci. **39**(3), 463–484 (2014)
18. Michailova, A., DelPrincipe, F., Egger, M., Niggli, E.: Spatiotemporal features of Ca^{2+} buffering and diffusion in atrial cardiac myocytes with inhibited sarcoplasmic reticulum. Biophys. J. **83**(6), 3134–3151 (2002)
19. Naik, P.A., Pardasani, K.R.: One dimensional finite element model to study calcium distribution in oocytes in presence of VGCC, RyR and buffers. J. Med. Imaging Health Inform. **5**(3), 471–476 (2015)
20. Naik, P.A., Pardasani, K.R.: Finite element model to study calcium distribution in oocytes involving voltage gated Ca^{2+} channel, ryanodine receptor and buffers. Alex. J. Med. **52**(1), 43–49 (2016)
21. Nivala, M., de Lange, E., Rovetti, R., Qu, Z.: Computational modeling and numerical methods for spatiotemporal calcium cycling in ventricular myocytes. Front. Physiol. **3**, 114 (2012)
22. Panday, S., Pardasani, K.R.: Finite element model to study effect of advection diffusion and Na^{+}/Ca^{2+} exchanger on Ca^{2+} distribution in oocytes. J. Med. Imaging Health Inform. **3**(3), 374–379 (2013)
23. Pathak, K., Adlakha, N.: Finite element model to study two dimensional unsteady state calcium distribution in cardiac myocytes. Alex. J. Med. **52**(3), 261–268 (2016)
24. Pathak, K.B., Adlakha, N.: Finite element model to study calcium signalling in cardiac myocytes involving pump, leak and excess buffer. J. Med. Imaging Health Inform. **5**(4), 683–688 (2015)
25. Pathak, K.B., Adlakha, N.: Finite element model to study one dimensional calcium dynamics in cardiac myocytes. J. Multiscale Model. **6**(02), 1550003 (2015)
26. Shannon, T.R., Wang, F., Puglisi, J., Weber, C., Bers, D.M.: A mathematical treatment of integrated ca dynamics within the ventricular myocyte. Biophys. J. **87**(5), 3351–3371 (2004)
27. Smith, G.D., Keizer, J.E., Stern, M.D., Lederer, W.J., Cheng, H.: A simple numerical model of calcium spark formation and detection in cardiac myocytes. Biophys. J. **75**(1), 15–32 (1998)
28. Sneyd, J., Sherratt, J.: On the propagation of calcium waves in an inhomogeneous medium. SIAM J. Appl. Math. **57**(1), 73–94 (1997)

29. Stewart, B.D., Scott, C.E., McCoy, T.P., Yin, G., Despa, F., Despa, S., Kekenes-Huskey, P.M.: Computational modeling of amylin-induced calcium dysregulation in rat ventricular cardiomyocytes. Cell Calcium **71**, 65–74 (2018)
30. Swaminathan, D.: Mathematical modeling of intracellular calcium signaling: a study of IP_3 receptor models. Ohio University (2010)
31. Tewari, S., Pardasani, K.: Finite difference model to study the effects of Na^+ influx on cytosolic Ca^{2+} diffusion. World Acad. Sci. Eng. Technol. **15**, 670–675 (2008)
32. Tripathi, A., Adlakha, N.: Finite volume model to study calcium diffusion in neuron cell under excess buffer approximation. Int. J. Math. Sci. Engg. Appl. (IJMSEA) **5**, 437–447 (2011)
33. Wagner, J., Fall, C.P., Hong, F., Sims, C.E., Allbritton, N.L., Fontanilla, R.A., Moraru, I.I., Loew, L.M., Nuccitelli, R.: A wave of ip_3 production accompanies the fertilization Ca^{2+} wave in the egg of the frog, xenopus laevis: theoretical and experimental support. Cell Calcium **35**(5), 433–447 (2004)
34. Watras, J., Ehrlich, B.E., et al.: Bell-shaped calcium-response curves of $ins(l, 4, 5)p_3$-and calcium-gated channels from endoplasmic reticulum of cerebellum. Nature **351**(6329), 751–754 (1991)

Wave Interaction with a Floating Circular Porous Elastic Plate

H. Behera, Siluvai Antony Selvan and Vinay Kumar Gupta

Abstract Wave interaction with a circular elastic porous plate, floating in a two-layer fluid is investigated under the linearized theory of water waves. The eigenfunction expansion method—along with the Darcy's law for wave past porous structure under the assumption that the flexible structure is having fine pores and is of homogeneous structural rigidity—is exploited to obtain an analytic solution of the problem. The plate deflection and flow distribution around the plate are further computed to study the effects of various wave and structural parameters. The study reveals that the porosity of the plate renders major part of the wave energy to dissipate.

Keywords Circular structures · Eigenfunction expansion method · Two-layer fluid · Plate deflection · Flow distribution

1 Introduction

The floating flexible structures are often used as the breakwater for its characteristics nature of reducing wave force on the structures by partial reflection and transmission of an incident wave energy. Wave interactions with various floating structures of different geometrical configurations have been widely studied over the past few decades. Garrett [1], using Galerkin's method, investigated the wave scattering by circular dock and analyzed the hydrodynamic force exerted due to incident wave. Meylan and Squire [2] analyzed the behavior of a solitary, circular, flexible ice

H. Behera (✉) · S. A. Selvan
Department of Mathematics, SRM Institute
of Science and Technology, Kattankulathur 603203, Tamil Nadu, India
e-mail: hkb.math@gmail.com

S. A. Selvan
e-mail: antony.selvan22@gmail.com

V. K. Gupta
Discipline of Mathematics, Indian Institute of Technology Indore, Indore 453552, Madhya Pradesh, India
e-mail: vkg@iiti.ac.in

S. Manna et al. (eds.), *Mathematical Modelling and Scientific Computing with Applications*, Springer Proceedings in Mathematics & Statistics 308,
https://doi.org/10.1007/978-981-15-1338-1_7

floe brought into motion by the action of long-crested sea waves and studied the strain field generated in the floe, its surge response and the energy initiated in the water encircling it. They found that the force on a circular floe or ice cake due to long crested ocean waves generally increases as the floe diameter or thickness is increased. Motivated by the design of the circular floating island along the Israeli coast, Zilman and Miloh [3] applied the shallow water approximation for studying the wave action on the circular plate. Tsubogo [4] employed the advanced boundary element method to analyze floating elastic plates subject to a train of plane waves and used Fourier series expansion instead of finite difference method for discretizing the free edge conditions of the plate. Surface wave scattering by ice floe of varying thickness was studied by Bennetts et al. [5] using the multi-mode approximation method. Sturova [6] studied the unsteady behavior of an elastic articulated floating flexible plate under the assumption of linear shallow water theory. Mondal et al. [7] analyzed the surface gravity wave interaction with circular flexible structures using the small amplitude wave theory and Fourier–Bessel series expansion.

Apart from impermeable floating/submerged horizontal structures, horizontal porous structures are preferred as effective breakwaters to reduce wave force on an existing structure for creating tranquility zone near ports and harbors. Cho and Kim [8] obtained a numerical solution using boundary element method along with the eigenfunction expansion method for wave scattering by submerged horizontal flexible porous membrane and validated their results with experiments. Yu [9] presented a comprehensive review on wave interaction with submerged horizontal porous plate. He analyzed the effects of porosity, length and submerged on reducing the wave force on the porous plate. Hu and Wang [10] studied the problem of wave interaction with system of submerged horizontal plate and a vertical permeable wall using the matched vertical eigenfunction expansion method. Liu et al. [11] studied wave interaction with submerged porous plates of different configurations. Behera and Sahoo [12] investigated the surface gravity wave interaction with a submerged horizontal flexible porous plate under the assumptions of the small amplitude water wave theory and small structural response. Moreover, the flexible porous plate in [12] was modeled using the thin plate theory and wave past porous structure was based on the generalized porous wavemaker theory. The authors observed that owing to the presence of a horizontal flexible porous plate, a significant amount of wave energy is dissipated when a wave interacts with the flexible porous plate and thus results in a very less wave transmission to the lee side of the submerged plate. Koley and Sahoo [13] analyzed the wave interaction with flexible membrane using Green's technique. Meylan et al. [14] studied the water wave scattering and energy dissipation by a floating porous elastic plate in three dimensions by using the eigenfunction expansion method and boundary and finite element methods. Recently, Behera and Ng [15] developed an analytic solution for the oblique wave scattering by floating and submerged porous elastic plates of different structural parameters using the eigenfunction expansion method. In [15], the authors found that a significant amount of wave energy is dissipated due to the presence of floating and submerged porous plates, thus resulting in less wave reflection and transmission.

In general, the ocean has several layers of fluid with varying density. The low-density fluid floats above the high-density one. Such stratification is mainly due to the mixing of oil/chemical and solar radiation. Most of the models are made more accurate by taking stratification into consideration. Using Green's second identity, Xu and Lu [16] derived the energy conservation relation for a floating plate in a two-layer fluid. Recently, Das et al. [17] found the dead water effect in a two-layer fluid when densities in both the layers are close enough.

All the aforementioned studies consider wave interaction with floating/submerged rigid/flexible impermeable/porous structures. However, wave interaction with circular floating flexible porous elastic plate in a single-/two-layer fluid has received much less attention, although the use of floating circular porous elastic structures as breakwater in coastal regions is ubiquitous in protecting various marine facilities. The present study focuses on wave interaction with flexible and porous floating circular plate in a two-layer fluid using the small amplitude wave theory and small structural response to different edge conditions. The problem is analyzed using the eigenfunction expansion method. The plate deflection and flow distribution around the plate are presented for various wave and structural parameters.

The rest of the paper is organized as follows. The detailed formulation and boundary conditions for the problem of wave interaction with a floating circular porous plate in a two-layer fluid are presented in Sect. 2. The method to solve the problem is described in Sect. 3. followed by discussion of the results in Sect. 4. The conclusion is given in Sect. 5.

2 Mathematical Formulation

The problem is analyzed in the cylindrical polar coordinates (r, θ, z) with $r\theta$-plane being the horizontal plane and vertical z-axis is taken positive in downward direction. Let a thin circular porous elastic plate of radius 'b' and thickness 'a' be floating on a two-layer fluid horizontally at $z = 0$ at finite water depth h_2 as shown in Fig. 1, and let the center of the plate be at origin. The upper layer $(0 \leq z \leq h_1)$ and lower layer $(h_1 \leq z \leq h_2)$ consists of non-homogeneous fluid having densities ρ_1 and ρ_2, respectively. The fluid interface between these two layers is denoted by $z = h_1$. Further, the incompressible and inviscid fluids exhibit the irrotational flow and follow the simple harmonic with respect to time. Hence, the velocity potential $\Phi_j(r, \theta, z, t)$ can be written as $\Phi_j(r, \theta, z) = \mathrm{Re}[\varphi_j(r, \theta, z)\mathrm{e}^{-\mathrm{i}\omega t}]$, where ω be the angular frequency and $\varphi_j(r, \theta, z)$ be the spatial velocity potential which satisfies the three-dimensional Laplace equation in the jth region is given by

$$\Delta\varphi_j = 0, \tag{1}$$

where $j = 1, 2$ and Δ is the Laplace operator, which is defined as

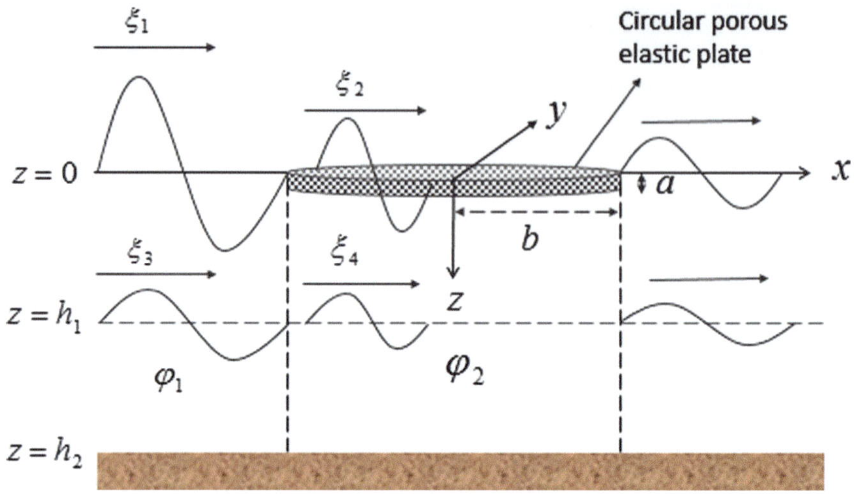

Fig. 1 Floating circular porous elastic plate in a two-layer fluid

$$\Delta \equiv \nabla_{r\theta}^2 + \partial_z^2, \text{ with } \nabla_{r\theta}^2 \equiv \left(\frac{\partial^2}{\partial r^2} + \frac{1}{r}\frac{\partial}{\partial r} + \frac{1}{r^2}\frac{\partial^2}{\partial \theta^2}\right) \text{ and } \partial_z^2 \equiv \frac{\partial^2}{\partial z^2}.$$

The normal velocity vanishes at the rigid bed, i.e.,

$$\partial_z \varphi_j = 0 \text{ on } z = h_2, \quad \text{for } j = 1, 2. \tag{2}$$

In both the modes, the elevations are given by

$$\xi_j(r, \theta, z, t) = \eta_j(r, \theta, z)\,\mathrm{e}^{-\mathrm{i}\omega t} \quad \text{for } j = 1, 3, \tag{3}$$

where η_1 and η_3 denote the amplitudes of elevations in surface and internal modes, respectively. Similarly, the elevation in the plate-covered region is given by

$$\xi_j(r, \theta, z, t) = \eta_j(r, \theta, z)\mathrm{e}^{-\mathrm{i}\omega t} \quad \text{for } j = 2, 4, \tag{4}$$

where η_2 and η_4 denote the amplitudes of plate deflection and internal mode elevation in the plate-covered region, respectively. The linearized kinematic and dynamic conditions at mean free surface in the open-water and plate-covered regions are given by

$$\mathrm{i}\omega\varphi_1 + g\eta_1 = 0, \tag{5}$$

$$\partial_z\varphi_1 + \mathrm{i}\omega\eta_1 = 0, \tag{6}$$

$$\partial_z \varphi_2 + \mathrm{i}\omega\eta_2 = -\mathrm{i}\alpha_I G\varphi_2, \tag{7}$$

where G is the porous-effect parameter and α_I is the surface mode progressive wave number. Similarly, the boundary conditions at the interface $z = h_1$ for the open-water and plate-covered regions are given by

$$\partial_z \varphi_j|_{z=h_1^-} = \partial_z \varphi_j|_{z=h_1^+}, \tag{8}$$

$$s(\mathrm{i}\omega\varphi_j + g\eta_i)|_{z=h_1^-} = (\mathrm{i}\omega\varphi_j + g\eta_i)|_{z=h_1^+}, \tag{9}$$

where $i = 3$ for $j = 1$ and $i = 4$ for $j = 2$. Substitution of Eqs. (2) and (6) in (7) yields

$$s\left(\partial_z \varphi_j + K\varphi_j\right)|_{z=h_1^-} = \left(\partial_z \varphi_j + K\varphi_j\right)|_{z=h_1^+}, \tag{10}$$

where $K = \omega^2/g$ with g being the magnitude of the acceleration due to gravity, $j = 1, 2$ and $s = \rho_1/\rho_2$ with $\rho_1 < \rho_2$ is the density ratio. The dynamical equation of flexible porous plate is given by

$$EI\nabla^4_{r\theta}\eta_2 + N\nabla^2_{r\theta}\eta_2 + \rho_2 g\eta_2 - \rho_p\omega^2 d\eta_2 = -\mathrm{i}\omega\rho_2\varphi_2, \tag{11}$$

where ρ_p is the density of the plate, E is the Young's modulus of the plate and $I = \frac{a^3}{12(1-\nu^2)}$ with ν being the Poisson's ratio. The far-field condition satisfies the Sommerfeld radiation condition and is given by

$$\lim_{r\to\infty} \sqrt{r}\left\{\frac{\partial(\varphi - \varphi_{\text{in}})}{\partial r} - \mathrm{i}\alpha_I(\varphi - \varphi_{\text{in}})\right\} = 0, \tag{12}$$

where

$$\varphi_{in} = \sum_{m=0}^{\infty}\sum_{n=I}^{II}\left(\frac{-\mathrm{i}gH_n}{2\omega}\right) f_{1n}(z)\varepsilon_m J_m(\alpha_n r)\cos(m\theta), \quad \text{for } 0 < \theta < 2\pi. \tag{13}$$

with H_n being the known incident wave amplitude, α_I and α_{II} be the progressive wavenumbers in both surface and interfacial modes, respectively, $\epsilon_m = 1$ for $m = 0$, $\epsilon_m = 2\mathrm{i}^m$ for $m = 1, 2, 3, \ldots$ and J_m being the Bessel function of first kind of order m. The continuity of pressure and velocity along the interface are given by

$$\varphi_1 = \varphi_2 \text{ and } \partial_r\varphi_1 = \partial_r\varphi_2, \text{ on } r = b \quad \text{for } 0 < z < h_2,\ 0 < \theta \le 2\pi. \tag{14}$$

It is assumed that the plate is having free-edge condition. Thus, the bending moment and shear strain vanishes along the edge of the floating circular flexible porous plate, and the edge conditions on $z = 0$ and $r = b$ for $0 < \theta \le 2\pi$ are given by

$$\left[\partial_r^2 + \nu\left(\frac{1}{r}\partial_r + \frac{1}{r^2}\partial_\theta^2\right)\right](\partial_z\varphi_2 - \mathrm{i}\alpha_0 G\varphi_2) = 0, \tag{15a}$$

$$\left[EI\left\{\partial_r\nabla_{r\theta}^2 + \frac{(1-\nu)}{r^2}\left(\partial_r - \frac{1}{r}\right)\partial_\theta^2\right\} + N\frac{\partial}{\partial r}\right](\partial_z\varphi_2 - \mathrm{i}\alpha_0 G\varphi_2) = 0. \tag{15b}$$

3 Method of Solution

The spatial velocity potentials $\varphi_j(r,\theta,z)$ in the open-water and plate-covered regions satisfy Eq. (2) along with linearized boundary condition in Eq. (5), and $\phi_j(r,\theta,z)$ in these regions are given by

$$\begin{aligned}\varphi_1(r,\theta,z) = \sum_{m=0}^{\infty}\Bigg(\sum_{n=I}^{II}\left(\frac{-\mathrm{i}gH_n}{2\omega}\right)\epsilon_m J_m(\alpha_n r) f_{1n}(z)\\ + \sum_{n=I,II,1}^{\infty} A_{mn}H_m^{(1)}(\alpha_n r) f_{1n}(z)\cos(m\theta),\end{aligned} \tag{16}$$

$$\varphi_2(r,\theta,z) = \sum_{m=0}^{\infty}\left[\sum_{n=I,II,III,IV,1}^{\infty} B_{mn}J_m(\beta_n r)\right] f_{2n}(z)\cos(m\theta), \tag{17}$$

where $H_m^{(1)}$ is the Hankel function of first kind of order m and A_{mn} are the constant coefficients to be determined. The eigenfunctions for the open-water region are given by

$$f_{1n}(z) = \begin{cases} -\frac{\sinh\{\alpha_n(h_2-h_1)\}[\alpha_n\cosh(\alpha_n z)-K\sinh(\alpha_n z)]}{\alpha_n\sinh(\alpha_n h_1)-K\cosh(\alpha_n h_1)} & \text{for } 0\le z\le h_1,\\ \cosh\{\alpha_n(h_2-z)\} & \text{for } h_1\le z\le h_2.\end{cases} \tag{18}$$

Similarly, the eigenfunctions for the plate-covered region are given by

$$f_{2n}(z) = \begin{pmatrix} F(\beta_n)[\beta_n\Im\cosh(\beta_n z) - (K-\mathrm{i}\alpha_0 G\Im)\sinh(\beta_n z)] & \text{for } \ 0\le z\le h_1,\\ \cosh\{\beta_n(h_2-z)\} & \text{for } \ h_1\le z\le h_2,\end{pmatrix} \tag{19}$$

where $F(\beta_n)$ and $\Im$ is given as,

$$F(\beta_n) = -\frac{\sinh\{\beta_n(h_2-h_1)\}}{\beta_n\Im\sinh(\beta_n h_1) - (K-\mathrm{i}\alpha_0 G)\Im\cosh(\beta_n h_1)}$$

and

$$\Im = \frac{EI}{\rho_2 g}\beta_n^4 - \frac{N}{\rho_2 g}\beta_n^2 - \frac{\rho_p\, a\, \omega^2}{\rho_2 g} + 1,$$

The eigenvalue α satisfies a dispersion relation in the open-water region, which is given by

$$\frac{\tanh(\alpha h_1) + \tanh\{\alpha(h_2 - h_1)\} + K^2[(s\tanh\{\alpha(h_2 - h_1)\}\tanh(\alpha h_1) + 1)]}{\left[(1-s)\alpha^2 \tanh\{\alpha(h_2 - h_1)\}\tanh(\alpha h_1)\right]} = \frac{1}{K}. \tag{20}$$

It may be noted that dispersion relation (20) has two real roots α_I and α_{II}, which correspond to surface and interface modes respectively, and infinitely many purely imaginary roots for $n = 1, 2, 3, \ldots$. Similarly, the eigenvalue β satisfies a dispersion relation in the plate-covered region, which is given by

$$\begin{aligned} &K^2[s\tanh(\beta h_1)\tanh\{\beta(h_2 - h_1)\} + 1] - \beta K[\Im\{s\tanh\{\beta(h_2 - h_1)\} + \tanh(\beta h_1)\} \\ &\quad + (1-s)\tanh\{\beta(h_2 - h_1)\}] + \beta^2\Im(1-s)\tanh\{\beta(h_2 - h_1)\}\tanh(\beta h_1) \\ &\quad + \mathrm{i}\alpha_I G[\beta\Im(1-s)\tanh\{\beta(h_2 - h_1)\} - \Im K\{1 + s\tanh\{\beta(h_2 - h_1)\}\tanh(\beta h_1)\}] = 0. \end{aligned} \tag{21}$$

It may be noted that all the roots of dispersion relation (21) are complex in nature. The eigenfunctions for the open-water region [as in Eq. (18)] satisfy the orthogonal property

$$Z_{\alpha_n} = s\int_0^{h_1} f_{1n}^2(z)\,\mathrm{d}z + \int_{h_1}^{h_2} f_{1n}^2(z)\,\mathrm{d}z.$$

Using velocity and pressure continuities as in Eq. (14) along with the above orthogonality relation, the following system of equations are obtained.

$$\left[\sum_{n=I}^{II}\left(\frac{-\mathrm{i}gH_n}{2\omega}\right)\epsilon_m J_m(\alpha_n b) + \sum_{n=I,II,1}^{\infty} A_{mn} H_m^{(1)}(\alpha_n b)\right] Z_{\alpha_n} - \sum_{n=I,II,III,IV,1}^{\infty} B_{mn} J_m(\beta_n b) Z_{\alpha_n\beta_n} = 0, \tag{22}$$

$$\left[\sum_{n=I}^{II}\left(\frac{-\mathrm{i}gH_n}{2\omega}\right)\epsilon_m J'_m(\alpha_n b) + \sum_{n=I,II,1}^{\infty} A_{mn} H_m^{(1)\prime}(\alpha_n b)\right] Z_{\alpha_n} - \sum_{n=I,II,III,IV,1}^{\infty} B_{mn} J'_m(\beta_n b) Z_{\alpha_n\beta_n} = 0, \tag{23}$$

where $Z_{\alpha_n\beta_n} = s\int_0^{h_1} f_{1n}(z) f_{2n}(z)\,\mathrm{d}z + \int_{h_1}^{h_2} f_{1n}(z) f_{2n}(z)\,\mathrm{d}z$. Further, using free-edge condition in Eqs. (15a) and (15b), the following system of equations are obtained,

$$\sum_{n=I,II,III,IV,1}^{\infty} B_{mn}\left[b^2 J''_m(\beta_n b) + \nu\left\{b J'_m(\beta_n b) - m^2 J_m(\beta_n b)\right\}\right](-K\beta_n F(\beta_n)) = 0, \tag{24}$$

$$\sum_{n=I,II,III,IV,1}^{\infty} B_{mn}\Big[EI\Big\{b^3 J_m'''(\beta_n b) + b^2 J_m''(\beta_n b) - \Big(1 + (2-\nu)m^2\Big) J_m'(\beta_n b) + (3-\nu)m^2 J_m(\beta_n b)\Big\}$$
$$+ b^3 N J_m^{'}(\beta_n b)\Big](-K\beta_n F(\beta_n)) = 0. \tag{25}$$

For some fixed value of m (say $\mathcal{M}$), and n (say $\mathcal{N}$), the above system of Eqs. (22–25) is solved to find $\mathcal{M}(2\mathcal{N}+6)$ unknowns.

4 Results and Discussion

In this Section, numerical computations have been developed to investigate the deformation of the circular porous plate and flow distribution around the flexible plate in a two-layer fluid. Similar to Ref. [18], it is found in the present work as well that the Fourier-Bessel series converges for $\mathcal{M} = 8$ and $\mathcal{N} = 8$. The following wave and structural parameters are fixed for the numerical computation irrespective of its change, which is mentioned accordingly. The water depth $h_2 = 50$ m, density of water $\rho_2 = 1025$ kg/m^3, density ratio $s = 0.8$, incident wave amplitude $H_I = 1$, depth ratio $h_1/h_2 = 0.5$, dimensionless radius of the plate $b/h_2 = 2$, porous-effect parameter $G = 0.25 + 0.25\mathrm{i}$, dimensionless thickness of the plate $a/h_2 = 0.01$, Poisson's ratio $\nu = 0.3$, Young's modulus $E = 1 \times 10^9$ Pa and compressive force $N = 0.5\sqrt{EI\rho_2 g}$.

Figure 2 illustrates the deflection of flexible porous plate for different values of the depth ratio h_1/h_2. As the value of the depth ratio h_1/h_2 decreases, the vertical wave force exerted on the porous plate increases due to more energy concentration near the free surface. Thus, most of the incident waves are transmitted through the porous plate which results in more deformation in the plate. As $h_1/h_2 \to 1$, the deflection of the porous plate in a two-layer fluid approaches the plate deflection in a single-layer fluid as studied by Mondal et al. [7] in case of wave interaction with a floating impermeable flexible plate.

In Fig. 3, the deflection in the flexible porous plate is plotted for different values of the density ratio s. It is found that the deflection of porous plate decreases with

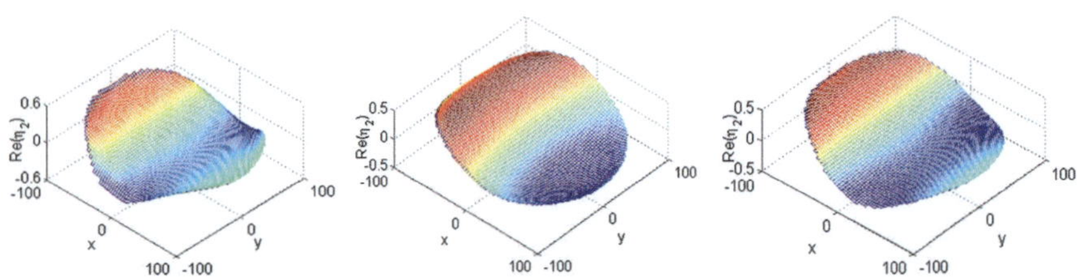

(a) $h_1/h_2 = 0.2$ (b) $h_1/h_2 = 0.5$ (c) $h_1/h_2 = 0.9$

Fig. 2 Deflection of the flexible porous plate for different value of depth ratio h_1/h_2 The fixed parameters are $G = 0.25 + 0.25\mathrm{i}$ and $s = 0.8$

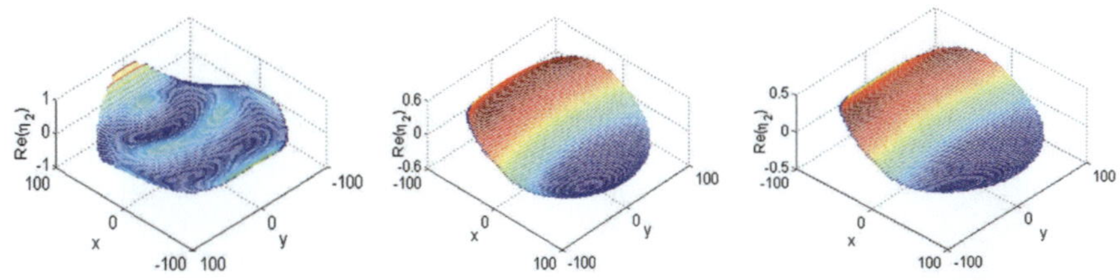

(a) $s = 0.5$ (b) $s = 0.8$ (c) $s = 0.9$

Fig. 3 Deflection of the flexible porous plate for different value of density ratio s. The fixed parameters are $G = 0.25 + 0.25\mathrm{i}$ and $h_1/h_2 = 0.5$

increase in the value of s. For higher values of s, the force acting on the plate decreases along with wave energy dissipation, which results in less deformation of the porous plate. As $s \to 1$, the two-layer becomes a homogeneous fluid and the deflection of plate approaches to that in the case of wave interaction with porous plate in a single-layer fluid as in Mondal et al. [7] in case of wave interaction with floating impermeable flexible plate. Figure 4 exhibits the deflection in the circular flexible porous plate for different values of porous-effect parameter G. It is found that the deflection decreases with increase in the absolute value of G. With increase in the absolute value of G, the wave energy dissipation increases which results in the less deformation in the porous plate.

Figure 5 displays the flow distribution around the circular flexible plate for different values of the porous-effect parameter G. In case of impermeable plate (Fig. 5a), the deformation is more which is due to more wave transmission from wind to lee side of the plate. However, in case of porous plate (Fig. 5b), there is less transmission in the lee side of the porous plate due to the wave energy dissipation. Thus, the surface amplitude in the lee side of plate is reduced significantly by a porous plate in a two-layer fluid.

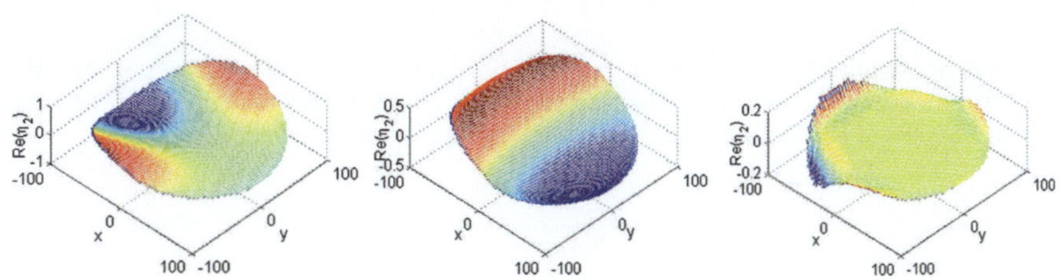

(a) $G = 0$ (b) $G = 0.25 + 0.25\mathrm{i}$ (c) $G = 0.5 + 0.5i$

Fig. 4 Deflection of the flexible porous plate for different value of porous-effect parameter G. The fixed parameters are $h_1/h_2 = 0.5$ and $s = 0.8$

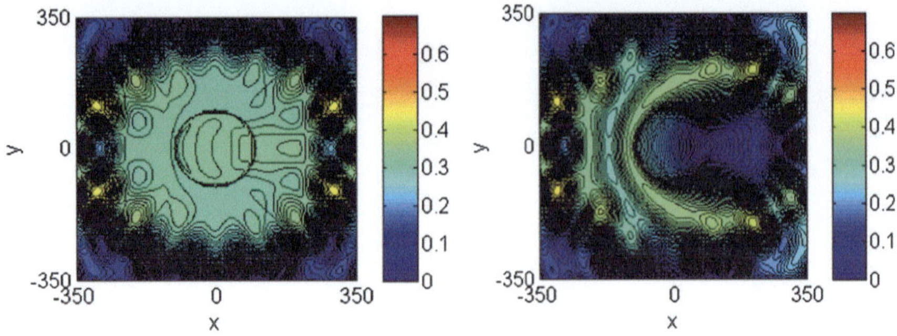

(a) $G = 0$ (b) $G = 0.25 + 0.25i$

Fig. 5 Flow distribution around the circular flexible plate for different value of porous-effect parameter G. The fixed parameters are $h_1/h_2 = 0.2$ and $s = 0.8$

5 Conclusion

Wave interaction with floating circular flexible porous plate at finite water depth in a two-layer fluid has been investigated using the eigenfunction expansion method. The deflection in the floating porous plate and flow distribution around it are computed and analyzed for various wave and structural parameters. The study reveals that in the case of a porous plate, a large amount of wave energy is dissipated, and hence the deflection in the porous plate is reduced significantly in comparison to that in the impermeable plate. Furthermore, the deflection in the porous plate decreases significantly while moving from the wind side to lee side of the plate. As $h_1/h_2 \to 0$, due to the more wave energy dissipation by the porous plate, the deflection of the plate decreases. Further, from the results of deflection of the structure and flow distribution around the plate (Figs. 4 and 5), it is concluded that with increasing in absolute vale of the porous-effect parameters, both deflection of the plate and amplitude of the free surface elevation in the transmitted region decrease.

Acknowledgements HB is thankful to SERB, Government of India (Award No. CRG/2018/004521) for the financial support. VKG gratefully acknowledges the financial support from the SERB, India through "MATRICS" project MTR/2017/000693.

References

1. Garrett, C.J.R.: Wave forces on a circular dock. J. Fluid Mech. **46**(1), 129–139 (1971)
2. Meylan, M.H., Squire, V.A.: Response of a circular ice floe to ocean waves. J. Geophys. Res. Oceans **101**(C4), 8869–8884 (1996)
3. Zilman, G., Miloh, T.: Hydroelastic buoyant circular plate in shallow water: a closed form solution. Appl. Ocean Res. **22**, 191–198 (2000)

4. Tsubogo, T.: The motion of an elastic disk floating on shallow water in waves. In: The Eleventh International Offshore and Polar Engineering Conference, Stavanger, Norway, June 17–22, 2001 (2001)
5. Bennetts, L.G., Biggs, N.R.T., Porter, D.: Wave scattering by an axisymmetric ice floe of varying thickness. IMA J. Appl. Math. **74**, 273–295 (2008)
6. Sturova, I.V.: Unsteady behavior of an elastic articulated beam floating on shallow water. J. Appl. Mech. Tech. Phys. **50**, 589–598 (2009)
7. Mondal, R., Mandal, S., Sahoo, T.: Surface gravity wave interaction with circular flexible structures. Ocean Eng. **88**, 446–462 (2014)
8. Cho, I.H., Kim, M.H.: Interactions of horizontal porous flexible membrane with waves. J. Waterway Port Coast. Ocean Eng. **126**(5), 245–253 (2000)
9. Yu, X.: Functional performance of a submerged and essentially horizontal plate for offshore wave control: a review. Coast. Eng. J. **44**(02), 127–147 (2002)
10. Hu, H.H., Wang, K.H.: Damping effect on waves propagating past a submerged horizontal plate and a vertical porous wall. J. Eng. Mech. **131**(4), 427–437 (2005)
11. Liu, Y., Li, Y., Teng, B.: Wave interaction with a perforated wall breakwater with a submerged horizontal porous plate. Ocean Eng. **34**(17), 2364–2373 (2007)
12. Behera, H., Sahoo, T.: Hydroelastic analysis of gravity wave interaction with submerged horizontal flexible porous plate. J. Fluids Struct. **54**, 643–660 (2015)
13. Koley, S., Sahoo, T.: Oblique wave scattering by horizontal floating flexible porous membrane. Meccanica **52**(1–2), 125–138 (2017)
14. Meylan, M.H., Bennetts, L.G., Peter, M.A.: Water-wave scattering and energy dissipation by a floating porous elastic plate in three dimensions. Wave Motion **70**, 240–250 (2017)
15. Behera, H., Ng, C.-O.: Oblique wave scattering by a system of floating and submerged porous elastic plates. In: Proceedings of the 32nd International Workshop on Water Waves and Floating Bodies (IWWWFB32), Dalian, China, Apr 23–26, 2017 (2017)
16. Xu, F., Lu, D.Q.: Wave scattering by a thin elastic plate floating on a two-layer fluid. Int. J. Eng. Sci. **48**(9), 809–819 (2010)
17. Das, S., Sahoo, T., Meylan, M.H.: Flexural-gravity wave dynamics in two-layer fluid: blocking and dead water analogue. J. Fluid Mech. **854**, 121–145 (2018)
18. Peter, M.A., Meylan, M.H., Chung, H.: Wave scattering by a circular elastic plate in water of finite depth: a closed form solution. Int. J. Offshore Polar Eng. **14**, 81–85 (2004)

Dynamics of SH Wave Propagation in Al/BaTiO$_3$ Composite Structure

Sonal Nirwal and Sanjeev A. Sahu

Abstract The present article manifests the transference of horizontally polarised shear (SH) wave in an aluminium (Al) plate overlying by a Functionally Graded Piezoelectric Material (FGPM) layer. The separation of variables method is adopted to find the solution. The phase velocity of the wave is calculated for both electrically open and short cases. The significant influence of various affecting parameters on the phase velocity curve is demonstrated through graphs. Findings may be applicable in the structural health monitoring, surface acoustic wave (SAW) devices and ultrasonic inspection techniques.

Keywords Piezoelectric · FGPM · SH wave · Electrically open and short cases

1 Introduction

Smart materials play a crucial role in the manufacturing of modern seismic devices. In the wide exploration of advanced material technologies, intelligent devices and smart electronic multifunctional materials which carry material gradients are needed. Theoretical and analytical studies of the smart materials contribute significantly to the manufacturing process for useful devices like transducers, Surface Acoustic Waves (SAW) devices and many more. Intelligent materials such as Functionally Graded Materials (FGMs) are helpful in several engineering applications which are created from more than one distinct constituent phases with continuous gradually varying properties and compositions. FGMs are thoroughly used in the electronic engineering area, aerospace engineering and spaceflight. Furthermore, to increase the efficiency work time and capacity of SAW devices and transducers, the gradient piezoelectric (PE) materials and plates are also taken under consideration for their extreme profound application and significance in SAW devices.

S. Nirwal (✉) · S. A. Sahu
Department of Mathematics and Computing, Indian Institute of Technology (Indian School of Mines), Dhanbad, Jharkhand 826004, India
e-mail: sonal.iitism@gmail.com; sonal.nirwal2012@gmail.com

S. Manna et al. (eds.), *Mathematical Modelling and Scientific Computing with Applications*, Springer Proceedings in Mathematics & Statistics 308,
https://doi.org/10.1007/978-981-15-1338-1_8

The piezoelectric effect is the capability of some materials to create an alternating current (AC) voltage when stress is subjected to the material surface. From the past few decades, the transference the different types of waves in composite layered-structure has received considerable attention due to its use in microwave communication, signal processing and intelligent structure design [1–3]. Kundu et al. studied the Love wave propagation in a piezoelectric layer overlying in an inhomogeneous elastic half-space [4]. Love-type wave propagation in a corrugated PE structure is analysed by Singh et al. [5].

Functionally graded piezoelectric material (FGPM) is a kind of smart PE material with material composition whose volume fraction changing gradually. The elastic properties of these smart composite material change continuously with increasing distance from the treated surface. Recently, Sahu et al. studied the surface wave propagation in an FGPM layered structure [6]. The approximation of surface wave frequency in the piezo-composite structure is studied by Singhal et al. [7]. Polarized SH wave in the FGPM layer sandwiched between corrugated piezomagnetic layer and the elastic substrate is also investigated by Sahu et al. [8]. Manoj et al. analyzed the approximation of surface wave velocity in smart composite structure using the Wentzel–Kramers–Brillouin method [9]. Various investigations have been undertaken to examine the characteristics of surface waves in the piezoelectric/piezomagnetic composite structures [10, 11].

The present article investigates the transference of the SH wave in Aluminium (Al)/$BaTiO_3$ composite structure (in which $BaTiO_3$ coated onto the Al plate). The influence of various affecting parameters on the phase velocity curve (for both electrically open and short circuit case) is examined graphically.

2 Formulation of the Problem

Considering an FGPM layer of thickness h_1 overlying by aluminium (Al) plate of thickness h_2 as illustrated in Fig. 1. The Cartesian coordinate system is taken in such a

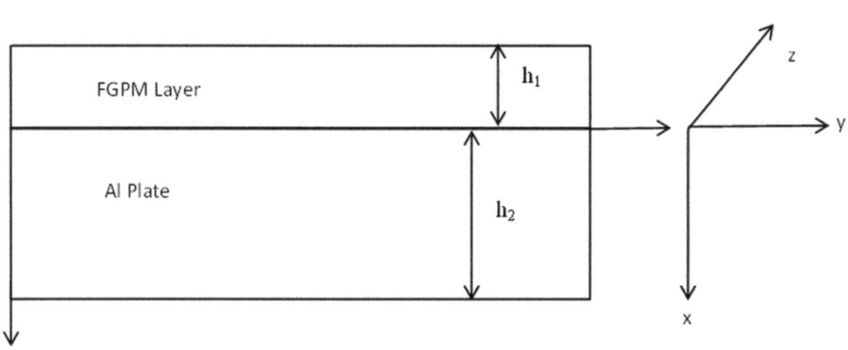

Fig. 1 Geometry of the problem

way that y-axis is in the direction of wave propagation and x-axis is positively pointing downwards. Therefore, the mechanical displacement components and electric potential function are given by

$$u(x, y, t) = v(x, y, t) = 0, w = w(x, y, t), \Theta = \Theta(x, y, t)$$

2.1 *For the Upper FGPM Layer*

The governing equation of motion for the piezoelectric layer (without body force and surface charge density) is given by [8]

$$\begin{bmatrix} \sigma_{ij}^{(p)} \\ D_i^{(p)} \end{bmatrix} = \begin{bmatrix} \bar{c}_{ijlk} & -\bar{e}_{kij} \\ \bar{e}_{ikl} & \bar{\varepsilon}_{ik} \end{bmatrix} \begin{bmatrix} S_{kl}^{(p)} \\ E_k^{(p)} \end{bmatrix}, \tag{1}$$

where, $\sigma_{ij}^{(p)}$ and $S_{kl}^{(p)}$ are the stress and strain tensors respectively; $E_k^{(p)}$ and $D_i^{(p)}$ are the electrical field and electrical displacement intensity respectively; $\bar{c}_{ijkl}$, $\bar{e}_{kij}$ and $\bar{\eta}_{ik}$ are the elastic, piezoelectric, and dielectric coefficients, respectively.

The relation between strain components and mechanical displacement are

$$S_{ij}^{(p)} = \frac{1}{2}\left(\frac{\partial w_i}{\partial x_j} + \frac{\partial w_j}{\partial x_i}\right), \tag{2}$$

where, $i, j = 1, 2, 3$.

According to the quasi-static Maxwell's equation, the relation between the electrical intensity and electrical potential is given by

$$E_i^{(p)} = -\frac{\partial \phi}{\partial x}, \tag{3}$$

where ϕ denotes the electrical potential function.

The governing equation of motion for a transversally isotropic piezoelectric layer is given by

$$\begin{aligned} \sigma_{ij,i}^{(p)} &= \rho_1 \frac{\partial^2 w}{\partial^2 t}, \\ D_{i,i}^{(p)} &= 0, \end{aligned} \tag{4}$$

where, w is the displacement component; ρ_1 is the mass density. The comma followed by the subscript "i" indicates space differentiation with respect to the corresponding coordinate.

On using Eqs. (1)–(4), the equilibrium equations for the FGPM layer are [8]

$$\bar{c}_{44}(x)\nabla^2 w_1 + \bar{e}_{15}(x)\nabla^2\phi_1 + \frac{\partial \bar{c}_{44}}{\partial x}\frac{\partial w_1}{\partial x} + \frac{\partial \bar{e}_{15}}{\partial x}\frac{\partial \phi_1}{\partial x} = \rho_1\frac{\partial^2 w_1}{\partial t^2}, \tag{5}$$

$$\bar{e}_{15}(x)\nabla^2 w_1 - \bar{\varepsilon}_{11}(x)\nabla^2\Theta_1 + \frac{\partial \bar{e}_{15}}{\partial x}\frac{\partial w_1}{\partial x} - \frac{\partial \bar{\varepsilon}_{11}}{\partial x}\frac{\partial \phi_1}{\partial x} = 0, \tag{6}$$

where, w_1 and Θ_1 represent the displacement and electric potential function for the upper layer.

Here we assume that all the material properties of the FGPM layer have the same exponential function distribution along the x direction

$$\bar{c}_{44}(x) = c_{44}^F e^{\sigma x}, \bar{e}_{15}(x) = e_{15}^F e^{\sigma x}, \bar{\varepsilon}_{11}(x) = \varepsilon_{11}^F e^{\sigma x}, \rho_1(x) = \rho^F e^{\sigma x} \tag{7}$$

where, σ is the exponential factor characterizing the degree of the material gradient in the x direction.

Using Eq. (7) in Eqs. (5) and (6)

$$c_{44}^F\left(\nabla^2 w_1 + \sigma\frac{\partial w_1}{\partial x}\right) + e_{15}^F\left(\nabla^2\phi_1 + \sigma\frac{\partial \phi_1}{\partial x}\right) = \rho^F\frac{\partial^2 w_1}{\partial^2 t}, \tag{8}$$

$$e_{15}^F\left(\nabla^2 w_1 + \sigma\frac{\partial w_1}{\partial x}\right) = \varepsilon_{11}^F\left(\nabla^2\phi_1 + \sigma\frac{\partial \phi_1}{\partial x}\right). \tag{9}$$

2.2 *For the Lower Plate*

The propagation of SH wave in the metal is governed as [11]

$$c_{44}^H\nabla^2 w_2 = \rho^H\frac{\partial^2 w_2}{\partial t^2}, \tag{10}$$

where, $c_{44}^H = \frac{Y}{2(1+v)}$ is the shear modulus with Young's modulus Y; ρ^H and v is the mass density and Poisson ratio; w_2 is the displacement function and $\nabla^2 = \frac{\partial^2}{\partial x^2} + \frac{\partial^2}{\partial y^2}$ is the two dimensional Laplacian operator.

3 Solution of the Problem

3.1 *For the Upper Layer*

From Eqs. (8) and (9), we can write

$$\nabla^2 w_1 + \sigma \frac{\partial w_1}{\partial x} = \frac{1}{\frac{1}{\rho^F}\left(c_{44}^F + \frac{\left(e_{15}^F\right)^2}{\eta_{11}^F}\right)} \frac{\partial^2 w_1}{\partial t^2}. \tag{11}$$

Let us introduce an auxiliary function

$$\varsigma_1 = \phi_1 - \frac{e_{15}^F}{\varepsilon_{11}^F} w_1, \tag{12}$$

on substituting the value from Eq. (12) into Eq. (9), we can write as

$$\left(\nabla^2 \varsigma_1 + \sigma \frac{\partial \varsigma_1}{\partial x}\right) = 0. \tag{13}$$

Assuming that the solution of Eqs. (11) and (13) in the harmonic form as

$$w_1(x, y, t) = f_1(x) e^{ik(y-ct)}, \tag{14}$$

$$\varsigma_1(x, y, t) = g_1(x) e^{ik(y-ct)}. \tag{15}$$

Substituting the value from Eqs. (14) and (15) in Eqs. (11) and (13) and using Eq. (12), the mechanical displacement and electric potential function are given by

$$w_1(x, y, t) = e^{rx}(Q_1 \mathrm{Cos}\mu_1 x + Q_2 \mathrm{Sin}\mu_1 x) e^{ik(y-ct)}, \tag{16}$$

$$\phi_1(x, y, t) = \left((Q_3 e^{m_1 x} + Q_4 e^{m_2 x}) + \frac{e_{15}^f}{\varepsilon_{11}^f} e^{rx}(Q_1 \mathrm{Cos}\mu x + Q_2 \mathrm{Sin}\mu x)\right) e^{ik(y-ct)}, \tag{17}$$

where, $r = -\frac{\sigma}{2}$, $m_{1,2} = \frac{-\sigma \pm \sqrt{\sigma^2 + 4k^2}}{2}$, $\mu_1 = \frac{\sqrt{4k^2(c^2/\beta_1^2 - 1) - \sigma^2}}{2}$ and $\beta_1^2 = \frac{1}{\rho^F}\left(c_{44}^F + \frac{\left(e_{15}^F\right)^2}{\varepsilon_{11}^F}\right)$.

3.2 *For the Lower Plate*

Assuming the solution of Eq. (10) in the harmonic form as

$$w_2(x, y, t) = f_2(x) e^{ik(y-ct)}, \tag{18}$$

on substituting the value from Eq. (18) in Eq. (10), we get

$$w_2(x, y, t) = \left(Q_5 e^{m_3 x} + Q_6 e^{-m_4 x}\right) e^{ik(y-ct)}, \tag{19}$$

where, $m_3 = \sqrt{k^2(-c^2/\beta_2^2 + 1)}$ with $\beta_2^2 = \frac{c_{44}^H}{\rho^H}$.

4 Boundary Conditions

(i) The mechanical traction free at the upper surface $(x = -h_1)$ is

- $\sigma_{zx}^{(p)}(-h_1, y) = 0.$ (20)

(ii) The electrical condition at the upper surface $(x = -h_1)$ is

- $D_x^{(p)}(-h_1, y) = 0,$ (Electrically open-circuit case) (21)

- $\phi_1(-h_1, y) = 0.$ (Electrically-short circuit case) (22)

(iii) Moreover, Continuity of stress, mechanical displacement and electrical potential at the common interface $(x = 0)$ are

- $\sigma_{zx}^{(p)}(0, y) = \sigma_{zx}^{(h)}(0, y), w_1(0, y) = w_2(0, y), \phi_1(0, y) = 0.$ (23)

(iv) The mechanical traction at the lower surface $(x = h_2)$ is

- $c_{44}^h \frac{\partial w_2}{\partial x} = 0.$ (24)

5 Dispersion Relation

5.1 Dispersion Relation for Electrically Open Case

Substituting the values of w_1, Θ_1 and w_2 from Eqs. (16), (17) and (19) in Eqs. (20), (21), (23) and (24) we obtain the algebraic equation with the unknown constants $\mathbf{Q_i}$ $\mathbf{i = 1, 2, \ldots, 6}$.

Now in order to get the non-trivial solution and hence the dispersion relation, we equate the determinant of these coefficients (Q_i) to zero.

$$\det\begin{bmatrix} Q_{11} & Q_{12} & Q_{13} & Q_{14} & Q_{15} & Q_{16} \\ Q_{21} & Q_{22} & Q_{23} & Q_{24} & Q_{25} & Q_{26} \\ Q_{31} & Q_{32} & Q_{33} & Q_{34} & Q_{35} & Q_{36} \\ Q_{41} & Q_{42} & Q_{43} & Q_{44} & Q_{45} & Q_{46} \\ Q_{51} & Q_{52} & Q_{53} & Q_{54} & Q_{55} & Q_{56} \\ Q_{61} & Q_{62} & Q_{63} & Q_{64} & Q_{65} & Q_{66} \end{bmatrix} = 0, \tag{25}$$

where the definition of $\mathbf{Q_i}$ $\mathbf{i = 1, 2, \ldots, 6}$. are given in Appendix 1.

5.2 Dispersion Relation for Electrically Short Case

Substituting the values of w_1, Θ_1 and w_2 from Eqs. (16), (17) and (19) in Eqs. (20), (22), (23) and (24), we obtain the algebraic equation with the unknown constants $\mathbf{Q_i}$ $\mathbf{i = 1, 2, \ldots, 6}$. Now in order to get the non-trivial solution and hence the dispersion relation, we equate the determinant of these coefficients (Q_i) to zero.

$$\det\begin{bmatrix} Q_{11} & Q_{12} & Q_{13} & Q_{14} & Q_{15} & Q_{16} \\ Q_{21} & Q_{22} & Q_{23} & Q_{24} & Q_{25} & Q_{26} \\ Q_{31} & Q_{32} & Q_{33} & Q_{34} & Q_{35} & Q_{36} \\ Q_{41} & Q_{42} & Q_{43} & Q_{44} & Q_{45} & Q_{46} \\ Q_{51} & Q_{52} & Q_{53} & Q_{54} & Q_{55} & Q_{56} \\ Q_{61} & Q_{62} & Q_{63} & Q_{64} & Q_{65} & Q_{66} \end{bmatrix} = 0, \tag{26}$$

where the definition of $\mathbf{Q_i}$ $\mathbf{i = 1, 2, \ldots, 6}$. are given in Appendix 2.

6 Numerical Results and Discussion

Based on the dispersion relation(s) [Eqs. (25 and 26)], numerical results are presented to show the propagation behavior of waves in the considered structure.

For the graphical illustration, the following physical quantities have been taken:

For the FGPM layer [8]

$$c_{44}^F = 4.4 \times 10^{10}\,\mathrm{N/m^2},\ e_{15}^F = 11.4\,\mathrm{c/m^2},\ \varepsilon_{11}^F = 9.82 \times 10^{-9}\,\mathrm{C^2/Nm},$$
$$\rho^F = 5.7 \times 10^3\,\mathrm{kg/m^3}$$

For the lower Al Plate [12]

$$\nu = 0.33, Y = 70 \times 10^3\,\mathrm{n/m^3},\ \rho^h = 2.8 \times 10^3\,\mathrm{kg/m^3}$$

Graphs have been sketched to depict the variation of non-dimensional phase velocity with respect to non-dimensional wave number. Figures 2, 3, and 4 represent the substantial impact of different parameters (as layer thickness, gradient factor and elastic coefficient) on the phase velocity curve for both electrically open and short cases. The solids line represents the curve for electrically open circuit case while the dotted line represents for electrically short circuit case, wherever they are.

Figure 2 represents the substantial impact of the upper layer thickness on the phase velocity curve for both electrically open and short circuit case. The figure approved that the increasing value of wave number leads to decreases in the phase velocity curve. It can also be noted that the phase velocity curve upswings as the thickness of the upper layer rise for both electrically open and short circuit cases.

In order to explore the influence gradient factor on the phase velocity curve, Fig. 3 is plotted. Phase velocity has been also calculated in case of homogeneous piezoelectric material ($\sigma = 0$). Subtle observation of the graphs adduced that, as we increase the wave number, the phase velocity curve follows a steep decreasing nature. It can also be noted that the phase velocity of the wave is influenced significantly by the functionally grading of the material. Increment in the coefficient of the material gradient raises the phase velocity curve for both open and short circuit cases.

The pronounced influence of elastic parameter on the phase velocity curve for both open and short circuit case is manifested through Fig. 4. As we continue to draw the results from Fig. 4, a similar influence is observed as compared to Fig. 3. Moreover, the impact of the elastic parameter is found less as compared to the gradient factor.

Furthermore, a comparison study from the graphs (Figs. 2, 3 and 4) yields that the phase velocity curve for the electrically open circuit case is always higher than that of the short circuit.

Fig. 2 Non-dimensional phase velocity versus non-dimensional wave number for different value of upper layer thickness

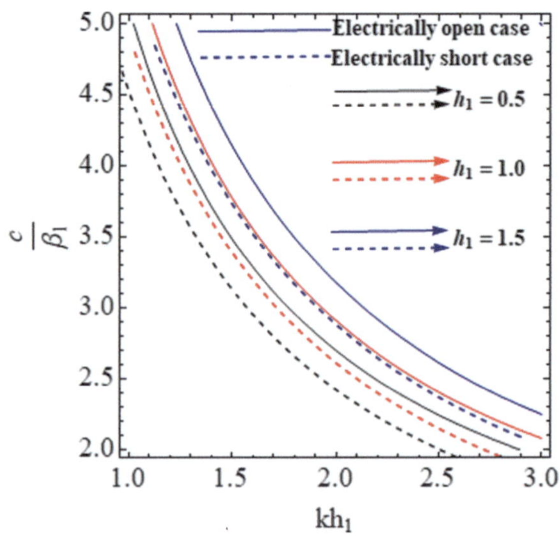

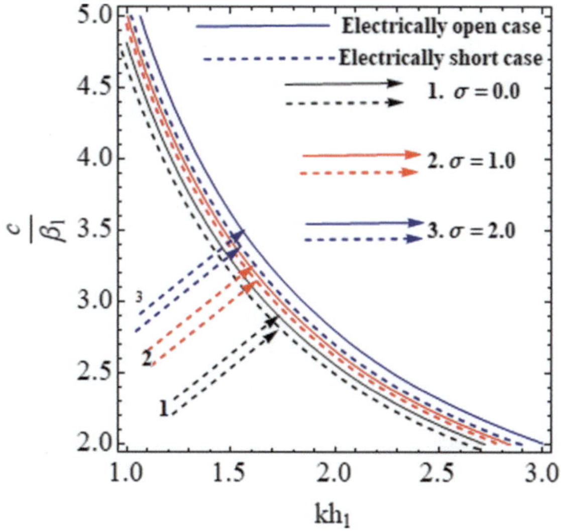

Fig. 3 Non-dimensional phase velocity versus non-dimensional wave number for different value of gradient factor

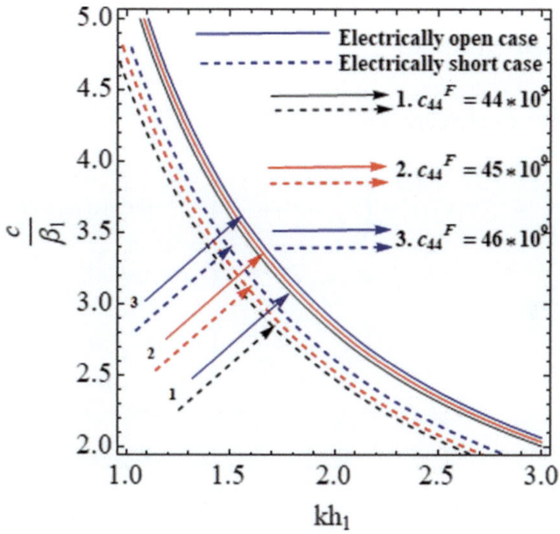

Fig. 4 Non-dimensional phase velocity versus non-dimensional wave number for different value of the elastic coefficient

7 Conclusions

Based on the graphical demonstration, the following remarks can be carried out.

1. The increasing value of the wavenumber leads to diminishing the phase velocity curves.

2. Through the study on the propagation behavior of Love-type waves, the results conclude that the FGM layer will lead to the increases in the phase velocity curves.
3. A leap value of gradient factor exerted a favourable impact on the phase velocity curve.
4. Broader the value of elastic coefficient elevates the phase velocity curve.
5. The phase velocity curves for the open-circuit case is found higher than that of the short-circuit case.

The present study comprehends mathematical modeling on a smart piezo-composite structure. This study finds its applications towards designing and optimization of devices and sensors. Such a study could be the bridge between theoretical computation and practical implication.

Acknowledgements The authors express their sincere thanks to the Science and Engineering Research Board (SERB), New Delhi, India for providing financial assistance through project No. YSS/2015/002057.

Appendix 1

For electrically open circuit case

$$Q_{11} = \left(c_{44}^F + \frac{\left(e_{15}^F\right)^2}{\varepsilon_{11}^F}\right)e^{-rh_1}[r\mathrm{Cos}(\mu_1 h_1) + \mu_1 \mathrm{Sin}(\mu_1 h_1)],$$

$$Q_{13} = m_1\, e_{15}^F\, e^{-m_1 h_1},\ Q_{14} = m_2\, e_{15}^F\, e^{-m_2 h_1},$$

$$Q_{12} = \left(c_{44}^F + \frac{\left(e_{15}^F\right)^2}{\varepsilon_{11}^F}\right)e^{-rh_1}[-r\,\mathrm{Sin}(\mu_1 h_1) + \mu_1 \mathrm{Cos}(\mu_1 h_1)],$$

$$Q_{15} = 0,\ Q_{16} = 0,\ Q_{21} = -m_1 \varepsilon_{11}^F e^{-m_1 h_1},$$

$$Q_{22} = -m_2 \varepsilon_{11}^F\, e^{-m_2 h_1},\ Q_{23} = Q_{24} = Q_{25} = Q_{26} = 0,$$

$$Q_{31} = r\left(c_{44}^F + \frac{\left(e_{15}^F\right)^2}{\varepsilon_{11}^F}\right),\ Q_{32} = \mu_1\left(c_{44}^F + \frac{\left(e_{15}^F\right)^2}{\varepsilon_{11}^F}\right),$$

$$Q_{33} = e_{15}^F m_1,\ Q_{34} = e_{15}^F\, m_2 Q_{35} = -m_3 c_{44}^H,\ Q_{36} = m_4 c_{44}^H,$$

$$Q_{41} = 1,\ Q_{42} = Q_{43} = Q_{44} = 0,\ Q_{45} = -1,$$

$$Q_{46} - 1,\ Q_{51} = \frac{e_{15}^F}{\varepsilon_{11}^F},\ Q_{52} = 0,\ Q_{53} = 1,\ Q_{54} = 1,$$

$$Q_{55} = Q_{56} = Q_{61} = Q_{62} = Q_{63} = Q_{64} = 0,$$

$$Q_{65} = m_3 c_{44}^H e^{m_3 h_2},\ Q_{66} = m_4 c_{44}^H e^{-m_4 h_2}.$$

Appendix 2

For electrically short circuit case

$$Q_{11} = \left(c_{44}^F + \frac{\left(e_{15}^F\right)^2}{\varepsilon_{11}^F} \right) e^{-rh_1} [r\text{Cos}(\mu_1 h_1) + \mu_1 \text{Sin}(\mu_1 h_1)],$$

$$Q_{13} = m_1 e_{15}^F e^{-m_1 h_1}, \; Q_{14} = m_2 e_{15}^F e^{-m_2 h_1},$$

$$Q_{12} = \left(c_{44}^F + \frac{\left(e_{15}^F\right)^2}{\varepsilon_{11}^F} \right) e^{-rh_1} [-r\,\text{Sin}(\mu_1 h_1) + \mu_1 \text{Cos}(\mu_1 h_1)],$$

$$Q_{15} = 0, \; Q_{16} = 0, \; Q_{25} = 0, \; Q_{26} = 0,$$

$$Q_{21} = \frac{e_{15}^F}{\varepsilon_{11}^F} e^{-rh_1} \,\text{Cos}(\mu_1 h_1), \; Q_{22} = -\frac{e_{15}^F}{\varepsilon_{11}^F} e^{-rh_1} \text{Sin}(\mu_1 h_1),$$

$$Q_{23} = e^{-m_1 h_1}, \; Q_{24} = e^{-m_2 h_1}$$

$$Q_{31} = r\left(c_{44}^F + \frac{\left(e_{15}^F\right)^2}{\varepsilon_{11}^F} \right), \; Q_{32} = \mu_1 \left(c_{44}^F + \frac{\left(e_{15}^F\right)^2}{\varepsilon_{11}^F} \right),$$

$$Q_{33} = e_{15}^F m_1, \; Q_{34} = e_{15}^F m_2 Q_{35} = -m_3 c_{44}^H,$$

$$Q_{36} = m_4 c_{44}^H, \; Q_{41} = 1, \; Q_{42} = Q_{43} = Q_{44} = 0,$$

$$Q_{45} = -1, \; Q_{46} - 1, \; Q_{51} = \frac{e_{15}^F}{\varepsilon_{11}^F}, \; Q_{52} = 0, \; Q_{53} = 1,$$

$$Q_{54} = 1, \; Q_{55} = Q_{56} = Q_{61} = Q_{62} = Q_{63} = Q_{64} = 0,$$

$$Q_{65} = m_3 c_{44}^H e^{m_3 h_2}, \; Q_{66} = m_4 c_{44}^H e^{-m_4 h_2}.$$

References

1. Bleustein, J.L.: A new surface wave in piezoelectric materials. Appl. Phys. Lett. **13**(12), 412–413 (1968)
2. Tiersten, H.F.: Linear Piezoelectric Plate Vibrations: Elements of the Linear Theory of Piezoelectricity and the Vibrations Piezoelectric Plates. Springer, Berlin (2013)
3. Wang, J., Yong, Y.K., Imai, T.: Finite element analysis of the piezoelectric vibrations of quartz plate resonators with higher-order plate theory. Int. J. Solids Struct. **36**(15), 2303–2319 (1999)
4. Kundu, S., Manna, S., Gupta, S.: Love wave dispersion in pre-stressed homogeneous medium over a porous half-space with irregular boundary surfaces. Int. J. Solids Struct. **51**(21–22), 3689–3697 (2014)
5. Singh, A.K., Parween, Z., Kumar, S.: Love-type wave propagation in a corrugated piezoelectric structure. J. Intel. Mater. Syst. Struct. **27**(19), 2616–2632 (2016)
6. Saroj, P.K., et al.: Love-type waves in functionally graded piezoelectric material (FGPM) sandwiched between initially stressed layer and elastic substrate. Waves Random Complex Media **25**(4), 608–627 (2015)

7. Singhal, A., Sahu, S.A., Chaudhary, Soniya: Approximation of surface wave frequency in piezo-composite structure. Compos. Part B: Eng. **144**, 19–28 (2018)
8. Sahu, S.A., Mondal, S., Dewangan, N.: Polarized shear waves in functionally graded piezoelectric material layer sandwiched between corrugated piezomagnetic layer and elastic substrate. J. Sandwich Struct. Mater. (2017). 1099636217726330
9. Singh, M.K., et al.: Approximation of surface wave velocity in smart composite structure using Wentzel–Kramers–Brillouin method. J. Intell. Mater. Syst. Struct. (2018). 1045389X18786464
10. Ezzin, H., Amor, M.B., Ghozlen, M.H.B.: Love waves propagation in a transversely isotropic piezoelectric layer on a piezomagnetic half-space. Ultrasonics **69**, 83–89 (2016)
11. Ezzin, H., Amor, M.B., Ghozlen, M.H.B.: Propagation behavior of SH waves in layered piezoelectric/piezomagnetic plates. Acta Mechanica **228**(3), 1071–1081 (2017)
12. Wang, Q., Wu, N., Quek, S.T.: Acoustic wave in piezoelectric coupled plates with open circuit. Int. J. Struct. Stab. Dyn. **10**(02), 299–313 (2010)

Predator-Prey Model with Prey Group Defense and Non-linear Predator Harvesting

Rajat Kaushik and Sandip Banerjee

Abstract This paper is concerned with a predator-prey system with a prey group defense and non-linear harvesting of the predator incorporating deterrence hypothesis for predators. Inclusion of predator deterrence rate makes the modelling approach more practicable and exhibits significant impact on the net predation. Taking all possible interactions into account, model equations are formulated. In brief qualitative analysis, existence of interior equilibrium and stabilities of all equilibrium points of the system are discussed to investigate the dynamical behavior of the ecosystem. Hopf, transcritical and saddle-node bifurcations are illustrated for various parameters. Numerical simulations are ecologically justified and supportive of theoretical results.

Keywords Predator-prey · Co-existence · Local stability · Hopf bifurcation · Stability switches

1 Introduction

Prey group defense has been an area of interest in predator-prey systems [17]. The defense approach of the preys is followed in the way that the preys at the boundary of the group hurt most, from the attacks of the predators. The number of preys remaining on the border of the group is proportional to the length of the perimeter of the ground region occupied by the group [6]. This perimeter is directly proportional to the square root of the area of that ground region. If we consider uniform biomass density throughout the region of the domain, then biomass of the population in any ground region is equal to the area of that region multiplying by the biomass density per unit area, which is a constant as per our assumption. Hence, it is logical to insert the square root term to portray the model with herd behavior.

R. Kaushik (✉) · S. Banerjee
Indian Institute of Technology Roorkee, Roorkee, Uttarakhand 247667, India
e-mail: rkaushik@ma.iitr.ac.in

S. Banerjee
e-mail: sandofma@iitr.ac.in

S. Manna et al. (eds.), *Mathematical Modelling and Scientific Computing with Applications*, Springer Proceedings in Mathematics & Statistics 308,
https://doi.org/10.1007/978-981-15-1338-1_9

In deterrence hypothesis, preys often deter predation by using certain signals to show that predators have been noticed and subsequent efforts for predations will be not likely to achieve the desired result [7]. Chemically defended preys can increase dietary wariness and advertise their unprofitability to predators [11]. The functions of aposematic signals such as animal coloration, sound, orders [8] are also to rescue the preys from predators attack, by warning potential predators that preys are unpalatable or poisonous. Camouflage is also one of the defense strategies used by undefended preys showing colour patterns to minimize detection [16]. These kinds of defensive deterrence mechanism are used to defend against the attacks and discourage predation [2].

Herd behavior of prey and predator models with the square root functional response has been studied by many other researchers. Ajraldi et al. [1], Belvisi and Venturino [3], Bera et al. [4], Braza [6], Gimmelli et al. [9], Ma et al. [10], Matia and Alam [12], Tang and Song [14, 15], Wang et al. [17], Xu et al. [18]. In [6], Braza introduced a predator-prey model in which prey population shows herd behavior to protect themselves from predators. The author revealed the fact that Hopf bifurcation can be experienced for some parameters in the model and some singularities must take place near the origin in this type of behavior. In [3], S. Belvisi and E. Venturino considered an eco-epidemic predator-prey model with the influence of predator-disease on prey group defense. Tang and Song [15] considered a predator-prey model with group defense taking hyperbolic mortality into account. Gimmelli and his co-authors [9] developed a predator-prey herd model incorporating the attack of specialist predators. In [10], Xiangmin Ma and his co-authors generalized the herd model taking stage-structure for predators with time delay in the model. In [18], Chaoqun Xu and his co-authors studied a more general predator-prey model with herd behavior in which global dynamics of the model is discussed. Therefore, we propose a stage structured predator-prey model, following herd behavior with the square root functional response.

In all these studies, predator-prey system with herd behavior of prey including non-linear harvesting of predators as well as prey's deterrence signals to discourage predation, is completely missing. Deterrence to predators being a natural phenomena, we have taken it into model formulation as a reduction factor for predation. Hence, considering predator harvesting and deterrence hypothesis, we propose a predator-prey model and then ask whether and under what conditions, there will be co-existence; whether the non-linear predator harvesting will effect the predator's survival; the prey's deterrence will prove a significant step against predator's attack or will there be any domination of the predator over prey and vice versa. The organization of the paper is as follows: In Sect. 2, we develop a predator-prey model with herd behavior and discuss positivity and boundedness of the system. A qualitative analysis of the model is discussed in Sect. 3, which includes existence of steady states and local stability of the predator-free as well as interior equilibrium point. Singularities of the trivial equilibrium point are observed and brief mathematical analysis is done to capture the local dynamics in Sect. 3.1, showing unusual behavior near the origin. Study of bifurcation is provided in Sect. 3.2. The analytical results are supported by numerical simulations in Sect. 4. The system exhibits Hopf bifurcation for many system parameters, which is discussed in detail in this section. The paper ends with a conclusion in Sect. 5.

2 Model Formulation and Boundedness

Based on the assumption mentioned above, the predator-prey system with prey group defense and non-linear harvesting of the predator incorporating predator deterrence hypotheses governed by following non-linear system of the ordinary differential equations:

$$\frac{dx}{dt} = rx\left(1 - \frac{x}{K}\right) - \left(\frac{\beta}{1 + \varphi}\right)\sqrt{x}y = f(x, y \tag{2.1}$$

$$\frac{dy}{dt} = \frac{n\beta}{1 + \varphi}\sqrt{x}y - dy - \frac{hy}{1 + y} = g(x, y) \tag{2.2}$$

where x(0) > 0, y(0) > 0 and all the parameters of the system viz. $r, K, \eta, \beta, \varphi, d, h$ are positive. Description of all parameters with units is shown in Table 1.

Theorem 1 *All solutions of the system* (2.1–2.2) *that start in* R_+^2 *remain positive* ∀ t > 0.

Proof From Eq. (2.2)

$$\frac{dy}{y} = \left\{\frac{n\beta}{1 + \varphi}\sqrt{x} - d - \frac{h}{1 + y}\right\}dt = \varphi(x\ ,y)dt, \tag{2.3}$$

where $\varphi(x, y) = \left\{\frac{n\ \beta}{1+\varphi}\sqrt{x} - d - \frac{h}{1+y}\right\}$.

Table 1 Parameters and variables used in the system with their description and units

Variable/parameter	Description	Unit
x	Biomass density of prey	Biomass per unit area
y	Biomass density of predator	Biomass per unit area
r	Intrinsic growth rate of prey	Per day
K	Carrying capacity of environment for prey	Biomass per unit area
η	Conversion efficiency from prey to predator biomass	Unit-less
β	Natural predation rate	Per day
φ	Deterrence rate that discourages predation	Per day
h	Non-linear predator harvesting	Per day
d	Starvation rate of predators	Per day

Integrated Eq. (2.3) in the region [0,t], we get

$$y(t) = y(0)e^{\int_0^t \varphi(x,y)dt} > 0.$$

Now from Eq. (2.1)

$$\frac{dx}{\sqrt{x}} = \left\{r\sqrt{x}\left(1 - \frac{x}{K}\right) - \frac{\beta y}{1+\delta}\right\}dt = \psi(x, y)dt \tag{2.4}$$

where

$$\psi(x, y) = \left\{r\sqrt{x}\left(1 - \frac{x}{K}\right) - \frac{\beta y}{1+\delta}\right\}.$$

Integrated Eq. (2.4) in the region [0,t], we get

$$x(t) = \left\{x(0) + \frac{1}{2}\int_0^t \psi(x, y)dt\right\}^2 > 0.$$

Moreover, at x(t) = 0, dx/dt = 0 and y(t) = 0, dy/dt = 0 which ceases x(t) and y(t) to become negative. Thus, x(t) and y(t) remain positive for all t > 0. This completes the proof of the theorem.

Theorem 2 *All the solutions of the system* (2.1–2.2) *which start in* R_+^2 *are ultimately bounded.*

Proof From Eq. (2.1), we have

$$\frac{dx}{dt} \le rx\left(1 - \frac{x}{K}\right),$$
$$\Rightarrow \lim_{t\to\infty} \sup x(t) \le K.$$

Let us define

$$\Omega = \eta x + y$$
$$\Rightarrow \frac{d\Omega}{dt} \le \eta rx\left(1 - \frac{x}{K}\right) - dy$$
$$\le \eta rx - dy$$
$$\le (\eta r + \eta d)K - d\Omega.$$

Hence

$$\frac{d\Omega}{dt} + d\Omega \le (\eta r + \eta d)K = \delta(say)$$

Applying a lemma on differential equation inequalities [5], we obtain

$$0 \leq \Omega(\mathrm{x}, \mathrm{y}) \leq \frac{\delta}{d} + \frac{(x(0), y(0))}{\mathrm{e}^{dt}},$$

and for $t \to \infty$,

$$0 \leq \Omega \leq \frac{\delta}{d}.$$

Thus, all the solutions of the system (2.1–2.2) enter into region

$$B = \left\{(x, y) : 0 \leq \Omega \leq \frac{\delta}{d} + \epsilon, \quad \text{for } any\, \epsilon > 0\right\}.$$

Hence the theorem.

3 Equilibrium State Analysis

System (2.1–2.2) has following equilibrium points:

(i) The trivial equilibrium point $\boldsymbol{E_0} = (\mathbf{0}, \mathbf{0})$;
(ii) Predator-free equilibrium point $\boldsymbol{E_1} = (\boldsymbol{K}, \mathbf{0})$;
(iii) Positive equilibrium point $\boldsymbol{E^*} = (\boldsymbol{x^*}, \boldsymbol{y^*})$ which satisfies the following system:

$$\mathrm{rx}\left\{1 - \frac{\mathrm{x}}{\mathrm{K}}\right\} - \frac{\beta}{1+\varphi}\mathrm{y}\sqrt{\mathrm{x}} = 0 \tag{3.1}$$

$$\frac{\eta\beta}{1+\varphi}\mathrm{y}\sqrt{\mathrm{x}} - \mathrm{dy} - \frac{\mathrm{hy}}{1+\mathrm{y}} = 0$$

Theorem 3 *If* $(\mathrm{d}+\mathrm{h})^2(1+\varphi)^2 < K\eta^2\beta^2$, *then system* (2.1–2.2) *has a unique equilibrium point* (x^*, y^*).

Proof Define

$$\mathrm{u(x)} = \frac{\mathrm{r}(1+\varphi)}{\beta}\sqrt{\mathrm{x}}\left\{1 - \frac{\mathrm{x}}{\mathrm{K}}\right\}, \quad v(\mathrm{x}) = \frac{(1+\varphi)h - \eta\beta\sqrt{x} + d(1+\varphi)}{\eta\beta\sqrt{x} - d(1+\varphi)}$$

The predator and prey null-cline curves can be given as y = u(x) and y = v(x). From system (3.1), we have that the interior equilibrium (x*, y*) must satisfy

$$y^* = u\left(x^*\right), \quad y^* = u\left(x^*\right).$$

By simple calculation, we obtain

$$u(0) = u(K) = 0,$$
$$u'(x) = \frac{r(1+\varphi)}{\beta}\frac{(K-3x)}{2K\sqrt{x}}, \tag{3.2}$$
$$u''(x) = -\frac{r(1+\varphi)}{\beta}\frac{(3x+K)}{4Kx^{3/2}} < 0 \text{ for } \quad x \in (0, K).$$

Thus, y = u(x) is a continuous concave downward function on (0, K) satisfying u (0) = u (K) = 0. Now,

$$v(x) = 0 \quad for \quad x = x_v = \frac{(d+h)^2(1+\varphi)^2}{\eta^2\beta^2},$$
$$v'(x) = -\frac{(1+\varphi)h\eta\beta}{2\left(\eta\beta\sqrt{x}-m\right)^2\sqrt{x}} < 0.$$

Therefore, y = v(x) is a decreasing function which becomes undefined $x = d^2$ $(1 + \varphi)^2/(\varphi^2\beta^2)$. Function is negative for x < d^2 $(1 + \varphi)^2/(\varphi^2\beta^2)$ and positive for = d^2 $(1 + \varphi)^2/(\varphi^2\beta^2)$ < x < $(d + h)^2(1 + \varphi)^2/(\varphi^2\beta^2)$. To obtain a unique equilibrium point both curves should intersect on a unique point for which intercept x_v $of y = v(x)$ on x-axis must be less than intercept K of $y = u(x)$ on x-axis (see Fig. 1) i.e.

$$\frac{(d+h)^2(1+\varphi)^2}{\eta^2\beta^2} < K.$$

This completes the proof of the theorem.

Theorem 4 *Predator-free equilibrium point E_1 is locally asymptotically stable if*

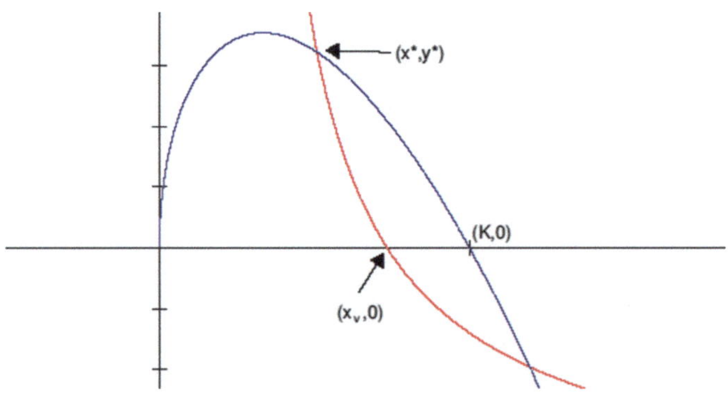

Fig. 1 Blue curve represents y = u(x) while red curve represents y = v(x). Intersection of both curves is (x^*, y^*)

$$(d+h)^2(1+\varphi)^2 > K\eta^2\beta^2.$$

Proof Jacobian matrix at equilibrium point E_1 is given as follows:

$$[J]_{E_1} = \begin{bmatrix} -r & \frac{-\beta\sqrt{K}}{(1+\varphi)} \\ 0 & \frac{\eta\beta\sqrt{K}}{(1+\varphi)} - d - h \end{bmatrix}.$$

Eigenvalues for matrix $[J]_{E_1}$ are

$$\lambda_1 = -r < 0, \quad \lambda_2 = \frac{\eta\beta\sqrt{K}}{(1+\varphi)} - d - h.$$

Therefore, condition for local stability for predator-free equilibrium point is $\lambda_2 < 0$.

i.e. $(d+h)^2(1+\varphi)^2 < K\eta^2\beta^2$. This completes the proof of the theorem.

Theorem 5 *Interior equilibrium point $E^*(x^*, y^*)$ is locally asymptotically stable if*

$$\left\{\frac{r}{2} - \frac{3rx^*}{2K} + \frac{hy^*}{(1+y^*)^2}\right\} < 0 \;\; \textit{and} \;\; \left(\frac{r}{2} - \frac{3rx^*}{2K}\right)\frac{hy^*}{(1+y^*)^2} + \frac{\eta\beta^2 y^*}{2(1+\varphi)^2} > 0.$$

Proof Jacobian matrix at E^* is given as follows:

$$[J]_{E^*} = \begin{bmatrix} r - \frac{2rx^*}{K} - \frac{\beta y^*}{2(1+\varphi)\sqrt{x^*}} & \frac{-\beta\sqrt{x^*}}{(1+\varphi)} \\ \frac{\eta\beta y^*}{2(1+\varphi)\sqrt{x^*}} & \frac{\eta\beta\sqrt{x^*}}{(1+\varphi)} - d - \frac{h}{(1+y^*)^2} \end{bmatrix} = \begin{bmatrix} \frac{r}{2} - \frac{3rx^*}{2K} & \frac{-\beta\sqrt{x^*}}{(1+\varphi)} \\ \frac{\eta\beta y^*}{2(1+\varphi)\sqrt{x^*}} & \frac{hy^*}{(1+y^*)^2} \end{bmatrix}$$

The corresponding characteristic equation is given as follows:

$$\lambda^2 - a\lambda + \mathrm{b} = 0, \tag{3.4}$$

where

$$a = \frac{r}{2} - \frac{3rx^*}{2K} + \frac{hy^*}{(1+y^*)^2},$$
$$b = \left(\frac{r}{2} - \frac{3rx^*}{2K}\right)\frac{hy^*}{(1+y^*)^2} + \frac{\eta\beta^2 y^*}{2(1+\varphi)^2}$$

For eigenvalues to have negative real part, we must have a < 0 and b > 0. Hence the theorem.

3.1 Steady State Analysis of Trivial Equilibrium Point

Since, $\boldsymbol{y}/\sqrt{\boldsymbol{x}}$ is not defined at origin, therefore we use the singular dynamics near the origin assuming $\boldsymbol{x} \ll 1, y \ll 1$. With these assumptions $(1 + \boldsymbol{y}) \approx \mathbf{1}$, $\sqrt{\boldsymbol{xy}} \ll y$ and $(\mathbf{1} - \boldsymbol{x}/\boldsymbol{K}) \approx \mathbf{1}$, system (2.1–2.2) reduces to

$$\frac{dx}{dt} \approx rx - \frac{\beta\sqrt{x}y}{(1+\varphi)} \tag{3.5}$$

$$\frac{dy}{dt} \approx -dy - hy. \tag{3.6}$$

From Eq. (3.6), $y(t) = y(0)e^{-(d+h)t}$, where $y(0) \ll 1$. Now consider the following cases:

i. $x = O(y_0)$, then $\frac{dx}{dt} \approx rx$ and $\frac{dy}{dt} \approx -dy - hy$ which implies that origin is a saddle point.
ii. $x = O\left(y_0^2\right)$, then solution of system (3.5–3.6) is given as follows:

$$\left\{\sqrt{x} - \frac{\beta y}{(1+\varphi)(r+2(d+h))}\right\} = \left\{\sqrt{x_0} - \frac{\beta y_0}{(1+\varphi)(r+2(d+h))}\right\}e^{rt/2}. \tag{3.7}$$

The dynamics near the origin is as follows:

(a) $\left\{\sqrt{x_0} - \frac{\beta y_0}{(1+\varphi)(r+2(d+h))}\right\}e^{rt/2} = 0 \Rightarrow$ solution tends to zero along Γ : $\left\{\sqrt{x} - \frac{\beta y}{(1+\varphi)(r+2(d+h))}\right\} = 0$ as shown in the Fig. 2.

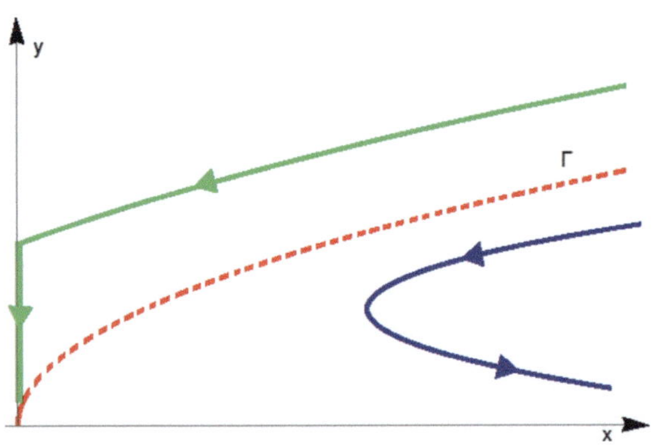

Fig. 2 The separatrix curve $y = (1+\varphi)(r+2d+2h)\sqrt{x}/\beta$ is a red dashed curve. Trajectories above the separatrix show extinct while below the separatrix show saddle behaviour

(b) $\left\{\sqrt{x_0} - \frac{\beta y_0}{(1+\varphi)(r+2(d+h))}\right\} e^{rt/2} < 0 \Rightarrow$ orbit of the model terminates at some positive y and then decreases to zero.

(c) $\left\{\sqrt{x_0} - \frac{\beta y_0}{(1+\varphi)(r+2(d+h))}\right\} e^{rt/2} < 0 \Rightarrow$ orbit of the model shows saddle behavior, as shown in Fig. 2. As y decreases, x decreases to a minimum value and then increases further. Hence, locally it looks like a saddle trajectory, but what happens when orbit goes away from the origin, local analysis cannot answer this question.

iii. $x = O\left(y_0^{\alpha}\right)$, then $x \ll \sqrt{x}y$ and system (3.5–3.6) reduces to, $\frac{dx}{dt} = \frac{-\beta\sqrt{x}y}{(1+\varphi)}$ and $\frac{dy}{dt} = -(d+h)y$. After solving we get,

$$y = y(t_0) + \frac{2(d+h)(1+\varphi)}{\beta}\left\{\sqrt{x} - \sqrt{x(t_0)}\right\}.$$

This curve terminates on y-axis at $y = y(t_0)\frac{2(d+h)(1+\varphi)}{\beta}\sqrt{x(t_0)} > 0$ (as $x \ll y$). This explains that trajectory terminates at $x = 0$ at some positive value of y and then predator y also tends to zero due to Eq. (3.6). Prey biomass is not sufficient enough to survive and go extinct. Ultimately, the predator also follows the same in absence of food.

3.2 *Study of Bifurcations*

3.2.1 Hopf Bifurcation

Let us consider φ as a bifurcation parameter first. Putting values of rest of the parameters, system (2.1–2.2) reduces to

$$\dot{x} = f(x, y, \varphi), \quad \dot{y} = g(x, y, \varphi) \tag{3.8}$$

Let the positive equilibrium point be $(x^*(\varphi), y^*(\varphi))$ and the eigenvalues for the linearised system about this equilibrium are $\lambda = \alpha_1(\varphi) \pm \alpha_2(\varphi)$. Then for a certain value of $\varphi = \varphi_0$, Hopf bifurcation takes place if the following conditions are satisfied.

I. **Non-hyperbolicity condition**: $\alpha_1(\varphi) = 0$, $\alpha_2(\varphi) = \omega \neq 0$, where $\mathrm{sgn}(w) = \mathrm{sgn}[(\partial g_{\varphi}/\partial x))|\varphi = \varphi_0(x_0, y_0)]$.

II. **Transversality condition**: $\frac{d\alpha_1(\varphi)}{d\varphi}|\varphi = \varphi_0 = d \neq 0$

III. **Genericity condition**: $\alpha \neq 0$, where

$$a = \frac{1}{16}\left(f_{xxx} + f_{xyy} + g_{xxy} + g_{yyy}\right) + \frac{1}{16\omega}\left(f_{xy}\left(f_{xx} + f_{yy}\right)\right.$$
$$\left. - g_{xy}\left(g_{xx} + g_{yy}\right) - f_{xx}g_{xx} + f_{yy}g_{yy}\right. \text{ with}$$

$$f_{xy} = \left((\partial^2 f_\varphi)/\partial x \partial y\right)\Big|_{\varphi=\varphi_0}(x_0, y_0), \text{ etc.}$$

The interesting observation in the system (2.1–2.2) is that Hopf bifurcation exists for not only the critical value of parameter φ, but for the critical value of every parameter present in the system.

3.2.2 Transcritical Bifurcation

Predator-free equilibrium point E_1 is locally asymptotically stable if $(d + h)^2(1 + \varphi)^2 > K\eta^2\beta^2$ and unstable if $(d + h)^2(1 + \varphi)^2 < K\eta^2\beta^2$. Therefore, system (2.1–2.2) exhibits a transcritical bifurcation for $(d + h)^2(1 + \varphi)^2 = K\eta^2\beta^2$, involving equilibria E^* and E_1. Also, it is easy to observe that feasibility condition for unique E^* and stability condition for E_1 are opposite of each other. System (2.1–2.2) shows transcritical bifurcation for all the parameters, except r.

Limit points (turning points) are called saddle-node bifurcation points which are associated with sudden creation of two fixed points. We can see saddle-node bifurcations also for many of the system parameters.

4 Numerical Simulation

The model is numerically simulated to analyse the dynamics of the predator-prey system after changing the values of the system parameters. Preselected values of parameters within biologically meaningful range for numerical simulation are $r = 1.5$, $K = 2$, $h = 0.3$, $\beta = 0.8$, $\varphi = 0.2$, $\eta = 0.9$, $d = 0.4$.

(i) **Effect of varying parameter r on the system**: Fig. 3 shows dynamics of the population for $K = 2$, $h = 0.3$, $\beta = 0.8$, $\eta = 0.9$, $d = 0.4$, $\varphi = 0.2$. We analyse the effect with different values for reproduction of preys. At $r = 1.5$, all species coexist with interior steady state (Fig. 3a). At $r = 2.307$, we obtain Hopf bifurcation as shown in Fig. 3c, d. Increasing prey reproduction rate destabilizes the equilibrium, giving rise to the occurrence of biomass oscillations (limit cycle) as shown in (Fig. 3b, d). Corresponding bifurcation diagram for the phenomenon can be shown in Fig. 7b

(ii) **Effect of varying parameter φ on the system**: Figs. 4 and 7a illustrate dynamics for $r = 1.5$, $K = 2$, $\beta = 0.8$, $\eta = 0.9$, $d = 0.4$, $h = 0.3$ and varying parameter φ. Prey's deference signals help in minimizing predation and establish ecological balance up to some extent. At low deterrence rate $\varphi = 0.14$, due to heavy predation unstable equilibrium with large oscillations is obtained (see Fig. 4a). When deterrence rate φ is increased to 0.2 and 0.52 respectively, it discourages unrestrained predation, fetch stability to the system and establishes an ecological balance with co-existence of all species (see Fig. 4b, c). Hence, stability

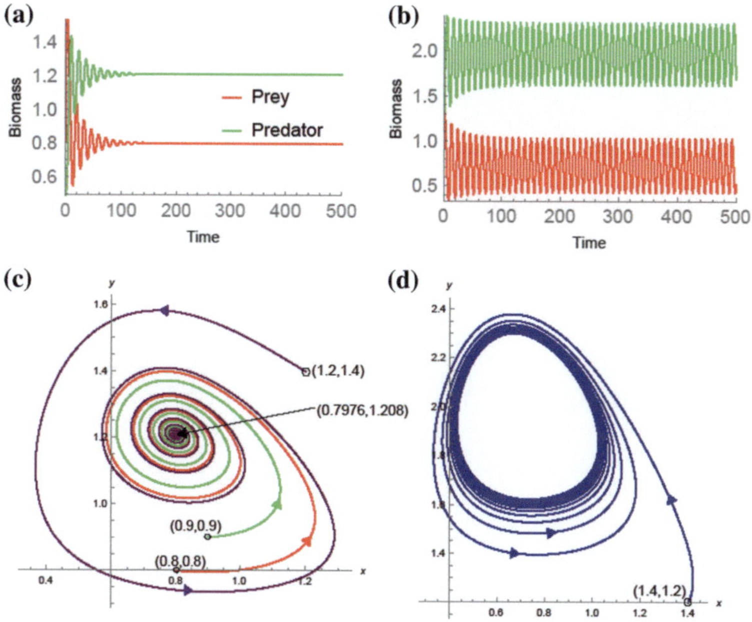

Fig. 3 Time evolution of biomass densities while varying r: parameter values K = 2, h = 0.3, β = 0.8, φ = 0.2, η = 0.9, d = 0.4 with initial conditions x(0) = 0.8, y(0) = 0.5 (for **a**) and x(0) = 1.4, y(0) = 1.2 (for **b**) remain as fixed and the values of parameter r for this simulation are taken as r = 1.5 (**a**), r = 2.5 (**b**). Corresponding diagrams for parametric plotting are shown in **c** and **d** respectively

switching takes place at $\varphi = 0.1444$ (Fig. 7a) and unstable equilibrium turns to the stable equilibrium state. We obtain a limit point (saddle-node bifurcation) at $\varphi = 0.5966$ (see LP in Fig. 7a), hence, at $\varphi = 0.52$, system shows different behavior with different initial conditions. One positive equilibrium and predator-free equilibrium, both are locally stable and hence an interesting parametric plotting diagram is obtained for different initial conditions (see Fig. 4e). A transcritical bifurcation is obtained at $\varphi = 0.4546$ as shown in Fig. 7a. On further increment in φ, sufficient deterrence of predation keeps the preys safe and ultimately life-dinner requirement for the predators remains unfulfilled, consequently, predators extinct from the environment as shown in Fig. 4f.

(iii) **Effect of varying parameter h on the system**: Fig. 5 deals with the dynamics for parameter values r = 1.5, K = 2, β = 0.8, η = 0.9, d = 0.4, φ = 0.2 and varying parameter h. This numerical dynamics illustrates that controlling the harvesting rate of predators, we can break the oscillating behavior of the model and drive it to the stable interior equilibrium of the system (see Fig. 5a, b). Stability switching with Hopf bifurcation takes place at h = 0.2425 (see HB in Fig. 5f). We find a limit point (saddle-node bifurcation) at h = 0.5659 (see LP in Fig. 5f), hence, at h = 0.5, system behaves differently for different initial

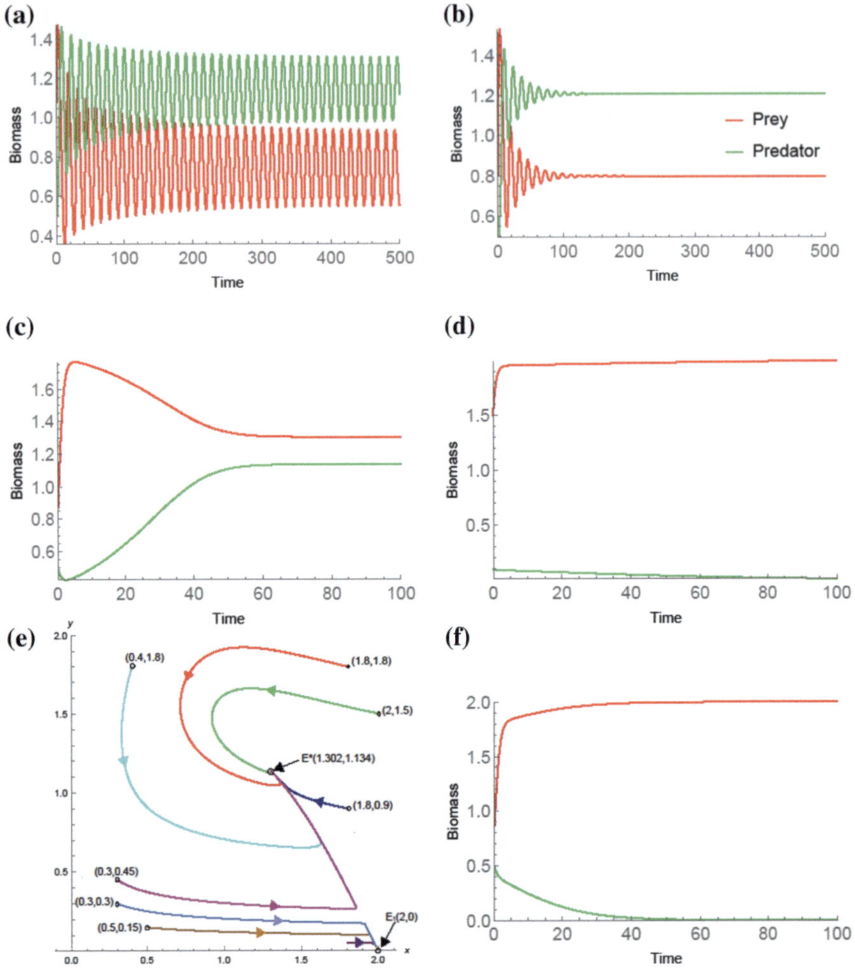

Fig. 4 Time evolution of biomass densities while varying φ: parameter values r = 1.5, K = 2, β = 0.8, η = 0.9, d = 0.4, h = 0.3 remain fixed and the values of parameter φ for this simulation are taken as φ = 0.14 with (x(0) = 0.8, y(0) = 0.5) (**a**), φ = 0.2 with (x(0) = 0.8, y(0) = 0.5) (**b**), φ = 0.52 with (x(0) = 0.8, y(0) = 0.5) (**c**), φ = 0.52 with (x(0) = 1.5, y(0) = 0.1) (**d**) and φ = 0.7 with (x(0) = 1.5, y(0) = 0.1) (**f**). Parametric plotting for φ = 0.52 with different initial conditions is shown in **e**

conditions as shown in Fig. 5c–e. Unlike in the previous case (Fig. 4e), trajectory starting from initial condition (0.4, 1.8) tends to predator-free equilibrium point (Fig. 5e). A transcritical bifurcation occurs at h = 0.4485 which biologically means that further increment of h can wipe out predation population completely, therefore, for sustainable development, we employ an economic threshold of the predator. Corresponding bifurcation diagram for the dynamics is shown in (Fig. 5f).

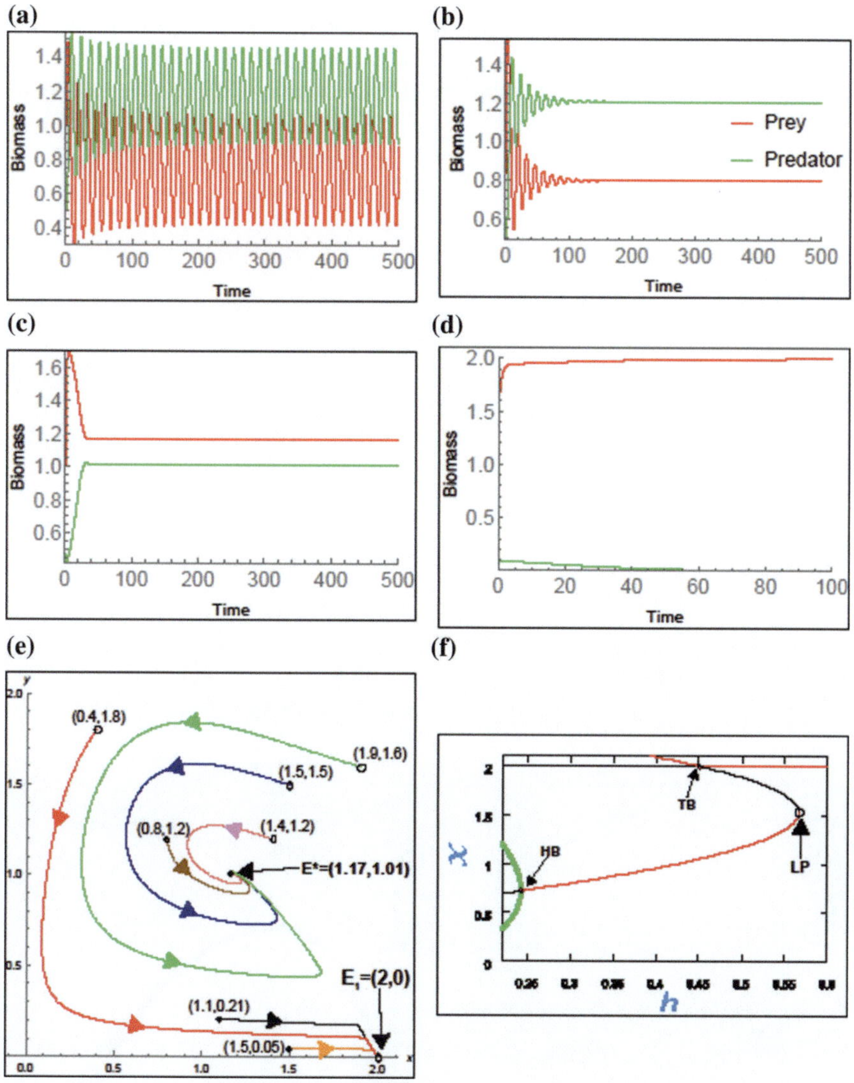

Fig. 5 Time evolution of biomass densities while varying h: parameter values $r = 1.5$, $K = 2$, $\beta = 0.8$, $\eta = 0.9$, $d = 0.4$, $\varphi = 0.2$ remain fixed and the values of parameter h for this simulation are taken as $h = 0.23$ with ($x(0) = 0.8$, $y(0) = 0.5$) (**a**), $h = 0.3$ with ($x(0) = 0.8$, $y(0) = 0.5$) (**b**), $h = 0.5$ with ($x(0) = 0.8$, $y(0) = 0.5$) (**c**) and $h = 0.5$ with ($x(0) = 1.5$, $y(0) = 0.1$) (**d**). Parametric plotting for $h = 0.5$ with different initial conditions is shown in **e** while corresponding bifurcation diagram is shown in **f**

(iv) **Effect of varying parameter K on the system**: Fig. 6 shows dynamics of the species for r = 1.5, h = 0.3, β = 0.8, η = 0.9, d = 0.4, φ = 0.2. We will investigate the effect with different carrying capacity on the system. We

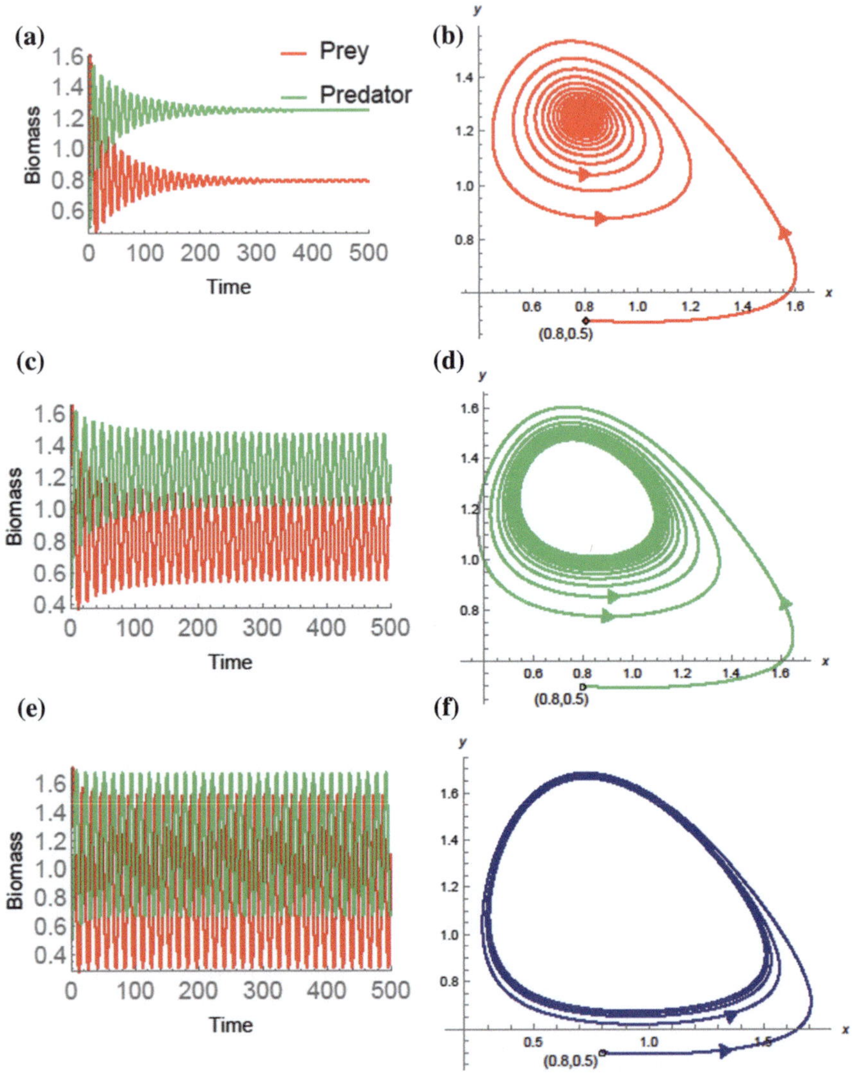

Fig. 6 6 Time evolution of biomass densities while varying K: parameter values r = 1.5, h = 0.3, β = 0.8, η = 0.9, d = 0.4, φ = 0.2. with initial conditions x(0) = 0.8, y(0) = 0.5 remain fixed and the values of parameter K for this simulation are taken as K = 2.1 (**a**), K = 2.17 (**c**), K = 2.25 (**f**). K = 2.15 is the point of Hopf bifurcation whose dynamics can be shown in **a** (K = 2.1 < 2.15) and **c** (K = 2.17 > 2.15). **b**, **d** and **f** show parametric plotting for the trajectories of **a**, **c** and **e** respectively. Limit cycle size is increasing gradually as shown in **b**, **d** and **f** respectively

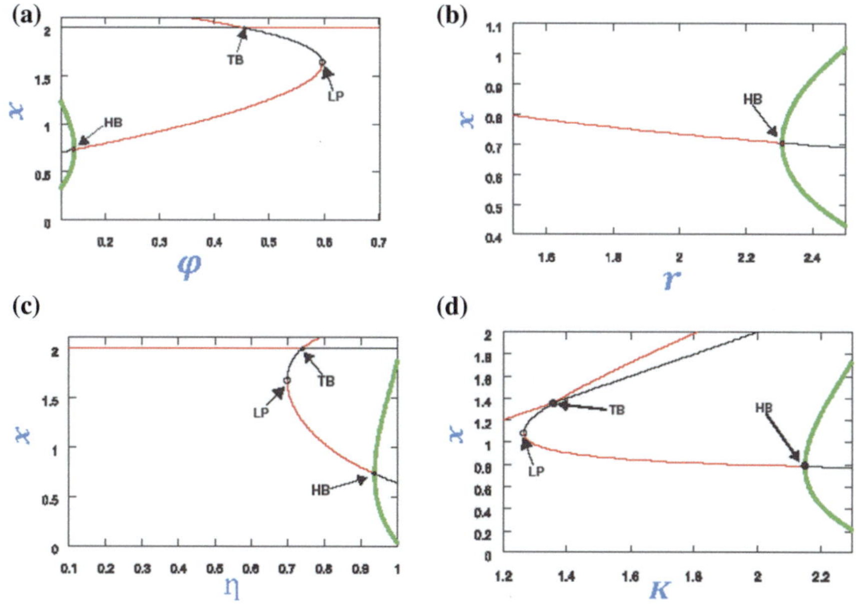

Fig. 7 Bifurcation diagrams for various parameters φ, r, η, and K: red, black and thick green lines represent stable steady state, unstable steady state and stable limit cycle respectively. In **a** HB (0.1444), TB (0.4546) and LP(0.5966) are Hopf, transcritical and saddle-node bifurcation points respectively. In **b** HB (2.307) is the point of Hopf bifurcation. In **c** HB (0.9377), TB (0.7425) and LP (0.6996) are Hopf, transcritical and saddle-node bifurcation points respectively. In **d** HB (2.15), TB (1.3611) and LP (1.265) are the points of Hopf, transcritical and saddle-node bifurcation

obtain transcritical bifurcation at $K = 1.361$ (see TB in Fig. 7d). Hence, for $K < 1.361$, predator-free equilibrium point is obtained with prey biomass K. At $K = 2.1 > 1.361$, all species coexist with positive equilibrium point (Fig. 6a). At $K = 2.15$, we get that supercritical Hopf bifurcation is obtained as shown in Fig. 6a ($K = 2.1 < 2.15$) and c ($K = 2.17 > 2.15$). As we change the carrying capacity to $K = 2.25$, we get a larger limit cycle because of the increased carrying capacity (Fig. 6e, f). We observe that on increasing carrying capacity, positive equilibrium state is shifted to a non-equilibrium state in the system. This phenomenon is known as the paradox of enrichment [13]. Corresponding bifurcation diagram for the phenomenon is illustrated in Fig. 7d.

(v) **Two parameter bifurcation diagram between parameters h and φ**: Nonlinear harvesting rate of predator h and deterrence rate to discourage predation φ are the key parameters in our system. We now discuss the feasibility of equilibria in the $h - \varphi$ parameter space (see Fig. 9). The boundary separating the region I and region II for E_1 and E^* respectively, is given by the curve $(0.4 + h)^2\,(1 + \varphi)^2 = 1.0368$. For small values of h and φ, uncontrolled predation gives rise to the oscillating behavior which can be shown by region III.

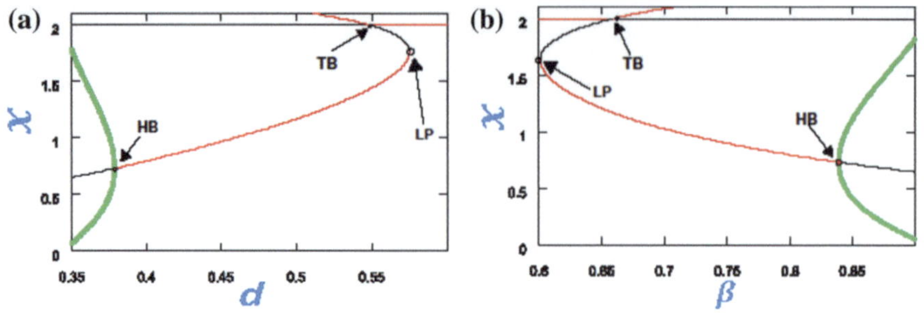

Fig. 8 Bifurcation diagrams for parameters d and β: In **a** HB (0.3785), TB (0.5485) and LP (0.5751) are Hopf, transcritical and saddle-node bifurcation points respectively. In **b** HB (0.8389), TB (0.66) and LP (0.6013) are Hopf, transcritical and saddle-node bifurcation points respectively

Fig. 9 Feasibility and stability regions for the equilibria of the system in h-φ parameter space

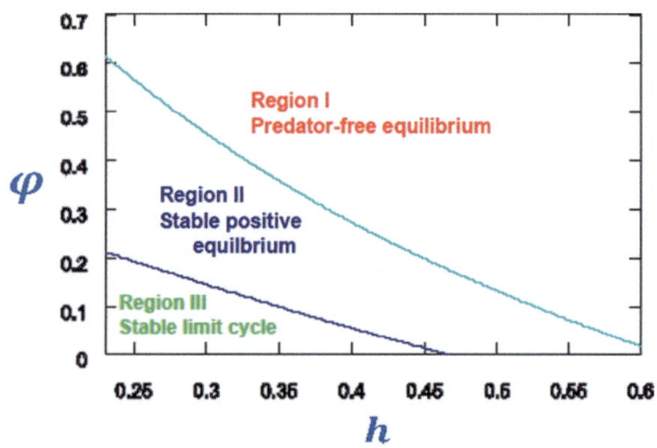

Remark 1 Bifurcation diagrams for rest parameters η, β and d are shown in Figs. 7 and 8. Steady state for $x > K = 2$ is biologically meaningless because predator equilibrium is negative for $x > 2$. However, mathematical significance of this is to show transcritical bifurcation (TB) at predator-free equilibrium point.

5 Conclusion

In this work, we have presented a predator-prey system with non-linear harvesting of predators and square root functional response of preys. We have investigated the equilibrium points and their stabilities. We have evaluated bifurcation analysis theoretically and that is perfectly justified by numerical simulations.

In the study, we can observe that non-linear harvesting of the predator as well as deterrence rate to minimize predation is a significant factor for the predator-prey interactions and affect the stability and coexistence for predator-prey system with herd behavior of prey. A sustainable harvesting and low prey's deterrence rate are important features in the perfect survival of all the species. Prey's deterrence phenomenon slows down the unrestricted predation which ultimately reduces periodic oscillations of the time-evolution and stabilizes the ecosystem. On other hand harvesting of predators does not allow predators to dominate on the system. At the same time, we conclude that uncontrolled harvesting rate or high deterrence rate to avoid predation, abolish predators completely from the system (see Figs. 5d and 4f) which is also ecologically justified. Therefore, we can make the prey and predator coexist through controlling the predator-harvesting rate. It is expected that the results illustrated in this study will be beneficial for the further study of complicated ecosystems.

References

1. Ajraldi, V., Pittavino, M., Venturino, E.: Modelling herd behavior in population system. Nonlinear Anal. RWA **12**, 2319–2338 (2011)
2. Bancala, F.: Function of mucus secretion by lamellose ormer, haliotis tuberculata lamellosa, in response to starfish predation. Anim. Behav. **78**, 1189–1194 (2009)
3. Belvisi, S., Venturino, E.: An ecoepidemic model with diseased predators and prey group defense. Simulat. Model. Pract. Theor. **34**, 144–155 (2013)
4. Bera, S.P., Maiti, A., Samanta, G.P.: Modelling herd behavior of prey: analysis of a prey-predator model. World J. Model. Simul. **11**, 3–14 (2015)
5. Birkhoff, G., Rota, G.C.: Ordinary Differential Equations. Needham Heights, Ginn (1982)
6. Braza, P.A.: Predator-prey dynamics with square root functional responses. Nonlinear Anal. RWA **13**(1837), 2012 (2012)
7. Clark, R.W.: Pursuit-deterrent communication between prey animals and timber rattlesnakes (*crotalus horridus*): the response of snakes to harassment displays. Behav. Ecol. Sociobiol. **59**, 258–261 (2005). https://doi.org/10.1007/s00265-005-0032-9
8. Eisner, T., Grant, R.P.: Toxicity, odor aversion, and 'olfactory aposematism. Science **213**(4506), 476 (1981). https://doi.org/10.1126/science.7244647
9. Gimmelli, G., Kooi, B.W., Venturino, E.: Ecoepidemic models with prey group defense and feeding saturation. Ecol. Complex. **22**, 50–58 (2015)
10. Ma, X., Shao, Y., Wang, Z., Luo, M., Fang, X., Ju, Z.: An impulsive two-stage predator-prey model with stage structure and square root functional responses. Math. Comput. Simulat. **119**, 91–107 (2016)
11. Maan, M.E., Cummings, M.E.: Poison frog colors are honest signals of toxicity, particularly for bird predators. Am. Nat. **179**(1), E1–E2 (2012). https://doi.org/10.1086/663197
12. Matia, S.N., Alam, S.: Prey-predator dynamics under herd behavior of prey. Univ. J. Appl. Math. **1**, 251–257 (2013)
13. Rosenzweing, M.: The paradox of enrichment. Science **171**, 385–387 (1971)
14. Tang, X., Song, Y.: Bifurcation analysis and turing instability in a diffusive predator-prey model with herd behavior and hyperbolic mortality. Chaos Solitons Fractals **81**, 303–314 (2015)
15. Tang, X., Song, Y.: Stability, hopf bifurcation and spatial patterns in a delayed diffusive predator-prey model with herd behavior. Appl. Math. Comput. **254**, 375–391 (2015)
16. Thomas, C., Candy, R.: Avian predators change their foraging strategy on defended prey when undefended prey are hard to find. Anim. Behav. **93**, 97–103 (2014)
17. Wang, X., Pan, Q., Kang, Y., He, M.: Predator group size distributions in predator-prey system. Ecol. Complex. **26**, 117–127 (2016)
18. Xu, C., Yuan, X., Zhang, T.: Global dynamics of a predator-prey model with defense mechanism for prey. Appl. Math. Lett. **62**, 42–48 (2016)

Case Wise Study of Heterogeneity on the Traversal Characteristics of Torsional Surface Waves

C. Kumari and S. Kundu

Abstract Within the framework of a heterogeneous anisotropic layer laid over a pre-stressed half-space, propagation of torsional surface wave is investigated. We have considered two different cases of heterogeneity function in the anisotropic layer, namely exponential and trigonometric hyperbolic functions. We have also obtained two different dispersion equations for both cases of heterogeneities on the torsional wave propagation in the said model. The effects of heterogeneities and pre-stress on the torsional wave of phase velocity have been shown graphically for their different numerical values.

Keywords Torsional · Anisotropic · Heterogeneity · Pre-stress · Phase velocity

1 Introduction

In seismology, we study about the seismic wave. Seismic waves are extremely noteworthy for ponder because of these wave contains so much vitality that they can make tremor, avalanches, volcanic emissions etc. Who studied about the seismic wave is called seismologists. We know, two kinds of seismic wave; Body wave and Surface wave. Love wave and Rayleigh wave are the part of surface wave, but in recently we know about another surface wave which is known as Torsional wave. It is also known as twisting wave because it behaves like twist. Torsional surface wave can just exist in solid medium rather than electromagnetic, gas, or fluid media. Ewing and Press [1] described lots of things in their book. They studied about many significant factors regarding seismic wave propagation, curvature influence of love wave in heterogeneous medium, refraction and reflection of waves at the interface of the rigid boundary, etc. Elastic wave propagation in layered medium is analyzed by Alterman

C. Kumari (✉) · S. Kundu
Department of Applied Mathematics, Indian Institute of Technology (Indian School of Mines), Dhanbad, Jharkhand 826004, India
e-mail: guptachandani002@gmail.com

S. Kundu
e-mail: kundu_santi@yahoo.co.in

S. Manna et al. (eds.), *Mathematical Modelling and Scientific Computing with Applications*, Springer Proceedings in Mathematics & Statistics 308,
https://doi.org/10.1007/978-981-15-1338-1_10

and Karal [2] using finite difference approach and studied the development, pulse width and depth of Rayleigh wave on the half space surface. Apart from it, they also presented a theoretical seismograms series that showed the variation of surface wave with the depth and density of the medium. About mechanical deformation and plate tectonics on Earth, there is a great information in the book of Love [3], Gubbins [4] Biot [5], and so on.

Torsional waves are the main attraction of researchers nowadays due to its exotic behavior and inadequate published data. While considering the analysis regarding wave propagation through layers it is necessary to gather information regarding rigidity and density of the medium because while earthquake due to generation of torsional wave rigidity and density variation can be observed. Propagation analysis of torsional wave is performed by Chattopadhyay et al. [6] and the solution for displacement along with the effect of inhomogeneity is obtained by utilizing the fact that material property is dependent on space variable harmonically. Another investigation of torsional wave is performed by Dey et al. [7] and analyzed the wave propagation possibility in the nonhomogeneous elastic medium with different types of rigidity variables such as exponential, polynomial. One point worthy finding of this investigation is that two torsional wavefronts may be there in certain types of nonhomogeneity. Chattaraj et al. [8] has showed keen interest towards torsional surface wave in the fluid saturated poroelastic layer and found that increment in phase velocity leads to decrease inhomogeneity factor. Another investigation of wave propagation is performed by Wang et al. [9] using cross-anisotropic inhomogeneous medium and described the analytical solution for wave vector and wave velocity. Gupta et al. [10] used heterogeneous half-space to analyze the torsional surface wave propagation and suggested the effect of rigid boundary on the torsional wave propagation in a homogeneous layer over a heterogeneous half-space. Some other related studies on torsional surface wave are due to articles Kundu et al. [12], Alam et al. [13], Gupta and Bhengra [14], Alam et al. [15].

In the present analysis our main attention is devoted to analyze the propagation of torsional surface waves passing through pre-stressed elastic half-space attached with heterogeneous anisotropic layers. In this paper, we have considered two different cases of heterogeneity function in the anisotropic layer, exponential and trigonometric hyperbolic functions. Anisotropic has directionally property, which suggests diverse properties in various ways, rather than isotropy. A case of anisotropy is light getting through a polarizer. Another is wood, which is less demanding part along its grain than the crosswise over it. In the half-space, we have taken pre-stressed elastic property. So we have obtained two different dispersion equations for both cases of heterogeneities on the torsional wave propagation. The effects of heterogeneities and pre-stress on the torsional wave of phase velocity have been shown graphically for their different numerical values. We observed the exceptional impact of initial pressure display on the wave phase speed. More precisely, an initial force leads to decrease the torsional wave phase speed. This investigation is done due to it has large potential to deal with many seismological and geological problems.

2 Formulation of the Problem

In considering this analysis we have assumed finite thickness (***H***) heterogeneous anisotropic media, which is attached over a pre-stressed elastic half-space. From Fig. 1 it can be seen that we have chosen r-axis in the direction of wave propagation and parallel to the layer, whereas z-axis is taken into lower half space in such a way that it is pointing positively downward.

We have considered two different cases of heterogeneity function in the anisotropic upper layer:

Case 1: $N = N_0 \mathrm{e}^{\alpha z},\ L = L_0 \mathrm{e}^{\alpha z},\ \rho_1 = \rho_1^0 \mathrm{e}^{\alpha z}$.
Case 2: $N = N_0 \cosh^2(\beta z),\ L = L_0 \cos h^2(\beta z),\ \rho_1 = \rho_1^0 \cos^2(\beta z)$.

where N and L are the rigidity moduli, ρ be the density of anisotropic layer and α, β are heterogeneities parameters.

The equations of motion are

$$\begin{aligned} &\frac{\partial \tau_{rr}}{\partial r} + \frac{1}{r}\frac{\partial \tau_{r\theta}}{\partial \theta} + \frac{\partial \tau_{rz}}{\partial z} + \frac{\tau_{rr} - \tau_{\theta\theta}}{r} = \rho \frac{\partial^2 u}{\partial t^2}, \\ &\frac{\partial \tau_{r\theta}}{\partial r} + \frac{1}{r}\frac{\partial \tau_{\theta\theta}}{\partial \theta} + \frac{\partial \tau_{\theta z}}{\partial z} + \frac{2\tau_{r\theta}}{r} = \rho \frac{\partial^2 v}{\partial t^2}, \\ &\frac{\partial \tau_{rz}}{\partial r} + \frac{1}{r}\frac{\partial \tau_{\theta z}}{\partial \theta} + \frac{\partial \tau_{zz}}{\partial z} + \frac{\tau_{rz}}{r} = \rho \frac{\partial^2 w}{\partial t^2}, \end{aligned} \tag{1}$$

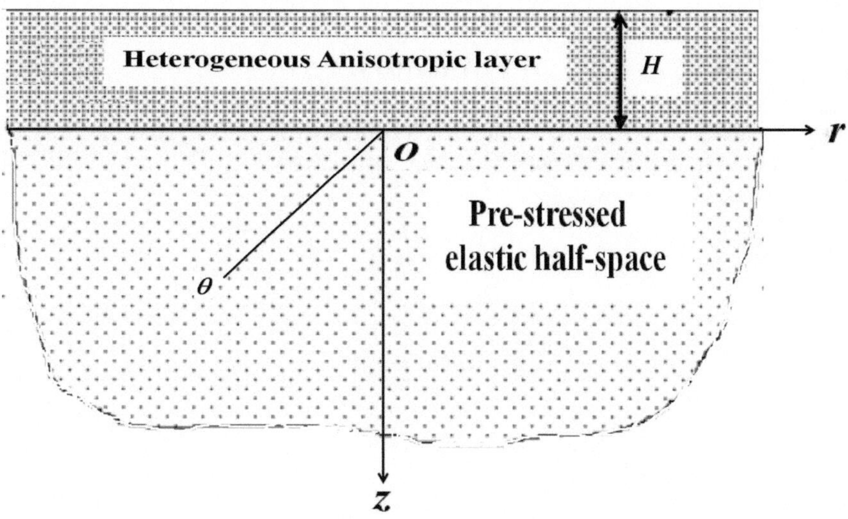

Fig. 1 Geometry of the problem

where stress components are $\tau_{rr}, \tau_{rz}, \tau_{r\theta}, \tau_{\theta\theta}, \tau_{z\theta}, \tau_{zz}$ respectively and displacement components are the (u, v, w). The condition of torsional wave propagation $u = 0$, $\omega = 0$, $v = v(r, z, t)$.

Now non-vanishing equation of motion in the absence of body force for the propagation of torsional wave is given by

$$\frac{\partial \tau_{r\theta}}{\partial r} + \frac{\partial \tau_{\theta z}}{\partial z} + \frac{2\tau_{r\theta}}{r} = \rho \frac{\partial^2 v}{\partial t^2}. \tag{2}$$

2.1 Upper Layer Solution

The stress-strain relation components for medium are given by

$$\tau_{r\theta} = 2Ne_{r\theta}, \tau_{\theta z} = 2Le_{z\theta}. \tag{3}$$

Using the above, Eq. (2) reduces to,

$$N\left(\frac{\partial^2}{\partial r^2} + \frac{1}{r}\frac{\partial}{\partial r} - \frac{1}{r^2}\right)v_1 + \frac{\partial L}{\partial z}\frac{\partial v_1}{\partial z} + L\frac{\partial^2 v_1}{\partial z^2} = \rho\frac{\partial^2 v_1}{\partial t^2}. \tag{4}$$

Let us consider the harmonic wave which propagates along the radial direction in the form of

$$v_1 = V_1(z)J_1(kr)e^{i\omega t}, \tag{5}$$

where k, c and $J_1(kr)$ are wave number, torsional wave velocity and the Bessel's function of first kind as well as of first order respectively.

From Eqs. (4) to (5) we get,

$$\frac{d^2V_1}{dz^2} + \frac{1}{L}\frac{dL}{dz}\frac{dV_1}{dz} - \frac{k^2N}{L}\left(1 - \frac{\rho c^2}{N}\right)V_1 = 0. \tag{6}$$

Now, substituting $V_1 = \frac{\phi_1(z)}{\sqrt{L}}$ in Eq. (6), we get

$$\frac{d^2\phi_1}{dz^2} - \frac{1}{2L}\left\{\frac{d^2L}{dz^2} - \frac{1}{2L}\left(\frac{dL}{dz}\right)\right\}\phi_1 = \frac{k^2N}{L}\left(1 - \frac{\rho c^2}{N}\right)\phi_1. \tag{7}$$

Case 1

$$N = N_0\exp(\alpha z),\ \ L = L_0\exp(\alpha z),\ \ \rho_1 = \rho_1^0\exp(\alpha z). \tag{8}$$

Using Eqs. (8) and (7) occurs

$$\frac{d^2\phi_1}{dz^2} + m_1^2\phi_1 = 0, \tag{9}$$

where $m_1^2 = \left[k^2 \frac{N_0}{L_0}\left(\frac{c^2}{c_1^2} - 1\right) - \frac{\alpha^2}{4}\right]$ and $c_1 = \sqrt{\frac{N_0}{\rho_0}}$ is the velocity.

Case 2

$$N = N_0 \cosh^2(\beta z),\ L = L_0 \cosh^2(\beta z),\ \rho_2 = \rho_2^0 \cosh^2(\beta z). \tag{10}$$

Using Eq. (10) in Eq. (7), becomes

$$\frac{d^2\phi_1}{dz^2} + m_2^2\phi_1 = 0, \tag{11}$$

where $m_2 = \left[k^2 \frac{N_0}{L_0}\left(\frac{c^2}{c_1^2} - 1\right) - \beta^2\right]^{\frac{1}{2}}$ and $c_2 = \sqrt{\frac{N_0}{\rho_0}}$ is the velocity.

Now the solution of Eqs. (9) and (11) is given by

$$\phi = A\cos(m_i z) + B\sin(m_i z), \quad i = 1 \text{ and } 2. \tag{12}$$

where for Case 1, $i = 1$ and for Case 2, $i = 2$.

Therefore the displacement component for the torsional wave propagation in the anisotropic layer (for Case 1 and Case 2) is given by following equation

$$v_1 = (A\cos(m_i z) + B\sin(m_i z))\frac{J_1(kr)}{\sqrt{L}}e^{i\omega t}, \quad i = 1, 2. \tag{13}$$

2.2 *Half-Space Solution*

$$L = N = \mu = \mu_0(1+\gamma z)^2,\ \rho = \rho_2^0(1+\gamma z)^2.$$

The stress-strain relation components for half-space are given by

$$\sigma_{r\theta} = \mu\left(\frac{\partial v_2}{\partial r} - \frac{v_2}{r}\right) \text{ and } \sigma_{\theta z} = \mu\frac{\partial v_2}{\partial z}. \tag{14}$$

Hence the non-vanishing equation of motion for the torsional wave propagation with initial stress is given by

$$\frac{\partial \sigma_{r\theta}}{\partial r} + \frac{\partial \sigma_{\theta z}}{\partial z} + \frac{2\sigma_{r\theta}}{r} - \frac{\partial}{\partial z}\left(\frac{P}{2}\frac{\partial v_\theta}{\partial z}\right) = \rho_2^0 \frac{\partial^2 v}{\partial t^2}. \tag{15}$$

Using Eq. (14) in (15), we have

$$\mu\left(\frac{\partial^2}{\partial r^2} + \frac{1}{r}\frac{\partial}{\partial r} - \frac{1}{r^2}\right)v_2 + \left(\mu - \frac{P}{2}\right)\frac{\partial^2 v_2}{\partial z^2} = \rho \frac{\partial^2 v_2}{\partial t^2}. \tag{16}$$

Again, we consider the solution of Eq. (16) as

$$v_2 = V_2(z) J_1(kr) e^{i\omega t}. \tag{17}$$

Using the above equation, Eq. (16) becomes

$$\frac{d^2 V_2}{dz^2} - n^2 V_2 = 0. \tag{18}$$

where, $n = k\left[\left(1 - \frac{c^2}{c_2^2}\right)/\left(1 - \frac{P}{2\mu}\right)\right]^{1/2}$.

The appropriate solution for the half-space is found as,

$$v_2 = De^{-nz} e^{i\omega t} J_1(kr), \tag{19}$$

where, D is an arbitrary constant.

3 Boundary Conditions and Dispersion Equation

The required boundary conditions for the analyzing this phenomenon include the continuity of the stress are as follows

1. Stress of the layer vanishes at $z = -H$ so that

$$\tau_{\theta z} = 0 \text{ at } z = -H.$$

2. The continuity of the displacement component gives

$$v_1 = v_2 \text{ at } z = 0.$$

3. At the interface z = 0, the continuity in the stress gives

$$\tau_{\theta z} = \delta_{\theta z} \text{ at } z = 0.$$

Where v_1 and v_2 the displacement in the upper layer and the half-space, respectively.

Using Eqs. (13) and (19) in the above boundary conditions, we obtain the following equations:

$$(i)\ A_1 - \sqrt{L}B_1 = 0$$
$$(ii)\ L_0 m A_2 + n\mu\sqrt{L}B_1 = 0$$
$$(iii)\ A_1 \sin(mH) + A_2 \cos(mH) = 0.$$

Eliminating the arbitrary constants from above eq's, the dispersion relation for both cases

$$\tan(m_i H) = \frac{\mu k}{m_i L_0}\sqrt{\frac{1 - c^2/c_2^2}{1 - P/2\mu}}, \quad i = 1 \text{ and } 2, \tag{20}$$

where μ, k and p are respectively rigidity of the elastic half-space, the wave number of the wave, pre-stress associated with the half-space.

$m_1 = k\left[\frac{N_0}{L_0}\left(\frac{c^2}{c_1^2} - 1\right) - \frac{\alpha^2}{4}\right]^{\frac{1}{2}}$ for Case 1 and $m_2 = k\left[\frac{N_0}{L_0}\left(\frac{c^2}{c_1^2} - 1\right) - \beta^2\right]^{\frac{1}{2}}$ for Case 2.

4 Numerical Calculations and Discussion

So as to get the effect of case wise inhomogeneities and initial stress on the torsional wave propagation, we have considered some numerical examples of material parameters as follow:

For layer:

$N = 1.87 * 10^9\,\text{N/m}^2$, $L = 2.64 * 10^9\,\text{N/m}^2$, $\rho = 1442\,\text{kg/m}^3$, [11].

For half-space:

$\mu = 78.4 * 10^9\,\text{N/m}^2$, $\rho = 3535\,\text{kg/m}^3$, [4].

All graphs are obtained between phase velocity (c/c_0) and wave number (kH) on the based on obtaining dispersion Eq. (20) for different values of respective parameters. Since we have considered two cases so Fig. 2a represents for Case 1 and Fig. 2b represents for Case 2 and similarly Fig. 3a represents for Case 1 and Fig. 3b represents for Case 2, each graph consists of two subfigures.

In Fig. 2a and b represent the variation of dimensionless phase velocity (c/c_0) against wave number (kH) for torsional surface wave associated with the layer. Figure 2a and b show the effect of phase velocity (c/c_0) and wave number (kH) for different values of heterogeneity parameters (0.1, 0.5, 0.9) for both Case 1 and Case 2. From Fig. 2a and b, we have observed that the phase velocity curves follow the same trend in both the cases with the variation in wave number and heterogeneity parameter. These figures show that the phase velocity of the torsional wave decreases with the increasing value of wave number, i.e., it is clear from figure that as the magnitude of heterogeneity associated with layer increases the phase velocity

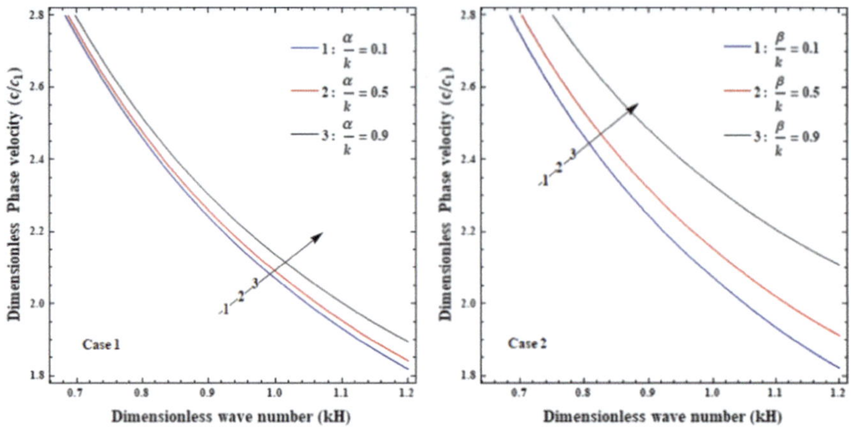

(a) *Case 1.* ***(b)*** *Case 2.*

Fig. 2 Variation of phase velocity (c/c0) and wave number (kH) for different values of heterogeneity parameters **a** Case 1 and **b** Case 2

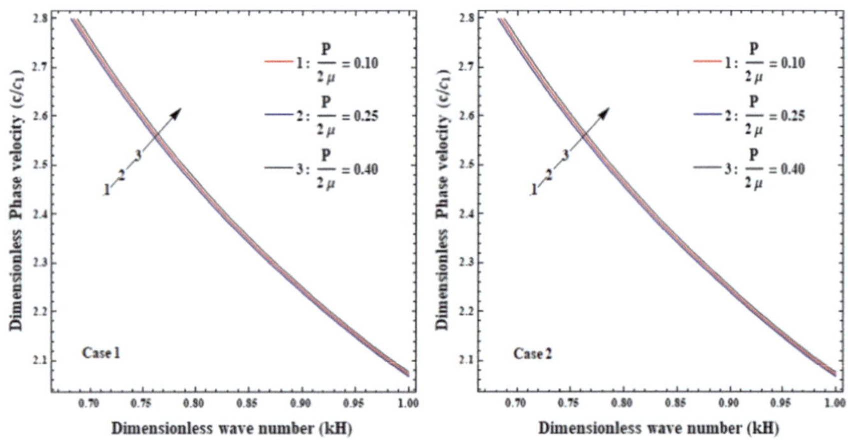

(a): *Case 1.* ***(b):*** *Case 2.*

Fig. 3 Variation of phase velocity (c/c0) and wave number (kH) for different values of initial stress parameters **a** Case 1 and **b** Case 2

also increases in both cases. The figure also suggests that the effect of heterogeneity on the phase velocity of torsional wave is more significant for Case 2 rather than Case 1.

In Fig. 3a and b also represent the variation of dimensionless phase velocity (c/c_0) against wave number (kH) for torsional surface wave. Figure 3a and b portrayed the influence of initial stress associated with the half-space with the phase velocity of torsional. Figure 3a and b show the variation effect of phase velocity (c/c_0) and wave number (kH) for different values of heterogeneity parameters (0.10, 0.25, 0.40) for both Case 1 and Case 2. Here we have same observed that these figures show that the phase velocity of the torsional wave decreases in the increasing value of wave number, i.e., It is clear from figure that as the magnitude of initial stress associated with half space increases the phase velocity also increases in both cases. We can also observe that the effect of initial stress on the phase velocity of torsional wave is almost similar for both cases; Case 1 and Case 2.

5 Conclusions

Propagation of torsional waves in a non-homogeneous anisotropic layer lying over a pre-stressed half-space. We have obtained the dispersion equation by using some assumption and boundary constraints and numeric calculation of velocity is performed with the variation in initial stress, elastic parameters, inhomogeneity. It can be seen from the graphical representation that increment in initial stress and inhomogeneity tends to increase in phase velocity of the torsional surface wave. Obtained result is illustrated using graphical representation by taking different anisotropic material. It is also observed that increment in heterogeneities as well as initial stress leads to increase the phase velocity of the torsional wave. For a graphical representation, Mathematica software is used to solve the dispersion relation. This investigation concludes that the presence of initial stress and heterogeneities enhances the phase velocity of the wave.

References

1. Ewing, W.M., Press, W.S.F.: Elastic Waves in Layered Media. McGraw-Hill, New York (1957)
2. Alterman, Z., Karal, F.C.: Propagation of waves in layered media by finite difference methods
3. Love, A.E.H.: The Mathematical Theory of Elasticity. Cambridge University Press, Cambridge, UK (1927)
4. Gubbins, D.: Seismology and Plate Tectonics. UK, Cambridge University Press, Cambridge (1990)
5. Biot, M.A.: Mechanics of Incremental Deformations. Wiley, New York (1965)
6. Chattopadhyay, A., Gupta, S., Kumari, P., Sharma, V.K.: Propagation of torsional waves in an inhomogeneous layer over an inhomogeneous half space. Meccanica **46**(4), 671–680 (2011)
7. Dey, S., Gupta, S., Gupta, A.K.: Torsional surface waves in non-homogeneous and anisotropic medium. J. Acoust. Soc. Am. **99**(5), 2737–2741 (1996)
8. Chattaraj, R., Samal, S.K., Mahanti, N.C.: Propagation of torsional surface wave in anisotropic poroelastic medium under initial stress. Wave Motion **48**(2), 184–195 (2011)
9. Wang, C., Lin, Y., Jeng, Y., Ruan, Z.: Wave propagation in an inhomogeneous cross-anisotropic medium. Int. J. Numer. Anal. Methods Geomech **34**(7), 711–732 (2010)
10. Gupta, S., Chattopadhyay, A., Kundu, S., Gupta, A.K.: Effect of rigid boundary on the propagation of Torsional wave in a homogeneous layer over a heterogeneous half-space. Arch. Appl. Mech. **80**(2), 143–150 (2010)

11. Singh, A.K., Das, A., Kumar, S., Chattopadhay, A.: Influence of corrugated boundary surfaces, reinforcement, hydrostatic stress, heterogeneity and anisotropy on love-type wave propagation. Meccanica **50**(12), 2977–2994 (2015)
12. Kundu, S., Saha, A., Gupta, S., Manna, S.: Propagation of torsional wave in a non-homogeneous crustal layer over a dry sandy mantle. Meccanica **50**(12), 3029–3040 (2015)
13. Alam, P., Kundu, S., Gupta, S., Saha, A.: Study of torsional wave in a poroelastic medium sandwiched between a layer and a half-space of heterogeneous dry sandy media. Waves Random Complex Media (2017). https://doi.org/10.1080/17455030.2017.1335915
14. Gupta, S., Bhengra, N.: Dispersion study of propagation of torsional surface wave in a layered structure. J. Mech. **33**(3), 303–315 (2017)
15. Alam, P., Kundu, S., Gupta, S.: Dispersion and attenuation of torsional wave in a viscoelastic layer bonded between a layer and a half-space of dry sandy media. Appl. Mathematics Mech. (English Edition) **38**(9), 1313–1328 (2017)

Effect of the Heterogeneity, Initial Stress and Viscosity on the Propagation Characteristics of Shear Wave

Raju Kumhar and S. Kundu

Abstract This paper irradiates the influence of initial stress, heterogeneity and viscoelasticity on the propagation behavior of shear wave in a heterogeneous orthotropic layer under initial stress overlying a heterogeneous viscoelastic half-space. A separation of variables method has been adopted to obtain the analytical solutions for both the layer and half-space separately. A complex frequency equation has been derived by using suitable boundary conditions. Thereafter, the complex frequency equation has been separated into dispersion and absorption equations. The effects of heterogeneities, initial stress and attenuation coefficient on the phase velocity and dissipation function of shear wave have been shown graphically.

Keywords Shear wave · Heterogeneity · Viscoelasticity · Orthotropic · Initial stress · Dispersion relation

1 Introduction

The study of seismic surface waves has been proved as most useful tool in order to understand the structure of the Earth as well as to comprehend causes of the earthquake and its consequent damages. Shear wave is a types of seismic surface waves, in which the movement of the particles is perpendicular to the direction of wave propagation. The investigation of shear waves propagating in different types of elastic materials under different physical properties is the topic of great interest to many seismologists, geologists and researchers. The basic information regarding the propagation of seismic waves in layered media is available in Ewing et al. [1]. Love [2], Gubbins [3] and Birch [4].

The constituent layers of finite width of the Earth considering them to be isotropic and homogeneous may not be sufficient to encapsulate most of the engineering problems including geological materials, response of soils and composites. Orthotropic

R. Kumhar (✉) · S. Kundu
Department of Mathematics and Computing, Indian Institute of Technology (Indian School of Mines), Dhanbad 826004, India
e-mail: raju.ism92@gmail.com

S. Manna et al. (eds.), *Mathematical Modelling and Scientific Computing with Applications*, Springer Proceedings in Mathematics & Statistics 308,
https://doi.org/10.1007/978-981-15-1338-1_11

materials are those in which the thermal or mechanical properties are unique and independent along three mutually perpendicular directions. The most suitable example of such types of material are wood, bone, ceramic and cold-rolled steel etc. Recently, Maity and Kundu [5] investigated the propagation of Love wave in a heterogeneous orthotropic layer under initial stress. Moreover, Abd-Alla and Ahmed [6] have published a paper on the traversal characteristics of Love waves in a heterogeneous orthotropic elastic medium influenced by an initial stress overlying semi-infinite medium. Wang et al. [7] pointed out their valuable research on the propagation behaviour of Love wave in an inhomogeneous orthotropic elastic layer obeying the both exponential and generalized Power Law Models. Taking into account the possibility and applications, the present problem has been formulated to study the propagation of shear wave (SH-wave) in a heterogeneous pre-stressed orthotropic layer over lying a heterogeneous viscoelastic half-space.

The ingredient layers of the Earth medium are usually found heterogeneous at each level with large alteration and acute changes in the material properties with respect to the vertical directions. The presence of heterogeneity has high influence on the traversal characteristics of seismic surface wave. Many researchers have shown great interest in studying the propagation of the seismic surface wave considering different type of heterogeneity in their problem. Some of whom are Manna et al. [8] studied the Love wave propagation in piezoelectric material resting over a heterogeneous elastic half-space with properties varying linearly. Saha et al. [9] examined the influence of heterogeneity on the propagation of SH-type wave in initially stressed composite structure. Craster et al. [10] discussed the Long-wave asymptotic theories: the connection between functionally graded waveguides and periodic media. Furthermore, the explicit asymptotic modelling of transient Love waves propagated along a thin coating was studied in detail by Ahmad et al. [11]. The complicated layers of the Earth owing to their viscous nature significantly affect the propagation behaviour of seismic wave. The material property of viscoelastic material is due to conjunction of two physical properties namely viscous and elastic and are primarily responsible for seismic attenuation. Therefore, the viscoelasticity present in a material medium has a great influence in propagation and attenuation of seismic waves and find extensive applications in many areas such as soil dynamics, fluid dynamics, earthquake engineering and seismology. Carcione [12] was the first to study the propagation of seismic waves in anisotropic linear viscoelastic media which has attracted various researchers around the globe. Recently, Kakar and Kakar [13] contributed their thoughts regarding the traversal of Love wave in a heterogeneous viscoelastic layered half-space with properties varying both exponentially and quadratically.

In this paper, the analysis of propagation characteristics of shear waves in a heterogeneous orthotropic medium under the initial stress resting over a heterogeneous viscoelastic half-space is presented. For the upper heterogeneous orthotropic medium, the hyperbolic variation in initial stress, shear moduli and density are taken as $P = P_1 \cosh^2(\alpha z)$, $Q_i = a_i \cosh^2(\alpha z)$ and $\rho_1 = \rho_1' \cosh^2(\alpha z)$ respectively, while in the heterogeneous viscoelastic half-space, the rigidity, density and viscosity varies exponentially as $\mu = \mu_2 e^{\varepsilon z}$, $\rho = \rho_2 e^{\varepsilon z}$ and $\mu' = \mu_2' e^{\varepsilon z}$ respectively. The effect of all affecting parameters associated with the layer and half-space on the traversal characteristics of shear waves have been shown graphically with the help of Wolfram Mathematica software. Classical relationship of Love wave has been received as a special case, which validates the undertaken problem. The outcomes and observations of the present study may play an important role in designing and constructing suitable kind of hazard resistant buildings as well as understanding the damage instigated due to earthquakes.

2 Formulation of the Problem

We have considered the traversal of shear waves in a heterogeneous orthotropic medium under the influence of initial stress P of finite width H resting over a heterogeneous viscoelastic half-space. The x-axis is taken in the direction of wave propagation and z-axis is taken vertically downwards, as shown in Fig. 1. Let (u_1, v_1, w_1) and (u_2, v_2, w_2) are the displacement components for the layer and half-space, respectively. Now for the propagation of shear wave

$$u_1 = 0 = w_1, \quad v_1 = v_1(x, z, t)$$
$$u_2 = 0 = w_2, \quad v_2 = v_2(x, z, t)$$

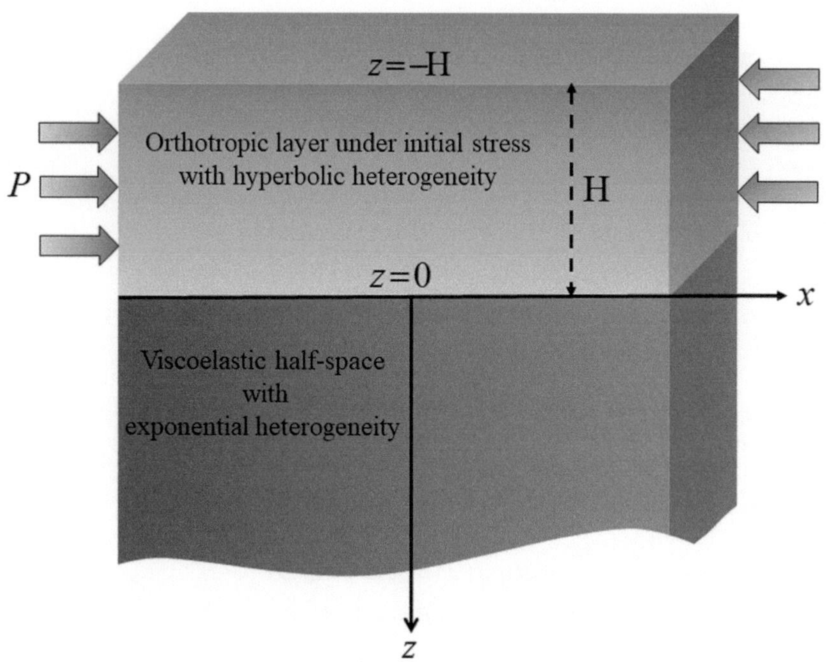

Fig. 1 Geometry of the problem

$$\text{and} \quad \frac{\partial}{\partial y} = 0 \tag{1}$$

In the upper heterogeneous orthotropic medium, the variation in initial stress, density and shear moduli are taken as:

$$P = P_1 \cosh^2(\alpha z), \quad \rho_1 = \rho_1' \cosh^2(\alpha z) \text{ and } Q_i = a_i \cosh^2(\alpha z) \tag{2}$$

where P is the initial stress, ρ_1 is density, Q_i are shear moduli, α is a heterogeneity constant having dimension that is inverse of length. Here $P_1,\ \rho_1',\ a_i$ are the values of $P,\ \rho_1,\ Q_i$ at $z = 0$.

The variation in rigidity, density and viscosity of the lower half-space are taken as:

$$\mu = \mu_2 e^{\varepsilon z}, \rho = \rho_2 e^{\varepsilon z} \quad \text{and} \quad \mu' = \mu_2' e^{\varepsilon z} \tag{3}$$

where ε is the heterogeneity for viscoelastic half-space, $\mu_2,\ \rho_2$ and μ_2' are constant values related with rigidity, density and internal friction, respectively.

3 Solution of the Upper Layer

In the absence of body force, the dynamical equation for the orthotropic medium under the influence of initial stress are given by Biot [14] as

$$\begin{aligned}
&\frac{\partial \sigma_{11}^{(1)}}{\partial x} + \frac{\partial \sigma_{12}^{(1)}}{\partial y} + \frac{\partial \sigma_{13}^{(1)}}{\partial z} - P\left(\frac{\partial w_z}{\partial y} - \frac{\partial w_y}{\partial z}\right) = \rho_1 \frac{\partial^2 u_1}{\partial t^2}\\
&\frac{\partial \sigma_{12}^{(1)}}{\partial x} + \frac{\partial \sigma_{22}^{(1)}}{\partial y} + \frac{\partial \sigma_{23}^{(1)}}{\partial z} - P\frac{\partial w_z}{\partial x} = \rho_1 \frac{\partial^2 v_1}{\partial t^2}\\
&\frac{\partial \sigma_{13}^{(1)}}{\partial x} + \frac{\partial \sigma_{23}^{(1)}}{\partial y} + \frac{\partial \sigma_{33}^{(1)}}{\partial z} - P\frac{\partial w_y}{\partial x} = \rho_1 \frac{\partial^2 w_1}{\partial t^2}
\end{aligned} \tag{4}$$

where $u_1,\ v_1,\ w_1$ are the displacement components and $w_x,\ w_y,\ w_z$ are the angular displacement along the direction $x,\ y,\ z$ respectively. $\sigma_{ij}^{(1)}(i, j = 1, 2, 3)$ are the incremental stress components.

The stress-strain relation for orthotropic medium are:

$$\begin{aligned}
\sigma_{11}^{(1)} &= B_{11}e_{11} + B_{12}e_{22} + B_{13}e_{33},\\
\sigma_{22}^{(1)} &= B_{21}e_{11} + B_{22}e_{22} + B_{23}e_{33},\\
\sigma_{33}^{(1)} &= B_{31}e_{11} + B_{32}e_{22} + B_{33}e_{33},\\
\sigma_{12}^{(1)} &= 2Q_3e_{12},
\end{aligned}$$

$$\sigma_{23}^{(1)} = 2Q_1 e_{23},$$
$$\sigma_{31}^{(1)} = 2Q_2 e_{31}, \tag{5}$$

where $B_{ij}(i, j = 1, 2, 3)$ are the incremental normal elastic coefficient and $Q_i(i = 1, 2, 3)$ are shear moduli. Here e_{ij} is the strain components, which is define by

$$e_{ij} = \frac{1}{2}\left(\frac{\partial u_i}{\partial x_j} + \frac{\partial u_j}{\partial x_i}\right), \qquad i, j = 1, 2, 3 \tag{6}$$

With the help of Eqs. (1), (5) and (6), the equation of motion (4) for the orthotropic medium becomes

$$Q_1 \frac{\partial^2 v_1}{\partial z^2} + \frac{\partial Q_1}{\partial z}\frac{\partial v_1}{\partial z} + \left(Q_3 - \frac{P}{2}\right)\frac{\partial^2 v_1}{\partial x^2} = \rho_1 \frac{\partial^2 v_1}{\partial t^2} \tag{7}$$

We may assume $v_1(x, z, t) = V_1(z)e^{i(\omega t - kx)}$, where k and c are the wave number and the common wave velocity respectively, we have

$$Q_1 \frac{d^2 V_1}{dz^2} + \frac{dQ_1}{dz}\frac{dV_1}{dz} + \left\{\rho_1\omega^2 + k^2\left(\frac{P}{2} - Q_3\right)\right\}V_1 = 0 \tag{8}$$

Using the Eq. (2) in Eq. (8), we obtain

$$\frac{d^2 V_1}{dz^2} + 2\alpha \tanh(\alpha z)\frac{dV_1}{dz} + \left\{\frac{\rho_1'\omega^2}{a_1} + k^2\left(\frac{P_1 - 2a_3}{2a_1}\right)\right\}V_1 = 0 \tag{9}$$

Substituting $V_1(z) = \frac{\phi(z)}{\cosh(\alpha z)}$ in Eq. (9) to eliminate the term $\frac{dV_1(z)}{dz}$ we obtain

$$\frac{d^2\phi(z)}{dz^2} + m^2\phi(z) = 0 \tag{10}$$

where $m^2 = \frac{\rho_1'\omega^2}{a_1} + k^2\left(\frac{P_1 - 2a_3}{2a_1}\right) - \alpha^2$.

The solution of Eq. (10) is

$$\phi(z) = F_1\cos(mz) + F_2\sin(mz) \tag{11}$$

where F_1 and F_2 are arbitrary constants.

Therefore, the displacement component of an orthotropic medium is obtained as

$$v_1(x, z, t) = \frac{F_1\cos(mz) + F_2\sin(mz)}{\cosh(\alpha z)}e^{i(\omega t - kx)}. \tag{12}$$

4 Solution of the Half-Space

The only non-vanishing governing equation without body forces for the traversal of shear waves in heterogeneous viscoelastic half-space is given by Bhattacharya [15]

$$\frac{\partial \sigma_{12}^{(2)}}{\partial x} + \frac{\partial \sigma_{23}^{(2)}}{\partial z} = \rho \frac{\partial^2 v_2}{\partial t^2} \tag{13}$$

where

$$\sigma_{12}^{(2)} = \left(\mu + \mu' \frac{\partial}{\partial t}\right) \frac{\partial v_2}{\partial x} \quad \text{and} \quad \sigma_{23}^{(2)} = \left(\mu + \mu' \frac{\partial}{\partial t}\right) \frac{\partial v_2}{\partial z} \tag{14}$$

From Eqs. (13) and (14), we have

$$\left(\mu + \mu' \frac{\partial}{\partial t}\right) \frac{\partial^2 v_2}{\partial x^2} + \frac{\partial}{\partial z}\left[\left(\mu + \mu' \frac{\partial}{\partial t}\right) \frac{\partial v_2}{\partial z}\right] = \rho \frac{\partial^2 v_2}{\partial t^2} \tag{15}$$

Assuming that

$$v_2(x, z, t) = V_2(z) e^{i(\omega t - kx)} \tag{16}$$

Substitute the Eq. (16) in Eq. (15), we get

$$\frac{d^2 V_2}{dz^2} + \frac{1}{\bar{\mu}} \frac{d\bar{\mu}}{dz} \frac{dV_2}{dz} + \left(\frac{\omega^2 \rho}{\bar{\mu}} - k^2\right) V_2 = 0 \tag{17}$$

where $\bar{\mu} = \mu + i\omega\mu'$

Now, we take the substituting $V_2(z) = \frac{\psi(z)}{\sqrt{\bar{\mu}}}$ in Eq. (17) to eliminate the term $\frac{dV_2(z)}{dz}$ we obtain

$$\frac{d^2 \psi(z)}{dz^2} - \left[-\frac{1}{4\bar{\mu}^2}\left(\frac{d\bar{\mu}}{dz}\right)^2 + \frac{1}{2\bar{\mu}} \frac{d^2 \bar{\mu}}{dz^2} - \left(\frac{\omega^2 \rho}{\bar{\mu}} - k^2\right)\right] \psi(z) = 0 \tag{18}$$

With the aid of Eqs. (3) and (18), Eq. (18) becomes

$$\frac{d^2 \psi}{dz^2} - D^2 \psi = 0 \tag{19}$$

where

$$D^2 = \frac{\varepsilon^2}{4} - \frac{\omega^2 \rho_2}{\bar{\mu}_2} + k^2 \text{ and } \bar{\mu}_2 = \mu_2 + i\omega\mu_2'$$

The solution of Eq. (19) is

$$\psi(z) = F_3 e^{Dz} + F_4 e^{-Dz} \tag{20}$$

where F_3 and F_4 are arbitrary constants.

Hence, the required displacement component for the viscoelastic half-space will be of the form

$$v_2(x, z, t) = F_4 \frac{1}{\sqrt{\bar{\mu}_2}} e^{-\left(\frac{\varepsilon}{2}+D\right)} e^{i(\omega t - kx)}. \tag{21}$$

5 Boundary Conditions

(i) At $z = -H$, the upper layer is free from external load, so the stress component $\sigma_{23}^{(1)} = 0$.
(ii) At $z = 0$,

  (a) The components of stress of both the orthotropic layer and viscoelastic half-space are continuous, i.e. $\sigma_{23}^{(1)} = \sigma_{23}^{(2)}$.
  (b) The displacement components of both the orthotropic layer and viscoelastic half-space are also continuous, i.e. $v_1 = v_2 o$.

Using Eqs. (12) and (21) in the above boundary conditions, we have

$$\begin{aligned} &[a_1 m \cosh(H\alpha)\sin(Hm) + a_1\alpha \sinh(H\alpha)\cos(Hm)]F_1 \\ &\quad + [a_1 m \cosh(H\alpha)\cos(Hm) - a_1\alpha\sinh(H\alpha)\sin(Hm)]F_2 = 0 \end{aligned} \tag{22}$$

$$m a_1 F_2 + \left(\frac{\varepsilon}{2} + D\right)F_4 = 0 \tag{23}$$

$$\sqrt{\bar{\mu}_2} F_1 - F_4 = 0 \tag{24}$$

Eliminating the arbitrary constants $F_1,\ F_2$ and F_4 by the Eqs. (22), (23) and (24), we get the frequency equation of shear waves as

$$\Upsilon(k_1, c) = \tan[Hm] - \frac{\mu_2\left(1 + iQ^{-1}\right)m(2D + \varepsilon) - 2\alpha m a_1 \tanh[H\alpha]}{2m^2 a_1 + \mu_2\left(1 + iQ^{-1}\right)\alpha(2D + \varepsilon)\tanh[H\alpha]} = 0 \tag{25}$$

where $Q^{-1} = \omega\frac{\mu_2'}{\mu_2}$ is dissipation function.

Equation (25) is a closed form of complex wave velocity equation of shear wave propagation in considered layered structure. In fact, the wave number is also complex and thus may be written as

$$k = k_1 + ik_2 = k_1(1 + i\delta) \tag{26}$$

where k_1 and k_2 are real and $\delta = k_2/k_1$ is attenuation coefficient.

Therefore, the Eq. (25) is separated into real and imaginary terms as

$$\Re[\Upsilon(k_1, c)] = 0 \tag{27}$$

$$\Im[\Upsilon(k_1, c)] = 0 \tag{28}$$

Equations (27) and (28) are the dispersion and absorption relations of shear waves, respectively.

6 Special Case

When $a_1 \rightarrow a_3 \rightarrow \mu_1$ (rigidity), $\alpha \rightarrow 0$ (heterogeneity associated with upper layer), $\varepsilon \rightarrow 0$ (heterogeneity associated with half-space), $P_1 \rightarrow 0$ (initial stress) and $\mu_2' \rightarrow 0$ (viscosity), then Eq. (28) vanishes and Eq. (27) reduces to

$$\tan\left[k_1 H\sqrt{\frac{c^2}{c_1^2} - 1}\right] = \frac{\mu_2\sqrt{1 - \frac{c^2}{c_2^2}}}{\mu_1\sqrt{\frac{c^2}{c_1^2} - 1}} \tag{29}$$

where $c_1 = \sqrt{\frac{\mu_1}{\rho_1'}}$ and $c_2 = \sqrt{\frac{\mu_2}{\rho_2}}$.

Equation (29) expresses the classical relation of Love wave, which coincides the results obtained by Love [2].

7 Numerical Results and Discussions

Numerical results are provided to show the effect of heterogeneities ($\alpha/k_1 \rightarrow$ upper layer, $\varepsilon/k_1 \rightarrow$ half-space), initial stress ($P_1/2a_1$) and attenuation coefficient (δ) on the traversal characteristics of shear wave in an orthotropic medium under initial stress overlying viscoelastic half-space. For the computational propose, we take the following relevant data:

(1) For the upper orthotropic medium [3]:

$$Q_1 = 5.82 \times 10^{10}\ \text{N/m}^2$$
$$Q_3 = 3.99 \times 10^{10}\ \text{N/m}^2$$
$$\rho_1 = 4.5 \times 10^3\ \text{kg/m}^3$$

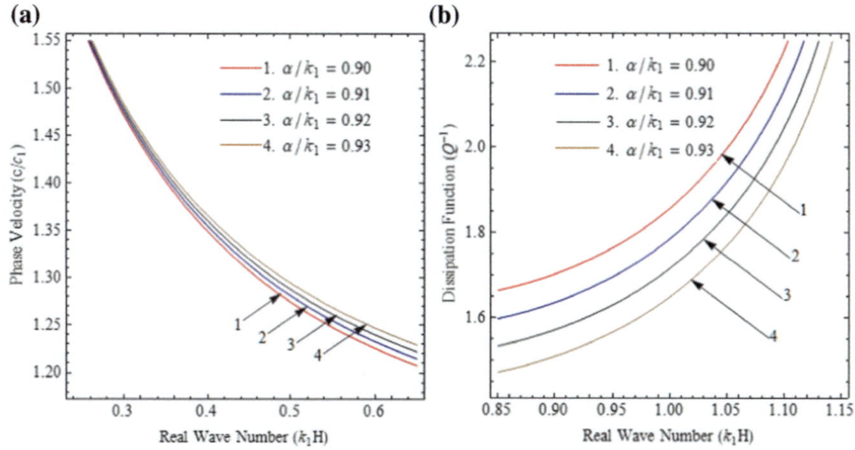

Fig. 2 Variations of **a** c/c_1 and **b** Q^{-1} against k_1H for different values of heterogeneity parameter α/k_1, when $\varepsilon/k_1 = 0.93$, $P_1/2a_1 = 0.15$ and $\delta = 0.75$

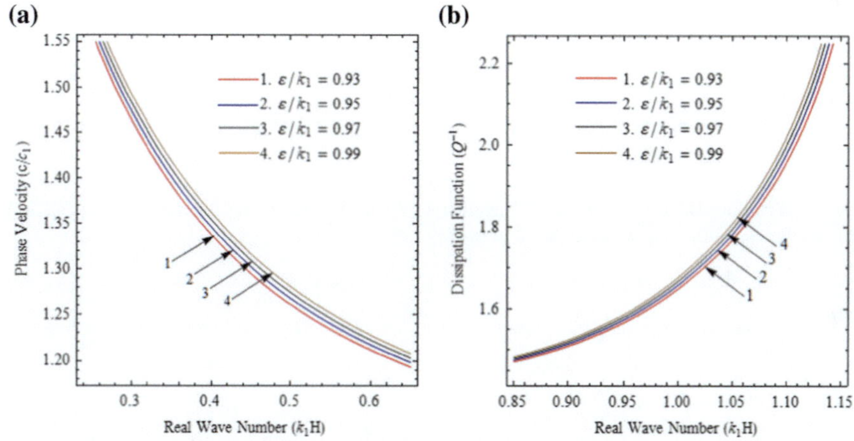

Fig. 3 Variations of **a** c/c_1 and **b** Q^{-1} against k_1H for different values of heterogeneity parameter ε/k_1, when $\alpha/k_1 = 0.90$, $P_1/2a_1 = 0.15$ and $\delta = 0.75$

(2) For the lower viscoelastic half-space [16]:

$$\mu_2 = 4.34 \times 10^{10}\,\text{N/m}^2$$
$$\rho_2 = 2217\,\text{kg/m}^3$$
$$\frac{\mu_2}{\mu_2'} = 60\,\text{s}^{-1}$$

The influence of all these parameters have been demonstrated in Figs. 2, 3, 4 and 5. All the figures have been plotted with vertical axis as dimensionless phase velocity c/c_1 and dissipation function Q^{-1} against horizontal axis as dimensionless real wave number $k_1 H$.

Figure 2 presents the influence of heterogeneity parameter (α/k_1) associated with the layer on the phase velocity and dissipation function of shear waves. In Figs. 2a, b, the increasing values of α/k_1 have been taken as 0.90, 0.91, 0.92 and 0.93 for curves 1, 2, 3 and 4 respectively. From these figures, it has been found that the heterogeneity

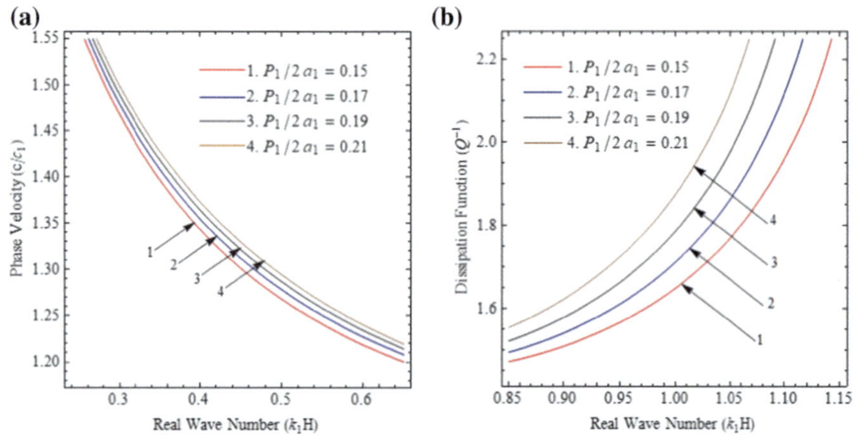

Fig. 4 Variations of **a** c/c_1 and **b** Q^{-1} against $k_1 H$ for different values of initial stress $P_1/2a_1$, when $\alpha/k_1 = 0.90$, $\varepsilon/k_1 = 0.93$ and $\delta = 0.75$

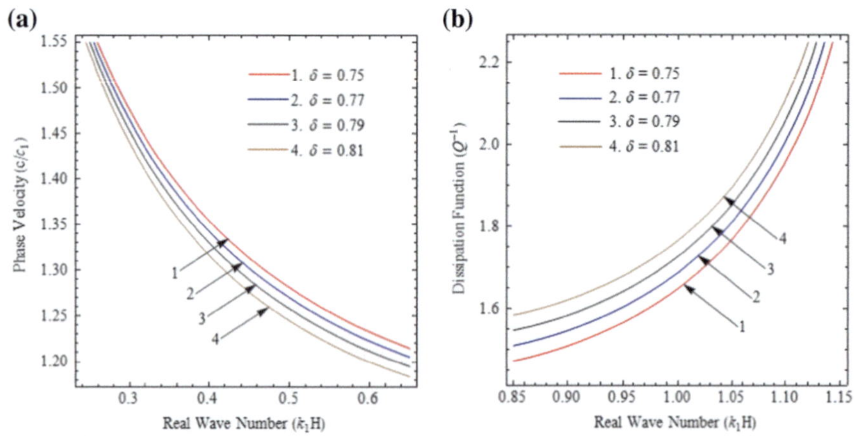

Fig. 5 Variations of **a** c/c_1 and **b** Q^{-1} against $k_1 H$ for different values of attenuation coefficient δ, when $\alpha/k_1 = 0.90$, $\varepsilon/k_1 = 0.93$ and $P_1/2a_1 = 0.15$

has a proportional influence on the increasing value of the phase velocity, but an inverse effect on the increasing value of the dissipation function.

The effect of heterogeneity parameter ε / k_1 along with the half-space on the phase velocity and dissipation function of propagation of shear wave are discussed in Fig. 3. The increasing values of ε / k_1 for curves 1, 2, 3 and 4 have been taken as 0.93, 0.95, 0.97 and 0.99 respectively in Figs. 3a, b. It can be seen from these figures that due to the increase in heterogeneity parameter, both the phase velocity and dissipation function are amplified.

Figures 4a, b can be mentioned as the impact of initial stress on the phase velocity and dissipation function for the propagation of shear waves. For curves 1, 2, 3 and 4, the values of $P_1/2a_1$ have been taken as 0.15, 0.17, 0.19 and 0.21 respectively. From this figure, it is clear that if the initial stress increases, the dimensionless phase velocity and dissipation function also get increased. It can be also seen that dissipation function is slightly affected at the lower frequency region compared with the higher frequency region.

In Figs. 5a, b, curves have been plotted for dimensionless phase velocity and dissipation function with respect to dimensionless real wave number by varying attenuation coefficient δ as 0.75, 0.77, 0.79 and 0.81 for curves 1, 2, 3 and 4 respectively. With the growing value of both wave number and attenuation coefficient, the phase velocity decreases but the dissipation function increases.

8 Conclusions

In this paper, we have studied the impacts of heterogeneity parameters, initial stress, and attenuation coefficient on the phase velocity and dissipation function against real wave number for the propagation of shear waves in a pre-stressed heterogeneous orthotropic layer lying over a heterogeneous viscoelastic half-space. The dispersion relation and absorption relation both are enumerated by using the technique of separation of variables, which coincides with the classical results of Love waves when the heterogeneity parameters, initial stress and viscoelastic parameter are neglected. Following significant outcomes can be drawn:

- The phase velocity of shear waves decreases when the real wave number increases, but the dissipation function also increases.
- The heterogeneity parameters (included in the layer and half-space) and initial stress exhibit the proportional impact on the phase velocity, but the increment in attenuation coefficient decreases the phase velocity of shear waves.
- The dissipation function increases with an increment in heterogeneity associated with the half-space, initial stress and attenuation coefficient, while it decreases with an increase in the heterogeneity associated with the layer.

Numerical computations of analytical results of the undertaken work may be useful to understand the nature of the propagation of shear waves in the introduced model. Moreover, the outcomes of the present study may find their potential application in the field of geology, geophysics, including civil engineering and construction sectors, and could be of interest to the scientific community.

Acknowledgements The authors would like to convey their deep sense of indebtedness to Indian Institute of Technology (Indian School of Mines) Dhanbad, India for providing all necessary resources to Mr. Raju Kumhar for this research work. The fertility of thoughts and the phenomena explained in the present research work can be expected to lead to extremely productive interactions across disciplines.

References

1. Ewing, W., Jardetzky, W., Press, F.: Elastic Waves in Layered Media. McGraw-Hill, New York (1957)
2. Love, A.E.H.: Mathematical Theory of Elasticity. Cambridge University Press, Cambridge (1920)
3. Gubbins, D.: Seismology and Plate Tectonics. Cambridge University Press, Cambridge (1990)
4. Birch, F.: Elasticity and constitution of the Earth's interior. J. Geophys. Res. **57**(2), 227–286 (1952)
5. Maity, M., Kundu, S.: Love wave propagation in a heterogeneous orthotropic layer under initial stress lying over an inhomogeneous half-space. In: AIP Conference Proceedings, vol. 1802, no. 1, p. 020009 (2017)
6. Abd-Alla, A.M., Ahmed, S.M.: Propagation of Love waves in a non-homogeneous orthotropic elastic layer under initial stress overlying semi-infinite medium. Appl. Math. Comput. **106**(2–3), 265–275 (1999)
7. Wang, C.D., Chou, H.T., Peng, D.H.: Love-wave propagation in an inhomogeneous orthotropic medium obeying the exponential and generalized power law models. Int. J. Geomech. **17**(7), 04017003 (2017)
8. Manna, S., Kundu, S., Gupta, S.: Love wave propagation in a piezoelectric layer overlying in an inhomogeneous elastic half-space. J. Vib. Control **21**(13), 2553–2568 (2015)
9. Saha, S., Chattopadhyay, A., Singh, A.K.: Impact of inhomogeneity on SH-type wave propagation in an initially stressed composite structure. Acta Geophys. **66**(1), 1–19 (2018)
10. Craster, R.V., Joseph, L.M., Kaplunov, J.: Long-wave asymptotic theories: the connection between functionally graded waveguides and periodic media. Wave Motion **51**(4), 581–588 (2014)
11. Ahmad, M., Nolde, E., Pichugin, A.V.: Explicit asymptotic modelling of transient Love waves propagated along a thin coating. Z. Angew. Math. Phys. **62**(1), 173–181 (2011)
12. Carcione, J.M.: Wave propagation in anisotropic linear viscoelastic media: theory and simulated wavefields. Geophys. J. Int. **101**(3), 739–750 (1990)
13. Kakar, R., Kakar, S.: Love wave in a Voigt-type viscoelastic heterogeneous layer overlying heterogeneous viscoelastic half-space. Int. J. Geomech. **17**(1), 06016009 (2016)
14. Biot, M.: Mechanics of Incremental Deformations. Wiley, New York (1965)
15. Bhattacharya, J.: The possibility of the propagation of Love type waves in an intermediate heterogeneous layer lying between two semi-infinite isotropic homogeneous elastic layers. Pure Appl. Geophys. **72**(1), 61–71 (1969)
16. Vashishth, A.K., Sharma, M.D.: Propagation of plane waves in poroviscoelastic anisotropic media. Appl. Math. Mech. **29**(9), 1141 (2008)

Haar Wavelet: History and Its Applications

Mahendra Kumar Jena and Kshama Sagar Sahu

Abstract In this paper, we have given a brief history of the Haar wavelet. Later the operational matrix which is obtained from Haar wavelet is used to find the numerical solutions of some differential equations. The solutions thus obtained from operational matrix method are compared with exact solution as well as solution from Runge-Kutta method and Modified Euler's method is presented.

Keywords Haar wavelet · Operational matrix · Initial value problem

AMS Classification 65L05 · 65L07

1 Introduction

Mathematical modeling of real life problems is bridge between the nature and us. Most of the models in science, engineering and economics are associated with the differential equations which may be ordinary differential equation (ODE) or partial differential equation (PDE). So it is challenging for us to formulate their mathematical model and obtain their solutions [4]. It is not always possible to find the exact solution where we take the help of numerical methods. Some of the methods are Euler's method, modified Euler's method, Runge-Kutta method and many more. Solving differential equation with better accuracy is appreciable.

One of the recent method in computational field is Haar wavelet operational matrix method (HWOM) [1, 6]. In [6], Jena and Sahu describe the method to solve the linear initial value problems (IVP) of first and second order. They also provide the method for first order nonlinear IVP.

This paper is divided into 5 sections. Section 1 is the introduction. History and definition of Haar wavelet is given in Sect. 2. HWOM method is also described in

M. K. Jena · K. S. Sahu (✉)
Department of Mathematics, Veer Surendra Sai University of Technology,
Burla, Sambalpur, Odisha 768018, India
e-mail: sahu.kshama@yahoo.in

S. Manna et al. (eds.), *Mathematical Modelling and Scientific Computing with Applications*, Springer Proceedings in Mathematics & Statistics 308,
https://doi.org/10.1007/978-981-15-1338-1_12

this section. Application of Haar wavelet to solve differential equation is given in Sect. 3. Some numerical examples are illustrated in Sect. 4. Conclusion and future work is provided in Sect. 5.

2 Haar Wavelet

There are different types of wavelet like Meyer wavelet, Morlet wavelet, Daubechies wavelet and Haar wavelet. Among them Haar wavelet is the oldest and simplest one. First, we present a historical background and then the definition is presented. Later operational matrix from Haar wavelet is constructed.

2.1 *Brief History*

The trigonometric functions $\sin x$ and $\cos x$ are of the period 2π. According to Fourier any periodic function of period 2π can be expressed as sum of *sine* and *cosine* functions. This series is called Fourier series and its theory is called frequency analysis [5]. This analysis came in 1807 and with this a new concept has been introduced in mathematics. Later, Fourier gave an idea about convergence of the Fourier series, orthogonal systems and a mathematical structure that vary with scale. High scale(low frequency) provides to a non-details global view of the signal and low scale(high frequency) corresponds to the detail information of hidden pattern of the signal.

The term wavelet was first found in 1909. It was in the appendix of the thesis of Alfred Haar. Unfortunately, it was unknown to other researchers. But they search of such transform which will provide local as well as global information of the signal. In 1981, same wavelet concept was proposed by geophysicist Jean Morlet. Again in 1984, he with the physicist Alex Grossmann invented the term wavelet. Till that, Haar wavelet was the only wavelet which has orthogonal property. However, Yves Meyer in 1985 constructed the second orthogonal wavelet which is called Meyer wavelet. In 1988, Stephans Mallat and Meyer introduced the concept of multiresolution analysis(MRA) [2]. At the same time Ingrid Daubechies explained the method to construct the compactly supported orthogonal wavelet [3]. Summary of the history of wavelet is listed below:

- 1909-Family of Haar wavelet
- 1981-Morlet Wavelet (Fig. 1)
- 1984-Morlet and Grossmann - wavelet
- 1985-Meyer orthogonal wavelet (Fig. 2)
- 1988-Mallat and Meyer- MRA
- 1988-Daubechies - Compactly supported orthogonal wavelet
- 1989-Mallat - Fast wavelet transform
- 1997-Chen and Hsiao - HWOM method

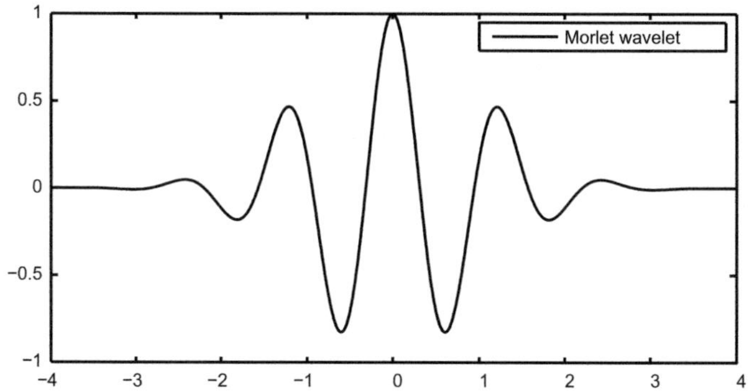

Fig. 1 Morlet wavelet

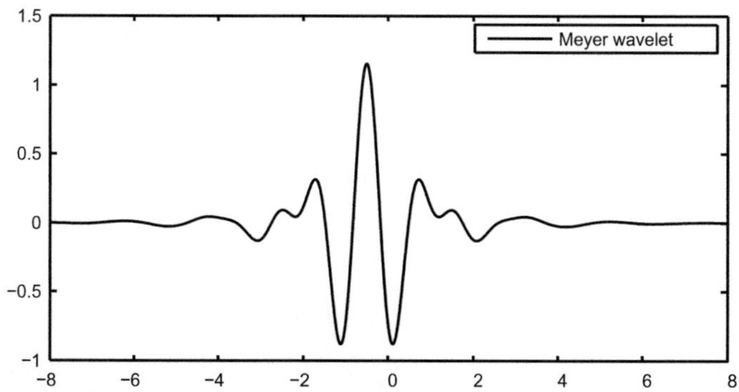

Fig. 2 Meyer wavelet

2.2 *Definition of Haar Wavelet*

Let $m = 2^j$, $j = 0, 1, \ldots, J$, $k = 0, 1, \ldots, m-1$ and $i = m + k + 1$. Here, i denotes wavelet number, j is wavelet level and k is the translation parameter. The maximum resolution is J. For the minimum $j = 0$ and $k = 0$, we have the minimum value of $i = 2$ and maximum value of $i = 2M = 2^{J+1}$. The Haar wavelet family for $t \in [0, 1]$ is defined as [6–8]

$$h_i(t) = \begin{cases} 1 & t \in \left[\frac{k}{m}, \frac{k+0.5}{m}\right] \\ -1 & t \in \left[\frac{k+0.5}{m}, \frac{k+1}{m}\right] \\ 0 & \text{otherwise} \end{cases}$$

For $i = 1$, $h_i = 1$ is the scaling function. The Haar wavelet are orthogonal to each other:

$$\int_0^1 h_i(t)\, h_l(t) = \begin{cases} 2^{-j}, & i = l \\ 0, & i \neq l. \end{cases} \tag{2.1}$$

Due to this orthogonal property these wavelets form a good basis. In particular, any function $y(t)$ which is square integrable in the interval [0, 1] can be expressed as infinite sum of Haar wavelet

$$y(t) = \sum_{i=1}^{2M} a_i h_i(t),$$

where $a_i = 2^j \int_0^1 y(t) h_i(t)\, dt$.

2.3 Operational Matrix

In 1997, Chen and Hsiao [1] described the integration of Haar wavelets which produces matrices called operational matrices. Let us define Haar wavelet matrix of order $2M \times 2M$ by $H_{2M\times 2M} = (h_i(t_l))_{i=1,l=1}^{2M\times 2M}$. Let us define matrices PH and QH by

$$(PH)_{il} = \int_0^{t_l} h_i(t)\, dt \tag{2.2}$$

and

$$(QH)_{il} = \int_0^{t_l} dt \int_0^{t} h_i(t)\, dt, \tag{2.3}$$

where the collocation points $t_l = (l - 0.5)/(2M)$. Then the $2M \times 2M$ matrices P and Q are called operational matrices. Taking $2M = 2$ and $2M = 4$, we have

$$H_{2\times 2} = \begin{bmatrix} 1 & 1 \\ 1 & -1 \end{bmatrix}, (PH)_{2\times 2} = \frac{1}{4}\begin{bmatrix} 1 & 3 \\ 1 & 1 \end{bmatrix}, P_{2\times 2} = \frac{1}{4}\begin{bmatrix} 2 & -1 \\ 1 & 0 \end{bmatrix},$$

$$H_{4\times 4} = \begin{bmatrix} 1 & 1 & 1 & 1 \\ 1 & 1 & -1 & -1 \\ 1 & -1 & 0 & 0 \\ 0 & 0 & 1 & -1 \end{bmatrix}, (PH)_{4\times 4} = \frac{1}{8}\begin{bmatrix} 1 & 3 & 5 & 7 \\ 1 & 3 & 3 & 1 \\ 1 & 1 & 0 & 0 \\ 0 & 0 & 1 & 1 \end{bmatrix}, P_{4\times 4} = \frac{1}{16}\begin{bmatrix} 8 & -4 & -2 & -2 \\ 4 & 0 & -2 & 2 \\ 1 & 1 & 0 & 0 \\ 1 & -1 & 0 & 0 \end{bmatrix}.$$

Similarly,

$$(QH)_{2\times 2} = \frac{1}{32}\begin{bmatrix} 1 & 9 \\ 1 & 15 \end{bmatrix}, Q_{2\times 2} = \frac{1}{32}\begin{bmatrix} 5 & -4 \\ 8 & -7 \end{bmatrix}$$

$$(QH)_{4\times4} = \frac{1}{128}\begin{bmatrix} 1 & 9 & 25 & 49 \\ 1 & 9 & 23 & 31 \\ 1 & 7 & 8 & 8 \\ 0 & 0 & 1 & 7 \end{bmatrix}, \quad (Q)_{4\times4} = \frac{1}{128}\begin{bmatrix} 21 & -16 & -4 & -12 \\ 16 & -11 & -4 & -4 \\ 6 & -2 & -3 & 0 \\ 2 & -2 & 0 & -3 \end{bmatrix}.$$

Chen and Hsiao [1] derived the following formula

$$P_{2m\times2m} = \frac{1}{4m}\begin{pmatrix} 4mP_{m\times m} & -H_{m\times m} \\ H_{m\times m}^{-1} & O \end{pmatrix}.$$

3 Applications

Haar wavelet can be localized in both time and scale. Due to this property it is being applied in many fields of engineering like digital image processing, image compression, decomposition of image. But here operational matrices are used to solve differential equations. Method to solve first order initial value problem (IVP) [6] is described below.

Consider the first order linear ordinary differential equation

$$\frac{dU}{dt} = [a(t)U + b(t)],\ t\epsilon\,[0, T],\ U(0) = U_0. \tag{3.1}$$

Let us divide the whole interval $[0, T]$ into n segments such that $t_{i+1} = t_i + d_i$. In terms of local coordinate $\tau = \frac{t-t_i}{d_i}$, we have $u(\tau) = U(t)$ and in the interval $[t_i, t_{i+1}]$, (3.1) becomes

$$\frac{d\mathbf{u}}{d\tau} = d_i\,[\mathbf{u}A(\tau) + B(\tau)] \tag{3.2}$$

where

$$\mathbf{u}(\tau) = [u(\tau_1), u(\tau_2), \ldots, u(\tau_{2M})]$$

is a row vector and collocation points in the interval [0, 1] are given by

$$\tau_j = \frac{\left(j - \frac{1}{2}\right)}{2M}, \quad j = 1, 2, \ldots, 2M.$$

$$A(\tau) = \begin{bmatrix} a(\tau_1 d_i + t_i) & 0 & 0 \cdots & 0 \\ 0 & a(\tau_2 d_i + t_i) & 0 \cdots & 0 \\ \vdots & \vdots & \vdots \cdots & \vdots \\ 0 & 0 & 0 \cdots & a(\tau_{2M} d_i + t_i) \end{bmatrix}$$

$$B(\tau) = (b(\tau_1 d_i + t_i), b(\tau_2 d_i + t_i), \ldots, b(\tau_{2M} d_i + t_i)).$$

Following [1], we take

$$\frac{d\mathbf{u}}{d\tau} = cH_{2M}^{(0)} \tag{3.3}$$

where $c = [c(1), c(2), \ldots, c(2M)]$.

Following [6], we have

$$c = d_i U_i Y S^{-1} + d_i B A^{-1} \left(H_{2M}^{(0)}\right)^{-1} S^{-1}$$

where

$$S = \left(H_{2M}^{(0)} A^{-1} \left(H_{2M}^{(0)}\right)^{-1} - d_i P_{2M}^{(1)}\right),$$

$$Y = E\left(H_{2M}^{(0)}\right)^{-1}, \quad E = [1, 1, 1, \ldots, 1]_{1\times 2M}$$

4 Numerical Examples

In this section some model problems are solved in HWOM method. The solution is also compared with the solution which is obtained from Runge-Kutta method, Modified Euler method and exact solution of the model problem.
Notation:

U_H: Haar approximate solution
U_R: Runge-Kutta solution
U_M: Modified Euler's solution
U_e: Exact solution

Example 4.1 Consider a first order linear IVP

$$U' = U, \quad U(0) = 1. \tag{4.1}$$

Exact solution for this model problem is $U(t) = e^t$. According to HWOM method $A = I$ and $B = 0$, where I is the identity matrix. Hence $S = I - d_i P_{2M}^{(1)}$ and $c = d_i U_i Y \left(I - d_i P_{2M}^{(1)}\right)^{-1}$. Comparison of error for Runge-Kutta method, Modified Euler method and HWOM method in terms of L_2 norm for this problem is presented in Table 1. The comparison of Haar approximate solution U_H, Runge-Kutta Solution U_R, Modified Euler solution U_M and exact solution U_e in graph is given in Fig. 3, where as the error plot for HWOM method is given in Fig. 4.

Example 4.2 Let us consider another model problem

$$U' = t + U, \quad U(0) = 1. \tag{4.2}$$

Table 1 L_2 Norm of the error $\left(U' = U, U(0) = 1\right.$, Number of points = np.$\left.\right)$

$h = 0.01$	R-K method	Modified Euler method	HWOM method
np = 100	1.0324×10^{-9}	2.0665×10^{-4}	6.6586×10^{-6}
np = 200	6.7600×10^{-9}	1.4000×10^{-3}	4.3294×10^{-5}
np = 300	2.9714×10^{-8}	5.9000×10^{-3}	1.8988×10^{-4}
np = 400	1.1208×10^{-7}	2.2400×10^{-2}	7.1546×10^{-4}
np = 500	3.9028×10^{-7}	7.8100×10^{-2}	2.5000×10^{-3}

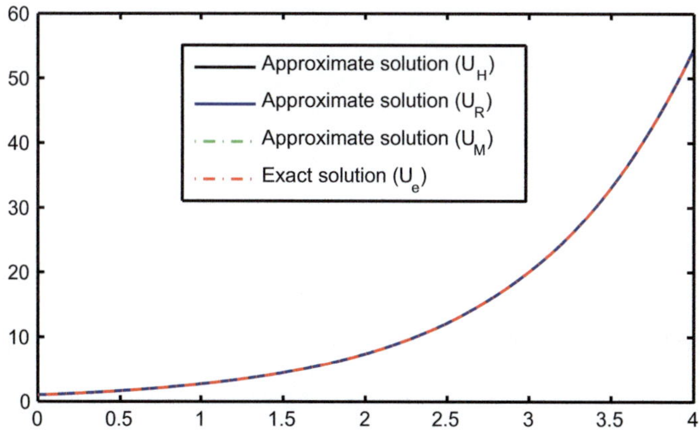

Fig. 3 Plot of U_e, U_H, U_R and U_M for Example 4.1

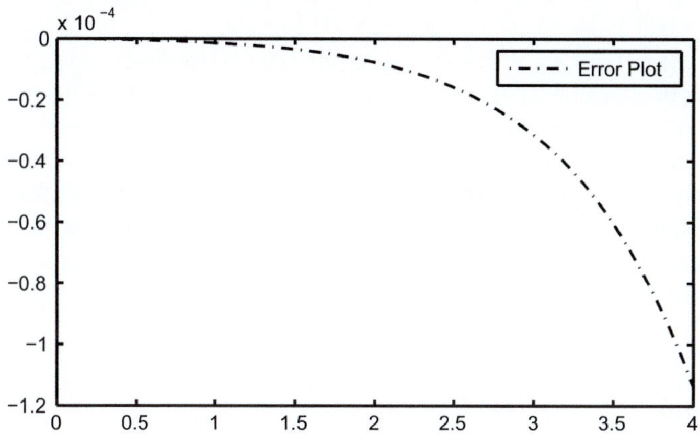

Fig. 4 Error plot for solution of $U^{'}(t) = U, U(0) = 1$

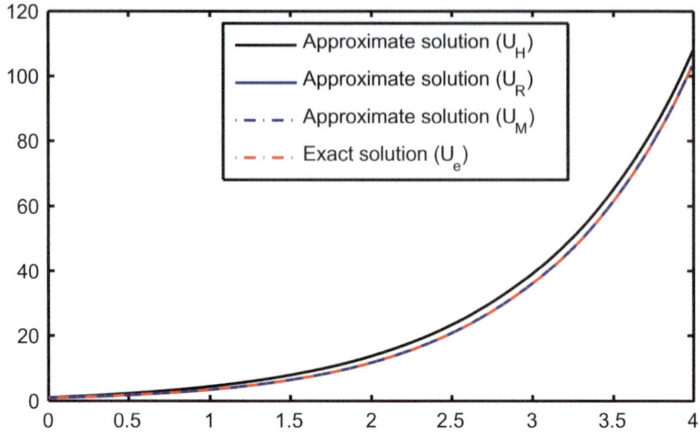

Fig. 5 Plot of U_e, U_H U_R and U_M for Example 4.2

and $U(t) = -t - 1 + 2e^t$ is the exact solution. Applying HWOM method we have $A = I$ and $B = [1, 1, 1, 1]$. So $S = I - d_i P_{2M}^{(1)}$ and $c = \left(d_i U_i Y + d_i B \left(H_{2M}^{0}\right)^{-1}\right) \left(I - d_i P_{2M}^{(1)}\right)^{-1}$. The comparison of solutions U_H, U_R, U_M and U_e is presented graphically in Fig. 5.

5 Conclusion and Future Work

In this paper, brief history of wavelets are given. HWOM method is briefly illustrated. Some numerical examples are provided to show efficiency of HWOM method. We know that HWOM method is of second order method [6]. Solution from this method provides good approximation although it is not better than Runge-Kutta method. Question arises "Is there any other wavelet like tools which yields operational matrices?" If it is, then "Is it possible to get order of approximation better than HWOM or Runge-Kutta method?" Answers to these questions may lead to some new operational matrix method.

References

1. Chen, C., Hsiao, C.H.: Haar wavelet method for solving lumped and distributed parameter system. IEE Proc. Control Theory Appl. **144**, 87–94 (1997)
2. Chui, C.K.: An Introduction to Wavelets. Academic Press, San Diego, CA (1992)
3. Daubechies, I.: Ten Lectures on Wavelet. SIAM Philadelphia (1992)
4. Davis, M.E.: Numerical Methods and Modeling for Chemical Engineers, pp. 54–63. Dover Publication, New York (2013)

5. Graps, A.: An Introduction to wavelet. IEEE Comput. Sci. Eng. **2**, 50–61 (1995)
6. Jena, M.K., Sahu, K.S.: Haar wavelet operational matrix method to solve initial value problems: a short survey. Int. J. Appl. Comput. Math. **3**, 3961–3975 (2017)
7. Lepik, U.: Numerical solution of differential equations using Haar wavelets. Math. Comput. Simul. **68**, 127–143 (2005)
8. Lepik, U., Hein, H.: Haar Wavelets With Applications. Springer, Berlin (2014)

Active Control with the Method of Receptances: Recent Progresses and Its Application in Active Aeroelastic Control

Kumar Vikram Singh

Abstract The design of active control for large and complex engineering structures requires accurate modeling and prediction of their dynamic response and instabilities. The performance of traditional model based control may be limited due to the errors in model approximation, size of the problem and/or availability of limited data for realizing active control. To overcome some of these challenges the method of receptances is developed. This method allows design and computation of controller gains based on a modest size of receptance matrices which can be extracted from transfer functions associated with available sensors and actuators. The area of active aeroelastic control deals with developing wing technology for the next generation aircrafts to achieve increased performance by controlling and manipulating the aeroelastic response by active means. In these applications, receptance based controller design is found to be promising as it eliminates the modeling of complex aeroelastic interaction between elastic structure and surrounding aerodynamics. In this paper, fundamental of the receptance based control is introduced and recent progress in this area is summarized. The effectiveness of the controller designed with on-board sensors (embedded) and actuators (control surfaces) for suppressing the flutter instabilities and flutter boundary extension is demonstrated with numerical examples. The performance of the controller such as its ability to control prescribed modes of interest without influencing the other is also presented. Ongoing research in this area is briefly summarized in this paper.

Keywords Active control · Aeroelastic · Frequency domain · Receptance transfer function

K. V. Singh (✉)
Mechanical & Manufacturing Engineering Department, Miami University,
Oxford, OH 45056, USA
e-mail: singhkv@miamioh.edu

S. Manna et al. (eds.), *Mathematical Modelling and Scientific Computing with Applications*, Springer Proceedings in Mathematics & Statistics 308,
https://doi.org/10.1007/978-981-15-1338-1_13

1 Introduction

Flexible structures in various engineering applications are susceptible to vibration due to the dynamic loading environment which needs to be minimized or eliminated to avoid catastrophic failure, early fatigue damage and reduced service life. These structural vibrations and related instabilities can be controlled by passive devices/materials or by applying feedback controlling force. Passive devices such as dynamic absorbers are widely used in several practical engineering applications [1, 2], because of their simplicity, low cost, and easiness to implement. However, they are limited by the operating range of frequencies and weight penalties. Their performance cannot be altered due to any change in operating conditions. Active control has a potential to overcome the disadvantages encountered with existing passive devices by supplying the desirable controlling forces in a dynamic environment (real time). The active vibration control (AVC) is implemented by means of sensors, actuators, and computers. The sensors detect vibration levels continuously, the computer then computes the appropriate amount of feedback control force needed to control vibrations at some given levels, and the calculated feedback gains are then applied to the structures by means of actuators. Recent advances in hardware and sensor technologies have allowed implementation of AVC in various engineering applications. These include integrated traction control, wheel-slip control and active suspension control [3] in automobile engineering, active and semi-active tendon control of building vibrations and cable stayed bridge vibration in civil engineering structures [4–6], active vibration isolation of helicopters in aerospace applications [7–9]. Developments of smart sensors and actuators (piezoelectric, magneto-restrictive, magneto-rheological fluids etc.) to achieve active and/or semi active vibration control in a range of engineering applications are widely available in literature [10–15].

Active control strategy requires efficient computation of feedback control gains in real time which can be computed either from the knowledge of the model (based on energy method, finite element and finite difference method etc.) or by using some time and/or frequency domain measurements. In general, the dynamic characteristic of the structure is defined by its eigenvalues (resonant frequencies or poles), and corresponding eigenvectors. The location of these eigenvalues or poles in the complex plane defines the response characteristics of the structure. One of the control objectives is to properly assign and place poles such that vibration suppression and desired dynamic response of the system can be achieved. By pole placement, the transient motion of the system can be manipulated, the damping of the system can be enhanced and natural frequencies of the system can be kept away from resonance with the time varying applied harmonic loads. Additionally, by using active control zero assignments or nodal control to suppress the vibration at chosen frequencies and locations on the structure can be achieved [16]. By both eigenvalues and eigenvectors (natural frequencies and mode shapes) active eigenstructure assignment problems can be formulated and solved for the precise shaping of the vibratory response of the system [17, 18].

In various engineering applications (buildings, structures, automobile etc.), the model size n is large and existing techniques for pole placement be applied to the reduced-order model through a first-order state-space realization [19, 20]. The transformation of a natural second-order model of a dynamical system to a first-order realization destroys some of the nice exploitable structures, such as the symmetry, positive definiteness, connectivity, etc., which are assets in a computational setting. A first-order realization might also require inversion of ill-conditioned matrix bringing numerical stability in computations. Moreover, in practice number of poles which can be measured and required by to be changed is comparatively small $p \ll 2n$. In such cases, the use of Ackermann's formula is impractical because for implementing the classical pole placement it is required to have a complete set of closed loop poles. Assigning a random (stable poles) $\{\mu_j\}_{j=1}^{2n}$ set may lead to a larger control force than really needed. Moreover, when the poles are moved to a desired location of designer's interest by the application of active controlling force, the energy is being transferred to the system including the higher modes. This may lead some of the poles corresponding to the higher modes move towards instability. Such a phenomenon in control is known as "spillover" [21]. These problems may become critical for large dimensional systems, which are characterized by many modes and associated large number of poles governing the dynamics of the system.

2 Traditional Modeling for Active Vibration Control

In general, vibrating structures are approximated by discrete models and their open loop dynamics is governed by the following set of differential equations,

$$\mathbf{M}\ddot{\mathbf{q}}(t) + \mathbf{C}\dot{\mathbf{q}}(t) + \mathbf{K}\mathbf{q} = \mathbf{f}(t), \tag{1}$$

where the dots denote derivatives with respect to time, $\mathbf{M}, \mathbf{C}, \mathbf{K} \in \Re^{n\times n}$, respectively, are the system matrices known as the mass (inertia), damping and stiffness matrices respectively, $\mathbf{q}(t), \dot{\mathbf{q}}(t), \ddot{\mathbf{q}}(t) \in \Re^{n\times 1}$, are the displacement, velocity and acceleration degrees of freedom vectors respectively, and $\mathbf{f}(t) \in \Re^{m\times 1}$ is the external excitation vector for a general multi-input case with $m \leq n$. The natural response of the system (1) due to external excitation and in the absence of controlling force is known as the *open loop response* of the system.

Active Vibration Control (AVC) strategy combines multiple sensors (to sense the vibration level continuously), computers (to compute the controlling force in real time) and actuators (to supply the controlling force) for suppressing the vibration corresponding wide range of dynamic operating conditions as shown in Fig. 1.

The feedback control laws to achieve AVC deals computing and supplying (feedback) a controlling force based upon the response (state) measurement such that the *closed loop response* of the system will have desired behavior, as defined by control designers or engineers. The governing closed loop system associated with (1) is

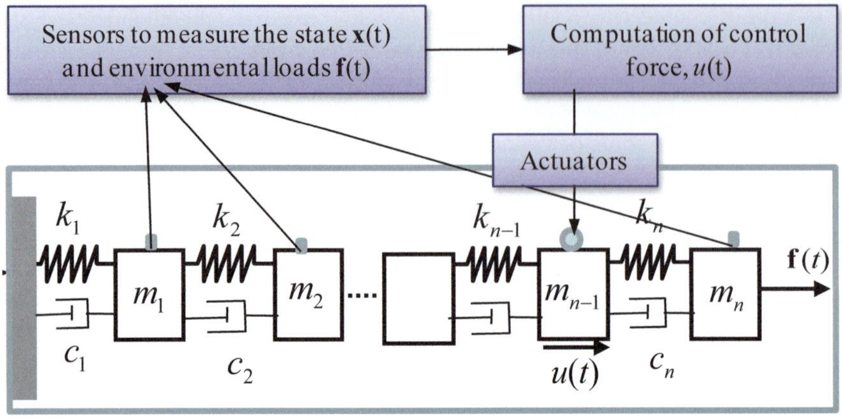

Fig. 1 Schematic of active vibration control strategy

expressed as,

$$\mathbf{M}\ddot{\mathbf{q}}(t) + \mathbf{C}\dot{\mathbf{q}}(t) + \mathbf{K}\mathbf{q}(t) = \mathbf{B}_c\mathbf{u}(t) + \mathbf{f}(t), \tag{2}$$

where, $\mathbf{u}(t) \in \Re^{m\times 1}$ are the controlling force, $\mathbf{B}_c \in \Re^{n\times m}$ is known as the control distribution matrix. The feedback controlling force $\mathbf{u}(t)$ can be scalar or vector and they can be distributed through the actuator/s by selecting $\mathbf{B}_c$ appropriately. The computation of the controlling force $\mathbf{u}(t)$ depends upon the measurement of state variables (displacement or velocity) or output measurements of response co-ordinates. The *state feedback control* law uses of control forces which are proportional to state (position and velocity) measurements as,

$$\mathbf{u}(t) = -\mathbf{F}_v^T\dot{\mathbf{q}} - \mathbf{G}_d^T\mathbf{q}, \tag{3}$$

where, $\mathbf{F}_v \in \Re^{n\times m}$ and $\mathbf{G}_d \in \Re^{n\times m}$ are constant velocity and displacement state feedback gain matrices respectively. The *single input state feedback* is a special case of (3), in which the scalar controlling force

$$u(t) = -\mathbf{f}_v^T\dot{\mathbf{q}} - \mathbf{g}_d^T\mathbf{q}, \;\; \mathbf{f}_v, \mathbf{g}_d \in \Re^{n\times 1}, \tag{4}$$

is distributed by selecting $\mathbf{B}_c = \mathbf{b}_c \in \Re^{n\times 1}$ vector. The non-zero entries of $\mathbf{b}_c$ defines the location of the actuator and its magnitude is often taken to be the gains of selected actuators. In practical point of view, it may happen that the scalar control force $u(t)$ required to achieve the desired control is so large that it cannot be implemented. In such cases, multi input control vector $\mathbf{u}(t)$ provides more flexibility (combination of sensors and actuators) in minimizing the controller gain norms.

By substituting (3) in (2), the following closed loop system,

$$\mathbf{M}\ddot{\mathbf{q}}(t) + \tilde{\mathbf{C}}\dot{\mathbf{q}}(t) + \tilde{\mathbf{K}}\mathbf{q}(t) = \mathbf{f}(t), \tag{5}$$

where,

$$\tilde{\mathbf{C}} = \mathbf{C} + \mathbf{B}_c\mathbf{F}_v^T,\ \tilde{\mathbf{K}} = \mathbf{K} + \mathbf{B}_c\mathbf{G}_d^T. \tag{6}$$

If there is no external disturbance $\mathbf{p}(t) = \mathbf{0}$, the dynamical behavior of the open loop system (1) and closed loop system (2) is characterized by the following eigenvalue problems,

$$\begin{cases} \left(\lambda_j^2\mathbf{M} + \lambda_j\mathbf{C} + \mathbf{K}\right)\boldsymbol{\psi}_j = \mathbf{0},\ \left(\lambda_j^2\mathbf{M} + \lambda_j\mathbf{C} + \mathbf{K}\right)^T\boldsymbol{\varphi}_j = \mathbf{0} \\ \left(\mu_j^2\mathbf{M} + \mu_j\tilde{\mathbf{C}} + \tilde{\mathbf{K}}\right)\tilde{\boldsymbol{\psi}}_j = \mathbf{0},\ \left(\mu_j^2\mathbf{M} + \mu_j\tilde{\mathbf{C}} + \tilde{\mathbf{K}}\right)^T\tilde{\boldsymbol{\varphi}}_j = \mathbf{0} \end{cases},\ \text{for } j = 1, 2, \ldots, 2n \tag{7}$$

where, λ_j are the open loop eigenvalues with corresponding right and left eigenvectors $\boldsymbol{\psi}_j$, $\boldsymbol{\varphi}_j$, respectively and μ_j are the closed loop eigenvalues and $\tilde{\boldsymbol{\psi}}_j$, $\tilde{\boldsymbol{\varphi}}_j$ are the corresponding closed loop eigenvectors. The eigenvalues of (7) are the roots of following characteristic polynomial

$$\det\left(\lambda_j^2\mathbf{M} + \lambda_j\mathbf{C} + \mathbf{K}\right) = 0,\quad \det\left(\mu_j^2\mathbf{M} + \mu_j\tilde{\mathbf{C}} + \tilde{\mathbf{K}}\right) = 0. \tag{8}$$

These eigenvalues and corresponding eigenvectors may be arranged to form the spectral and modal matrices as,

$$\boldsymbol{\Lambda} = \mathrm{diag}\left(\lambda_j\right),\ \boldsymbol{\Phi} = \left[\boldsymbol{\varphi}_1 \ldots \boldsymbol{\varphi}_{2n}\right],\ \boldsymbol{\Psi} = \left[\boldsymbol{\psi}_1 \ldots \boldsymbol{\psi}_{2n}\right] \in \mathbb{C}^{n\times 2n}. \tag{9}$$

When the system matrices are symmetric, $\mathbf{M} = \mathbf{M}^T,\ \mathbf{C} = \mathbf{C}^T,\ \mathbf{K} = \mathbf{K}^T$ the left and right eigenvectors will be same implying $\boldsymbol{\varphi}_j = \boldsymbol{\psi}_j$ and leading to spectral and modal matrices defined by (9) with $\boldsymbol{\Phi} = \boldsymbol{\Psi} \in \mathbb{C}^{n\times 2n}$.

By using the first order realization of the governing differential equations, the open loop system (1) is often converted to the following state space form,

$$\begin{aligned} \dot{\mathbf{x}}_{ss} &= \mathbf{A}_{ss}\mathbf{x}_{ss} + \mathbf{B}_{ss}\mathbf{u}_{ss} \\ \mathbf{y}_{ss} &= \mathbf{C}_{ss}\mathbf{x}_{ss} + \mathbf{D}_{ss}\mathbf{u}_{ss} \end{aligned}, \tag{10}$$

where, $\mathbf{x}_{ss} = \left(\mathbf{q}\ \dot{\mathbf{q}}\right)^T$ is the state vector, $\mathbf{A}_{ss} \in \Re^{2n\times 2n}$ is the state matrix, $\mathbf{B}_{ss} \in \Re^{2n\times m}$ is the input matrix, $\mathbf{u}_{ss} \in \Re^{m\times 1}$ is the input vector (excitation and/or controlling force), $\mathbf{C}_{ss} \in \Re^{s\times 2n}$ is the output matrix, and $\mathbf{D}_{ss} \in \Re^{s\times m}$ is transition (feed-forward) matrix. In the absence of control, the eigenvalues of the state space system (10) are known as the poles $\left\{\lambda_j\right\}_{j=1}^{2n}$ of the open loop system.

Design of a linear and modern control system deals with computing and/or estimating the controller force in time and frequency domain associated with the state space system (10) and several tools for controller design are available in literature [18–20]. For example, the classical eigenvalue or pole assignment problem deals with finding the control gain vectors/matrices in (3) from the known system matrices/vectors **M**, **C**, **K**, $\mathbf{B}_c$, such that the controlled loop system (7) has the desired closed loop poles $\{\mu_j\}_{j=1}^{2n}$. The solution of the classical pole placement problem and the robust algorithm for such a pole placement is developed in [22].

The partial pole assignment (PPA) for quadratic eigenvalue problems associated with (2) is first developed in [23] which computes the controller gains without the state space transformation, such that a partial set of eigenvalues $\{\lambda_j\}_{j=1}^{p}$ are assigned to a desired set $\{\mu_j\}_{j=1}^{p}$, while keeping all other eigenvalues unchanged $\{\lambda_j = \mu_j\}_{j=p+1}^{2n}$. This method ensures that the spillover phenomenon does not occur and higher modes are not affected by the placement of few open loop poles by implying that the characteristic equation for the closed loop system satisfies

$$\begin{cases} \det\left(\mu_j^2\mathbf{M} + \mu_j\tilde{\mathbf{C}} + \tilde{\mathbf{K}}\right) = 0, \ j = 1, 2, \ldots, p \\ \det\left(\lambda_j^2\mathbf{M} + \lambda_j\tilde{\mathbf{C}} + \tilde{\mathbf{K}}\right) = 0, \ \ j = p+1, p+2, \ldots, 2n \end{cases}. \tag{11}$$

They have computed the explicit solution for the control gain vectors associated with the symmetric quadratic pencil which only requires the system matrices, desired closed loop poles, and those open loop eigenpairs (eigenvalues and corresponding eigenvectors) which are to be invariant. Subsequent progress in this area could be found in [24, 25].

It is important to note that for complex structures mathematical models defining (1) are obtained by using the finite element modeling technique which approximate the structural properties such as mass and stiffness, joints, boundary conditions, etc. Moreover, due to the unavailability of accurate damping values, they are always assumed to be negligible or considered proportionally damped for the purposes of modal or generalized co-ordinate transformation. These approximation errors in modeling can be detrimental while applying controlling force in real time. Hence, in order to implement the state space control or PPA algorithms for large and complex structures, an observer needs to be designed for estimating the unmeasured states needed for full state feedback control [18–20].

3 Active Control by the Method of Receptances

To overcome the challenges of the traditional state space modeling approach, or any model dependent controller-design approach, a new method of receptances is developed for active vibration control [26]. This approach does not require the knowledge of models defining the large and complex structures and completely relies on available frequency response transfer functions which can be extracted from vibration experiments. The size of the matrices involved in the controller design is small and they depend on the number of sensors and actuators available on the structure for controller design. This method is generic and can be used for wide applications. It is not limited to particular types of structures and is not limited by physical size or complexity of the structure. The formulation of this method and subsequent progress made in this area are highlighted here.

3.1 Formulation and Progress for Active Vibration Control

The dynamic analysis of a structure often lead to the following second order system of equations in the Laplace domain associated with the open loop system (1),

$$\left(s^2\mathbf{M} + s\mathbf{C} + \mathbf{K}\right)\mathbf{q}(s) = \mathbf{f}(s). \tag{12}$$

The above dynamical system (12) can also be expressed in terms of receptance matrix $\mathbf{H}(s)$ as follows,

$$\mathbf{H}(s)\mathbf{f}(s) = \mathbf{q}(s), \tag{13}$$

where the matrix,

$$\mathbf{H}(s) = \left(s^2\mathbf{M} + s\mathbf{C} + \mathbf{K}\right)^{-1}, \tag{14}$$

is known as the receptance matrix for the open loop system. The receptance matrix may also be expressed in terms of the eigenvalues and eigenvectors as follows,

$$\mathbf{H}(s) = \left(s^2\mathbf{M} + s\mathbf{C} + \mathbf{K}\right)^{-1} = \sum_{j=1}^{2n} \frac{\boldsymbol{\psi}_j \boldsymbol{\varphi}_j^T}{\left(s - \lambda_j\right)} + \frac{\boldsymbol{\psi}_j^* \boldsymbol{\varphi}_j^{*T}}{\left(s - \lambda_j^*\right)}, \tag{15}$$

It is important to note that the receptance matrix $\mathbf{H}(s)$ in (15) can be extracted purely from the measurement and are dominated by the eigenvalues (or poles) closest to the frequency of excitation as shown in (15) where n is the number of eigenvalues and eigenvectors describing the system, $\boldsymbol{\varphi}_j$ and $\boldsymbol{\psi}_j$ are the jth left and right eigenvector respectively corresponding to the jth eigenvalue λ_j satisfying the eigenvalue problem (7) for the open loop system.

By substituting the state feedback single input control (4), which is expressed in the following Laplace domain form,

$$u_c(s) = -\left(s\mathbf{f}_v^T + \mathbf{g}_d^T\right)\mathbf{q}(s) \tag{16}$$

in the closed loop system (2) the following closed loop system is obtained,

$$\left(s^2\mathbf{M} + s\mathbf{C} + \mathbf{K} + \mathbf{b}_c(\mathbf{g}_d + s\mathbf{f}_d)^T\right)\mathbf{q}(s) = \mathbf{f}(s). \tag{17}$$

Hence the closed loop receptance can be represented as,

$$\hat{\mathbf{H}}(s) = \left(s^2\mathbf{M} + s\mathbf{C} + \mathbf{K} + \mathbf{b}_c(\mathbf{g}_d + s\mathbf{f}_d)^T\right)^{-1}. \tag{18}$$

It is shown in [26] that a rank-1 modification of (18) can be obtained in terms of the open loop receptances $\mathbf{H}(s)$ as follows,

$$\hat{\mathbf{H}}(s) = \mathbf{H}(s) - \frac{\mathbf{H}(s)\mathbf{b}_c(\mathbf{g}_d + s\mathbf{f}_v)^T\mathbf{H}(s)}{1 + (\mathbf{g}_d + s\mathbf{f}_v)^T\mathbf{H}(s)\mathbf{b}_c}. \tag{19}$$

The characteristics denominator roots of (19) represents the closed loop poles. Hence for any set of chosen closed loop poles $\{s_j = \mu_j \quad j = 1, 2, \ldots, 2n\}$ the following linear system of equations can be obtained,

$$\left(\mathbf{g}_d + \mu_j\mathbf{f}_v\right)^T\mathbf{r}_j = -1 \text{ with } \mathbf{r}_j = \mathbf{H}(\mu_j)\mathbf{b}_c. \tag{20}$$

By solving (20) the control gain vectors $\mathbf{f}_v$, $\mathbf{g}_d$ can be computed for a desired pole placement $\left\{\mu_j\right\}_{j=1}^{2n}$ with the knowledge of available receptance frequency response transfer function $\mathbf{r}_j$ while circumventing the knowledge of system matrices $\mathbf{M}$, $\mathbf{C}$, and $\mathbf{K}$. It is shown in [26] that if $\left\{\mu_j\right\}_{j=1}^{2n}$ is closed under conjugation, then the controller gains $\mathbf{f}_v$, $\mathbf{g}_d$ are real. Similar to describing the denominator dynamics of (19) in linear system of Eq. (20) for the pole placement, it is also shown that the zero assignment can be realized by developing the linear systems of equations from the numerator of (19). The theory and associated examples for poles, zero and simultaneous assignment of poles and zeros are also shown in [26].

For the case of multi-input control (3) for the system (2), the following closed loop receptance as a function of open loop receptances are obtained in [27],

$$\hat{\mathbf{H}}(s) = \frac{\operatorname{adj}\left(\mathbf{I} + \overline{\mathbf{H}}(s)(\mathbf{G}_d + s\mathbf{F}_v)^T\right)}{\det\left(\mathbf{I} + \overline{\mathbf{H}}(s)(\mathbf{G}_d + s\mathbf{F}_v)^T\right)}\mathbf{H}(s). \tag{21}$$

The control problem for pole placement in this case deals in finding the feedback control gains defining (21) by solving the following nonlinear set of equations

$$\det\left(\mathbf{I} + \overline{\mathbf{H}}(\mu_j)\hat{\mathbf{Q}}(\mu_j)\right) = 0, \quad \text{for } j = 1, 2, \ldots, 2n. \tag{22}$$

Experiments were conducted on T-shaped plate to demonstrate and verify the receptance based control in [27]. It is shown that by using collocated accelerometers and inertial actuators, poles and zeros of a system can be assigned for the case of output feedback control. This approach has also been extended to analyze the sensitivity of the individual poles by computing their sensitivities with respect to the controller gains in [28]. The realization of active/hybrid control by this method in friction induced vibration problems is demonstrated in [29, 30].

The closed loop system for the case of single input control (17) can have the following state space form,

$$\left(\lambda_j \boldsymbol{\varphi}_j^T \ \boldsymbol{\varphi}_j^T\right)\left(\lambda_j\begin{bmatrix}\mathbf{0} & \mathbf{M}\\ \mathbf{M} & \mathbf{C}\end{bmatrix}+\begin{bmatrix}-\mathbf{M} & \mathbf{0}\\ \mathbf{0} & \mathbf{K}\end{bmatrix}\right)\begin{pmatrix}\lambda_j\boldsymbol{\psi}_j\\ \boldsymbol{\psi}_j\end{pmatrix}=-(\boldsymbol{\varphi}_j^T\mathbf{b}_c)((\mathbf{g}_d^T+\lambda_j\mathbf{f}_v^T)\boldsymbol{\psi}_j), \tag{23}$$

defining the model uncontrollability and unobservability condition whenever the right-hand-side of (23) vanishes. Hence, when $\left(\boldsymbol{\varphi}_j^T\mathbf{b}_c\right) = 0$, it is known as the *uncontrollability* condition or when $\left(\mathbf{g}_d^T+\lambda_j\mathbf{f}_v^T\right)\boldsymbol{\psi}_j = 0$, it is known to be the *unobservability* condition, and under either of these conditions the eigenvalue λ_j remains unchanged by control action.

The uncontrollability condition $\left(\mathbf{b}_c^T\boldsymbol{\varphi}_j\right) = 0$ requires that the control distribution vector $\mathbf{b}_c$ must be perpendicular to both the real and imaginary parts of left eigenvectors $\boldsymbol{\varphi}_j$. It is shown in [31] that the following selection of $\mathbf{b}_c$

$$\mathbf{b}_c(s)=\mathbf{b}_1+\frac{\mathbf{b}_2}{s}, \tag{24}$$

makes the system uncontrollable for closed loop poles $\{\lambda_j\}_{j=p+1}^{2n}$. This allowed the computation of controller gains $\mathbf{f}_v$, $\mathbf{g}_d$ for partial pole assignment such that selected number of open loop poles are assigned to a new closed loop values $\{\mu_j\}_{j=1}^{p}$, $p < 2n$ and $\{\lambda_j\}_{j=p+1}^{2n}$ number of open loop poles remain invariant. By using two different test structures, one light fiber composite beam controlled by piezoelectric patches and other a heavy modular structure which is controlled by electromagnetic actuators, it is shown in [31] that selected poles of these structures can be controlled without influencing other poles associated with higher modes.

Subsequently, the effect of noise in measured receptance is addressed in [32] and a sequential multi-input feedback control approach is developed for robust pole placement. Both passive modifications in system matrices and design of active control by the method of receptances with experimental verification are investigated in [33]. This approach has been implemented for control of flexible links in [34] and for systems involving asymmetric matrices in [35]. The experimental demonstration of receptance based control for AgustaWestland W30 helicopter airframe is detailed in [36].

In order to overcome the challenges of solving nonlinear systems (22) associated with multi-input and output feedback control, eigenvalue and eigenvector assignment by multi-input-multi-output control using experimentally measured receptances is developed in [37]. The control distribution matrices and feedback gain matrices were partitioned as a combination of single input control as follows,

$$\mathbf{B}_c=\left[\mathbf{b}_{c_1}\ \mathbf{b}_{c_2}\ \cdots\ \mathbf{b}_{c_m}\right],\ \mathbf{F}_v=\left[\mathbf{f}_{v_1}\ \mathbf{f}_{v_2}\ \cdots\ \mathbf{f}_{v_m}\right],\ \mathbf{G}_d=\left[\mathbf{g}_{d_1}\ \mathbf{g}_{d_2}\ \cdots\ \mathbf{g}_{d_m}\right]. \tag{25}$$

Now by substituting (3) and (25) in (2) the eigenvalue problem associated with the closed loop system (without external excitation $\mathbf{f}(t) = \mathbf{0}$) can be represented as,

$$\left(\begin{array}{l} \mu_j^2 \mathbf{M} + \mu_j \mathbf{C} + \mathbf{K} - \\ \left(\mathbf{b}_{c_1}\left(\mu_j \mathbf{f}_{v_1}^T + \mathbf{g}_{d_1}^T\right) + \mathbf{b}_{c_2}\left(\mu_j \mathbf{f}_{v_2}^T + \mathbf{g}_{d_2}^T\right) + \cdots + \mathbf{b}_{c_r}\left(\mu_j \mathbf{f}_{v_m} + \mathbf{g}_{d_m}\right)\right) \end{array} \right) \mathbf{v}_j = \mathbf{0}. \quad (26)$$

where, $\left\{\mu_j\right\}_{j=1}^{2n}$ is the desired closed loop eigenvalues of for each velocity and $\mathbf{v}_j$ for $j = 1, 2, \ldots, 2n$ are the associated eigenvectors.

The system (26) can be rearranged as,

$$\mathbf{v}_j = \mathbf{H}(\mu_j)\left(\mathbf{b}_{c_1}\left(\mu_j \mathbf{f}_{v_1}^T + \mathbf{g}_{d_1}^T\right) + \mathbf{b}_{c_2}\left(\mu_j \mathbf{f}_{v_2}^T + \mathbf{g}_{d_2}^T\right) + \cdots + \mathbf{b}_{c_m}\left(\mu_j \mathbf{f}_{v_m}^T + \mathbf{g}_{d_m}^T\right)\right)\mathbf{v}_j. \quad (27)$$

Now by denoting,

$$\mathbf{r}_{\mu_{j,r}} = \mathbf{H}(\mu_j)\, \mathbf{b}_{c_r}, \text{ for } j = 1, 2, \ldots, 2n, \text{ and } r = 1, 2, \ldots, m \quad (28)$$

and,

$$e_{\mu_{j,r}} = \left(\mu_j \mathbf{f}_{v_r}^T + \mathbf{g}_{d_r}^T\right)\mathbf{v}_j, \text{ for } j = 1, 2, \ldots, 2n, \text{ and } r = 1, 2, \ldots, m, \quad (29)$$

the eigenvectors $\mathbf{v}_j$ of the closed loop system of Eqs. (26) can be expressed in terms of the linear combination available transfer functions $\mathbf{r}_{\mu_{j,r}}$ as,

$$\mathbf{v}_j = e_{\mu_{j,1}} \mathbf{r}_{\mu_{j,1}} + e_{\mu_{j,2}} \mathbf{r}_{\mu_{j,2}} + \cdots + e_{\mu_{j,m}} \mathbf{r}_{\mu_{j,m}}. \quad (30)$$

Equation (30) can be now expressed in the following linear systems of equations for each velocity of interest,

$$\begin{bmatrix} \mu_j \mathbf{v}_j^T & \mathbf{0} & \cdots & \mathbf{0} & \mathbf{v}_j^T & \mathbf{0} & \cdots & \mathbf{0} \\ \mathbf{0} & \mu_j \mathbf{v}_j^T & \cdots & \mathbf{0} & \mathbf{0} & \mathbf{v}_j^T & \cdots & \mathbf{0} \\ \vdots & \vdots & \vdots & \vdots & \vdots & \vdots & \vdots & \vdots \\ \mathbf{0} & \mathbf{0} & \cdots & \mu_j \mathbf{v}_j^T & \mathbf{0} & \mathbf{0} & \cdots & \mathbf{v}_j^T \end{bmatrix} \begin{pmatrix} \mathbf{f}_{v_1} \\ \vdots \\ \mathbf{f}_{v_m} \\ \mathbf{g}_{d_1} \\ \vdots \\ \mathbf{g}_{d_m} \end{pmatrix} = \begin{pmatrix} e_{\mu_{j,1}} \\ e_{\mu_{j,2}} \\ \vdots \\ e_{\mu_{j,m}} \end{pmatrix}. \quad (31)$$

As the eigenvectors $\mathbf{v}_j$ can be scaled arbitrarily, by choosing a random self-conjugate pair $\left(e_{\mu_{j,r}}, e_{\mu_{j+1,r}} = e_{\overline{\mu}_{j,r}}\right)$ associated with the self-conjugate closed loop pair of poles $\left(\mu_{j,r}, \mu_{j+1,r} = \overline{\mu}_{j,r}\right)$ in (31) the control gain matrices $\mathbf{F}_v = \left[\mathbf{f}_{v_1}\ \mathbf{f}_{v_2} \cdots \mathbf{f}_{v_m}\right]$, $\mathbf{G}_d = \left[\mathbf{g}_{d_1}\ \mathbf{g}_{d_2} \cdots \mathbf{g}_{d_m}\right]$ can be obtained. It is shown in [37] that selection of self-conjugate pairs of poles and $e_{\mu_{j,r}}$ ensures the real control gain and multiple of solutions for control gains can be obtained for various scaling choices of

$e_{\mu_{j,k}}$. In order to ensure the partial pole assignment, some of the open loop eigenvalues $(2n - p)$ can be made invariant in the closed loop system (26), by ensuring the following unobservability condition

$$\left(\mathbf{b}_{c_1}\left(\lambda_k \mathbf{f}_{v_1}^T + \mathbf{g}_{d_1}^T\right) + \cdots + \mathbf{b}_{c_m}\left(\lambda_k \mathbf{f}_{v_m} + \mathbf{g}_{d_m}\right)\right)\mathbf{w}_k = \mathbf{0}, \tag{32}$$

for $k = p+1,\ p+2,\ \ldots, 2n,\ p < 2n$, where, λ_k and $\mathbf{w}_k$ are eigenvalues and eigenvectors of the open loop system respectively, which remains unchanged. In this case the $(2n - p)$ equations in (32) can be expressed in the following linear system of equation form,

$$\begin{bmatrix} \lambda_k \mathbf{w}_k^T & \mathbf{0} & \cdots & \mathbf{0} & \mathbf{w}_k^T & \mathbf{0} & \cdots & \mathbf{0} \\ \mathbf{0} & \lambda_k \mathbf{w}_k^T & \cdots & \mathbf{0} & \mathbf{0} & \mathbf{w}_k^T & \cdots & \mathbf{0} \\ \vdots & \vdots & \vdots & \vdots & \vdots & \vdots & \vdots & \vdots \\ \mathbf{0} & \mathbf{0} & \cdots & \lambda_k \mathbf{w}_k^T & \mathbf{0} & \mathbf{0} & \cdots & \mathbf{w}_k^T \end{bmatrix} \begin{pmatrix} \mathbf{f}_{v_1} \\ \vdots \\ \mathbf{f}_{v_m} \\ \mathbf{g}_{d_1} \\ \vdots \\ \mathbf{g}_{d_m} \end{pmatrix} = \begin{pmatrix} 0 \\ 0 \\ \vdots \\ 0 \end{pmatrix}. \tag{33}$$

By combining (31) and (33) the controller gains to assign the new closed loop poles $\{\mu_j\}_{j=1}^{p}$ while ensuring the unaffected open loop poles $\{\lambda_k\}_{k=p+1}^{2n}$ can be obtained by solving the following linear system of equations,

$$\begin{bmatrix} \mu_j \mathbf{v}_j^T & \mathbf{0} & \cdots & \mathbf{0} & \mathbf{v}_j^T & \mathbf{0} & \cdots & \mathbf{0} \\ \mathbf{0} & \mu_j \mathbf{v}_j^T & \cdots & \mathbf{0} & \mathbf{0} & \mathbf{v}_j^T & \cdots & \mathbf{0} \\ \vdots & \vdots & \vdots & \vdots & \vdots & \vdots & \vdots & \vdots \\ \mathbf{0} & \mathbf{0} & \cdots & \mu_j \mathbf{v}_j^T & \mathbf{0} & \mathbf{0} & \cdots & \mathbf{v}_j^T \\ \lambda_k \mathbf{w}_k^T & \mathbf{0} & \cdots & \mathbf{0} & \mathbf{w}_k^T & \mathbf{0} & \cdots & \mathbf{0} \\ \mathbf{0} & \lambda_k \mathbf{w}_k^T & \cdots & \mathbf{0} & \mathbf{0} & \mathbf{w}_k^T & \cdots & \mathbf{0} \\ \vdots & \vdots & \vdots & \vdots & \vdots & \vdots & \vdots & \vdots \\ \mathbf{0} & \mathbf{0} & \cdots & \lambda_k \mathbf{w}_k^T & \mathbf{0} & \mathbf{0} & \cdots & \mathbf{w}_k^T \end{bmatrix} \begin{pmatrix} \mathbf{f}_{v_1} \\ \vdots \\ \mathbf{f}_{v_m} \\ \mathbf{g}_{d_1} \\ \vdots \\ \mathbf{g}_{d_m} \end{pmatrix} = \begin{pmatrix} e_{\mu_{j,1}} \\ e_{\mu_{j,2}} \\ \vdots \\ e_{\mu_{j,m}} \\ 0 \\ 0 \\ \vdots \\ 0 \end{pmatrix}. \tag{34}$$

In this way the complete spectrum may be separated into assigned modes and retained modes to achieve partial pole placement (PPA). The assignment of selected finite number of modes, which usually can be measured in practice, stability of the system can be guaranteed without worrying about the other measures being taken to solve the problem of spillover in the uncontrolled/unobserved modes.

In recent years this receptance based approach has gained more attention and it is being used for designing different types of controllers and for various engineering applications. For example, method of receptances is also used for designing control for the systems with inherent time delay in the feedback loop. Partial pole assignment with time delay by combining the receptance matrix and system matrices are shown in [38]. For the case of multi-input various methods are investigated by Bai et al. to compute efficient and robust controller gains for time delayed system in [39–41]. The effect of time delay in friction induced vibration problem [42], output feedback problem [43], higher order systems [44] and the stability of the closed loop system [45] using the receptance method are also investigated. Receptance based control is also developed for nonlinear systems [46, 47], distributed parameter systems [48], viscoelastic system [49], structural modification for rotor-bearing system [50] and fluid loaded systems [51]. Partial pole assignment with inaccessible degrees of freedom is demonstrated in [52]. Eigenstructure assignment by this method is shown in [53, 54]. The role of controllability and observability fir the receptance method is discussed in [55]. Robust stabilization of eigenvalue assignment and their sensitivity for the friction induced vibration problems are shown in [56]. A multi-step method for higher order system is developed in [57], passive modification for partial frequency assignment is demonstrated in [58] and partial quadratic eigenvalue assignment and development the strategy for minimum controller norm by this approach is shown in [59, 60].

4 Receptance Based Active Aeroelastic Control

Active aeroelastic control deals with the mutual interaction between aerodynamic forces, structural vibration, stability and the closed loop control to suppress dynamic flutter instabilities and gust load alleviation. Modeling and design of aeroelastic and aeroservoelastic control has been investigated in the past and their details can be found in several monographs [61–63]. Traditionally these problems are often modeled and solved in state-space form and they are derived in generalized or modal co-ordinates which may require design of observers/estimators to realize the active control. The aircraft wings and UAVs are generally embedded with several in-board sensors and actuators such as strain gages, accelerometers, pitch gyros and servo-actuators to manipulate the leading and trailing edge control surfaces. The method of receptance is a natural fit to design the controller for such structures, which allows the controller design based upon receptances available from the combination of available sensors and actuators on the aircraft. In this section, with the help of a flexible wing model [64], the development of receptance based control for the aeroelastic control application is summarized with several numerical examples. The discussion is limited to the pole assignment problems for flutter suppression and flutter boundary extension.

Let us consider a flexible wing model defined by two bending and two torsion modes, with multiple control surfaces, as shown in Fig. 2. The parameters defining the wing model is summarized in Table 1.

The closed loop dynamics of the linear aeroelastic system is governed by (2) with the following system matrices,

$$\mathbf{M} = m \begin{bmatrix} \bar{s}^5 c/5 & \bar{s}^6 c/6 & \bar{s}^4 \bar{c}/4 & \bar{s}^5 \bar{c}/5 \\ \bar{s}^6 c/6 & \bar{s}^7 c/7 & \bar{s}^5 c\bar{c}/5 & \bar{s}^6 c\bar{c}/6 \\ \bar{s}^4 \bar{c}/4 & \bar{s}^5 \bar{c}/5 & \bar{s}^3 \tilde{c}/3 & \bar{s}^4 \tilde{c}/4 \\ \bar{s}^5 \bar{c}/5 & \bar{s}^6 \bar{c}/6 & \bar{s}^4 \tilde{c}/4 & \bar{s}^5 \tilde{c}/5 \end{bmatrix}, \tag{35}$$

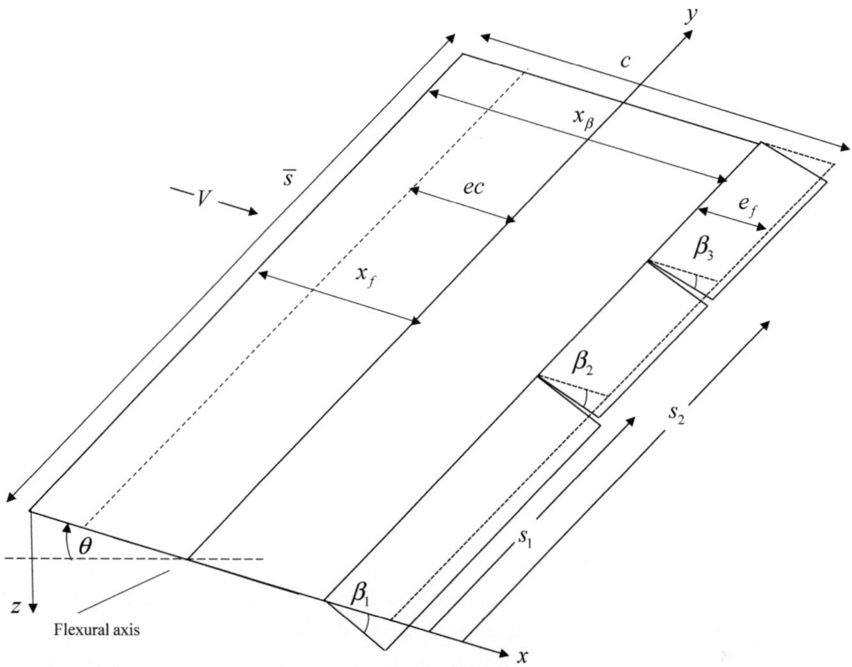

Fig. 2 A flexible wing model with multiple control surfaces

Table 1 Parameters for the flexible wing model shown in Fig. 4

Parameter	Symbol	Values
Semi-span (m)	$\bar{s}$	6
Chord (m)	c	1
Mass per unit area (kg/m^2)	m	10
Air density (kg/m^3)	ρ	1.225
2D lift curve slope	a_w	2π
Unsteady torsional velocity term	$M_{\dot{\theta}}$	$-1.2c$
Fraction of chord	e_f	0.1
Flexural rigidity	EI	400×10^3
Torsional rigidity	GJ	200×10^3
Position of flexural axis	x_f	$0.4c$
Eccentricity ratio	e	0.15

$$\mathbf{C}(V) = \mathbf{C}_s + \rho V \underbrace{\begin{bmatrix} -\frac{ca_w\bar{s}^5}{10} & -\frac{ca_w\bar{s}^6}{12} & 0 & 0 \\ -\frac{ca_w\bar{s}^6}{12} & -\frac{ca_w\bar{s}^7}{14} & 0 & 0 \\ \frac{c^2ea_w\bar{s}^4}{8} & \frac{c^2ea_w\bar{s}^5}{10} & \frac{c^3M_{\dot{\theta}}\bar{s}^3}{24} & \frac{c^3M_{\dot{\theta}}\bar{s}^4}{32} \\ \frac{c^2ea_w\bar{s}^5}{10} & \frac{c^2ea_w\bar{s}^6}{12} & \frac{c^3M_{\dot{\theta}}\bar{s}^4}{32} & \frac{c^3M_{\dot{\theta}}\bar{s}^5}{40} \end{bmatrix}}_{\mathbf{C}_a}, \tag{36}$$

$$\mathbf{K}(V) = \underbrace{\begin{bmatrix} 4EI\bar{s} & 6\bar{s}^2EI & 0 & 0 \\ 6\bar{s}^2EI & 12\bar{s}^3EI & 0 & 0 \\ 0 & 0 & GJ\bar{s} & GJ\bar{s}^2 \\ 0 & 0 & GJ\bar{s}^2 & 4GJ\bar{s}^3/3 \end{bmatrix}}_{\mathbf{K}_s} + \rho V^2 \underbrace{\begin{bmatrix} 0 & 0 & -ca_w\bar{s}^4/8 & -ca_w\bar{s}^5/10 \\ 0 & 0 & -ca_w\bar{s}^5/10 & -ca_w\bar{s}^6/12 \\ 0 & 0 & c^2ea_w\bar{s}^3/6 & c^2ea_w\bar{s}^4/8 \\ 0 & 0 & c^2ea_w\bar{s}^4/8 & c^2ea_w\bar{s}^5/10 \end{bmatrix}}_{\mathbf{K}_a}, \tag{37}$$

and,

$$\mathbf{B}_c = \rho V^2 c \begin{bmatrix} -a_c s_1^3/6 & -a_c\left(s_2^3 - s_1^3\right)/6 & -a_c\left(\bar{s}^3 - s_2^3\right)/6 \\ -a_c s_1^4/8 & -a_c\left(s_2^4 - s_1^4\right)/8 & -a_c\left(\bar{s}^3 - s_2^4\right)/8 \\ 4cb_c s_1^2/4 & cb_c\left(s_2^2 - s_1^2\right)/4 & cb_c\left(\bar{s}^3 - s_2^2\right)/4 \\ cb_c s_1^3/6 & cb_c\left(s_2^3 - s_1^3\right)/6 & cb_c\left(\bar{s}^3 - s_2^3\right)/6 \end{bmatrix} \tag{38}$$

where, $\mathbf{q} = \left(q_1\ q_2\ q_3\ q_4\right)^T$ are the generalized coordinates. The stiffness and damping matrices composed of structural stiffness $\mathbf{K}_s$ and structural damping $\mathbf{C}_s = \mathbf{0}$ as well as the aerodynamic stiffness $\mathbf{K}_a$ and damping $\mathbf{C}_a$ matrices which are velocity (V) dependent due to the contribution of the aerodynamic forces surrounding the wing plan form. The control is distrusted using the rotation $\mathbf{u}(t) = \left(\beta_1\ \beta_2\ \beta_3\right)^T$ of the control surfaces which are hinged at the trailing edge of the wing planform. The control distribution matrix $\mathbf{B}_c$ (38) represents the aerodynamic loads on the control surfaces. The receptance frequency response transfer function between the control surface rotation and the response degrees of freedom can be represented by the following matrix,

$$\mathbf{r}(s) = \mathbf{H}(s)\mathbf{B}_c = \begin{bmatrix} h_{\beta_1 q_1}(s) & h_{\beta_2 q_1}(s) & h_{\beta_3 q_1}(s) \\ h_{\beta_1 q_2}(s) & h_{\beta_2 q_2}(s) & h_{\beta_3 q_2}(s) \\ h_{\beta_1 q_3}(s) & h_{\beta_2 q_3}(s) & h_{\beta_3 q_3}(s) \\ h_{\beta_1 q_4}(s) & h_{\beta_2 q_4}(s) & h_{\beta_3 q_4}(s) \end{bmatrix}. \tag{39}$$

Hence the number of control input ($m = 3$) and measured degrees of freedom ($n = 4$) defines the size of receptance matrix which are needed to design the control. For the case of single control surface $m = 1$ at the trailing end, the control distribution matrix (38)and single input control in (2) will have the following form,

$$\mathbf{B}_c = \mathbf{b}_c = \rho V^2 c \left(-\frac{a_c s_1^3}{6}\ -\frac{a_c s_1^4}{8}\ \frac{cb_c s_1^2}{4}\ \frac{cb_c s_1^3}{6} \right)^T, \quad u(t) = \beta_1(t). \tag{40}$$

The control surface parameters are defined as follows: $e_d = \sqrt{e_f(1-e_f)}$, $b_c = a_w/\pi(1-e_f)e_d$, and $a_c = a_w/\pi\left(\cos^{-1}(1-2e_f)+2e_d\right)$. Alternatively, it is shown in [65] that the wing model can be defined with control surface rotation as additional degrees of freedom in the model.

4.1 Example: Pole Placement with Single Control Surface

Consider the case of single input control with the trailing edge control surface having the length of $s_1 = \bar{s}$. For the air speed $V = 60$ m/s the system matrices (35)–(38) are obtained with the following open loop poles:

$$\begin{aligned} \lambda_{1,2} &= -6.10 \pm 111.04i, \ \ \lambda_{3,4} = -10.00 \pm 459.11i \\ \lambda_{5,6} &= -11.31 \pm 192.66i, \ \ \lambda_{7,8} = -15.11 \pm 14.36i \end{aligned} \tag{41}$$

The closed loop poles are chosen to be:

$$\mu_j = \left(\mathrm{Re}(\lambda_j) - 15\right) + \left(\mathrm{Im}(\lambda_j) + 1\right)i, \quad \text{for } j = 1, 2, \ldots, 8. \tag{42}$$

Linear system of equations (20) is solved and the following controller gains are computed with the available receptance transfer functions $\mathbf{r}(\mu_j)$ obtained from the system matrices and control surface input matrix (40) for the chosen closed loop poles (42),

$$\mathbf{F}_v = \mathbf{f}_v = \begin{pmatrix} -1.4007 \\ -5.1935 \\ -0.0677 \\ 0.2193 \end{pmatrix}, \ \mathbf{G}_d = \mathbf{g}_d = \begin{pmatrix} -18.0772 \\ 92.5955 \\ 5.7555 \\ 86.7900 \end{pmatrix}. \tag{43}$$

With the control gains (43) the closed loop transfer function (19) can be computed and they are plotted with open loop receptance transfer functions (39) in Fig. 3 in the frequency domain ($s = j\omega$). It is evident from Fig. 3 that more modal damping at each corresponding natural frequencies are added, as defined by the chosen closed loop poles in (42). This process can be repeated for each velocity and flutter instability can be suppressed by pole placement. This will allow the design of the controller such that the open loop flutter velocity can be increased and the flutter boundary for a given aircraft can be extended for a given aircraft. These aspects are highlighted in the subsequent examples which consider the general solution form (31) for the pole and partial pole placement.

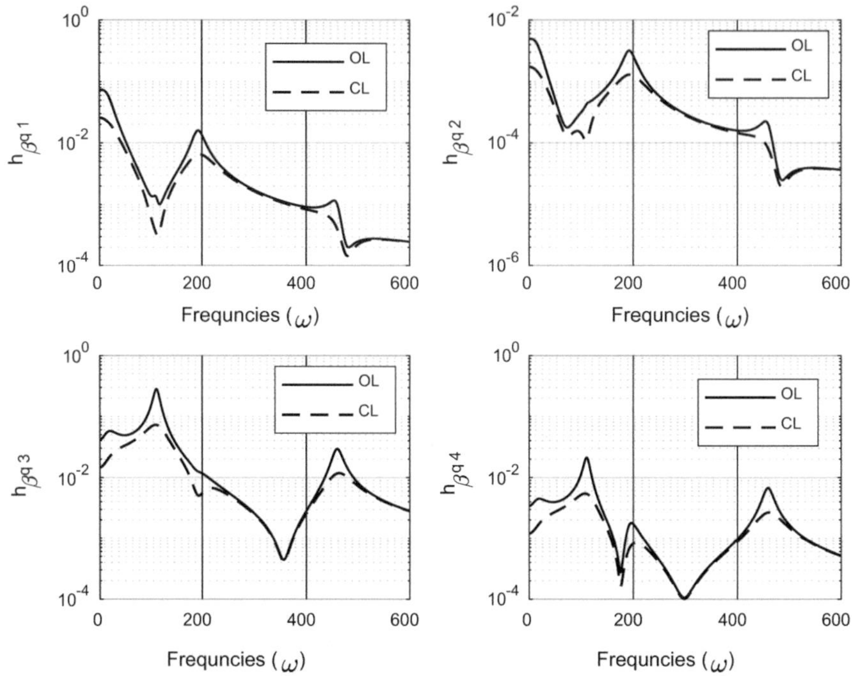

Fig. 3 Open loop (OL) and closed loop (CL) transfer functions at $V = 60$ m/s

4.2 Example: Flutter Boundary Extension via Pole and Partial Pole Placement

Now we consider the case of multi-input control where the control force is defined by (3) for the case of three control surfaces ($m = 3$) with $s_1 = 0.5\bar{s}$ and $s_2 = 0.75\bar{s}$. We reconsider the pole assignment defined in (42).

The system of Eqs. (31) is constructed with the knowledge of (27) which requires the knowledge of closed loop pole in (42) and the receptance transfer functions obtained between the partitioned control distribution matrix $\mathbf{B}_c = \left(\mathbf{b}_{c_1}\ \mathbf{b}_{c_2}\ \mathbf{b}_{c_3}\right)$ as defined in (38) and degrees of freedom. By choosing a random conjugate pair $\left(e_{\mu_{j,k}}, e_{\mu_{j+1,k}} = \pm i\right)$ the system of Eq. (31) is solved and the following gains $\mathbf{F}_v = \left[\mathbf{f}_{v_1}\ \mathbf{f}_{v_2}\ \mathbf{f}_{v_3}\right]$ and $\mathbf{G}_d = \left[\mathbf{g}_{d_1}\ \mathbf{g}_{d_2}\ \mathbf{g}_{d_3}\right]$ are computed:

$$\mathbf{F}_v = \begin{pmatrix} 2.0640 & 2.0640 & 2.0640 \\ 7.7255 & 7.7255 & 7.7255 \\ 0.0567 & 0.0567 & 0.0567 \\ -0.2380 & -0.2380 & -0.2380 \end{pmatrix}, \quad \mathbf{G}_d = \begin{pmatrix} 27.53 & 27.53 & 27.53 \\ -127.23 & -127.23 & -127.23 \\ -4.91 & -4.91 & -4.91 \\ -82.99 & -82.99 & -82.99 \end{pmatrix}. \tag{44}$$

These feedback gains satisfies the desire pole placement (42) by ensuring,

$$\det\left(\mu_j^2\mathbf{M}+\mu_j\mathbf{C}(V)+\mathbf{K}(V)-\begin{pmatrix}\mathbf{b}_{c1}\left(\mu_j\mathbf{f}_{v1}^T+\mathbf{g}_{d1}^T\right)\\ +\mathbf{b}_{c2}\left(\mu_j\mathbf{f}_{v2}+\mathbf{g}_{d2}\right)+\mathbf{b}_{c3}\left(\mu_j\mathbf{f}_{v3}+\mathbf{g}_{d3}\right)\end{pmatrix}\right)=0. \tag{45}$$

Controller gains are now computed for each velocity ($V = 1, 2, 3, \ldots, 120$) m/s defining the entire flight envelop. By using controller gains for each velocity the closed loop poles are computed. In Fig. 4, the real part of open and closed loop poles are plotted together, demonstrating that by active aeroelastic control the open loop flutter velocity ~87 m/s is increased to a new closed loop flutter velocity of ~115 m/s. This extends the flutter boundary by more than 30% for this wing configuration. Note that the selection of closed loop poles is arbitrary here for introducing more damping in the modes and it may place additional requirement for the selection of actuators driving the control surfaces. Moreover the computation of controller gain is needed for each velocity.

Alternatively, it is shown in [66] that a single controller gain can be computed for entire flight envelop by formulating and solving a least square problem associated with (31). For example, by denoting the system (31) for each velocity V_i as,

$$\mathbf{G}_j(V_i)\mathbf{z}=\tilde{\mathbf{e}}(V_i),\ \text{for}\ j=1,2,\ldots,2n,\ k=1,2,\ldots,r. \tag{46}$$

Then for the entire flight envelop the following system of equations can be assembled as follows,

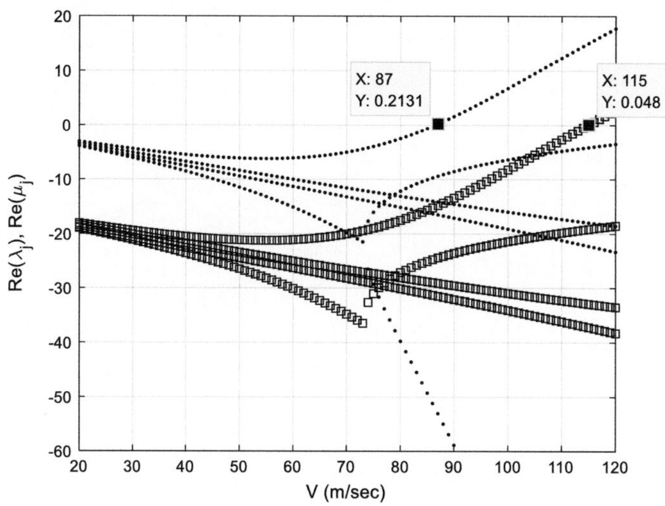

Fig. 4 Real part of eigenvalues of the open loop system (Dots) and closed loop system (Squares) with velocity dependent gains as a function of velocity

$$\begin{bmatrix} \mathbf{G}_j(V_1) \\ \vdots \\ \mathbf{G}_j(V_k) \\ \vdots \\ \mathbf{G}_j(V_r) \end{bmatrix} \begin{pmatrix} \mathbf{f}_{v_1} \\ \vdots \\ \mathbf{f}_{v_r} \\ \mathbf{g}_{d_1} \\ \vdots \\ \mathbf{g}_{d_r} \end{pmatrix} = \begin{pmatrix} \mathbf{e}_{j_r}(V_1) \\ \vdots \\ \mathbf{e}_{j_r}(V_k) \\ \vdots \\ \mathbf{e}_{j_r}(V_r) \end{pmatrix}, \text{for } k = 1, 2, \ldots, r. \tag{47}$$

A unique set of controller gains now can be obtained by solving (47) in the least square sense. By choosing the flight envelop with $V_1 = 1$ m/s and $V_r = 120$ m/s with the increment of 1 m/s, following set of unique controller gains are computed in the least square sense by solving (47) for the pole placement (42),

$$\mathbf{F}_v = \begin{pmatrix} -0.3443 & -0.3443 & -0.3443 \\ -2.6942 & -2.6942 & -2.6942 \\ 0.0605 & 0.0605 & 0.0605 \\ 0.0526 & 0.0526 & 0.0526 \end{pmatrix}, \quad \mathbf{G}_d = -\begin{pmatrix} 15.23 & 15.23 & 15.23 \\ 111.77 & 111.77 & 111.77 \\ 1.47 & 1.47 & 1.47 \\ 12.68 & 12.68 & 12.68 \end{pmatrix}. \tag{48}$$

With the controller gains in (48) the closed loop poles are computed and their real part are compared with their open loop counterparts in Fig. 5. It is important to note that for the entire flight envelop this unique controller gain matrices ensure the flutter boundary extension from ~87 to ~114 m/s. However, not all the closed loop poles are placed at the desired location as chosen in (42).

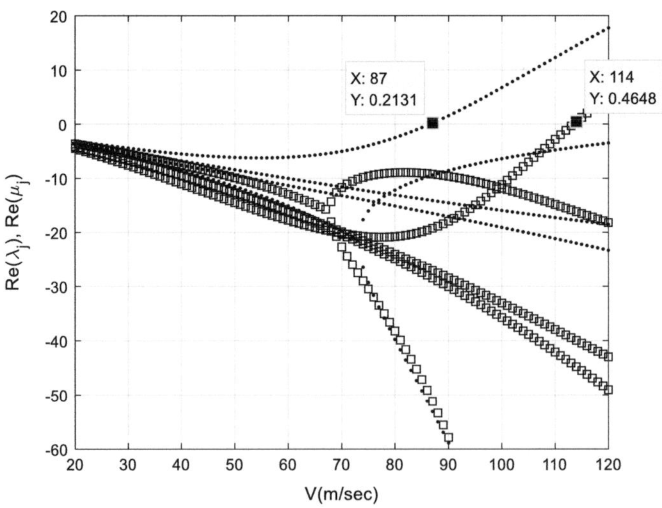

Fig. 5 Real part of eigenvalues of the open loop system (Dots) and closed loop system (Squares) with unique control gain (48) as a function of velocity

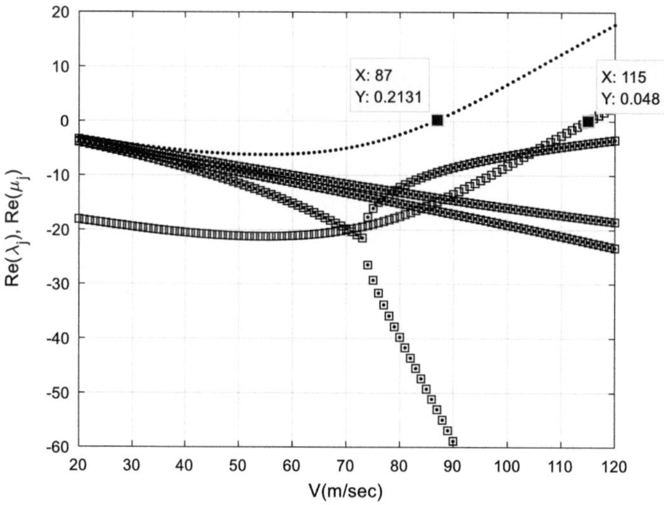

Fig. 6 Real part of eigenvalues of the open loop system (Dots) and closed loop system (Squares) as a function of velocity for the case of PPA

For the case of PPA the following closed loop poles are chosen

$$\mu_j = (\mathrm{Re}(\lambda_j) - 15) + (\mathrm{Im}(\lambda_j) + 1)i, \quad \text{for } j = 1, 2 \quad \mu_j = \lambda_j, \quad \text{for } j = 3, 4, \ldots, 8, \tag{49}$$

which allows the control of the mode contributing to the flutter instability without affecting any other higher modes of the aeroelastic system. By choosing the pole placement (49) the controller gains for the case of three control surfaces are computed for each velocity ($V = 1, 2, 3, \ldots, 120$) m/s in the flight envelop by solving (34). The real part of the open loop eigenvalues are plotted with the closed loop poles obtained from the computed controller gains in Fig. 6.

It can be seen in Fig. 6 that the controller introduces the damping at only one mode which is contributing to the flutter. It achieves the same flutter boundary extension for the case of full pole assignment as shown is Fig. 4. However, the remaining eigenvalues were invariant. By constructing the least square system (47) associated with (34), a unique controller gains can also be computed. Note that the performance of the controller with unique control gain for an entire flight envelop will depend on the selection of the closed loop poles and the selection of the velocity range for which least square system (47) is constructed.

These examples demonstrated the implementation of linear aeroelastic control by the method of receptances. The selection of unsteady aerodynamic term $M_{\dot{\theta}}$ in (36) allowed the receptance of aeroelastic systems to be similar to (14). The receptances for more complex unsteady aeroelastic interaction will have different form which is not demonstrated here for brevity. However, the progress in this area is summarized briefly in the following section.

4.3 Recent Progress in Receptance Based Active Aeroelastic Control

The receptance method is promising in the area of aeroelastic control as the receptance transfer function needed for controller design can be obtained from open loop wind tunnel testing during the design/testing phase or during its flight operation. In practice the input/output data may be perfectly general, making use of the available excitation and response signals to and from actuators and sensors respectively. For example, the input might be the voltage supplied to a motor in order to move a control surface and the output obtained from accelerometers placed on a wing. Moreover, measured receptance data (from wind tunnel test or inflight) truly represent the aeroelastic interactions and do not require the approximation of aeroelastic interaction between the structure and unsteady aerodynamic forces. This allowed the several progresses in recent years for designing the active aeroelastic control by the receptance method.

It is shown in [66] that by designing a single input control force active pole assignment for flutter suppression/boundary extension can be achieved from numerical receptances for a range of wing configurations subjected to unsteady aerodynamic loads. The usage of single input feedback control by piezo-actuated ailerons has been also verified in a wind tunnel testing for flutter suppression [67]. For high performance flexible wings having multiple leading and trailing edge control surfaces, multi-input state feedback control problems developed in [68] for achieving flutter boundary augmentation in fighter wing models having two and three control surfaces. It is shown that on the onset of flutter the closed loop system can be stabilized by designing the prescribed controller output or control surface rotation. State feedback control problems are developed with actuator dynamics and partial pole assignment is carried out in [69], which overcame the problem of potential actuator spillover and facilitated simultaneous control for aeroelastic and actuator modes. The study of optimal sizing and placement of control surfaces using the receptance based control were investigated in [70–72] such that minimal controller efforts can be achieved for enhanced flutter boundaries. To deal with flutter suppression of wings with structural nonlinearities receptance based control is developed in [73] to assign the limit cycles to stable oscillations. Receptance method for controlling nonlinear aeroelastic system is investigated in [74] and for active flutter suppression with prescribed flutter margin is shown in [75]. Feedback linearization and control of a nonlinear aeroelastic system is demonstrated with experimental validation in [76]. Research is currently underway to mature the receptance based control strategy in various engineering applications including in the area of active aeroelastic control.

References

1. Soong, T.T., Dargush, G.F.: Passive energy dissipation systems in structural engineering (1997)
2. Piersol, A.G., Harris, C.M.: Harri's Shock and Vibration Handbook, 5th edn. McGraw-Hill (2017)
3. Janocha, H.: Adaptronics and smart structures. Springer, Berlin (1999)
4. Achkire, Y., Preumont, A.: Active tendon control of cable-stayed bridges. Earthq. Eng. Struct. Dyn. **25**(6), 585–597 (1996)
5. Housner, G.W., et al.: Structural control: past, present, and future. J. Eng. Mech. **123**(9), 897–971 (1997)
6. Soong, T.T., Costantinou, M.C. (eds.): Passive and Active Structural Vibration Control in Civil Engineering, vol. 345. Springer, Berlin (2014)
7. Balas, M.: Trends in large space structure control theory: fondest hopes, wildest dreams. IEEE Trans. Autom. Control **27**(3), 522–535 (1982)
8. Sutton, T.J., Elliott, S.J., Brennan, M.J., Heron, K.H., Jessop, D.A.C.: Active isolation of multiple structural waves on a helicopter gearbox support strut. J. Sound Vib. **205**(1), 81–101 (1997)
9. Kessler, C.: Active rotor control for helicopters: motivation and survey on higher harmonic control. CEAS Aeronaut. J. **1**(1–4), 3 (2011)
10. Fuller, C.C., Elliott, S., Nelson, P.A.: Active Control of Vibration. Academic Press, London (1996)
11. Clark, R.L., Saunders, W.R., Gibbs, G.P.: Adaptive Structures: Dynamics and Control, vol. 28, No. 2. Wiley, New York (1998)
12. Symans, M.D., Constantinou, M.C.: Semi-active control systems for seismic protection of structures: a state-of-the-art review. Eng. Struct. **21**(6), 469–487 (1999)
13. Giurgiutiu, V.: Review of smart-materials actuation solutions for aeroelastic and vibration control. J. Intell. Mater. Syst. Struct. **11**(7), 525–544 (2000)
14. Irschik, H.: A review on static and dynamic shape control of structures by piezoelectric actuation. Eng. Struct. **24**(1), 5–11 (2002)
15. Preumont, A., Seto, K.: Active Control of Structures. Wiley, Chichester (2008)
16. Mottershead, J.E.: On the zeros of structural frequency response functions and their sensitivities. Mech. Syst. Sig. Process. **12**(5), 591–597 (1998)
17. White, B.A.: Eigenstructure assignment: a survey. Proc. Inst. Mech. Eng. Part I J. Syst. Control Eng. **209**(1), 1–11 (1995)
18. Inman, D.J.: Vibration with Control. Wiley, Chichester (2017)
19. Ogata, K., Yang, Y.: Modern Control Engineering, vol. 4. Prentice Hall, India (2002)
20. Datta, B.: Numerical Methods for Linear Control Systems, vol. 1. Academic Press, Cambridge (2004)
21. Balas, M.J.: Active control of flexible systems. J. Optim. Theory Appl. **25**(3), 415–436 (1978)
22. Kautsky, J., Nichols, N.K., Van Dooren, P.: Robust pole assignment in linear state feedback. Int. J. Control **41**(5), 1129–1155 (1985)
23. Datta, B.N., Elhay, S., Ram, Y.M.: Orthogonality and partial pole assignment for the symmetric definite quadratic pencil. Linear Algebra Appl. **257**, 29–48 (1997)
24. Datta, B.N., Elhay, S., Ram, Y.A., Sarkissian, D.R.: Partial eigenstructure assignment for the quadratic pencil. J. Sound Vib. **230**(1), 101–110 (2000)
25. Datta, B.N.: Finite-element model updating, eigenstructure assignment and eigenvalue embedding techniques vibrating systems. Mech. Syst. Sig. Process. **16**(1), 83–96 (2002)
26. Ram, Y.M., Mottershead, J.E.: Receptance method in active vibration control. AIAA J. **45**(3), 562–567 (2007)
27. Mottershead, J.E., Tehrani, M.G., James, S., Ram, Y.M.: Active vibration suppression by pole-zero placement using measured receptances. J. Sound Vib. **311**(3–5), 1391–1408 (2008)
28. Mottershead, J.E., Tehrani, M.G., Ram, Y.M.: Assignment of eigenvalue sensitivities from receptance measurements. Mech. Syst. Sig. Process. **23**(6), 1931–1939 (2009)

29. Ouyang, H.: Pole assignment of friction-induced vibration for stabilisation through state-feedback control. J. Sound Vib. **329**(11), 1985–1991 (2010)
30. Ouyang, H.J.: A hybrid control approach for pole assignment to second-order asymmetric systems. Mech. Syst. Sig. Process. **25**(1), 123–132 (2011)
31. Tehrani, M.G., Elliott, R.N.R., Mottershead, J.E.: Partial pole placement in structures by the method of receptances: theory and experiments. J. Sound Vib. **329**(24), 5017–5035 (2010)
32. Tehrani, M.G., Mottershead, J.E., Shenton, A.T., Ram, Y.M.: Robust pole placement in structures by the method of receptances. Mech. Syst. Sig. Process. **25**(1), 112–122 (2011)
33. Mottershead, J.E., Tehrani, M.G., Ram, Y.M.:. Eigenvalue assignment problems in vibration using measured receptances: passive modification and active control. In: Gladwell, G.M.L., Morassi, A. (eds.) Dynamical Inverse Problems: Theory and Application, pp. 179–202 (2011)
34. Ouyang, H., Richiedei, D., Trevisani, A.: Pole assignment for control of flexible link mechanisms. J. Sound Vib. **332**(12), 2884–2899 (2013)
35. Tehrani, M.G., Ouyang, H.: Receptance-based partial pole assignment for asymmetric systems using state-feedback. Shock Vib. **19**(5), 1135–1142 (2012)
36. Mottershead, J.E., Tehrani, M.G., James, S., Court, P.: Active vibration control experiments on an AgustaWestland W30 helicopter airframe. Proc. Inst. Mech. Eng. Part C J. Mech. Eng. Sci. **226**(C6), 1504–1516 (2012)
37. Ram, Y.M., Mottershead, J.E.: Multiple-input active vibration control by partial pole placement using the method of receptances. Mech. Syst. Sig. Process. **40**(2), 727–735 (2013)
38. Ram, Y.M., Mottershead, J.E., Tehrani, M.G.: Partial pole placement with time delay in structures using the receptance and the system matrices. Linear Algebra Appl. **434**(7), 1689–1696 (2011)
39. Bai, Z.-J., Chen, M.-X., Yang, J.-K.: A multi-step hybrid method for multi-input partial quadratic eigenvalue assignment with time delay. Linear Algebra Appl. **437**(7), 1658–1669 (2012)
40. Bai, Z.J., Chen, M.X., Datta, B.N.: Minimum norm partial quadratic eigenvalue assignment with time delay in vibrating structures using the receptance and the system matrices. J. Sound Vib. **332**(4), 780–794 (2013)
41. Bai, Z.J., Yang, J.K., Datta, B.N.: Robust partial quadratic eigenvalue assignment with time delay using the receptance and the system matrices. J. Sound Vib. **384**, 1–14 (2016)
42. Singh, K.V., Ouyang, H.: Pole assignment using state feedback with time delay in friction-induced vibration problems. Acta Mech. **224**(3), 645–656 (2013)
43. Xiang, J.W., Zhen, C., Li, D.C.: Partial pole assignment with time delay by the receptance method using multi-input control from measurement output feedback. Mech. Syst. Signal Process. **66–67**, 743–755 (2016)
44. Wang, X.T., Zhang, L.: Partial eigenvalue assignment with time delay in high order system using the receptance. Linear Algebra Appl. **523**, 335–345 (2017)
45. Santos, T.L.M., Araujo, J.M., Franklin, T.S.: Receptance-based stability criterion for second-order linear systems with time-varying delay. Mech. Syst. Signal Process. **110**, 428–441 (2018)
46. Tehrani, M.G., Wilmshurst, L., Elliott, S.J.: Receptance method for active vibration control of a nonlinear system. J. Sound Vib. **332**(19), 4440–4449 (2013)
47. Zhen, C., Jiffri, S., Li, D.C., Xiang, J.W., Mottershead, J.E.: Feedback linearisation of nonlinear vibration problems: a new formulation by the method of receptances. Mech. Syst. Sig. Process. **98**, 1056–1068 (2018)
48. Maha, A., Ram, Y.M.: The method of receptances for continuous rods. J. Appl. Mech. Trans. ASME **81**(7), 071009 (2014)
49. Singh, K.V., Ling, X.X.: Active control of viscoelastic systems by the method of receptance. J. Vib. Acoust. Trans. ASME **140**(2) (2018)
50. Tsai, S.-H., Ouyang, H., Chang, J.-Y.: Inverse structural modifications of a geared rotor-bearing system for frequency assignment using measured receptances. Mech. Syst. Signal Process. **110**, 59–72 (2018)
51. Xia, M.L., Li, S.: A combined approach for active vibration control of fluid-loaded structures using the receptance method. Arch. Appl. Mech. **88**(10), 1683–1694 (2018)

52. Wei, X.J., Mottershead, J.E., Ram, Y.M.: Partial pole placement by feedback control with inaccessible degrees of freedom. Mech. Syst. Signal Process. **70–71**, 334–344 (2016)
53. Liu, Z.H., Li, W.Y., Ouyang, H.J., Wang, D.H.: Eigenstructure assignment in vibrating systems based on receptances. Arch. Appl. Mech. **85**(6), 713–724 (2015)
54. Richiedei, D., Trevisani, A.: Simultaneous active and passive control for eigenstructure assignment in lightly damped systems. Mech. Syst. Signal Process. **85**, 556–566 (2017)
55. Wei, X., Maha, A., Ram, Y., Mottershead, J.E.: The role of controllability and observability in partial pole placement by the method of receptance. In: International Conference on Noise and Vibration Engineering (ISMA) (2014)
56. Liang, Y., Yamaura, H., Ouyang, H.J.: Active assignment of eigenvalues and eigen-sensitivities for robust stabilization of friction-induced vibration. Mech. Syst. Sig. Process. **90**, 254–267 (2017)
57. Liu, H., Xu, J.: A multi-step method for partial eigenvalue assignment problem of high order control systems. Mech. Syst. Signal Process. **94**, 346–358 (2017)
58. Belotti, R., Ouyang, H.J., Richiedei, D.: A new method of passive modifications for partial frequency assignment of general structures. Mech. Syst. Sig. Process. **99**, 586–599 (2018)
59. Bai, Z.-J., Wan, Q.-Y.: Partial quadratic eigenvalue assignment in vibrating structures using receptances and system matrices. Mech. Syst. Sig. Process. **88**, 290–301 (2017)
60. Bai, Z.J., Lu, M., Wan, Q.Y.: Minimum norm partial quadratic eigenvalue assignment for vibrating structures using receptances and system matrices. Mech. Syst. Sig. Process. **112**, 265–279 (2018)
61. Dowell, E.H.: A Modern Course in Aeroelasticity, vol. 3. In: Curtiss, H.C., Scanlan, R.H., Sisto, F. (eds.). Kluwer Academic Publishers, Dordrecht, The Netherlands (1989)
62. Rodden, W.P.: Theoretical and Computational Aeroelasticity. Crest Publication (2011)
63. Tewari, A.: Aeroservoelasticity: Modeling and Control. Springer, Berlin (2015)
64. Wright, J.R., Cooper, J.E.: Introduction to Aircraft Aeroelasticity and Loads, vol. 20. Wiley, Chichester (2008)
65. Singh, K.V., McDonough, L.A., Mottershead, J., Cooper, J.: Active aeroelastic control using the receptance method. In: ASME 2010 International Mechanical Engineering Congress and Exposition, pp. 137–146 (2010) (January)
66. McDonough, L.A., Singh, K.V., Kolonay, R.: Active control for coupled unsteady aeroelastic models. In: Proceedings of the International Forum on Aeroelasticity and Structural Dynamics, IFASD-2011-076, Paris, France, pp. 26–30 (2011)
67. Papatheou, E., Wei, X., Jiffri, S., Tehrani, M.G., Bode, S., Singh, K.V., Cooper, J.E.: Flutter control using vibration test data: theory, rig design and preliminary results. In: International Conference on Noise and Vibration Engineering, Leuven, Belgium, September, pp. 17–19 (2012) (September)
68. Singh, K.V., McDonough, L.A., Kolonay, R., Cooper, J.E.: Receptance-based active aeroelastic control using multiple control surfaces. J. Aircr. **51**(1), 335–342 (2014)
69. Singh, K.V., Brown, R.N., Kolonay, R.: Receptance-based active aeroelastic control with embedded control surfaces having actuator dynamics. J. Aircr. **53**(3), 830–845 (2016)
70. Brown, R.N., Singh, K.V.: Optimal sizing and configurations of the control surfaces for active aeroelastic control. In: AIAA Atmospheric Flight Mechanics Conference, p. 2241 (2015)
71. Brown, R., Singh, K.V.: Optimal selection of control surfaces for active aeroelastic control of prescribed modes. In: 57th AIAA/ASCE/AHS/ASC Structures, Structural Dynamics, and Materials Conference, p. 0712 (2016)
72. Brown, R.N., Singh, K.V., Kolonay, R.M.: Optimal sizing and placement of control surfaces for active aeroservoelastic control. In: 58th AIAA/ASCE/AHS/ASC Structures, Structural Dynamics, and Materials Conference, p. 0569 (2017)
73. Wei, X., Mottershead, J.E.:. Limit cycle assignment in nonlinear aeroelastic systems using describing functions and the receptance method. In: Topics in Modal Analysis, vol. 7, pp. 701–713. Springer, New York, NY (2014)
74. Zhen, C., Li, D., Xiang, J.: A modified receptance method for active control of a nonlinear aeroelastic system. In: 56th AIAA/ASCE/AHS/ASC Structures, Structural Dynamics, and Materials Conference, p. 1855 (2015)

75. Wu, Z., Cooper, J.E.: Active flutter suppression combining the receptance method and flutter margin. In: 57th AIAA/ASCE/AHS/ASC Structures, Structural Dynamics, and Materials Conference, p. 1227 (2016)
76. Jiffri, S., Fichera, S., Mottershead, J.E., Da Ronch, A.: Experimental nonlinear control for flutter suppression in a nonlinear aeroelastic system. J. Guid. Control Dyn. **40**(8), 1925–1938 (2017)

Computational Fluid Dynamics

Numerical Study on the Efficiency of Magnetophoresis in Human Vasculature like Conditions

S. Rekha and Sarbari Bhattacharya

Abstract Magnetic Nanoparticles (MNPs) have been proposed as therapeutic agents for treatment of medical conditions like cancer by using them either as drug carriers to specific locations in the body or as agents for localized heating. For these treatments to be effective, a substantial quantity of MNPs need to reach the target site from the site of injection. However, this may not be the case if blood flow alone is relied on to transport the MNPs as losses to branching blood vessels that lead to transport in directions away from the target site can occur. We have carried out numerical simulations on the flow of a dilute concentration of magnetic particles (MPs), where inter particle interactions can be ignored, in channels that mimic human vasculature like conditions. This has been done using Open Field Operation And Manipulation (OpenFOAM). Blood has been treated as a Newtonian fluid and only laminar flows are considered. The size of channels and direction of flow, which is always towards the target site, mimics the situation in arteries and arterioles. We have only taken into consideration situations where the main channel gives rise to a single daughter channel at different branching angles to the direction of flow. We find that a substantial number of MPs are lost to the daughter channel when the branching angle is less than 45° while a moderate loss is seen for branching angles greater than 45°. Application of an optimum external magnetic field gradient significantly improves the fraction of particles moving towards the target site in the case of branching angles less than 90° and ensures practically no loss for branching angles greater than 90°.

Keywords Magnetic guiding · Branching channels · OpenFOAM · Lagrangian particles

---

S. Rekha · S. Bhattacharya (✉)
Department of Physics, Bangalore University, Bengaluru 560056, India
e-mail: sarbari.bhattacharya@bub.ernet.in

S. Manna et al. (eds.), *Mathematical Modelling and Scientific Computing with Applications*, Springer Proceedings in Mathematics & Statistics 308,
https://doi.org/10.1007/978-981-15-1338-1_14

1 Introduction

Magnetic Nano Particles (MNPs) of diameter less than 50 nm are generally superparamagnetic (SPM) in nature and have the potential for a large variety of applications [1, 2]. The synthesis and characterization of MNPs has become both easier and cheaper over the years [3]. Today it is possible to synthesize MNPs with size, shape, magnetic volume and surface modifications to address a particular problem [4, 5]. MNPs can be designed to be biodegradable and low in cytotoxicity when administered in moderate quantities thus enabling applications, both diagnostic and therapeutic, in the field of medicine. MNPs are already being used to enhance contrast in Magnetic Resonance Imaging [6]. MNPs embedded in micron sized beads are used for separating specific cells from an assay by first magnetically tagging them and then guiding them using magnetic field gradient [7, 8]. MNPs have been proposed as therapeutic agents for treatment of cancer either as drug carriers or as local heating agents (Magnetic Fluid Hyperthermia) [9].

The targeted delivery of MNPs within a living system is in its early stage and much more research has to be carried out before this can be put into regular practice [10] for the treatment of human subjects. The transport of the MNPs from the site of injection to the target site would rely mainly on the blood flow. If the full benefits of a targeted delivery treatment are to be realized, a significant fraction of the MNPs injected need to reach the target site. Owing to the complex nature of human vasculature, if one relies on blood flow alone, the efficacy of treatment will be compromised as loss of MNPs can be expected at every branch point encountered during flow towards the target site. In order to overcome or minimize this, guiding MNPs to the target site using external magnetic field gradients (EMFG) is essential. Detailed knowledge and understanding of the human vasculature and the nature of blood flow within it is necessary to design an optimum EMFG. The strength of the applied EMFG should be such that the velocity acquired by the MNPs is not significantly greater than the fluid velocity as this may disturb the flow pattern. Further, as these particles also interact with each other in the presence of an EMFG, the field strength cannot be increased arbitrarily as aggregation of MNPs would increase significantly leading to an increased risk of blocking the flow channels. A trade-off between minimizing loss of MNPs and EMFG strength has to be made to ensure that the treatment is of maximum efficacy with minimal risk.

Several theoretical [11, 12], numerical [13–17] and experimental studies [18–23] have been carried out to study the magnetic guiding of MNPs in the form of ferrofluids or magnetic microspheres (MNPs incorporated in a non-magnetic matrix). There are also some studies which have looked at the magnetophoresis of MNPs in Y-shaped channels [24–27]. In this study, we simulate the flow of SPM beads through a channel in the presence of a daughter channel of uniform width branching out at various angles to the direction of flow in the primary channel, both without as well as in the presence of an EMFG along the primary flow direction. Using Open Field Operation and Manipulation (OpenFOAM)—a Computational Fluid Dynamics (CFD) software, we study how the guiding efficiency improves in channels with branching angles from 15° to 120°, with the application of an optimum EMFG.

2 Methods

2.1 Description of the Model

The MNPs, in order to be transported to the target tissue, have to be injected into an artery. It is known that arteries found in human vasculature vary in diameter from a few centimeters to a few hundred microns. The main elastic arteries branch out into several muscular arteries (small arteries) of diameter 0.1–10 mm that branch out further into arterioles that ultimately deliver blood to specific organs/tissues through a network of capillaries [28]. We have carried out 2D simulations of magnetophoresis of SPM beads of 6 μm diameter (*d*) in the channel set up described in the earlier section. The guiding efficiency, which is the ratio of the number of particles that exit from the main channel (MC) to the total number of particles injected at the inlet (in percentage), was computed in each case.

Two different values of primary channel width (*D*) have been considered: (i) 500 μm which mimics an artery and (ii) 50 μm which mimics an arteriole. The total channel length along the x-axis (*L*) was taken to be 9D while the length of the daughter channel was taken to be 4.5D. The daughter channel was assigned a constant width (D_d) throughout its length and the opening of the daughter channel was positioned along L in such a way that the length of MC was always 4.5D.

2.2 Mesh Parameters and Boundary Conditions

Gmsh (version 2.15.0), an open software was used to generate finite element mesh of the channel according to the above description. A 2D mesh was created and extended in z-axis (by about 1% of L) in order to satisfy OpenFOAM criteria [29]. The channels were meshed using prism elements of uniform size throughout. The grid size (Δx) was kept greater than the particle size in all cases (see Table 1).

Table 1 Parameters used for the simulations

D (μm)	a (m)	NI (At)	Mesh size (μm)	Time step (ms)
500	0.01	1000	24	0.1[a]
	0.01	3000		
50	0.001	3.2	6.25	0.005

[a]For simulations with $\Theta = 15°$, a time step of 0.015 ms was used

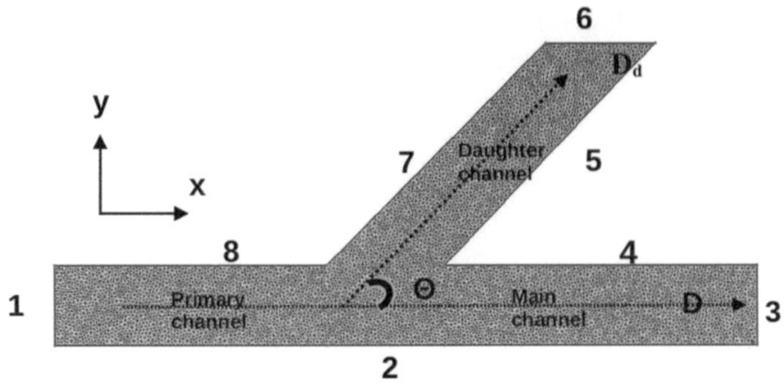

Fig. 1 Finite element mesh of the channel with $\Theta = 45°$ (with $D_d = D$) with all the physical surfaces labeled. Arrows indicate the flow direction. 1 is the inlet to the primary channel, 3 is the outlet for the main channel in the direction of the flow at the inlet and 6 is the outlet for the daughter channel. 2, 4, 5, 7 and 8 are walls

Surfaces 1–8 were declared as physical surfaces with the enclosed volume to be filled with fluid while the rest (front and back) were default faces (Fig. 1). Physical surfaces 1 and 3 were treated as the inlet (*In*) and outlet of MC ($O1$) respectively while physical surface 6 was taken as the outlet of the daughter channel ($O2$). These three surfaces were assigned to be patches while the other five surfaces were taken to be walls [with a no slip boundary condition (BC)] [30]. The default faces were declared to be empty faces (which ensures that the simulation is indeed in 2D). Uniform velocity was provided at *In*, zeroGradient at outlet 1 and 2 and noSlip conditions at the walls [30]. The pressure (p) at *In* was set as zeroGradient and at $O1$ and $O2$, it was kept a constant ($p = 0$). Note, that the channels here are mimicking blood vessels in the human vasculature. This being a closed system, with all arteries finally branching down to a capillary bed in the tissues, usage of identical pressure at both outlets is justifiable. However, this may not always be the case.

2.3 Blood flow

We have treated blood as a Newtonian fluid with a density (ρ_f) of 1086 kg/m^3 and a kinematic viscosity (η_f) of 10^{-6} m^2/s. The value of kinematic viscosity used here would match that of a dilute suspension of red blood cells in plasma [31]. Note that kinematic viscosity of whole human blood depends strongly on the red blood cell content. Additionally, the value of density taken matches that of the SPM beads used. This is done so that the effect of gravity can be ignored as the beads become neutrally buoyant in this fluid. Further, the blood flow was assumed to be laminar for both cases considered. The velocity of the blood flow in case (i) was assigned a value of 2 mm/s while for case (ii) it was assigned a value of 0.5 mm/s. These are towards the lower end of the range of velocities found in channels of similar size in vivo.

Navier-Stokes equation is discretized (finite volume method) and solved using PISO (Pressure—Implicit with Splitting of Operators) algorithm to obtain the flow field [32].

2.4 Super Paramagnetic Beads

The characteristics of SPM bead used in the simulations have been taken from [33]. As mentioned earlier, SPM beads have been assigned a diameter of 6 μm with a density (ρ_b) of 1086 kg/m^3. SPM beads contains SPM nanoparticles embedded in a non-magnetic matrix and the magnetic volume susceptibility of the entire bead (χ_b) = 0.17 is used to compute the magnetic force acting on it. Magnetization of the beads have been reported to reach saturation values for magnetic field strengths greater than 0.5T. The particles were injected randomly across the inlet patch at a rate of 15 particles/second for a duration of 4 s. Note, SPM beads can be considered as a spherical aggregate of MNPs with an aggregate diameter of 6 μm and an effective magnetic susceptibility of χ_b.

2.5 OpenFOAM Specifications

2.5.1 Solver Modifications and Compilation of a New Particle Force

OpenFOAM—CFD (version 5.0) software was used to perform the numerical simulations. Euler-Lagrangian approach was used where the blood flow was modeled as an Eulerian phase with the beads as discrete Lagrangian particles. An existing solver called *nonNewtonianicoFoam* was modified to add the Lagrangian particles to the flow [34]. A two way coupling is envisaged where the fluid and particles are coupled to each other (Eq. 2). We use a dilute concentration of SPM beads thereby ignoring any kind of particle-particle interactions. Magnetic field (***B***) is specified as a boundary condition along with pressure (*p*) and Velocity (***U***).

The equations governing the fluid flow are

$$\nabla \circ \vec{U} = 0 \tag{1}$$

$$\frac{\partial \vec{U}}{\partial t} + \nabla \circ \vec{U}\vec{U} - \Delta\left(\eta_f \vec{U}\right) - \nabla\vec{U} \circ \nabla\eta_f = -\nabla p - \vec{S_b} \tag{2}$$

where $\mathbf{S}_b$ is the momentum source correction due to the presence of Lagrangian particles [34].

The forces acting on the SPM bead are drag (F_d), gravity (F_g) and magnetic force (F_m) due to the presence of EMFG. As this is a case of neutrally buoyant spheres, gravitational force plays no role, none the less it is included for extension of the simulation beyond the case of neutrally buoyant beads.

The expressions used are given below,

$$F_d = \frac{00.75 C_d m_b \mu_f Re}{\rho_d d^2} \tag{3}$$

where

$$C_d = 24\left(1 + \frac{1}{6} Re^{\frac{2}{3}}\right) \tag{4}$$

for particle Reynolds number (Re) < 1000

$$F_g = m_b g\left(1 - \frac{\rho_f}{\rho_b}\right) \tag{5}$$

$$\overrightarrow{F_m} = \frac{m_b \chi_b}{\rho_b \mu_0}\left(\overrightarrow{B}.\nabla\right)\overrightarrow{B} \tag{6}$$

where m_b is the mass of SPM bead, χ_b, the magnetic susceptibility of the bead, g, the acceleration due to gravity and μ_o, the permeability of free space.

Lagrangian particles (SPM beads in our case) are governed by the equation

$$m_b \frac{d\overrightarrow{U_b}}{dt} = \Sigma \overrightarrow{F} \tag{7}$$

A new particle force was compiled to include F_m using a standard paramagnetic force model as a template. This new particle force model called *magneticForce* requires specification of $\boldsymbol{B}$ in the initial boundary condition and χ_b as an input parameter. The *magneticForce* particle model calculates the force on the SPM beads using the expression in Eq. (6).

2.5.2 External Magnetic Field (B)

The expression for magnetic field along the axis of a coil (B_x) of radius a with N turns of wire carrying a current I whose center is at the origin is given by [35]

$$B_x(\rho, x) = \frac{\mu_o NI}{2\pi} \frac{1}{\rho\sqrt{(\rho + a)^2 + x^2}}\left[\frac{(a^2 - \rho^2 - x^2)E(k)}{(a - \rho)^2 + x^2} + K(k)\right] \tag{8}$$

where $\rho = \sqrt{y^2 + z^2}$ is the radial distance from the axis of the coil, *K(k)* and *E(k)* are the elliptic integrals of first and second kind respectively with k being the elliptic modulus given as

$$k(\rho, x) = \sqrt{\frac{4\rho a}{(\rho + a)^2 + x^2}} \tag{9}$$

In order to guide SPM beads into the MC, we need an EMFG along the x-axis. We use a gradient field created by an anti-Helmholtz pair of coils which are essentially symmetric coils with centers located at $x = -a/2$ and $+a/2$ and the current flows in one coil in a direction opposite to that in the other.

$$B_x^{anti} = B_x\left(\rho, x - \frac{a}{2}\right) - B_x\left(\rho, x + \frac{a}{2}\right) \tag{10}$$

It is clear from Eq. 6 that F_m depends on $\left(\vec{B} \cdot \nabla\right)\vec{B}$. In order to increase the magnitude of F_m and to saturate the magnetization of the SPM beads, we superpose a uniform magnetic field of 0.5T along with a suitable gradient field produced from the anti-Helmholtz coil pair. Note that the coordinates of the grid points are obtained using *funkySetFields* [36] and are used for the calculation of B_x^{anti} field (using *Qt Octave*). This is specified as **B** in the initial boundary condition and the resultant B field is shown in Fig. 2a, b for two different values of *NI* for the same anti-Helmholtz coil pair.

The simulation was executed for a total time of 5 s using the largest time step (Δt) that allowed for a simulation (see Table 1) with a Courant number always less than 1, Courant number being defined as $Co = U\frac{\Delta t}{\Delta x}$. The guiding efficiency was computed for all the channel configurations considered. A screen shot of the simulation carried out in the case of $\Theta = 30°$ at $t = 2.5$ s is shown in Fig. 2, both in the presence and without the applied EMFG.

2.6 Validation of Code

The code developed is based on the solver "nonNewtonianicoFoam" that already exists in the OpenFOAM library. This code is similar to the "icoFoam" solver, which is used to solve incompressible laminar flows of Newtonian fluids except that it is modified here to handle non Newtonian fluids [37]. The simulations carried out in this current work is for Newtonian fluids. Bayraktar et al. [38] have verified the icoFoam solver for bench mark test cases and validated the code. In our study, we have modified the "nonNewtonianicoFoam" solver to incorporate Lagrangian particles and have also introduced a new particle force model which has been discussed in the earlier section. In order to validate the modifications made, we have chosen a straight channel of D = 500 μm and L = 4500 μm with no branches and injected particles manually at the inlet along various streamlines. The time taken by a particle, for example, on the axis of the channel (where the fluid velocity is maximum) to reach the outlet of the channel (t_o) in the presence and absence of EMFG from the simulations were

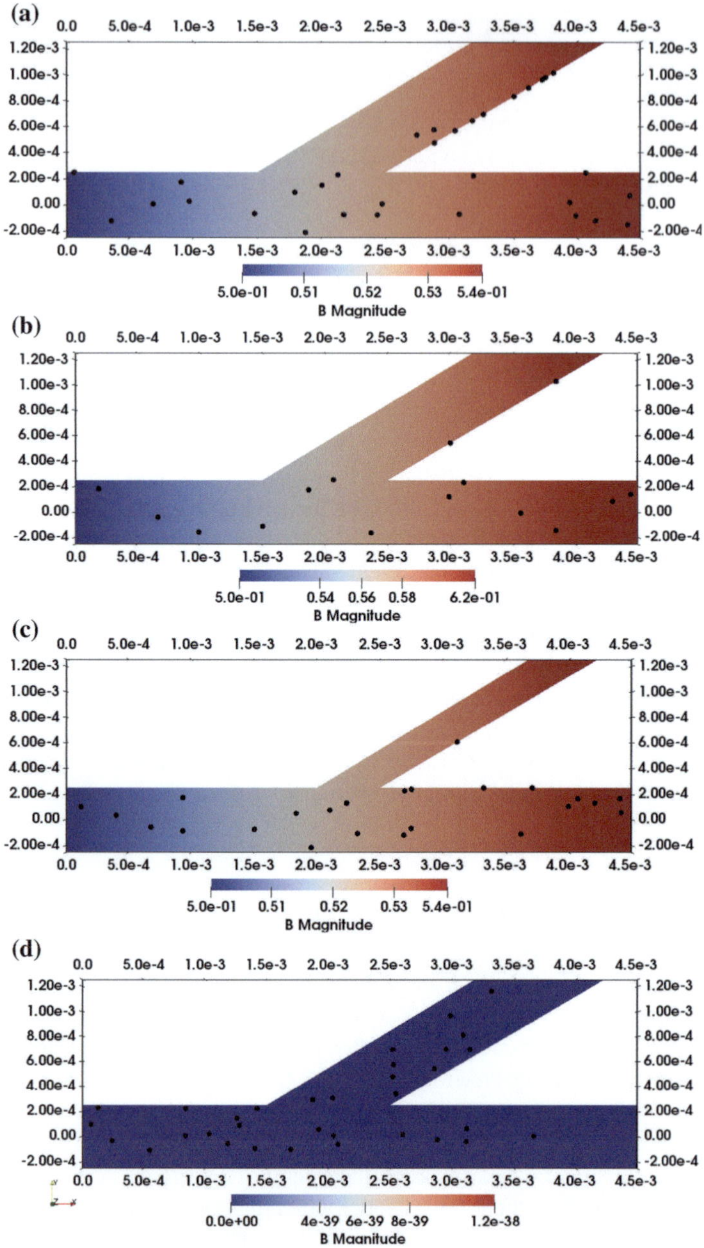

Fig. 2 A screenshot of D = 500 μm, simulation at t = 2.5 s, Θ = 30° (Units: x and y are in metres while magnitude of B is in tesla). The particle positions are represented by filled circles that are scaled to 8 times larger than the particle size for purposes of visibility. **a** NI = 1000 At (D_d = D), **b** NI = 3000 At (D_d = D), **c** NI = 1000 At (D_d = D/2) and **d** NI = 0 (D_d = D). Note that the number of particles flowing along the daughter channel reduces as the strength of the EMFG along the X-axis increases

Table 2 Comparison of particle exit time "t_o" computed from OpenFoam simulations with modified solver and new particle force model and numerical solutions of Eq. 11 for a test case of channel with D = 500 μm without any branch

NI (At)		t_o (s)	
		OpenFoam simulations	Numerical calculations
Coil radius = 0.01 m	0	1.535	1.510
	1000	1.095	1.090
	3000	0.665	0.680

compared with the values obtained by solving the equation of motion of the magnetic bead in the fluid flow.

The equation of motion of the bead that was solved numerically is given by

$$\rho_b V_b \frac{d^2x(t)}{dt^2} = -6\pi\mu_f(d/2)|v_f - \frac{dx(t)}{dt}| + \frac{V_b \chi_b}{\mu_0}\frac{\partial B_x^2(x)}{\partial x} \tag{11}$$

where V_b is the volume of the bead, v_f is the maximum velocity of the fluid (v_f= 0.003 m/s) and B_x is given by Eq. 10 with $\rho = 0$ (as the bead is considered to be moving along the axis of the channel) and a superposed uniform field of 0.5T.

It is evident from Table 2 that the simulated and numerically computed values agree well with each other. The modified solver code as well as the *magneticForce* model are thus validated and can be used for further studies now also involving channel branches.

2.7 Grid Dependency Tests

Test simulations were carried out with varying grid sizes (Δx) both in the case of D = 500 and 50 μm. The particles were injected manually at the same positions in all the cases and the time taken to reach either of the outlet (t_o) was found out from the simulation results. Results of test simulations for a channel having $\Theta = 30°$ branch are reported below.

Case 1: D = 500 μm

The grid size was varied from 18 to 48 μm and 16 particles were injected manually at the "inlet" patch. The t_o value of the particle injected at the 5th position from the upper wall of the channel for all the cases are shown in Table 3. The t_o value decreases from 1.845 to 1.835 s when Δx is decreased from 48 to 36 μm and then remains unaltered for further reduction down to 18 μm. This indicates that the results are independent of the grid size in this range. A reduction of Δx below 18 μm was not attempted to avoid large increase in computation time. We have chosen a Δx of 24 μm for the rest of the simulations with $D = 500$ μm channel.

Table 3 Variation of particle exit time t_o in channels of D = 500 and 50 μm with branch at Θ = 30° and meshes of varying grid sizes

D (μm)	Δx (μm)	t_o (s)
500	48	1.845
	36	1.835
	30	1.835
	24	**1.835**
	18	1.835
50	10	1.075
	8.0	1.003
	6.25	**1.000**

Highlighted grid sizes are the ones used for the simulation studies presented in this current work

Case 2: D = 50 μm

Three different meshes with grid sizes of 6.25, 8 and 10 μm were used for the test simulations. Reduction in Δx below 6.25 μm was not attempted as then the grid size Δx becomes smaller than the particle size d. Out of the 5 particles injected manually at the inlet, the t_o values of the particle injected at 3rd position from the upper wall of the channel computed for various grid sizes are shown in Table 3 for comparison. Given that there is practically no variation in the particle exit times t_o with the variation in Δx, we conclude that our results are independent of this parameter.

3 Results and Discussion

In the absence of EMFG, the SPM beads migrated into the daughter channel irrespective of the branching angle for both the cases. The SPM beads were injected randomly to mimic actual experiments. It was found that the beads which were injected into the stream lines in the upper half of the primary channel (closer to the opening of the daughter channel) had a higher probability of entering the daughter channel. This is because these stream lines are far more affected by the presence of the daughter channel than the stream lines in the lower half of the primary channel. The flow which starts out with only an x component of velocity in the primary channel, begins to develop a y-component around the branching point in order to flow through the daughter channel. We find that only about 50% of the SPM beads flow through the MC in the absence of EMFG.

Figure 3a shows the guiding efficiency for the case of the wider channel of D = 500 μm with the daughter channel of the same width, as a function of branch angle in the absence of an EMFG as well with two different EMFG strengths whose specifications are given in Table 1. A substantial increase in the guiding efficiency is obtained for $\Theta < 90°$ in the presence of an EMFG. While SPM beads flowing in

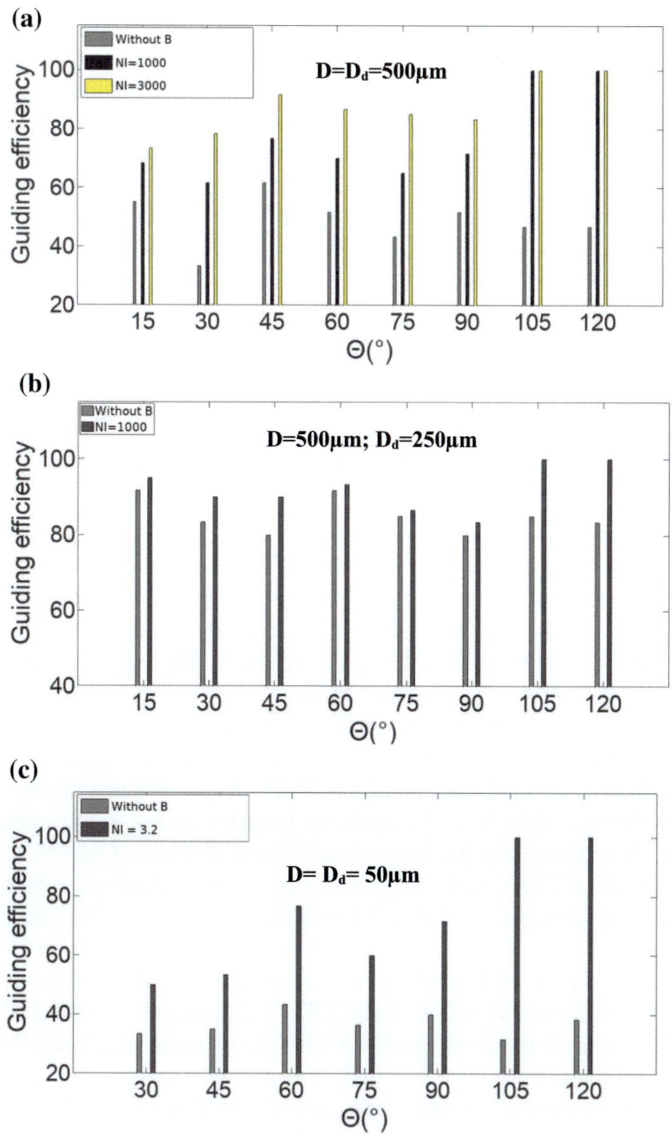

Fig. 3 Guiding efficiencies computed for channels of various branching angles with and without EMFG **a** $D = D_d = 500$ μm, **b** $D = 500$ μm, $D_d = D/2$ and **c** $D = D_d = 50$ μm. *Note* The simulations with $\Theta = 15°$ were not carried out as it required a $\Delta t \sim 10$ ns

the lower half of the channel (away from the branch opening) continue along the MC, beads flowing in the upper half of the channel with $F_m > F_d$ remain in the main channel increasing the guiding efficiency as compared to the situation without an EMFG. Moreover, the efficiency is better for the stronger gradient field, as is evident from Fig. 3a. Further, when branching angles are greater than 90°, we obtain 100% guiding efficiency for both field gradient strengths unlike the situation without the EMFG. This is because SPM beads moving into the daughter channel are attracted back into the MC when the EMFG, which is along the positive X direction, is present as this determines the direction of F_m.

Now, given the above results, we conclude that the lower *NI* value of 1000 At is a better choice between the two practically realizable anti-Helmholtz coil configurations that we have considered as complications like heating effects and aggregation will increase with the higher value.

As an extension, we have also carried out simulations with $D = 500$ µm and $D_d = 250$ µm (Fig. 2c), to see how the guiding efficiency is affected by the width of the daughter channel. While there is a definite loss of SPM beads into the daughter channel at all angles in the absence of EMFG (Fig. 3b), the loss is smaller when compared to the case of $D_d = D$ (Fig. 3a). We find that only the beads in stream lines very close to the upper wall of the primary channel enter the daughter channel unlike the earlier case where almost all SPM beads in the upper half of the channel were affected by the presence of the daughter channel. We find that in the presence of EMFG (*NI* = 1000 At) the guiding efficiency is improved. We conclude that the guiding efficiency in the absence of EMFG increases with decrease in the width of the daughter channel. At the same time, to achieve maximum guiding efficiency in channels with $\Theta < 90°$, the strength of the EMFG required is practically similar to that of the case with $D_d = D$ as F_d acting on particles closest to the walls are also the highest. Note that while the loss of beads in the absence of an EMFG appears reasonable when only a single branch is considered, this could prove to be a problem as the number of branches encountered during the flow of the beads to the target site increases.

Figure 3c shows the guiding efficiency computed for SPM beads in a channel of $D = D_d = 50$ µm, with and without EMFG. These simulations were carried out with the idea of gauging the practically realizable EMFG strengths that would be required to improve the guiding efficiency at such dimensions. The general trend does not seem to be qualitatively any different from the case of $D = D_d = 500$ µm.

4 Conclusion

It is evident from the simulations that a significant fraction of magnetic particles will invariably be lost from the primary channel into a branching daughter channel. These losses can be reduced by the use of an optimally designed EMFG. Using these simulations, it may be possible to estimate the total loss of magnetic particles due to branching channels encountered en route to the target site. This will enable better estimation of the appropriate dosage of magnetic particles that needs to be administered at the site of injection in order to achieve effective treatment. Choosing a proper dosage is a major challenge faced in such therapeutic techniques. Thus, these kind of simulations can be extremely helpful if one has to customize a magnetophoresis procedure for an individual undergoing treatment. We plan to extend these simulations to a more realistic three dimensional situation with differently sized branching channels and further also incorporate the situation where particles are no longer neutrally buoyant.

Acknowledgements RS would like to thank University Grants Commission (Govt. of India) for her fellowship under the NET—JRF/SRF scheme. The authors are grateful to the organizers of the "International Conference on Mathematical Modelling and Scientific Computing" for providing an opportunity to present this work and interact with experts working in a similar field leading to constructive critical input for this study.

References

1. Pankhurst, Q.A., Connolly, J., Jones, S.K., Dobson, J.: Applications of magnetic nanoparticles in biomedicine. J. Phys. D Appl. Phys. **36**, R167–R181 (2003)
2. Mohammed, L., Gomaa, H.G., Ragab, D., Zhu, J.: Magnetic nanoparticles for environmental and biomedical applications: a review. Particuology **30**, 1–14 (2017)
3. Andujar, C.B., Tung, L.D., Thanh, N.T.K.: Synthesis of nanoparticles for biomedical applications. Annu. Rep. Prog. Chem., Sect. A **106**, 553–568 (2010)
4. Bolden, N.W., Rangari, V.K., Jeelani, S., Boyoglu, S., Singh, S.R.: Synthesis and evaluation of magnetic nanoparticles for biomedical applications. J. Nanopart. Res **2013**, Article ID 370812 (2013)
5. Patsula, V., Kosinová, L., Lovrić, M., Hamzić, L.F., Rabyk, M., Konefal, R., Paruzel, A., Šlouf, M., Herynek, V., Gajović, S., Daniel Horák, D.: Superparamagnetic Fe_3O_4 nanoparticles: synthesis by thermal decomposition of iron[III] glucuronate and application in magnetic resonance imaging. ACS Appl. Mater. Interfaces **8**(11), 7238–7247 (2016)
6. Estelrich, J., Sánchez-Martín, M.J., Busquets, M.A.: Nanoparticles in magnetic resonance imaging: from simple to dual contrast agents. Int J Nanomedicine. **10**, 1727–1741 (2015)
7. Sinha, A., Ganguly, R., Puri, I.K.: Magnetic separation from superparamagnetic particle suspensions. J. Magn. Magn. Mater. **32**, 2251–2256 (2009)
8. Forbes, T.P., Forry, S.P.: Microfluidic magnetophoretic separations of immunomagnetically labeled rare mammalian cells. Lab Chip **12**, 1471 (2012)
9. Pham, X.N., Nguyen, T.P., Pham, T.N., Tran, T.T.N., Tran, T.V.T.: Synthesis and characterization of chitosancoated magnetite nanoparticles and their application in curcumin drug delivery. Adv. Nat. Sci. Nanosci. Nanotechnol. **7**(4), 045010 (2016)
10. Chomoucka, J., Jana, D., Huska, D., Adam, V., Kizek, R., Hubalek, J.: Magnetic nanoparticles and targeted drug delivering. Pharmacol. Res. **62**(2), 144–149 (2010)
11. Grief, A.D., Richardson, G.: Mathematical modelling of magnetically targeted drug delivery. J. Magn. Magn. Mater. **293**, 455–463 (2005)
12. Afkhami, S., Renardy, Y.: Ferrofluids and magnetically guided superparamagnetic particles in flows: a review of simulations and modeling. J. Eng. Math. **107**, 231–251 (2017)
13. Liu, Y., Shah, S., Tan, J.: Computational modeling of nanoparticle targeted drug deliver. Rev. Nanosci. Nanotechnol. **1**, 66–83 (2012)
14. Shi, Z., Sun, J., Jia, S., Zhang, P.: Simulation of magnetophoresis of magnetic nanoparticles in liquids. J. Phys. D: Appl. Phys. **49**(33), 335005 (2016)

15. Alam, M., Golozar, M., Darabi, J.: Modelling and simulation of particle-particle interaction in a magnetophoretic bioseparation chip. Phys. Fluids **30**, 042001 (2018)
16. Sharma, S., Kumar, R., Gaur, A.: A model for magnetic nanoparticles transport in a channel for targeted drug delivery. Procedia Mater. Sci. **10**, 44–49 (2015)
17. Uthra, C.R., Vasanthakumari, R.: Mathematical model for nano magnetic drug delivery and targeting using computational methods. WASJ **34**(8), 1083–1089 (2016)
18. Marcus, M., Karni, M., Baranes, K., Levy, I., Alon, N., Margel, S., Shefi, O.: Iron oxide nanoparticles for neuronal cell applications: uptake study and magnetic manipulations. J. Nanobiotechnol. **14**, 37 (2016)
19. Holligan, D.L., Gillies, G.T., Dailey, J.P.: Magnetic guidance of ferrofluidic nanoparticles in an in vitro model of intraocular retinal repair. Nanotechnology **14**, 661–666 (2003)
20. Li, W., Yangyang Liu, Y., Zhiyu Qian, Z., Yang, Y.: Evaluation of tumor treatment of magnetic nanoparticles driven by extremely low frequency magnetic field. Sci. Rep **7**, 46287 (2017)
21. Janikowska, A., Matuszak, J., Lyer, S., Schreiber, E., Unterweger, H., Zaloga, J., Groll, J., Alexiou, C., Cicha, I.: A novel human artery model to assess the magnetic accumulation of SPIONs under flow conditions. Sci. Rep. **7**, 42314 (2017)
22. Cui, Y., Zhang, M., Zeng, F., Jin, H., Xu, Q., Huang, Y.: Dual-targeting magnetic PLGA nanoparticles for codelivery of paclitaxel and curcumin for brain tumor therapy. ACS Appl. Mater. Interfaces **8**(47), 32159–32169 (2016)
23. Kim, S.-M., Seo, H.-S., Kim, Y.-J.: A study on the flow characteristics of oil-based ferrofluid with magnetic nanoparticles in the patterned microchannel under agnetophoresis. J. Nanosci. Nanotechnol. **16**, 10951–10958 (2016)
24. Hoshiar, A.K., Le, T.A., Amin, F.U., Kim, M.O., Yoon, J.: A novel magnetic actuation scheme to disaggregate nanoparticles and enhance passage across the blood-brain barrier. Nanomaterials **8**, 3 (2018)
25. Khashan, S.A., Dagher, S., Alazzam, A., Mathew, B., Alnaqbi, A.H.: Microdevice for continuous flow magnetic separation for bioengineering applications. J. Micromech. Microeng. **27**, 055016 (2017)
26. Do, T.D., Noh, Y., Kim, M.O., Yoon, J.: An optimized field function scheme for nanoparticle guidance in magnetic drug targeting systems. In Proceedings of the 2015 IEEE/RSJ International Conference on Intelligent Robots and Systems (IROS), Hamburg, Germany, pp. 4388–4393, 28 Sept–2 Oct 2015
27. Hamdipoor, V., Afzal, M.R., Le, T.-A., Yoon, J.: Haptic-based manipulation scheme of magnetic nanoparticles in a multi-branch blood vessel for targeted drug delivery. Micromachines **9**(1), 14 (2018)
28. Anatomy and physiology, Chapter 20. In The Cardiovascular System: Blood Vessels and Circulation, BC Open text book
29. D Mesh Tutorial using GMSH, Openfoamwiki
30. Chapter 5.2 Boundaries, OpenFOAM v5: User Guide
31. Késmárky, G., Kenyeres, P., Rábai, M., Tóth, K.: Plasma viscosity: a forgotten variable. Clin. Hemorheol. Microcirc. **39**, 243–246 (2008)
32. Weller, H.G., Tabor, G., Jasak, H., Fureby, C.: A tensorial approach to computational continuum mechanics using object-oriented techniques. Comput. Phys. **12**(6), 620–631 (1998)
33. Shevkoplyas, S.S., Siegel, A.C., Westervelt, R.M., Prentiss, M.G., Whitesides, G.M.: The force acting on a superparamagnetic bead due to an applied magnetic field. Lab Chip **7**, 1294–1302 (2007)
34. Tutorial icoLagrangianFoam/solidParticle. Author, Aurelia Vallier
35. Lin, D., Chen, X.: Mathematical models of 3D magnetic field and 3D positioning system by magnetic field. Appl. Math. Inf. Sci. **8**(4), 1647–1654 (2014)
36. Contrib/funkySetFields—OpenFOAMWiki
37. Hamedi, N.: Non-newtonian models in OpenFOAM—Implementation of a non-Newtonian model
38. Bayraktar, E., Mierka, O., Turek, S.: Benchmark computations of 3D laminar flow around a cylinder with CFX, OpenFOAM and FeatFlow. Int. J. Comput. Sci. Eng. **7**, 253–266 (2012)

Numerical Study on Fluid Flow Through Collapsible Channels

Vedant Dhruv, Ujwal Mishra and Ranjith Maniyeri

Abstract The fluid flow in collapsible channels or tubes is an interesting problem with several physiological applications; for example, blood flow in veins, air flow in lungs and wheezing. In this paper, we present a fluid-structure interaction based model for single-phase fluid flow through a microchannel containing two elastic walls. A two-dimensional model is developed and simulations have been performed using a commercial software. The deforming geometry is analyzed using moving mesh. The flow field and deformation of the elastic walls for different boundary loads and inlet flow conditions are presented and discussed.

Keywords Collapsible channel · Fluid-structure interaction · Boundary load · Microchannel

1 Introduction

The theory of internal fluid flow is nearly two centuries old. The flow through rigid tubes and between rigid plates has been understood to a great extent. However, several real life phenomena, especially in the physiological domain involve internal flow of fluid through elastic and deformable tubes. In all such cases, employing the hydrodynamics of flow through rigid tubes is not sufficient as it would not accurately model the situation. Such tubes or channels which consist of elastic boundaries tend to 'collapse' based on the transmural pressure and thus are known as 'collapsible channels.' The transmural pressure determines the shape of the collapsible channel. When the transmural pressure is positive and has a large value, the cross-sectional

V. Dhruv · U. Mishra · R. Maniyeri (✉)
Department of Mechanical Engineering, National Institute of Technology,
Surathkal, Mangalore, Karnataka 575025, India
e-mail: mranji1@nitk.edu.in

V. Dhruv
e-mail: vedantdhruv96@gmail.com

U. Mishra
e-mail: ujwalmishra86@gmail.com

S. Manna et al. (eds.), *Mathematical Modelling and Scientific Computing with Applications*, Springer Proceedings in Mathematics & Statistics 308,
https://doi.org/10.1007/978-981-15-1338-1_15

area of the channel is circular and the fluid flow can be considered as that of pipe flow. When the transmural pressure is reduced, a critical value is reached upon which the tubular channel begins to buckle and attains an elliptical cross-section. At low values of transmural pressure, the tube is said to be completely collapsed [1]. An important aspect of flow through collapsible tubes in the self-induced oscillations and the instabilities created in the flow when transmural pressure drops below certain value. These oscillations play an important role in several biological processes. Collapsible channel models can be used to study flow-induced collapse of blood vessels to carry out auto-regulation of blood supply to internal organs, flow of urine in urethras, airflow in the bronchial airways, flow-induced deformation of soft palate and pharyngeal wall which causes snoring, spontaneous flow-induced oscillations in the cardiovascular system such as oscillations of the external jugular vein, wheezing in lung airways. Hence, a better understanding of the flow through collapsible tubes is essential for the study and prevention of cardiovascular diseases (heart stroke) and lung diseases (asthma and emphysema).

The majority of the work carried out in the area of collapsible channel has been in the experimental regime. Due to the complexity of the problem, in recent years, numerical studies have been performed to better understand the flow through collapsible channels. Rosar and Peskin [2] developed a three-dimensional model to simulate the flow through a collapsible tube. The numerical method used to solve the equations was immersed boundary method. Cai and Luo [3] developed a plain-strained elastic beam model with large deflection and incremental linear extension and utilized finite element method (FEM). Jensen and Heil [4] performed a combination of asymptotic analysis and numerical simulation to describe high Reynolds number(Re) unsteady, pressure-driven flow in a finite length channel of which one wall contains a section of the membrane under longitudinal tension (modelling a Starling Resistor). Pihler-Puzovic and Pedley [5] have studied high-Re flow for an incompressible Newtonian fluid in a two-dimensional collapsible channel.

Motivated by these works, in our present paper we develop a fluid-structure interaction model to understand the flow through a collapsible channel and study the deformation of the elastic walls by considering different inlet conditions and boundary loads. The primary application of such an analysis would be to get a better understanding of the blood flow in human arteries [6].

2 Mathematical Modelling

Figure 1 shows the two-dimensional schematic diagram of the microchannel containing two elastic walls along with the boundary conditions. The outer walls are fixed and hence the no-slip condition is imposed. The elastic walls are fixed at the two extreme ends of the channel and are allowed to deform only within the specified domain.

The total channel length is 300 μm and the spacing between the elastic walls is 98 μm. The outer walls are at a distance of 50 μm from the elastic walls when

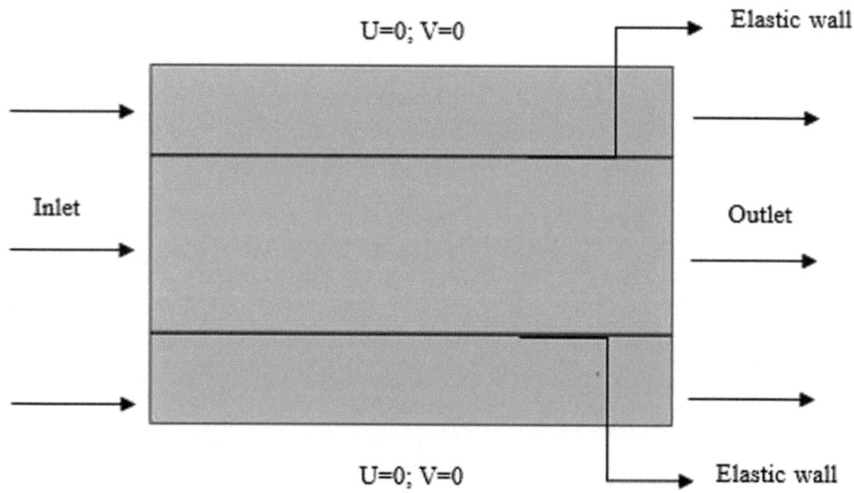

Fig. 1 Schematic diagram of the microchannel with the elastic walls

they are not deformed. The elastic walls are 1 μm thick. For the fluid domain, the boundaries on the left are specified as inlet for fluid flow and the boundaries on the right are specified as outlet with pressure equal to 0 Pa.

The properties of the deformable walls and the incompressible fluid that flows through the microchannel are shown in Tables 1 and 2 respectively.

Two cases for inlet fluid flow are considered: (i) constant flow velocity of 12 cm/s, which is the average blood flow speed in the aorta and (ii) a time dependent function with fully developed laminar characteristics and parabolic velocity profile whose amplitude changes with time. This function reaches a peak value at 0.215 s and thereafter gradually decreases to a steady-state amplitude of 0.033 m/s at inlet. Three cases are considered for the boundary load applied to the elastic walls: (i) zero boundary load, (ii) constant boundary load of 5 N/m and (iii) sinusoidal boundary load with maximum amplitude of 5 N/m. Hence we have overall six cases which are specified in Table 3.

Table 1 Physical properties of elastic medium

Property	Variable	Value
Young's modulus	E (N/m^2)	2e5
Poisson's ratio	Nu	0.33
Density	Rho (kg/m^3)	7850

Table 2 Physical properties of fluid medium (water)

Property	Variable	Value
Density	Dens (kg/m^3)	1000
Dynamic viscosity	Visc (Pa s)	0.001

Table 3 Different boundary loads and inlet flow conditions

Case	Boundary load (N/m)	Inlet velocity(m/s)
1a	0	u = 0.12
1b	0	$u = u_{mean}$
2a	5	u = 0.12
2b	5	$u = u_{mean}$
3a	5*sin(t)	u = 0.12
3b	5*sin(t)	$u = u_{mean}$

Here, the function u_{mean} is defined as,

$$u_{mean} = \frac{\mathbf{0.0333} * t^2}{\sqrt{t^4 - \mathbf{0.07}[s^2] * t^2 + \mathbf{0.0016}[s^4]}} \tag{1}$$

Since the present analysis consists of a deformable structure surrounded by fluid flow, the governing equations of fluid-structure interaction (FSI) are used. Equations (2) and (3) are the structural and fluid-flow equations respectively. The dynamics of the boundary is governed by Eq. (4). The commercial software employs an arbitrary Lagrangian-Eulerian (ALE) method to solve these equations for a continuously deforming geometry.

$$\rho^s \frac{D^2\boldsymbol{u}}{Dt^2} - \nabla . (\boldsymbol{F} . \boldsymbol{S}(\boldsymbol{u})) = \rho^s b^s \tag{2}$$

where $\boldsymbol{u}$ represents the displacements of the structure, $\boldsymbol{b}^s$ represents the body forces applied on the structure, $\boldsymbol{S}$ represents the second Piola-Kirchoff stress tensor, ρ^s represents the density of the structure and $\boldsymbol{F}$ represents the deformation gradient tensor.

$$\rho^F \frac{d\boldsymbol{v}}{dt}|_\chi + \rho^F . \boldsymbol{c} . \nabla \boldsymbol{v} - 2\mu \nabla . \boldsymbol{\varepsilon}(\boldsymbol{v}) + \nabla p = \rho^F \boldsymbol{b}^F; \quad \nabla . \boldsymbol{v} = 0 \tag{3}$$

where $\boldsymbol{v}$ denotes the fluid velocity, $|_\chi$ denotes the derivative is carried out while the referential domain is constant, p denotes the physical pressure and ρ^F and μ denote the fluid density and viscosity respectively. The fluid body forces are represented by $\boldsymbol{b}^F$ and $\boldsymbol{\varepsilon(v)}$ represents the strain rate tensor. It can be observed that the ALE formulation comes into the above equation in fluid acceleration and convective terms.

$$\boldsymbol{u}_\Gamma(t) = \boldsymbol{d}_\Gamma^F(t); \; \dot{\boldsymbol{u}}_\Gamma(t) = \boldsymbol{v}_\Gamma(t); \; \ddot{\boldsymbol{u}}_\Gamma(t) = \dot{\boldsymbol{v}}_\Gamma(t) \tag{4}$$

where Γ denotes the interface and $\boldsymbol{d}_\Gamma^F(t)$ represents the displacement of the fluid mesh nodes at the interface.

3 Results and Discussion

We use a time-dependent solver to simulate the fluid flow through the microchannel consisting of elastic walls. The simulations are carried out from t = 0 s to t = 5 s with a time step of 0.1 s for all the cases as mentioned in Table 3.

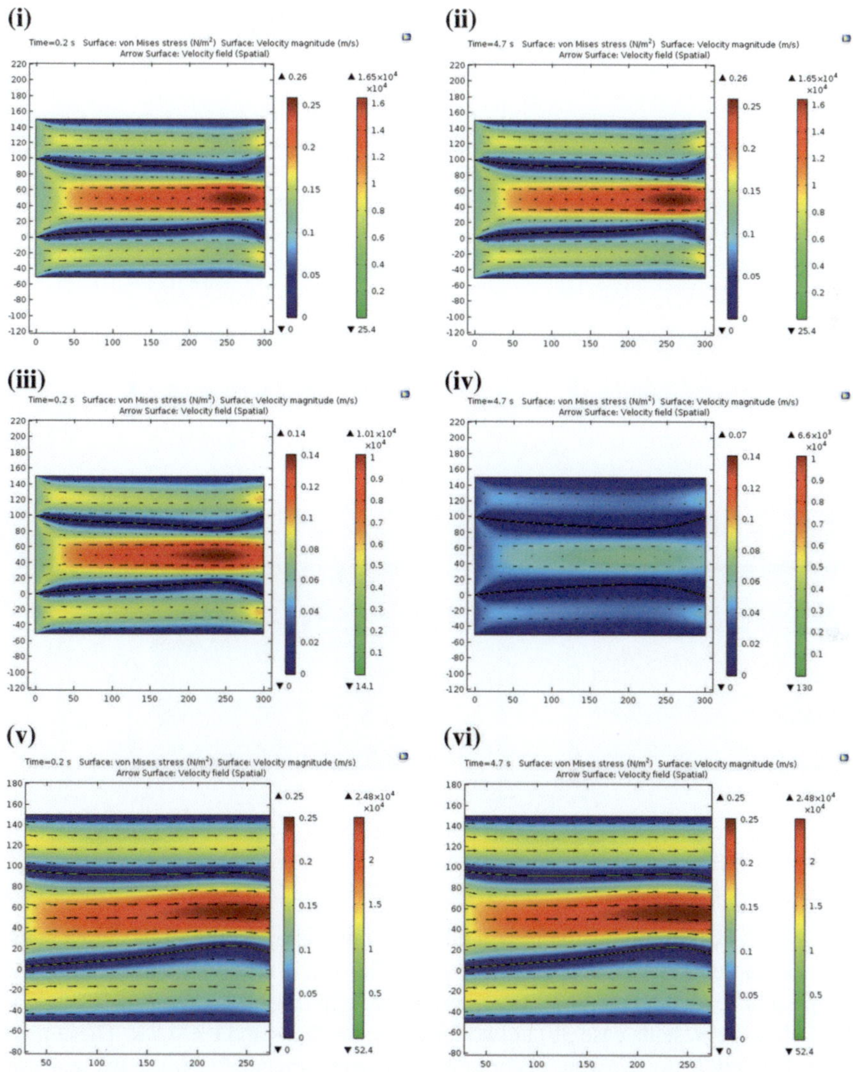

Fig. 2 i–xii Illustration of the flow field, maximum velocity achieved, stresses on the elastic walls and their deformation for the six cases. i, ii—case 1a; iii, iv—case 1b; v, vi—case 2a; vii, viii—case 2b; ix, x—case 3a; xi, xii—case 3b as shown in Table 3 at t = 0.2 s and t = 4.7 s

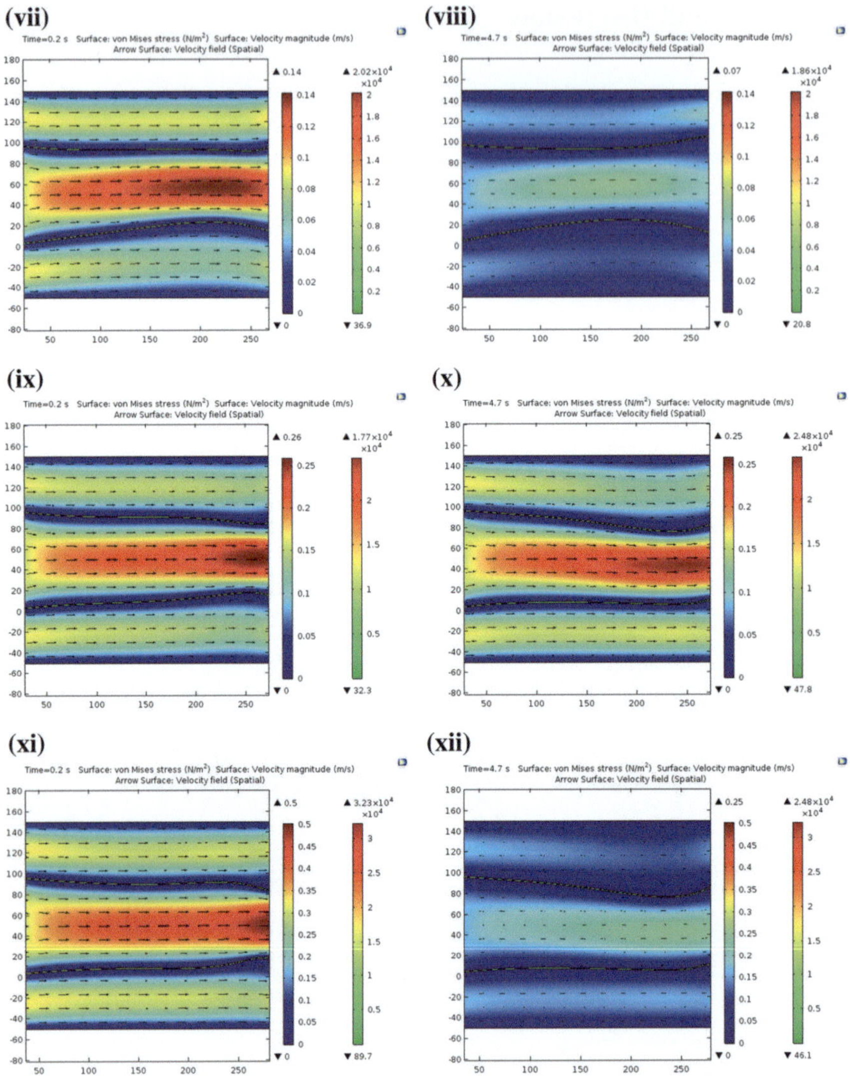

Fig. 2 (continued)

Figure 2i–xii highlights the flow field in the microchannel for all the cases mentioned in Table 3 (Fig. 2i, ii represents case 1a; Fig. 2iii, iv represents case 1b; Fig. 2v, vi represents case 2a; Fig. 2vii, viii represents case 2b; Fig. 2ix, x represents case 3a; Fig. 2xi, xii represents case 3b) at two intervals; t = 0.2 s and t = 4.7 s. These time intervals allow us to see what occurs inside the microchannel near the two ends of the time range, i.e., at initial stages and when the flow has settled to a constant value. The stresses acting on the elastic walls and their deformation due to the flow of the fluid are also depicted in Fig. 2.

Figure 2i, ii, v, vi shows that for zero boundary load or constant boundary load, the fluid velocity remains the same over the entire time range. The maximum velocity that is achieved in the channel once the fluid becomes fully developed remains the same over the entire duration of the simulation time (0.26 m/s for zero boundary load and 0.25 m/s for constant boundary load). For zero boundary load i.e., cases 1(a) and 1(b) (Fig. 2i–iv) we observe that the elastic walls of the channel collapse towards the outlet of the domain. This resembles the throat section that is present in nozzles and consequently the fluid velocity rises and reaches a value much higher than the inlet velocity and the pressure drops to a value much lower than the surrounding region. For example, in case 1a, the fluid velocity at the collapse region ($x = 260\ \mu m$, $y = 50\ \mu m$) is 0.26 m/s and pressure is 7.545 Pa.

We observe in Fig. 2i–viii that once the fluid velocity attains steady state, the elastic walls do not undergo further deformation. Hence the deformation present in cases 3(a) and 3(b) at later time intervals is clearly due to the sinusoidal time-varying boundary load that has been imposed.

To better understand the deformation of the elastic walls, we consider two points on each of the walls at a length of 50 and 250 μm. The deflection of each of these points for all the cases at time intervals of 1 s is presented in Fig. 3.

The data points in Fig. 3a, c are more clustered than those in Fig. 3b, d. This shows that the elastic walls deformation is more pronounced at the outlet than at the inlet and that the collapse of the channel occurs near the outlet. This is in agreement with the results of the study performed by Zhu and Wang [7].

4 Conclusion

A time-dependent study is performed on a two-dimensional microchannel consisting of two elastic walls to understand the phenomenon of fluid flow through collapsible tubes. Six test cases are considered with different inlet flow velocities and boundary loads acting on the elastic walls. Numerical simulation results for different cases show that the walls undergo deformation primarily at the outlet region. Moreover, these deformed walls do not undergo further change once the flow has attained a steady state. Only upon inducing a time-varying boundary load on the walls, they undergo further deformation. It is to emphasize here that, the developed mathematical model and the scientific computing done using commercial software in understanding the fluid dynamic behaviour of collapsible channels will help the research community to deduce physical results which are difficult to explore through experimentation. Hence, suitable mathematical model and the associated numerical simulations are need of the hour bringing detailed scientific results to solve the real life engineering problems. Also, in our future work, we plan to improve this primitive model so that it can simulate physiological processes with better accuracy.

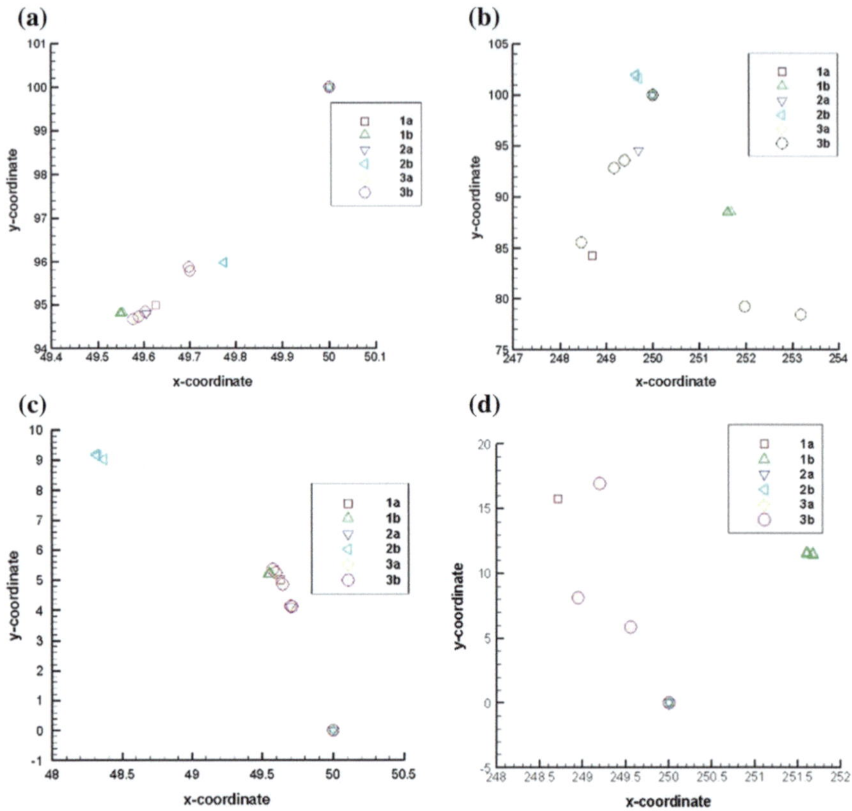

Fig. 3 **a** Deformation of point on top wall near inlet. **b** Deformation of point on top wall near outlet. **c** Deformation of point on bottom wall near inlet and **d** Deformation of point on bottom wall near outlet

References

1. Luo, X.Y., Pedley, T.J.: Modelling flow and oscillation in collapsible tubes. Theoret. Comput. Fluid Dyn. **10**, 277–294 (1998)
2. Rosar, M.E., Peskin, C.S.: Fluid flow in collapsible elastic tubes: a three-dimensional numerical model. New York J. Math. **7**, 281–302 (2001)
3. Cai, Z.X., Luo, X.Y.: A fluid–beam model for flow in a collapsible channel. J. Fluids Struct. **17**(1), 125–146 (2003)
4. Jensen, O.E., Heil, M.: High-frequency self-excited oscillations in a collapsible-channel flow. J. Fluid Mech. **481**, 235–268 (2003)
5. Pihler-Puzović, D., Pedley, T.J.: Stability of high-Reynolds-number flow in a collapsible channel. J. Fluid Mech. **714**, 536–561 (2013)
6. Tuema E.V., Ilegbusi O.: Unsteady integrodifferential equation of fluid-structure interaction in constricted collapsible tube model of diseased human coronary artery. Int. J. Differ. Equ. 1–21 (2012)
7. Zhu, J., Wang, X.: A study of steady and unsteady flow in a collapsible channel. In Computational Fluid and Solid Mechanics, pp. 1597–1600 (2003)

Water Boundary Layer Flow over an Exponentially Permeable Stretching Sheet with Variable Viscosity and Prandtl Number

Abhishek Kumar Singh, N. Govindaraj and S. Roy

Abstract The present work focus on water boundary layer flow over an exponential permeable stretching sheet in the presence of suction/injection with variable viscosity and prandtl number. The nonlinear partial differential equations governing flow and thermal fields are presented in non-dimensional form using suitable non-similar transformation. Finally non dimensional partial differential the equations are solved by the implicit finite difference method in combination with the Quasi-linearization technique. The numerical results for skin-friction and local Nusselt number are shown graphically to display effects of physical parameters.

Keywords Exponential permeable stretching sheet · Water boundary layer · Variable Prandtl number · Quasilinearization technique · Variable viscosity

Nomenclature

Pr	Prandtl number
f	Dimensionless streamfunction
C_p	Specific heat at constant pressure
T	Temperature
g	Acceleration due to gravity
U_W	Moving plate velocity
Nu_x	Nusselt number

A. K. Singh (✉) · N. Govindaraj
Mathematics Division, School of Advanced Science,
VIT University, Chennai Campus, Chennai 600127, India
e-mail: abhishek.iitm1@gmail.com

N. Govindaraj
e-mail: govindaraj07@yahoo.co.in

S. Roy
Department of Mathematics, Indian Institute of Technology Madras, Chennai 600036, India
e-mail: sjroy@iitm.ac.in

S. Manna et al. (eds.), *Mathematical Modelling and Scientific Computing with Applications*, Springer Proceedings in Mathematics & Statistics 308,
https://doi.org/10.1007/978-981-15-1338-1_16

Cf_x	Skin friction coefficient
U_∞	Free stream velocity
v	Velocity component in the y direction
u	Velocity component in the x direction
ν	Kinematic viscosity
ρ	Density
x, y	Cartesian coordinates
Re_L	Local Reynolds number
μ	Dynamic viscosity

1 Introduction

The phenomenon of mixed convection water boundary layer flow over an exponential permeable stretching surface has more practical application in the engineering field. The well know examples of permeable stretching surface are follow as paper production, glass spinning, metal spinning, fiber and extrusion processes, the cooling of an infinite metallic plate in a cooling bath, the aerodynamic extrusion of plastic sheets, condensation processes in film, roofing shingles, extraction of polymer and rubber sheets, wire drawing and glass-fiber. In these engineering industries, the quality of final product depends on the rate of heat transfer at the stretching surface which is presented in term of Nusselt number in the current work. Therefore this study may useful for improvement of quality of final product. Recently the permeable exponential stretching surface and temperature has been getting special attention due to above mentioned applications in that, surface undergoes stretching due to geophysical situation.

A few earlier studied on boundary layer flow over the surface may be outlined as follows. Elbashbeshy [1] investigated laminar boundary flow in quiescent fluid over an exponentially stretching surface subject to the suction. Zaimi et al. [2] studied boundary layer flow and heat transfer over a permeable stretching/shrinking sheet in viscous fluid. Thermal radiation and chemical reaction on MHD flow of a nanofluid past a permeable sheet in the presence of suction has been done by Sandeep [3]. Hafidzuddin [4] discussed boundary layer flow and heat transfer over a permeable exponential stretching sheet. Ishak [5] performed the steady boundary layer flow over a moving permeable sheet with various effects of thermal field. Patil [6] analyzed double diffusive mixed convection boundary layer flows over an exponentially stretching sheet. Olusoji [7] discussed the heat and mass transfer effect of the electrically conducting micropolar fluid flow past a stretching sheet with influence of velocity and thermal slip condition. Hayat et al. [8] investigated the MHD boundary layer of a nanofluid flow moving over a permeable stretching sheet with convective boundary condition. Srinvasulu et al. [9] discussed MHD boundary layer flow of a nanofluid over a stretching sheet with the effects of non-uniform heat source and chemical reaction.

Few numerical studies report similarity solutions on boundary layer flows over stretching sheet [10–12]. However, water boundary layer study with permeable exponentially stretching sheet on moving vertical plate for variable viscosity and Prandtl number did not attempted upto till date.

The aim of the present work is to investigate the mixed convection water boundary layer flow over an exponentially permeable stretching sheet continuously moving over a vertical plate with variable viscosity and Prandtl number. Consider the water is a most common application in the engineering field. The coupled non-linear partial differential equations have been solved numerically using an implicit finite difference scheme in combination with the Quasi-linear technique used by Singh et al. [13]. Results are compared with the results reported earlier by Tsou et al. [14], Soundalgekar and Murthy [15], Ali [16], Moutsoglou and Chen [17], Chen [18] and are found to be in excellent agreement (Table 2).

2 Mathematical Formulation

Let us consider the incompressible mixed convection water boundary layer flow over an exponentially permeable stretching surface (Fig. 1). The x-axis is taken along the permeable exponentially stretching sheet in the vertically upward direction and y-axis is taken to the normal to it. The stretching surface velocity and permeable exponentially free stream velocity are defined as follows (Table 1): $U_w(x)=U exp\left(\frac{x}{L}\right)$, $U_e(x)=U_\infty exp\left(\frac{x}{L}\right)$

$$\mu = \frac{1}{b_1 + b_2 T} \quad \text{and} \quad Pr = \frac{1}{c_1 + c_2 T}$$

where $b_1 = 53.41$, $b_2 = 2.43$, $c_1 = 0.068$, $c_2 = 0.004$

Governing equations:

$$\frac{\partial u}{\partial x} + \frac{\partial v}{\partial y} = 0 \tag{1}$$

$$u\frac{\partial u}{\partial x} + v\frac{\partial u}{\partial y} = U_e\left(\frac{dU_e}{dx}\right) + \frac{1}{\rho}\frac{\partial}{\partial y}\left(\mu\frac{\partial u}{\partial y}\right) + g[\beta(T - T_\infty)] \tag{2}$$

$$u\frac{\partial T}{\partial x} + v\frac{\partial T}{\partial y} = \frac{1}{\rho}\frac{\partial}{\partial y}\left(\frac{\mu}{Pr}\frac{\partial T}{\partial y}\right) \tag{3}$$

The boundary conditions are given by:

$$y = 0: \quad u(x, 0) = U_w(x), \quad v = v_w, \quad T = T_\infty + (T_w - T_\infty) exp\left(\frac{2x}{L}\right)$$

$$y \to \infty: \quad u \to U_e(x), \quad T \to T_\infty \tag{4}$$

Applying the following transformations:

$$\xi = \frac{x}{L};\ \eta = \left(\frac{U}{x\nu}\right)^{1/2} exp\left(\frac{x}{2L}\right) y;\ \ \psi(x, y) = (Ux\nu)^{1/2} exp\left(\frac{x}{2L}\right) f(\xi, \eta);$$
$$u = \frac{\partial\psi}{\partial y};\ v = -\frac{\partial\psi}{\partial x};\ u = U \exp\left(\frac{x}{L}\right) F;\ f_\eta(\xi; \eta) = F(\xi, \eta);$$
$$v = -\left(\frac{\nu U}{x}\right)^{\frac{1}{2}} exp\left(\frac{x}{2L}\right)\left\{\frac{f}{2}(1+\xi) + \xi f_\xi + \frac{\eta}{2}(\xi - 1)F\right\};\ Re_L = \frac{UL}{\nu};$$
$$T = T_\infty + (T_w - T_\infty) exp\left(\frac{2x}{L}\right)\theta(\xi, \eta);\ Gr = \frac{g\beta_T(T_w - T_\infty)L^3}{\nu^2} \tag{5}$$

to Eqs. (1)–(3), then Eq. (1) is identically satisfied, and Eqs. (2)–(3) reduce to:

$$(NF_\eta)_\eta + \frac{(\xi+1)}{2} fF_\eta - \xi F^2 + \xi\lambda\theta + \xi\epsilon^2 = \xi(FF_\xi - f_\xi F_\eta) \tag{6}$$

$$(NPr^{-1}\theta_\eta)_\eta + \frac{(\xi+1)}{2} f\theta_\eta - 2\xi F\theta = \xi(F\theta_\xi - f_\xi\theta_\eta) \tag{7}$$

where $\mu = \frac{1}{b_1 + b_2 T};\ Pr = \frac{1}{c_1 + c_2 T} = \frac{1}{a_2 + a_3\theta};\ N = \frac{\mu}{\mu_\infty} = \frac{b_1 + b_2 T_\infty}{b_1 + b_2 T} = \frac{1}{1 + a_1\theta};\ a_1 = \frac{b_2(T_\infty - T_w)}{b_1 + b_2 T_\infty};\ a_2 = c_1 + c_w;\ a_3 = c_2(T_\infty - T_w);\ \Delta T_w = (T_\infty - T_w);\ b_1 = 53.41,\ \ b_2 = 2.43,\ \ c_1 = 0.068,\ \ c_2 = 0.004$
where

$$f = \int_0^\eta F d\eta + f_w$$

f_w can be obtained from the following equation
$v_w = -\left(\frac{\nu U}{x}\right)^{\frac{1}{2}} exp\left(\frac{x}{2L}\right)\left\{\frac{f_w}{2}(1+\xi) + \xi(f_\xi)_w\right\}$
i.e, $\{f_w(1+\xi) + 2\xi(f_\xi)_w\} = -2v_0\xi^{\frac{1}{2}}\left(\frac{L}{\nu U}\right)^{\frac{1}{2}} = a\xi^{\frac{1}{2}}$ where $a = -2v_0\left(\frac{L}{\nu U}\right)^{\frac{1}{2}} =$ constant, is the surface mass transfer parameter with $a > 0$ for the suction, $a < 0$ for the injection and a = 0 for an impermeable. Further, $v_w = v_0 e^{\frac{x}{2L}}$, is the surface mass transfer which is also varies with x, v_0 is the initial value of the surface mass transfer. Thus, $v > 0$ corresponds to the case of suction while $v < 0$ corresponds to the case injection.

The non-dimensional boundary conditions become:

$$F = 1,\ \ \theta = 1\ \ at\ \ \eta = 0$$

$$F = \varepsilon,\ \ \theta = 0\ \ at\ \ \eta = \eta_\infty \tag{8}$$

where $\epsilon = \dfrac{U_\infty}{U}$ corresponds to the ratio of free stream velocity to the reference velocity. The skin friction (C_{fx}) and Nusselt number (Nu_x) are defined as:

$$C_{fx} = \mu \frac{2\left(\frac{\partial u}{\partial y}\right)_{y=0}}{\rho U^2} = 2\left(Re_L \xi exp(\xi)\right)^{-1/2} F_\eta(\xi, 0) \\ i.e.\ C_{fx}\left(Re_L \xi exp(\xi)\right)^{1/2} = 2F_\eta(\xi, 0) \tag{9}$$

$$Nu_x = -x \frac{\left(\frac{\partial T}{\partial y}\right)_{y=0}}{(T_w - T_\infty)} = \left(Re_L \xi exp(\xi)\right)^{1/2} \theta_\eta(\xi, 0) \\ i.e.\ Nu_x\left(Re_L \xi exp(\xi)\right)^{-1/2} = -\theta_\eta(\xi, 0) \tag{10}$$

The boundary value problem represented by Eqs. (6)–(7) with boundary condition [Eq. (8)] has been solved numerically using an implicit finite difference scheme in combination with the Quasi-linearization technique. The non linear partial differential [Eqs. (6)–(7)] is converted to linear partial differential equations:

$$X_1^k F_{\eta\eta}^{k+1} + X_2^k F_\eta^{k+1} + X_3^k F^{K+1} + X_4^k F_\xi^{k+1} + X_5^k \theta_\eta^{k+1} + X_6^k \theta^{k+1} = X_7^k \tag{11}$$

$$Y_1^k \theta_{\eta\eta}^{k+1} + Y_2^k \theta_\eta^{k+1} + Y_3^k \theta^{k+1} + Y_4^k \theta_\xi^{k+1} + Y_5^k F^{k+1} = Y_6^k \tag{12}$$

with boundary condition

$$F^{k+1} = 1,\ \theta^{k+1} = 1\ at\ \eta = 0 \\ F^{k+1} = \epsilon,\ \theta^{k+1} = 0\ at\ \eta = \eta_\infty \tag{13}$$

The coefficient function with iteration index k are know and the functions with iterative index (k+1) are to be determined. The coefficients in Eqs. (11) and (12) are given by:
$X_1^k = N;\ X_2^k = -a_1 N^2 \theta_\eta + (\xi + 1)\frac{f}{2} + \xi f_\xi;\ X_3^k = -\xi F_\xi - 2\xi F; X_4^k = -\xi F;\ X_5^k = -a_1 N^2 F_\eta;\ X_6^k = -a_1 N^2 F_{\eta\eta} + 2a_1^2 N^3 F_\eta \theta_\eta + \lambda\xi;\ X_7^k = -a_1 N^2 F_\eta \theta_\eta - a_1 N^2 F_{\eta\eta}\theta + 2a_1^2 N^3 F_\eta \theta_\eta \theta - \xi F^2 - \xi\epsilon^2 - \xi F F_\xi;\ Y_1^k = N pr^{-1};\ Y_2^k = -2a_1 N^2 pr^{-1}\theta_\eta + 2a_3\theta_\eta N + \frac{f}{2}(\xi + 1) + \xi f_\xi;\ Y_3^k = a_3 N \theta_{\eta\eta} - a_1 N^2 pr^{-1}\theta_{\eta\eta} - 2a_1 a_3 N^2 \theta_\eta^2 + 2a_1^2 N^3 pr^{-1}\theta_\eta^2;\ Y_4^k = -\xi F;\ Y_5^k = -\xi\theta_\xi;\ Y_6^k = a_3\theta_\eta^2 N - a_1\theta_\eta^2 pr^{-1} N^2 + a_3\theta_{\eta\eta}\theta N - a_1\theta_{\eta\eta}\theta pr^{-1} N^2 - 2a_1 a_3 \theta_\eta^2 \theta N^2 + 2pr^{-1} a_1^2 \theta_\eta^2 \theta N^3 - \xi F \theta_\xi$
The system is reduced to a system of linear algebraic equations with block tri-diagonal matrix, which is solved by Varga's algorithm [19]. To ensure the convergence of numerical solutions to the exact solutions.

3 Results and Discussion

The numerical results are displayed in terms of velocity profile (F), temperature profile (θ), skin friction coefficient (C_{fx} $(Re_L\xi exp(\xi))^{1/2}$) and heat transfer coefficients ($Nu_x(Re_L\xi exp(\xi))^{-1/2}$) for various values of $\epsilon(0.2 \leq \epsilon \leq 2)$, $\lambda(1 \leq \lambda \leq 3)$, $\xi(0 \leq \xi \leq 2)$ and $a(-1 \leq a \leq 1)$. Figures 2 and 3 displays the velocity profile and Fig. 4 display the temperature profile. The skin friction coefficient and Nusselt number are shown in Figs. 5 and 6 exponentially permeable stretching surface (see Fig. 1) to be in excellent agreement (see Table 2) velocity are defined as follows (see Table 1).

The effects of ε (= 0.5, 1.0 and 1.5) and a (= −1 and 1) on the velocity profile (F) is presented in Fig. 2. Results indicate that ε and a parameters have high impact on the appearance of velocity profiles. It is noted that the velocity profile exponentially increasing trends in F is observed for $\varepsilon = 1.5$ whereas symmetric profile is observed at $\varepsilon = 1.0$ at a = 1, for all λ. On the other hand, back flow is observed at $\varepsilon = 0.5$, hence velocity profile decreases with η for all λ. For example, at $\lambda = 1$, $\xi = 0.5$, velocity increases approximately by 14% with the increases from $\epsilon = 0.5$ to $\epsilon = 1.0$ at a = 1.

Figure 3 displays the effects of λ (= 1 and 3) and a(= −1, 0 and 1) on the velocity profile (F) at $\varepsilon = 0.5$ and $\xi = 1.0$. The velocity profile exponentially decreasing trends in observed for all values λ within $\eta(0 \leq \eta \leq 4)$ and after that constant pattern is observed. It is seen that the magnitude of velocity overshoot is observed for higher buoyancy force $\lambda(= 3)$ and injection parameter ($a < 0$). The physical reason is that

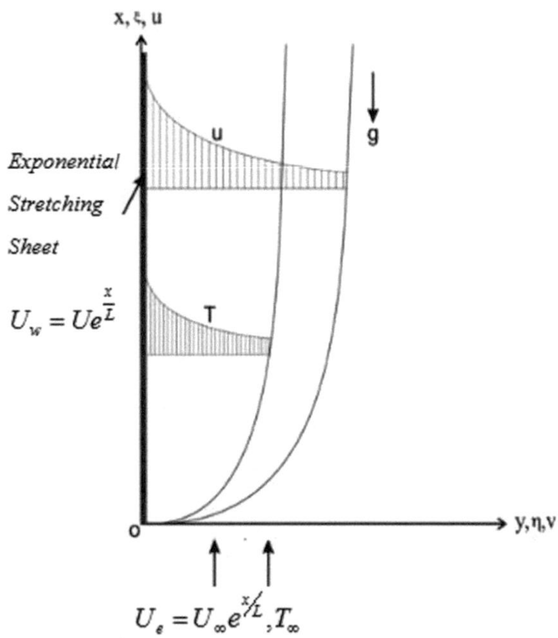

Fig. 1 Physical model and coordinate system

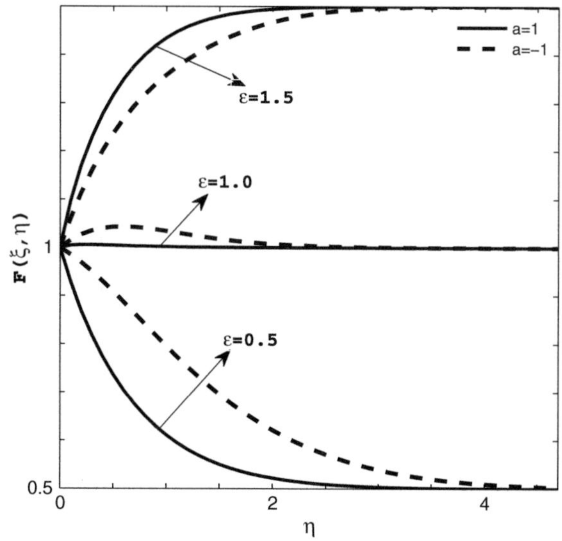

Fig. 2 Effects of ϵ and a on velocity profile for $\xi = 0.5$ and $\lambda = 1$

the buoyancy force effects are more significant in injection parameter. In particular, increase in ξ from $\xi = 0$ to $\xi = 1$ results in an decreases of the velocity profile approximately 24% for a = 0, at = 0.5 and $\lambda = 3.0$.

The effects of ε (= 0.2 and 0.8) and ξ (= 0 and 1) on temperature profile (Θ) are presented in Fig. 4 at $\lambda = 2.0$ and $a = 0$. Decreasing trend in Θ downstream direction η $(0 \leq \eta \leq 1)$, irrespective of ε and ξ. Thereafter constant term is presented for higher range of η $(1.6 \leq \eta \leq 2)$. The temperature profile decreases with η indicates that an increase in stream wise coordinate ξ acts as a decelerating pressure gradient and thus fluid flows slower. In particular for $\epsilon = 0.8$ and $\lambda = 2.0$, the temperature profile decreases approximately by 99% as ξ increases from 1.0 to 2.0

Figure 5 shows the influence of λ (= 2 and 4) and ε (= 0.5 and 1.5) on the skin friction coefficient $[C_{fx}\,(Re_L\xi exp(\xi))^{1/2}]$. The skinfriction $C_{fx}\,(Re_L\xi exp(\xi))^{1/2}$ coefficient increases with increases of λ and ϵ. This is due to the fact that the increase in ε enhances the fluid acceleration and hence the skin friction coefficient increases. The skin-friction coefficient increases approximately by 182% due to increase of ϵ from $\epsilon = 0.5$ to $\epsilon = 1.5$ at $\lambda = 2.0$ and a = 0. Figure 6 displays the variation of the $\lambda(= 1$ and 2) and $\epsilon(= 1$ and 2) on heat transfer coefficient $[Nu_x(Re_L\xi exp(\xi))^{-1/2}]$. The heat transfer rate $(Nu_x(Re_L\xi exp(\xi))^{-1/2})$ increases with buoyancy parameter (λ) and ϵ. In particular, for $\epsilon = 1.0$ heat transfer rate $(Nu_x(Re_L\xi exp(\xi))^{-1/2})$ increases about 61% as ξ from $\xi = 0$ to $\xi = 1$ at $\lambda = 1$ and a = 0.

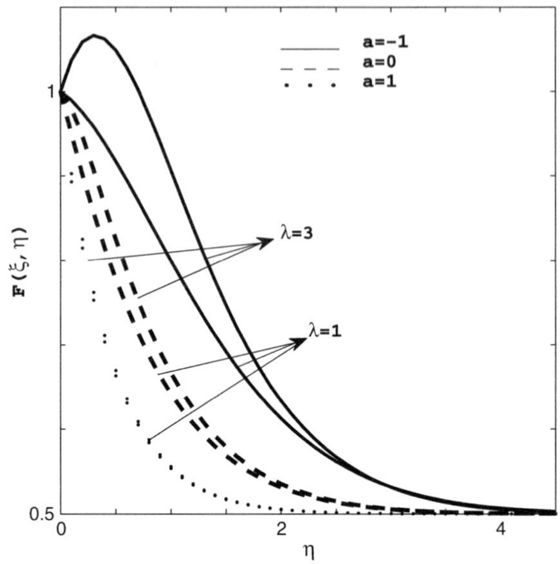

Fig. 3 Effects of λ and a on velocity profile for and $\epsilon = 0.5$ and $\xi = 1.0$

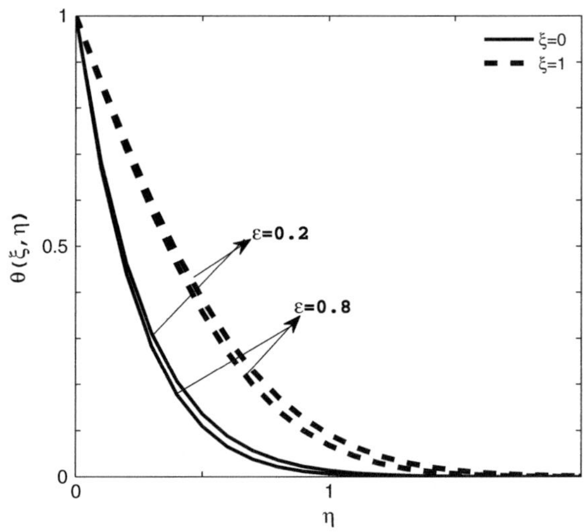

Fig. 4 Effects of ϵ and ξ on temperature profile for $\lambda = 2$ and $a = 0$

4 Conclusions

In this study the following conclusions may be remarked from the finding of this numerical investigation.

- The velocity profile overshoot near the wall region increases with buoyancy parameter (λ) and velocity ratio (ϵ). In particular, as ξ increases from $\xi = 2$ to $\xi = 3$ velocity profile (F) increases from 11% for $\lambda = 1$, $\epsilon = 0.5$ and a = 1.

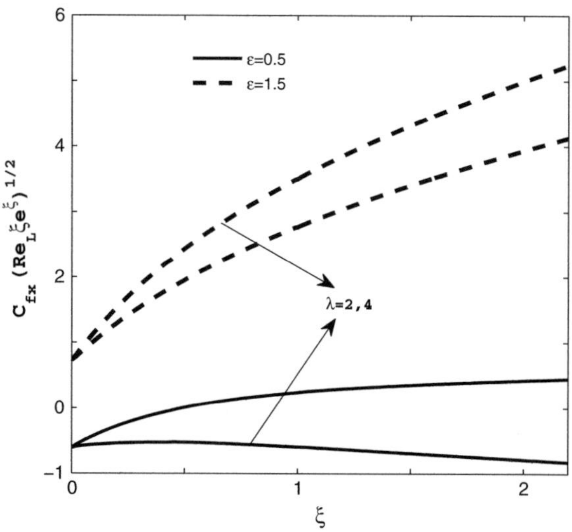

Fig. 5 Effects of ϵ and λ on skinfriction coefficient

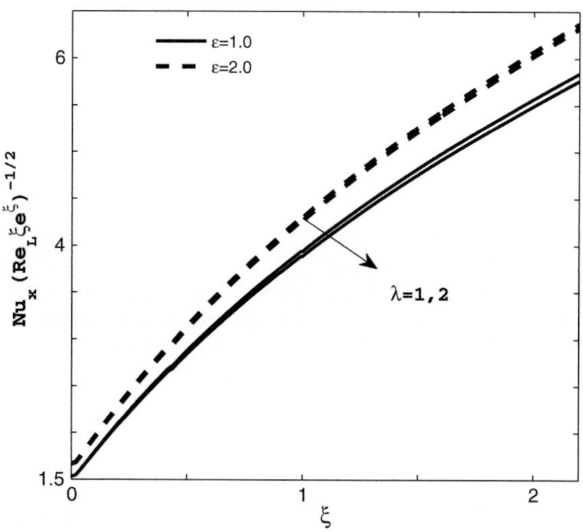

Fig. 6 Effects of ϵ and λ on heat transfer coefficient

- The temperature profile increases with increase in stream wise coordinate ξ and decreases velocity ratio (ϵ). In particular, as ξ increases from $\xi = 0$ to $\xi = 1$ temperature profile θ decreases 90%.
- The skin friction coefficient $C_{fx}\,(Re_L\xi exp(\xi))^{1/2}$ increases with velocity ratio (ϵ) and buoyancy parameters (λ). In particular at $\epsilon = 0.5$, as ξ increase from $\xi = 1$ to $\xi = 2$, the skin friction increases by about 2% at m = 0 and $\lambda = 2$.
- The heat transfer coefficient $Nu_x(Re_L\xi exp(\xi))^{-1/2}$ decreases with buoyancy parameter (λ) and composite velocity (ϵ). In particular, for $\lambda = 2$, as ϵ increases from $\epsilon = 0.5$ to $\epsilon = 1.5$, the heat transfer rate decrease by about 8% at a = 0 and $\xi = 0$.

Table 1 Values of thermophysical properties of water at different temperature [23]

Temperature (T)(°C)	Density (g/cm^3)	Specific heat ($J\,10^7/kg\,K$)	Thermal conductivity ($erg\,10^5/cm\,s\,K$)	Viscosity ($g\,10^{-2}/cm\,s$)	Prandtl no Pr
0	1.00228	4.2176	0.5610	1.7930	13.48
10	0.99970	4.1921	0.5800	1.3070	9.45
20	0.99821	4.1818	0.5984	1.0060	7.03
30	0.99565	4.1784	0.6154	0.7977	5.12
40	0.99222	4.1785	0.6305	0.6532	4.32
50	0.98803	4.1806	0.6435	0.5470	3.55

Table 2 Comparison of the steady state results $-\theta_\eta(0)$ for $\lambda = 0, \xi = 0, \varepsilon = 0$, and selected value of Pr to previously published works

Pr	2	5	7	10	100
Tsou et al. [20]	–	–	–	1.6804	5.545
Soundalgekar and Murty [15]	0.6831	–	–	1.6808	–
Ali [16]	–	–	–	1.6713	–
Moutsoglou and Chen [21]	–	–	1.38703	–	–
Chen [22]	0.68324	–	1.38619	1.68008	5.54450
Present work	0.6830	1.151	1.386	1.6801	5.5450

Acknowledgements Let me thanks organization team of International Conference on Mathematical Modelling and Scientific Computing who brought the platform to express our idea about mathematical modelling and simulation in applied mathematics. The current work is completely based on modelling of fluid dynamics problem and solution has been obtained by using finite difference method and given in terms of velocity profile (F), temperature profile (θ), skin friction coefficient ($C_{fx}\,(Re_L\xi exp(\xi))^{1/2}$) and heat transfer coefficients ($Nu_x(Re_L\xi exp(\xi))^{-1/2}$).

References

1. ElbElbashbeshy, E.M.A.: Heat transfer over an exponentially stretching continuous surface with suction. Arch. Mech. **53**, 643–651 (2001)
2. Zaimi, K., Ishak, A.: Boundary layer flow and heat transfer over a permeable stretching /shrinking sheet with convective boundary condition. J. Appl. Fluid Mech. **8**, 499–505 (2015)
3. Naramgari, S., Sulochana, C.: MHD flow over a permeable stretching sheet of a nanofluid with suction/injection. Alexandria Eng. J. **55**, 819–827 (2016)
4. Hafidzuddin, E.H., Nazar, R., Arifin, N.M., Pop, I.: Boundary layer flow and heat transfer over a permeable exponentially stretching/shrinking sheet with generalized slip velocity. J. Appl. Fluid Mech. **9**, 2025–2036 (2016)
5. Ishak, I., Nazar, R., Pop, I.: The effects of transpiration on the flow and heat transfer over a moving permeable surface in a parallel stream, chem. Eng. J. **148**, 63–67 (2009)
6. Patil, P.M., Latha, D.N., Roy, S., Momoniat, E.: Non similar solutions of mixed convection flow from an exponentially stretching surface. Ain Shams Eng. J. **8**, 697–705 (2015)

7. Olusoji, E.: Heat and mass transfer in MHD micropolar fluid flow over a stretching sheet with velocity and thermal slip conditions (2018)
8. Hayat, T., Imtiaz, M., Alsaedi, A.: MHD flow of Nanofluid over permeable stretching sheet with convective boundary conditions. Open J. Fluid Dyn. **8**(2), 195 (2014)
9. Srinvasulu, T., Bandari, Shankar: MHD boundary layer flow of nanofluid over a nonlinear stretching sheet with effect of non-uniform heat source and chemical reaction. J. Nanofluids **6**(4), 637–646 (2017)
10. Hayat, T., Shafiq, A., Alsaedi, A., Shahzad, S.A.: Unsteady MHD flow over exponentially stretching sheet with slip conditions. Appl. Math. Mech. **37**(2), 193–208 (2016)
11. Bidin, Biliana, Nazar, Roslinda: Numerical solution of the boundary layer flow over an exponentially stretching sheet with thermal radiation. Eur. J. Sci. Res. **33**, 710–717 (2009)
12. Cortel, R.: Fluid flow and radiative non linear heat transfer over a stretching sheet. J. King Saud Univ. Sci. **26**, 161–167 (2014)
13. Singh, P.J., Roy, S., Pop, I.: Unsteady mixed convection from a rotating vertical slender cylinder in an axial flow. Int. J. Heat Mass Transf. **51**, 1423–1430 (2008)
14. Tsou, F.K., Sparrow, E.M., Goldstein, R.J.: Flow and heat transfer in the boundary layer on a continuous moving surface. Int. Heat Mass Transfer **10**, 219–235 (1967)
15. Soundalgekar, V.M., Murty, T.V.R.: Heat transfer in flow past a continuous moving plate with variable temperature. Warme-und Stoffubertragung **14**, 91–93 (1980)
16. Ali, M.E.: On thermal boundary layer on a power-law stretched surface with suction or injection. Int. J. Heat Fluid Flow **16**, 280–290 (1995)
17. Moutsoglou, T.S.Chen: Buoyancy effects in boundary layers on inclined, continuous, moving sheets. ASME J. Heat Transf. **102**, 371–373 (1980)
18. Chen, C.H.: Laminar mixed convection adjacent to vertical, continuously stretching sheets. Heat Mass Transf. **33**, 471–476 (1998)
19. Varga, R.S.: Matrix Itrative Analysis. Prentice-Hall, Englewood Cliffs, NJ (2000)
20. Tsou, F.K., Sparrow, E.M., Goldstein, R.J.: Flow and heat transfer in the boundary layer on a continuous moving surface. Int. J. Heat Mass Transf. **10**, 219–235 (1967)
21. Moutsoglou, A., Chen, T.S.: Buoyancy effects in boundary layers on inclined continuous moving sheets. ASME J. Heat Transf. **102**, 371–373 (1980)
22. Chen, C.H.: Laminar mixed convection adjacent to vertical continuously stretching sheets. Heat Mass Transf. **33**, 471–476 (1998)
23. Lide, D.R. (ed.): CRC Handbook of Chemistry and Physics, 71st edn. CRC Press, BocaRaton, FL (1990)

Finite Element Analysis of MHD Blood Flow in Stenosed Coronary Artery with the Suspension of Nanoparticles

Ankita Dubey and B. Vasu

Abstract The numerical study presents a two dimensional mathematical modelling and computational simulation of blood flow in a stenosed coronary artery in the presence of magnetic field. Blood flow model is considered based on second grade fluid flow and heat transfer with the suspension of nanoparticles. Vogel's model is employed for viscosity of blood as a function of temperature. In order to complete our model, the variability in design and size of stenosis is considered. The finite element method is used to solve the transformed conservation equations numerically in conjunction of variational approach and FreeFEM++. The results show that an increase in the thermophoresis parameter (N_t) decreases the velocity while the increment in the Brownian motion parameter (N_b) increases the velocity in the whole domain. An increase in N_t and Brownian motion parameter (N_b), there is an increase in temperature values and nanoparticles concentration at the throat of the stenosis and as well as in the remaining domain. These properties changes in the domain by changing the shapes and designs of the stenosis in the domain.

Keywords Blood flow · Vogel's model · Nanoparticles · Magnetohydrodynamics · Thermophoresis · Brownian motion · Coronary artery · Stenosis

Nomenclature

A, B	Constants in viscosity function
B_r	Brownian diffusion constant
D_b	Brownian diffusion coefficient
D_T	Thermophoretic diffusion coefficient
g	Gravitational vector

A. Dubey (✉) · B. Vasu
Department of Mathematics, MNNIT, Allahabad, Prayagraj 211004, India
e-mail: dubey.ankita179@gmail.com

B. Vasu
e-mail: bvasu@mnnit.ac.in

S. Manna et al. (eds.), *Mathematical Modelling and Scientific Computing with Applications*, Springer Proceedings in Mathematics & Statistics 308,
https://doi.org/10.1007/978-981-15-1338-1_17

α_1, α_2	Material modules
G_r	Grashof Number
B_0	Magnetic field
ρ_f	Density of the base fluid
ρ_p	Density of the nanoparticles
M	Magnetohydrodynamics parameter
N_b	Brownian motion parameter
N_t	Thermophoresis parameter
V	Velocity vector
A_1, A_2	Rilvin Erickson Tensors
J	Electric current density
θ	Temperature
κ	Thermal conductivity
Φ	Nanoparticle volume fraction
λ_1	Viscoelastic parameter

1 Introduction

Non-Newtonian fluid flows have many important roles in the chemical processing industry, plastics processing industry, and in other engineering and industrial areas. Non-Newtonian behaviour is also encountered in the mining industry, where slurries and muds are often handled, and in applications such as lubrication and biomedical flows. 7% of the human body weight has accounts for blood. In 1981, Fung [1] represents blood as suspension of various tiny particles, in a continuous saline solution plasma in which plasma behaves as a Newtonian fluid whereas the whole blood is a suspension of cells and highly viscous in nature, exhibits the property of a non-Newtonian fluid (Biswas [2] and Fung [1]). At low shear rate blood shows a non-Newtonian character but at high shear rate, blood behaves like a Newtonian fluid which is usually available in large arteries [3, 4]. In the present study, blood is considered as non-Newtonian fluid as discussed by Baieth [5].

The arterial blood flow has some important role due to its application in engineering and medical field. The presence of arterial stenosis influence the hemodynamic behaviour of blood flow. The characteristics of blood flow hemodynamics has great significance in assessment and diagnosis of cardiovascular disease, testing the hypothesis of disease formation, helping in plan of vascular surgery, modelling the transport of drugs through the circulatory systems and also determining nanoparticle concentrations. The study helps to predict the performance of cardiovascular equipments or instruments that have not yet been built such as heart valves, stents, probes etc.

In human vascular system, it is thought that the stenosis is to be a significant factor in the onset of coronary artery disease (CAD), a progressive disease characterized [6] by the accumulation of plaques on the artery walls, which is one of the major causes of deaths in all the society as discussed in [7]. It is the most common diseases, affecting the cardiovascular system by the deposition of fatty materials in the coronary artery which thickens the plaque on the arterial wall. The obstruction of blood flow in artery due to thickening of wall with plaque is called stenosis. According to Ali et al. [8] study it is clearly visible that the development of stenosis in an artery results in the reduction of the blood flow rate or ischemia. In the coronary artery, ischemia may the constriction of the arterial lumen grows inward and restricts the normal movement of blood where the transport of blood to the region beyond the narrowing is reduced. Moreover, under normal physiological condition, the transport of blood in the human circulatory system depends entirely on the pumping action of the heart which produces a pressure gradient, throughout the arterial system. Carlos and Kiran [6] and Mathur and Jain [9] have discussed the blood flow in stenotic artery whereas Akbar [10] discussed the blood flow modelling in tapered arteries with stenosis.

The new class of fluids which consist of a base fluid with nano-sized particles (1–100 nm) suspended within them, called nanofluids [11, 12], which appear to be applicable in heat exchanger, cooling of electronics and diesel electric generator, nuclear reactor cooling, refrigeration (domestic refrigerator, chillers) solar water heating, boiler flue gas temperature reduction, diesel combustion cooling, and heating in buildings, biomedical application, fiber petroleum reservoirs, and nuclear waste repositories, cameras, micro devices, and displays. Nanofluids have specific property to enhance thermo physical properties such as thermal conductivity, thermal diffusivity, viscosity, and convective heat transfer coefficients compared to those of base fluids like oil or water. Pak and Cho [13] discussed that the chemically stable metals (e.g., gold, copper), metal oxides [14] (e.g., alumina, silica, zirconia, titania), oxide ceramics (e.g., Al_2O_3, CuO), metal carbides (e.g., SiC), metal nitrides and functionalized nanoparticles are commonly used as nanoparticles. Kumar et al. [15] and Giljohann et al. [16] also investigated the effect of the gold nanoparticles suspension in blood, which states that in biomedical sciences and different biomedical applications the small size of gold nanoparticles is very important. Nowadays for certain diseases, various nano sized particle drugs are used to increase or decrease the blood capillary growth and most of them are very effective.

Magnetohydrodynamics (MHD), study the motion of highly conducting fluid under the influence of the magnetic field. Some of the researches which have been done on the effects of magnetization on the arterial vessel of the blood flow are such as Darcy [17], studied fluid flow through porous media whereas Korchevskii and Marochnik [18], discussed the magneto-hydrodynamic version of blood flow. In 1974, Sud et al. [19] gave an idea about the effect of magnetic field on oscillating blood flow in arteries. In their study, authors have given a brief report that increment in the value of the magnetic field shows the increment in flow resistance of the blood flow in the stenosed artery.

In the present study a small segment of stenosed coronary artery is considered in which the non-Newtonian blood flow and heat transfer effect due the nano particles is modelled and studied. The aim of this paper is to simulate the MHD flow and heat transfer for blood in the stenosed coronary artery with the suspension of nanoparticles by using the finite element method with variational approach in FreeFem++. And also to investigate influence of some parameters such as thermophoresis parameter and Brownian motion parameter on velocity, nanoparticle concentration and temperature profile.

2 Mathematical Model for Blood Flow

A 2-D mathematical model for blood flow in a coronary artery is considered where blood flow is modeled as homogeneous fluid flow with the suspension of nanoparticles. The blood is modelled as second grade fluid (Hayat et al. [20]). For the simulation, the stationary case of the coronary artery is assumed i.e., the pulsatile nature of blood is neglected. The velocity is taken zero at the arterial wall. Coronary artery is chosen as cylindrical tube and considered the cylindrical coordinates (r, φ, z), where r is the radial direction, the z as axial direction and φ as azimuthal direction. Since the flow is in two dimension so we can neglect the flow and heat transfer in azimuthal (φ) direction i.e. the flow is in only radial (r) and axial (z) direction [7] as shown in Fig. (1).

Considering blood to be an incompressible second grade non-Newtonian fluid and the viscosity of blood is taken as the function of temperature as in Vogel's method [21].

$$\mu = \mu_0 e^{((A/B+\theta)-\theta_L)} \tag{1}$$

Since the natural convection is taken here then the nanofluid's density is will be [22]:

$$\begin{aligned} \rho &= \phi\rho_p + (1-\phi)\rho_{f_0} \\ &\cong \phi\rho_p + (1-\phi)[\rho_f(1-\beta_T(\theta-\theta_L))] \end{aligned} \tag{2}$$

where ρ_{f_0} is base fluid (blood) density and ρ_f is the base fluid's density at the reference temperature θ_L. Yadav et al. [23] adopted the base fluid's density as nanofluid's density. Hence the base fluid's density will be as in Eq. (2). The conservation of mass, momentum, thermal energy and total mass equation of the nanofluid will be as:

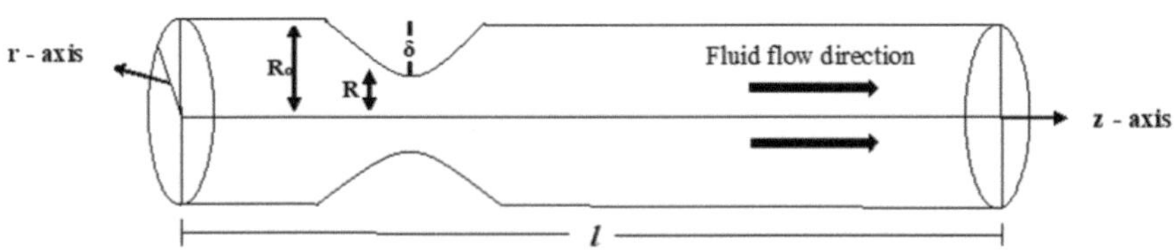

Fig. 1 Schematic diagram of stenosed coronary artery

$$\nabla \cdot V(r, z) = 0 \tag{3}$$

$$\rho_f\left(\frac{\partial V}{\partial t} + (V \cdot \nabla)V\right) = divT_s + [\phi\rho_p + (1-\phi)[\rho_f(1-\beta_T(\theta-\theta_L))]]g + J \times B \tag{4}$$

$$(\rho c)_f\left(\frac{\partial \theta}{\partial t} + (V \cdot \nabla)\theta\right) = k\nabla^2\theta + (\rho c)_p[D_b(\nabla\phi \cdot \nabla\theta) + \frac{D_T}{\theta_L}(\nabla\theta \cdot \nabla\theta)] \tag{5}$$

$$\left(\frac{\partial \phi}{\partial t} + (V \cdot \nabla)\phi\right) = D_b\nabla^2\phi + \frac{D_T}{\theta_L}\nabla^2\theta \tag{6}$$

where ϕ is the nanoparticles volume fraction. In Eq. (4) the third term on the right hand side is considered due to the consideration of electrically conducting fluids. Hence the following equation, presents the Ohm's law by which we can find the relation between J and B [24]:

$$J = \sigma(E + V \times B) \tag{7}$$

where E represents the electric field and in the last term of above equation the right-hand side $\boldsymbol{B}$=$\boldsymbol{B}_0$+$\boldsymbol{b}$ represents the total magnetic field, σ the electrical conductivity, $\boldsymbol{V}$, the velocity vector and $\boldsymbol{J}$ represents the electric current density.

For small magnetic Reynolds number [23, 24], the induced magnetic field will be neglected. Hence:

$$J \times B = -\sigma B_0^2 V \tag{8}$$

Stress for second grade fluid will be

$$T_s = -pI + \mu A_1 + \alpha_1 A_2 + \alpha_2 A_1^2 \tag{9}$$

where α_1, α_2 are material modules and in general considered to be functions of temperature.

In Eq. (9), due to the restraint of incompressibility $-pI$ represents the spherical stress, and the kinematical tensors A_1, A_2 can be defined by following equations [25],

$$A_1 = \nabla V + (\nabla V)^t \tag{10}$$

$$A_n = \frac{dA_{n-1}}{dt} + A_{n-1}(\nabla V) + (\nabla V)^t A_{n-1} \tag{11}$$

where the fluid velocity is taken only in radial and axial direction and represented as $V = [u(r, z), 0, w(r, z)]$

For the model it is required to be compatible with thermodynamics in the sense that all motions satisfy the Clasius-Duhem inequality and it is assumed that the specific

Helmholtz free energy is minimum in equilibrium, then all the material derivatives taken in above equations must meet the following conditions [26, 27].

$$\mu \geq 0, \alpha_1 \geq 0 \quad and \quad \alpha_1 + \alpha_2 = 0 \tag{12}$$

The corresponding boundary conditions are:

$$\begin{aligned} &u = 0, w = U_0, \frac{\partial w}{\partial r} = 0, \frac{\partial \theta}{\partial r} = 0, \frac{\partial \phi}{\partial r} = 0 \quad at\, r = 0 \\ &u = 0, w = 0, \theta = \theta_w, \phi = \phi_w \qquad at\, r = R(z) \end{aligned} \tag{13}$$

Introducing the non-dimensional parameters

$$\begin{aligned} &\bar{u} = \frac{uL_0}{U_0\delta}, \bar{w} = \frac{w}{U_0}, \bar{z} = \frac{z}{L_0}, \bar{r} = \frac{r}{R_0}, \bar{\mu} = \frac{\mu}{\mu_0} \\ &\bar{\theta} = \frac{\theta - \theta_L}{\theta_w - \theta_L} \quad and\, \bar{\phi} = \frac{\phi - \phi_L}{\phi_w - \phi_L} \end{aligned} \tag{14}$$

The nondimensional geometric parameters appearing in the terms defined above are stenosis height parameter ($\delta^* = \delta/R_0 << 1$) and the vessel aspect ratio ($\varepsilon = R_0/L_0$).

For the subsequent analysis, we shall assume that $\delta^* \ll 1$ and $\varepsilon = O(1)$, i.e., the maximum height of the stenosis is small in comparison with the radius of the artery and also that the radius of the artery and length of the stenotic region are of comparable magnitude. Hence these parameters should be neglected.

Hence the modified non-dimensional form of the governing equations: as follows

$$\frac{\partial \bar{\mu}}{\partial \bar{r}} \frac{\partial \bar{w}}{\partial \bar{r}} + \lambda_1 \left[-\frac{1}{\bar{r}} \left(\frac{\partial \bar{w}}{\partial \bar{r}} \right)^2 + 2 \frac{\partial \bar{w}}{\partial \bar{r}} \frac{\partial^2 \bar{w}}{\partial \bar{r}^2} \right] + \left(B_r \bar{\phi} + G_r \bar{\theta} \right) = 0 \tag{15}$$

$$\begin{aligned} &\frac{\partial \bar{\mu}}{\partial \bar{z}} \frac{\partial \bar{w}}{\partial \bar{r}} + \lambda_1 \begin{pmatrix} -\frac{\partial \bar{w}}{\partial \bar{z}} \frac{\partial^2 \bar{w}}{\partial \bar{r}^2} - \bar{w} \frac{\partial^3 \bar{w}}{\partial \bar{r}^2 \partial \bar{z}} + \frac{\bar{w}}{\bar{r}} \frac{\partial^2 \bar{w}}{\partial \bar{r} \partial \bar{z}} \\ -\frac{1}{\bar{r}} \frac{\partial \bar{w}}{\partial \bar{r}} \frac{\partial \bar{w}}{\partial \bar{z}} + \frac{\partial \bar{w}}{\partial \bar{r}} \frac{\partial^2 \bar{w}}{\partial \bar{r} \partial \bar{z}} + 2 \frac{\partial \bar{w}}{\partial \bar{r}} \frac{\partial^2 \bar{w}}{\partial \bar{z} \partial \bar{r}} \end{pmatrix} \\ &+ G_r \bar{\theta} + B_r \bar{\phi} - M \bar{w} = 0 \end{aligned} \tag{16}$$

$$\nabla^2 \bar{\theta} + N_b (\nabla \bar{\phi} \cdot \nabla \bar{\theta}) + N_t (\nabla \bar{\theta} \cdot \nabla \bar{\theta}) = 0 \tag{17}$$

$$\nabla^2 \bar{\phi} + \frac{N_t}{N_b} (\nabla^2 \bar{\theta}) = 0 \tag{18}$$

In this study viscosity of nanofluid is considered as a function of temperature as given in Vogel's method [21].

We can rewrite Eq. (1) as [28]

$$\mu = \frac{c}{S}\left[1 - \frac{A\bar{\theta}}{B^2}\right], \quad \text{where } S = \mu_0 e^{((A/B)-\theta_L)} \tag{19}$$

where A, B, c, μ_0, θ_0 are the different parameters.

After applying the above equation in system of Eqs. (15)–(18), we get

$$C\frac{\partial\bar{\theta}}{\partial\bar{r}}\frac{\partial\bar{w}}{\partial\bar{r}} + \lambda_1\left[-\frac{1}{\bar{r}}\left(\frac{\partial\bar{w}}{\partial\bar{r}}\right)^2 + 2\frac{\partial\bar{w}}{\partial\bar{r}}\frac{\partial^2\bar{w}}{\partial\bar{r}^2}\right] + \left(B_r\bar{\phi} + G_r\bar{\theta}\right) = 0 \tag{20}$$

$$C\frac{\partial\bar{\theta}}{\partial\bar{z}}\frac{\partial\bar{w}}{\partial\bar{r}} + \lambda_1\begin{pmatrix} -\frac{\partial\bar{w}}{\partial\bar{z}}\frac{\partial^2\bar{w}}{\partial\bar{r}^2} - \bar{w}\frac{\partial^3\bar{w}}{\partial\bar{r}^2\partial\bar{z}} + \frac{\bar{w}}{\bar{r}}\frac{\partial^2\bar{w}}{\partial\bar{r}\partial\bar{z}} \\ -\frac{1}{\bar{r}}\frac{\partial\bar{w}}{\partial\bar{r}}\frac{\partial\bar{w}}{\partial\bar{z}} + \frac{\partial\bar{w}}{\partial\bar{r}}\frac{\partial^2\bar{w}}{\partial\bar{r}\partial\bar{z}} + 2\frac{\partial\bar{w}}{\partial\bar{r}}\frac{\partial^2\bar{w}}{\partial\bar{z}\partial\bar{r}} \end{pmatrix}$$
$$+ G_r\bar{\theta} + B_r\bar{\phi} - M\bar{w} = 0 \tag{21}$$

$$\nabla^2\bar{\theta} + N_b(\nabla\bar{\phi}\cdot\nabla\bar{\theta}) + N_t(\nabla\bar{\theta}\cdot\nabla\bar{\theta}) = 0 \tag{22}$$

$$\nabla^2\bar{\phi} + \frac{N_t}{N_b}(\nabla^2\bar{\theta}) = 0 \tag{23}$$

With non-dimensional boundary conditions:

$$\begin{aligned} &\bar{u} = 0, \bar{w} = 1, \frac{\partial\bar{w}}{\partial\bar{r}} = 0, \frac{\partial\bar{\theta}}{\partial\bar{r}} = 0, \frac{\partial\bar{\phi}}{\partial\bar{r}} = 0 \quad at\,\bar{r} = 0 \\ &\bar{u} = 0, \bar{w} = 0, \bar{\theta} = 1, \bar{\phi} = 1 \quad at\,\bar{r} = \bar{R}(z) \end{aligned} \tag{24}$$

where $U_0, L_0, R_0, \mu_0, \theta_w, \theta_L$ and ϕ_L denote the reference velocity, reference length of the blood vessel, reference diameter, reference viscosity, pipe temperature, reference fluid temperature, and reference nanoparticle concentration in the arterial tube model respectively.

$$\begin{aligned} &M = \frac{\sigma B_0^2 L_0 R_0}{\mu_0}, N_b = \frac{D_b(\phi_w - \phi_L)(\rho c)_p}{\kappa}, N_t = \frac{D_T(\theta_w - \theta_L)(\rho c)_p}{\theta_0\kappa}, \\ &C = \frac{cA}{SB^2}, Gr = \frac{(\theta_w - \theta_L)\rho_{f,w}g(1-\phi_L)L_0R_0\beta_T}{U_0\mu_0}, \text{Re} = \frac{\rho_f U_0 L_0}{\mu_0}, \\ &Br = \frac{(\rho_p - \rho_f)L_0R_0(\phi_w - \phi_L)g}{U_0\mu_0}, \lambda_1 = \frac{\alpha_1 U_0}{\mu_0 R_0}, \text{Pr} = \frac{c_p\mu_0}{\kappa}, Le = \frac{\mu_0}{\rho_f D_b} \end{aligned} \tag{25}$$

3 Applied Numerical Solution

To solve the system of governing equations we have used finite element method of variational approach with FreeFEM++.

For weak formulation of the system of differential equations, we define the function spaces

$$X = \{\bar{u} \in (H_1(\Omega))|\bar{u} = a\, on\, \Gamma_{in}, \bar{u} = 0\, on\, \Gamma_{wall}\}$$
$$Q = \{\bar{u} \in (H_1(\Omega))|\bar{u} = 0\, on\, \Gamma_{in} \cup \Gamma_{wall}\} \tag{26}$$

The weak form of Eqs. (20)–(23) is to find u ϵ X and ϕ, θ ∈ X such that $u_h, v_h, w_h, q_h \in Q$. Here X and Q are finite dimensional spaces.

So the weak formulation of Eqs. (20)–(23) will be

$$\int_\Omega C\frac{\partial\bar{\theta}}{\partial\bar{r}}\frac{\partial\bar{w}}{\partial\bar{r}}\cdot u_h\, d\Omega - \int_\Omega \lambda_1\frac{1}{\bar{r}}\left(\frac{\partial\bar{w}}{\partial\bar{r}}\right)^2\cdot u_h\, d\Omega$$
$$+\int_\Omega 2\frac{\partial\bar{w}}{\partial\bar{r}}\frac{\partial^2\bar{w}}{\partial\bar{r}^2}\cdot u_h\, d\Omega + \int_\Omega B_r\bar{\phi}\cdot u_h\, d\Omega + \int_\Omega G_r\bar{\theta}\cdot u_h\, d\Omega = 0 \tag{27}$$

$$\int_\Omega C\frac{\partial\bar{\theta}}{\partial\bar{z}}\frac{\partial\bar{w}}{\partial\bar{r}}\cdot v_h\, d\Omega - \int_\Omega \lambda_1\frac{\partial\bar{w}}{\partial\bar{z}}\frac{\partial^2\bar{w}}{\partial\bar{r}^2}\cdot v_h\, d\Omega - \int_\Omega \lambda_1\bar{w}\frac{\partial^3\bar{w}}{\partial\bar{r}^2\partial\bar{z}}\cdot v_h\, d\Omega$$
$$+\int_\Omega \lambda_1\frac{\bar{w}}{\bar{r}}\frac{\partial^2\bar{w}}{\partial\bar{r}\partial\bar{z}}\cdot v_h\, d\Omega - \int_\Omega \lambda_1\frac{1}{\bar{r}}\frac{\partial\bar{w}}{\partial\bar{r}}\frac{\partial\bar{w}}{\partial\bar{z}}\cdot v_h\, d\Omega + \int_\Omega \lambda_1\frac{\partial\bar{w}}{\partial\bar{r}}\frac{\partial^2\bar{w}}{\partial\bar{r}\partial\bar{z}}\cdot v_h\, d\Omega$$
$$+\int_\Omega 2\lambda_1\frac{\partial\bar{w}}{\partial\bar{r}}\frac{\partial^2\bar{w}}{\partial\bar{z}\partial\bar{r}}\cdot v_h\, d\Omega + \int_\Omega G_r\bar{\theta}\cdot v_h\, d\Omega + \int_\Omega B_r\bar{\phi}\cdot v_h\, d\Omega$$
$$-\int_\Omega M\bar{w}\cdot v_h\, d\Omega = 0 \tag{28}$$

$$\int_\Omega \nabla^2\bar{\theta}\cdot w_h\, d\Omega + N_b\int_\Omega (\nabla\bar{\phi}\cdot\nabla\bar{\theta})\cdot w_h\, d\Omega + N_t\int_\Omega (\nabla\bar{\theta}\cdot\nabla\bar{\theta})\cdot w_h\, d\Omega = 0 \tag{29}$$

$$\int_\Omega (\nabla^2\bar{\phi})\cdot q_h\, d\Omega + \frac{N_t}{N_b}\int_\Omega (\nabla^2\bar{\theta})\cdot q_h\, d\Omega = 0 \tag{30}$$

Equations (27–30) are solved numerically using FreeFEM++ [29] in fixed mesh with 6462 triangular elements and number of vertices are 3392. For the mesh, the minimum step size ($h_{\min}$) is 0.0001 and the tolerance for computation is $<10^{-6}$. Classic Taylor-Hood element (P_1, P_2) technique is adopted to approximate the velocity, temperature and nanoparticles volume fraction concentration.

4 Results and Discussion

In this section the behaviour of the velocity magnitude, temperature and the mass concentration of nanoparticles for different shape and designs of stenosis is presented. Figure 2a, b, c d shows plots of the velocity magnitude for two different values of thermophoresis parameter (N_t) and Brownian motion parameter(N_b) for domain Ω that resembles the case of stenosis (here the blood flow direction is taken from left to right). Similarly for temperature distribution and nanoparticles concentration for the different values of thermophoresis parameter (N_t) and the Brownian motion parameter(N_b) are shown in Figs. 3a, b, c, d and 4a, b, c, d respectively. For computation of the model, the default values of different parameters are given in Table 1.

It is observed from Figs. 2a–d and 3a–b, increasing the values of thermophoresis parameter (N_t) decreases the value of velocity in whole domain while increasing the Brownian motion parameter (N_b), the velocity magnitude increases and again decreases by increasing thermophoresis parameter (N_t). The value of temperature decreases in whole domain when Brownian motion parameter (N_b) increases (at the stenosis and the remaining boundary of vessel before and after stenosis) and increases with increment of thermophoresis parameter (N_t). And the increase in value of N_t increase the concentration in whole domain with the increase in Brownian motion parameter (N_b) as one can find in Fig. 4a–d.

The three Figs. 5, 6, and 7 are showing the graphs of velocity profile, temperature distribution and nanoparticles concentration respectively. Figures 5 is showing velocity changes at $z = 0.005$ and variation between r-axis and velocity profile which clearly depicts that the velocity decreases with an increase in thermophoresis parameter. To study the effect of Brownian motion parameter (N_b) in above plotted Figs. 5, 6 and 7, the increase in N_b improves the velocity in the mid-way (centreline) of the artery but reduces at the walls. Brownian motion tries to distribute the particles as uniformly as possible, which reduces the nanoparticle concentration gradient and diminishes the regional variations in fluid characteristics. Hence the temperature and nanoparticles concentration lower at mid-way but improves at the vessel walls and decreases by increasing N_b.

Two different rates of stenosis are simulated by comparing the magnitudes of the velocity fields for two cases: diameter of stenosis $= 0.4\, l_0$(in Fig. 2a–d) and diameter of stenosis $= 0.3\, l_0$ (in Fig. 8a–d). Here we can see that in the case where the artery is narrower due to stenosis, the velocity magnitude is increased. Increase in the velocity magnitude by increasing the value of Brownian motion parameter (N_b) is shown in Fig. 8a–c, but there is a slight decrement in velocity due to increase in thermophoresis parameter (N_t) which is clearly shown in Fig. 8b–d. Similarly the temperature distribution and concentration of nanoparticles are shown in Figs. 9a–d and 10a–d for another shape of stenosis respectively. We see that in the second case the magnitudes of the temperature in the narrowed part is higher than the first case at the stenosis. Figures 8a–b depicts that there is an increment in velocity after increasing the thermophoresis parameter (N_t) and somehow the temperature decreases due to increment

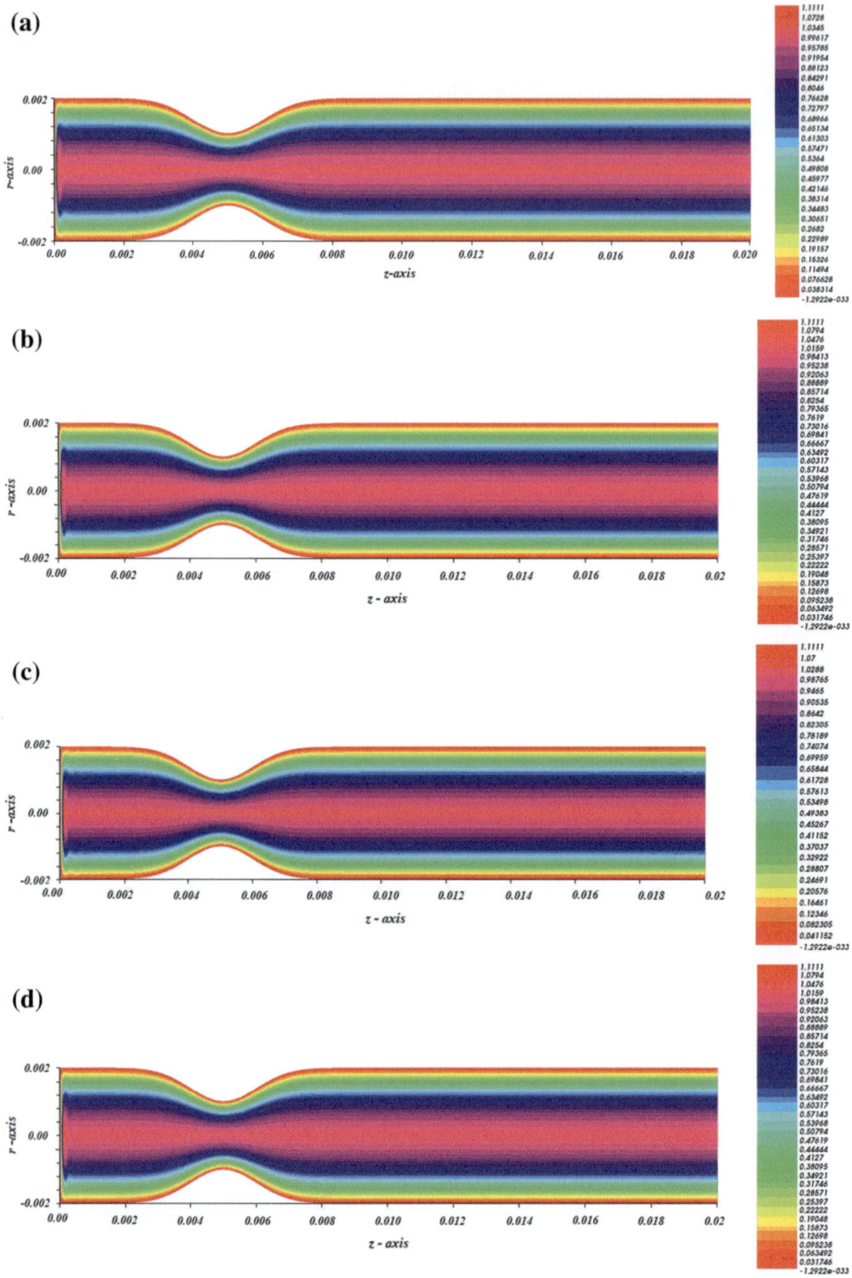

Fig. 2 Velocity field at **a** $N_b = 0.3$ and $N_t = 0.3$, **b** $N_b = 0.3$ and $N_t = 0.6$, **c** $N_b = 0.6$ and $N_t = 0.3$, **d** $N_b = 0.6$ and $N_t = 0.6$

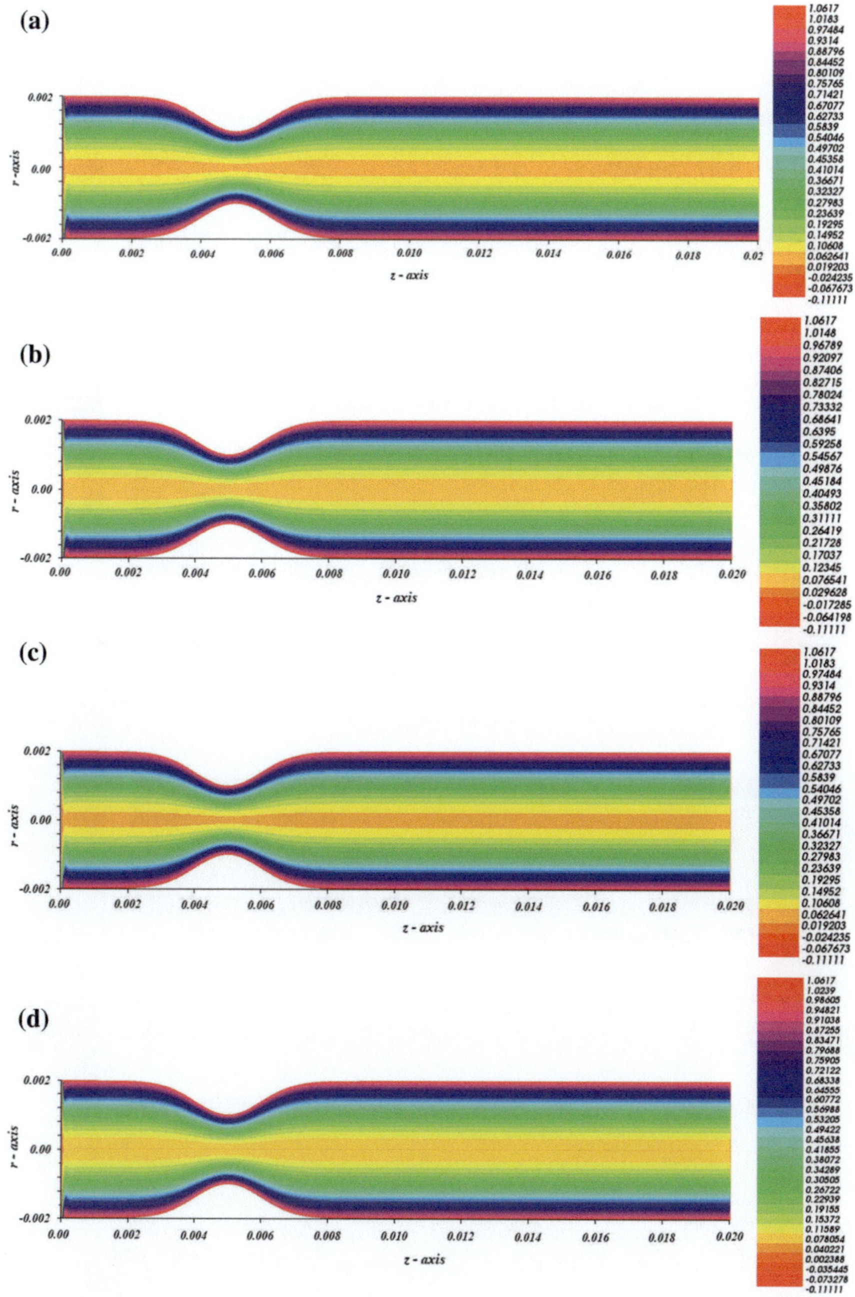

Fig. 3 Temperature distribution at **a** $N_b = 0.3$ and $N_t = 0.3$, **b** $N_b = 0.3$ and $N_t = 0.6$, **c** $N_b = 0.6$ and $N_t = 0.3$, **d** $N_b = 0.6$ and $N_t = 0.6$

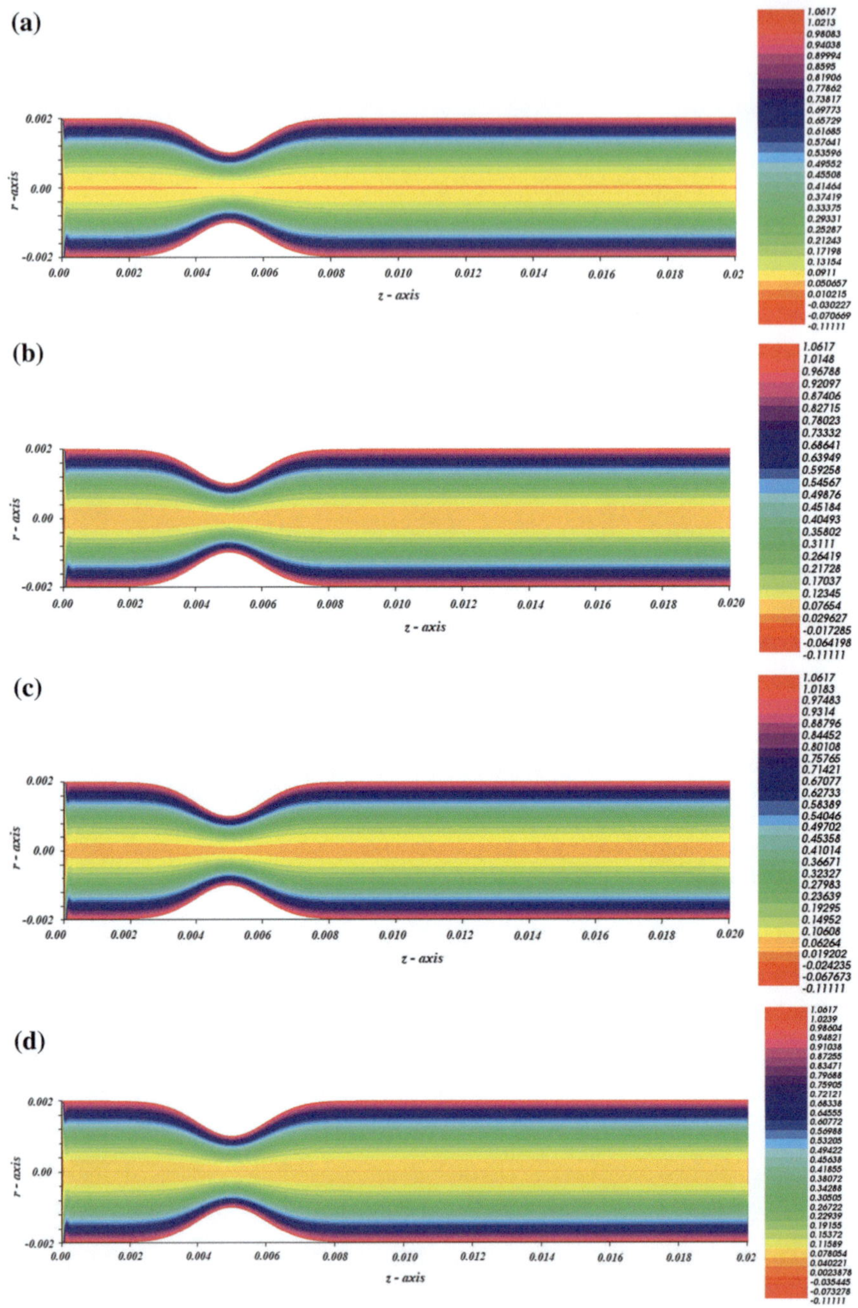

Fig. 4 Nanoparticle concentration at **a** $N_b = 0.3$ and $N_t = 0.3$, **b** $N_b = 0.3$ and $N_t = 0.6$, **c** $N_b = 0.6$ and $N_t = 0.3$, **d** $N_b = 0.6$ and $N_t = 0.6$

Table 1 Default values of thermo physical parameters

Parameter	S	c	M	μ_0	θ_0	λ_1	G_r	B_r	l	l_o
Values	0.824360	0.0727	0.3	0.5	0.5	0.5	1.0	1.0	0.020	0.005

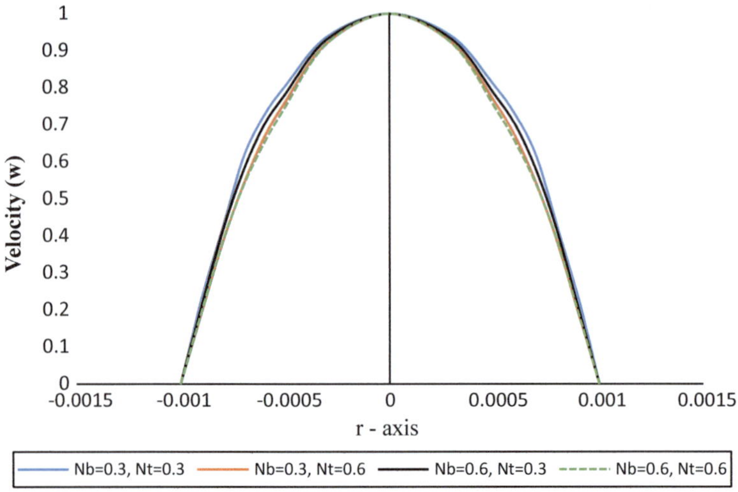

Fig. 5 Velocity profile for diameter of stenosis = 0.4 l_0 at M = 0.3, $z = 0.005$

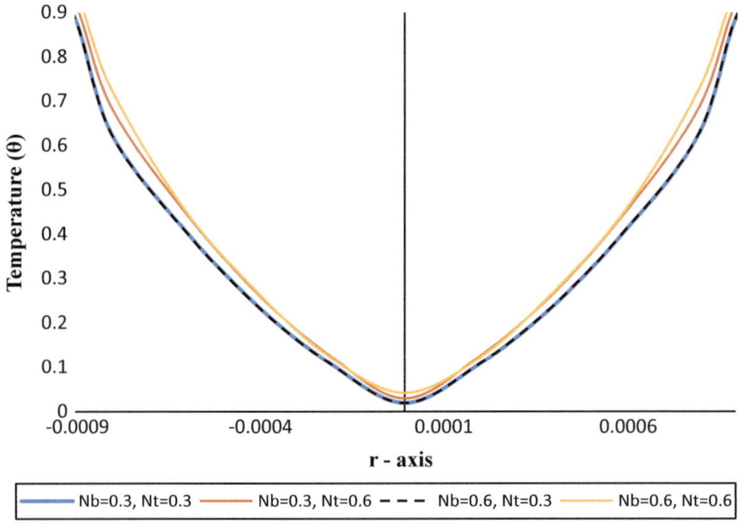

Fig. 6 Temperature profile for diameter of stenosis = 0.4 l_0 at M = 0.3, $z = 0.005$

in Brownian motion parameter (N_b). It is found that the nonuniformity of nanoparticle concentration increases when considering the thermophoresis hence the concentration profiles increased significantly with increase in thermophoresis parameter (N_t) and Brownian motion parameter (N_b). It is observed that concentration and temperature distributions are more noticeable with increase in thermophoresis parameter

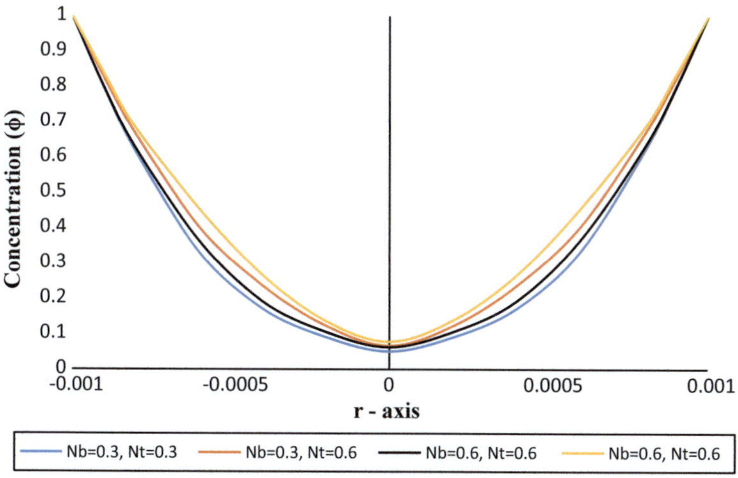

Fig. 7 Nanoparticles concentration for diameter of stenosis = 0.4 l_0 at M = 0.3, $z = 0.005$

(N_t) and hence these results are considerably increased within the boundary layer region for both cases of stenosis.

Figures 11, 12 and 13 are depicts that velocity profile, temperature distribution and nano-particle concentration for another size of stenosis respectively. From the above plots one can see that the velocity decreases by increasing thermophoresis parameter while in contrary the temperature and nanoparticles concentration increases by increasing the value of thermophoresis parameter. It is due to the reason that thermophoresis induces a force on the suspended nanoparticles in the fluid to accumulate them in the opposite direction of the temperature gradient, generating a greater nanoparticle concentration gradient and accentuating the variation in regional properties. It can be shown that considering nanoparticle migration is consequential to predict the heat transfer rate with high accuracy.

5 Conclusion

In the present paper, 2D mathematical model of MHD-blood flow, as second grade non-Newtonian nanofluid through coronary artery is studied numerically using finite element analysis. The study investigated the behaviour of the blood flow for the most common cardiovascular disease—stenosis and simulated the flow behaviour and heat transfer of MHD blood conveying nanoparticles by applying the finite element method of variational approach and obtained the results for symmetric case of stenosis. The main outcomes from the present investigation are as follows:

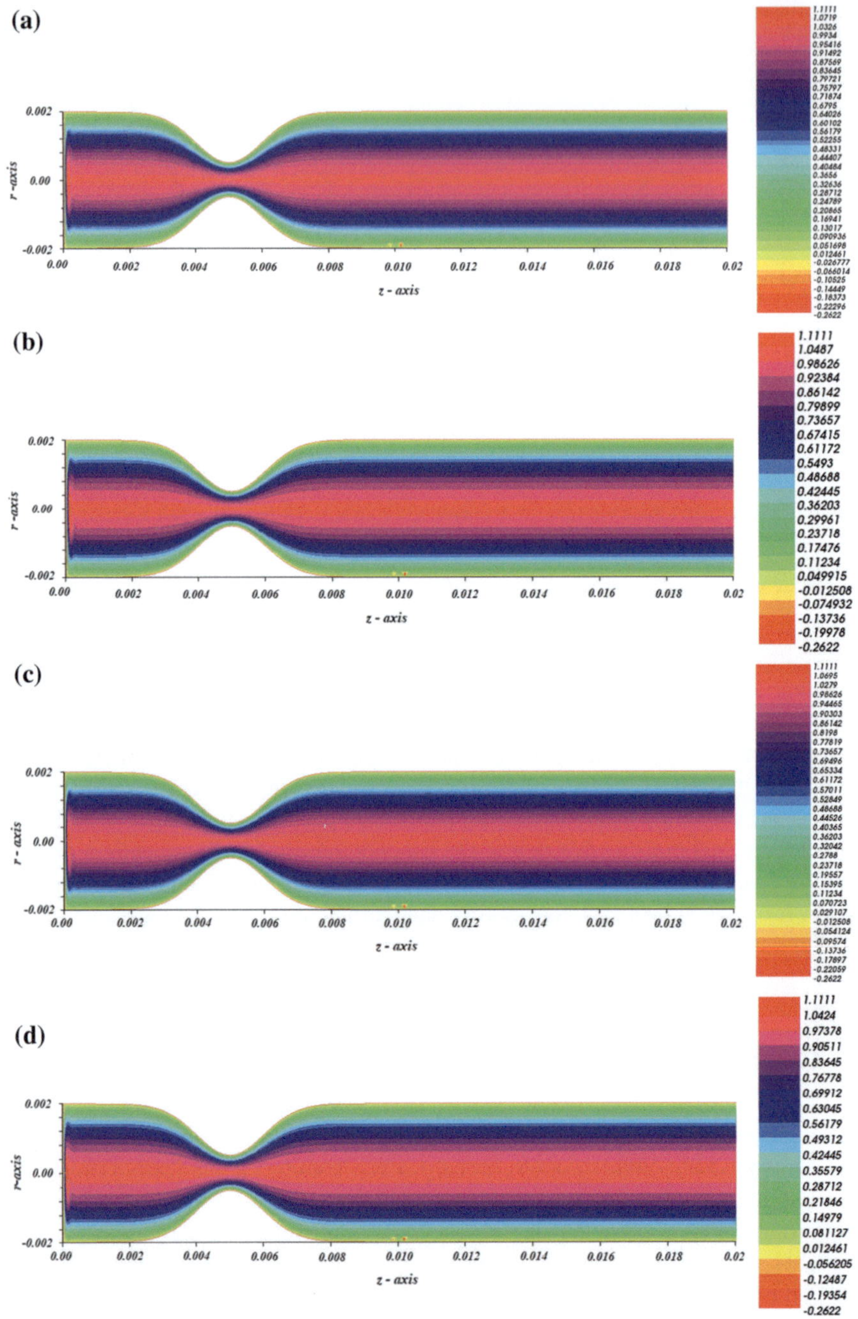

Fig. 8 Velocity field at **a** $N_b = 0.3$ and $N_t = 0.3$, **b** $N_b = 0.3$ and $N_t = 0.6$, **c** $N_b = 0.6$ and $N_t = 0.3$, **d** $N_b = 0.6$ and $N_t = 0.6$

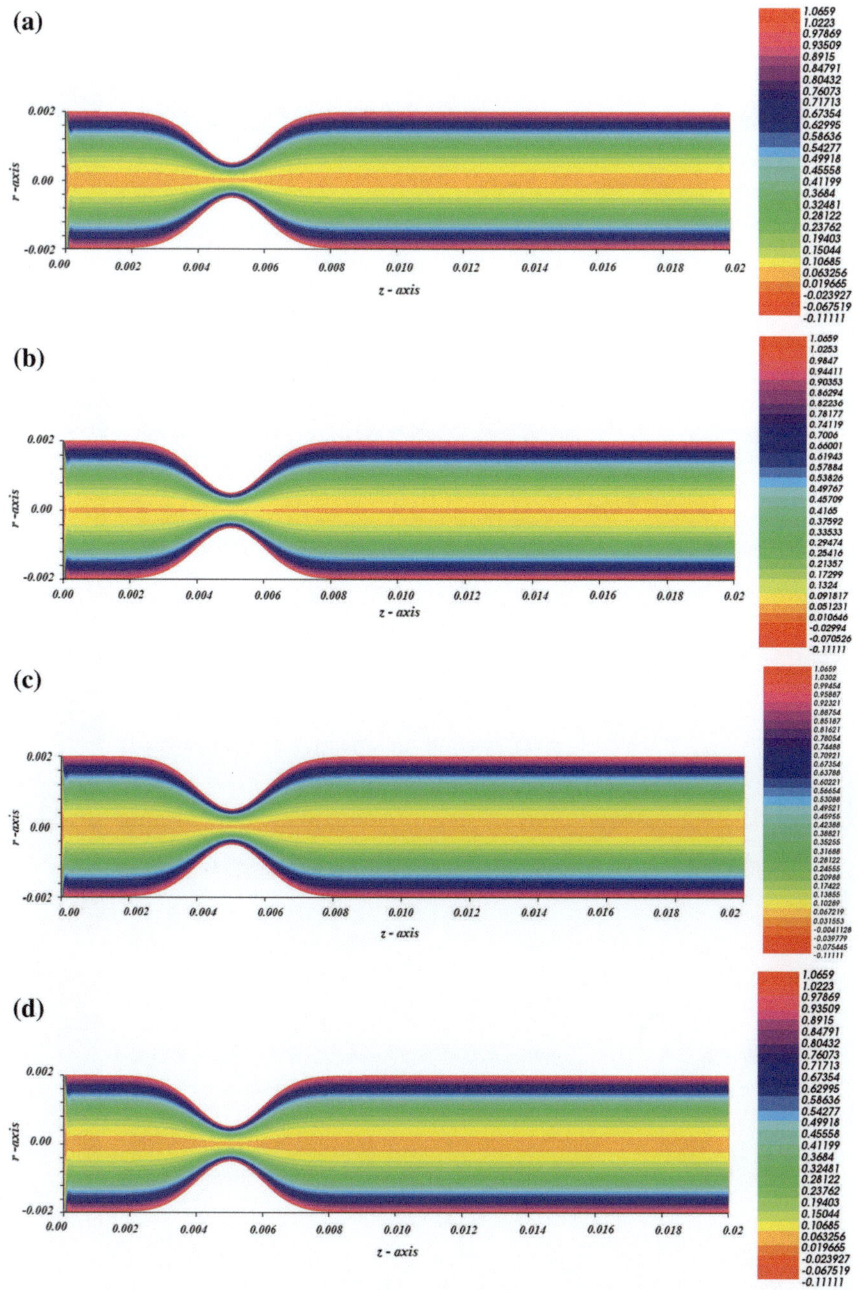

Fig. 9 Temperature distribution at **a** $N_b = 0.3$ and $N_t = 0.3$, **b** $N_b = 0.3$ and $N_t = 0.6$, **c** $N_b = 0.6$ and $N_t = 0.3$, **d** $N_b = 0.6$ and $N_t = 0.6$

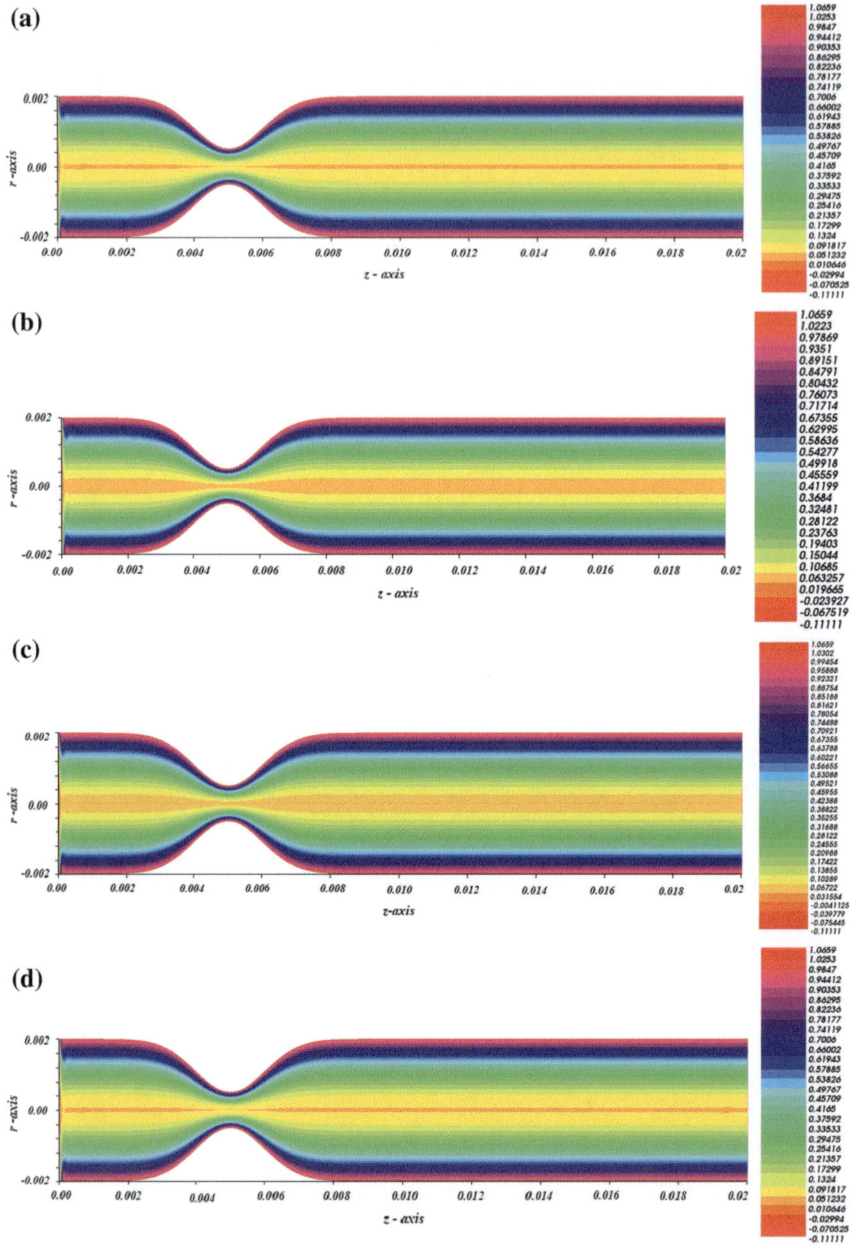

Fig. 10 Nanoparticles concentration at **a** $N_b = 0.3$ and $N_t = 0.3$, **b** $N_b = 0.3$ and $N_t = 0.6$, **c** $N_b = 0.6$ and $N_t = 0.3$, **d** $N_b = 0.6$ and $N_t = 0.6$

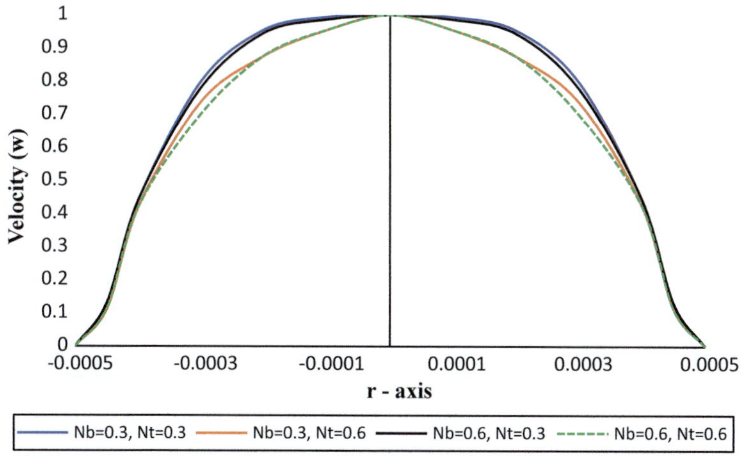

Fig. 11 Velocity profile for diameter of stenosis = 0.3 l_0 at M = 0.3, $z = 0.005$

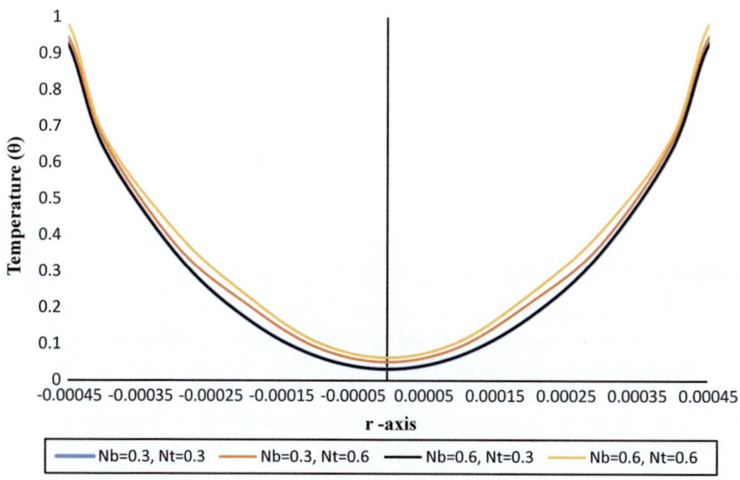

Fig. 12 Temperature profile for diameter of stenosis = 0.3 l_0 at M = 0.3, $z = 0.005$

- It is observed that in the first case of stenosis (diameter of stenosis = 0.4 l_0), increasing the thermophoresis parameter (N_t) caused an increase in nanoparticles concentration and temperature values in whole domain.
- An increase in Brownian motion parameter (N_b) leads to increase in velocity at the stenosis and in whole domain.
- For the case where diameter of stenosis = 0.3 l_0, in later case, by increasing stenosis depth the velocity is increased from the earlier case and also by increasing the Brownian motion parameter (N_b) velocity increases in the whole domain.

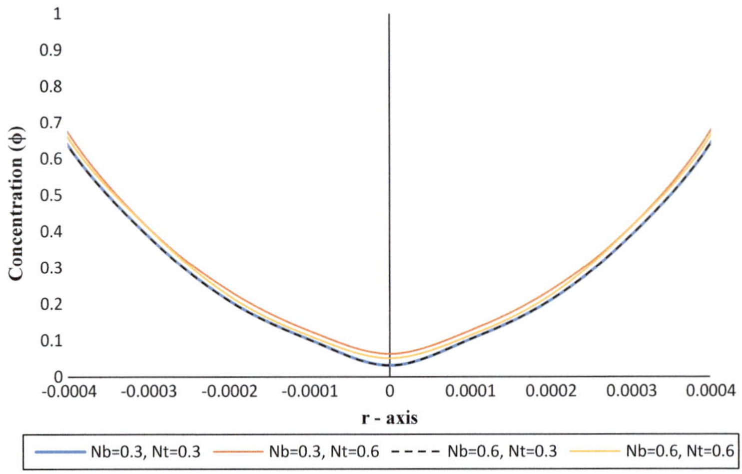

Fig. 13 Nanoparticles concentration for diameter of stenosis $= 0.3\ l_0$ at M $= 0.3$, $z = 0.005$

- Also from both the cases of stenosis, it can be concluded from the simulations shows for both the cases of stenosis the temperature and nanoparticles is increased at the stenosis by increasing the thermophoresis parameter (N_t).
- The velocity at the stenotic region is more in second case as compared to first one.

References

1. Fung, Y.C.: The flow properties of blood. In Biomechanics: Mechanical properties of living tissues, Springer, Berlin, pp. 62–98 (1981)
2. Biswas, D.: Blood flow models: a comparative study. Mittal Publications, Delhi (2000)
3. Merrill, E.W.: Rheology of human blood and some speculations on its role in vascular homeostasis. In: Biomechanical Mechanisms in Vascular Homeostasis and Intravascular Thrombus (1965)
4. Taylor, M.G.: The influence of the anomalous viscosity of blood upon its oscillatory flow. Phys. Med. Biol. **3**(3), 273 (1959)
5. Baieth, H.E.A.: Physical parameters of blood as a non-Newtonian fluid. Int. J. Biomed. Sci. IJBS **4**(4), 323 (2008)
6. Moreno, C., Bhaganagar, K.: Modeling of Stenotic coronary artery and implications of plaque morphology on blood flow. Model. Simul. Eng. **2013**, 14 (2013)
7. Sauvage, E.: Patient-specific blood flow modelling. Ph.D. diss., Ph. D. Thesis, Université catholique de Louvain (2014)
8. Ali, N., Zaman, A., Sajid, M., Anwar Bég, O., Shamshuddin, M.D., Kadir, A.: Numerical simulation of time-dependent non-newtonian nanopharmacodynamic transport phenomena in a tapered overlapping stenosed artery. Nanosci. Technol. Int. J. **9**(3), 247–282 (2018)
9. Mathur, P., Jain, S.: Mathematical modelling of non-Newtonian blood flow through artery in the presence of stenosis. Appl. Math. Biosci. **4**(1), 1–12 (2013)
10. Akbar, N.S.: Non-Newtonian model study for blood flow through a tapered artery with a stenosis. Alexandria Eng. J. **55**(1), 321–329 (2016)

11. Godson, L., Raja, B., Mohan Lal, D., Wongwises, S.: Enhancement of heat transfer using nanofluids—an overview. Renew. Sustain. Energy Rev. **14**(2), 629–641 (2010)
12. Li, Q., Xuan, Y.: Convective heat transfer and flow characteristics of Cu-water nanofluid. Sci. China Ser. E Technol. Sci. **45**(4), 408–416 (2002)
13. Pak, B.C., Cho, Y.I.: Hydrodynamic and heat transfer study of dispersed fluids with submicron metallic oxide particles. Exp. Heat Transf. Int. J. **11**(2), 151–170 (1998)
14. Lee, S., Choi, S.U.S., Li, S., Eastman, J.A.: Measuring thermal conductivity of fluids containing oxide nanoparticles. J. Heat Transf. **121**(2), 280–289 (1999)
15. Kumar, K.P., Paul, W., Sharma, C.P.: Green synthesis of gold nanoparticles with Zingiber officinale extract: characterization and blood compatibility. Process Biochem. **46**(10), 2007–2013 (2011)
16. Giljohann, D.A., Seferos, D.S., Daniel, W.L., Massich, M.D., Patel, P.C., Mirkin, C.A.: Gold nanoparticles for biology and medicine. Angew. Chem. Int. Ed. **49**(19), 3280–3294 (2010)
17. Darcy, H.: The flow of fluids through porous media. Mc-Graw Hill, NewYork, NY, USA (1937)
18. Korchevskii, E.M., Marochnik, L.S.: Magneto-hydrodynamic version of movement of blood. Biophysics **10**(2), 411–414 (1965)
19. Sud, V.K., Suri, P.K., Mishra, R.K.: Effect of magnetic field on oscillating blood flow in arteries. Stud. Biophys. **46**(3), 163–171 (1974)
20. Hayat, T., Khan, M., Ayub, M.: Some analytical solutions for second grade fluid flows for cylindrical geometries. Math. Comput. Model. **43**(1–2), 16–29 (2006)
21. Hatami, M., Hatami, J., Ganji, D.D.: Computer simulation of MHD blood conveying gold nanoparticles as a third grade non-Newtonian nanofluid in a hollow porous vessel. Comput. Methods Programs Biomed. **113**(2), 632–641 (2014)
22. Sheikholeslami, M., Gorji-Bandpy, M., Soleimani, S.: Two phase simulation of nanofluid flow and heat transfer using heatline analysis. Int. Commun. Heat Mass Transf. **47**, 73–81 (2013)
23. Yadav, D., Agrawal, G.S., Bhargava, R.: Thermal instability of rotating nanofluid layer. Int. J. Eng. Sci. **49**(11), 1171–1184 (2011)
24. Aziz, A., Aziz, T.: MHD flow of a third grade fluid in a porous half space with plate suction or injection: an analytical approach. Appl. Math. Comput. **218**(21), 10443–10453 (2012)
25. Asghar, S., Hanif, K., Hayat, T., Khalique, C.M.: MHD non-Newtonian flow due to non-coaxial rotations of an accelerated disk and a fluid at infinity. Commun. Nonlinear Sci. Numer. Simul. **12**(4), 465–485 (2007)
26. Dunn, J.E., Fosdick, R.L.: Thermodynamics, stability, and boundedness of fluids of complexity 2 and fluids of second grade. Arch. Ration. Mech. Anal. **56**(3), 191–252 (1974)
27. Dunn, J.E., Rajagopal, K.R.: Fluids of differential type: critical review and thermodynamic analysis. Int. J. Eng. Sci. **33**(5), 689–729 (1995)
28. Ellahi, R., Raza, M., Vafai, K.: Series solutions of non-Newtonian nanofluids with Reynolds' model and Vogel's model by means of the homotopy analysis method. Math. Comput. Model. **55**(7–8), 1876–1891 (2012)
29. Hecht, F.: New development in FreeFem++. J. Numer. Math. **20**(3–4), 251–266 (2012)
30. Abraham, F., Behr, M., Heinkenschloss, M.: Shape optimization in steady blood flow: a numerical study of non-Newtonian effects. Comput. Methods Biomech. Biomed. Eng. **8**(2), 127–137 (2005)
31. Formaggia, L., Quarteroni, A., Veneziani, A. (eds.): Cardiovascular mathematics: modeling and simulation of the circulatory system, vol. 1. Springer Science & Business Media (2010)

Comparative Study of Boundary Conditions in LBM for Incompressible Laminar Flow

Alankar Agarwal and Akshay Prakash

Abstract In this paper, we conduct a comparative study amongst different boundary conditions with two dimensional single-relaxation time lattice Boltzmann method (SRT-LBM), for incompressible laminar flow. Three types of boundary condition are considered for the simulation: including full-way bounce-back, half-way bounce-back, and modified bounce-back for the implementation of no-slip boundary condition on the wall with, pressure (density) boundary condition proposed by Zuo and He (Phys Fluids 9(6):1591–1598, 1997 [1]) for inlet/oulet. The benchmark fluid flow problem of steady plane Poiseuille's flow with Reynolds number, Re = 75 is choosen. The numerical simulations are validated with the analytical solution, and grid convergence test are performed to compare accuracy of different boundary conditions.

Keywords Lattice Boltzmann method (LBM) · Single-relaxation time · Incompressible flow · Boundary conditions

1 Introduction

In the past few decades, lattice Boltzmann method (LBM) has been adopted by a large number of researchers for the simulations of incompressible Navier-Stokes (NS) equations [2, 3]. The method differs from conventional computational fluid dynamics (CFD) approaches (such as finite difference method, finite volume method and finite element method) in the sense, that the traditional CFD approaches based on macroscopic continuum equations, where LBM is based on microscopic laws of collison-streaming and mesoscopic kinetic equations. The method has gained significant success over other numerical Navier-Stokes solutions due to its easy

A. Agarwal (✉)
Indian Institute of Technology Jodhpur, Jodhpur, Rajasthan 342037, India
e-mail: agarwal.1@iitj.ac.in

A. Prakash
Indian Institute of Technology Kharagpur, Kharagpur, West Bengal 721302, India
e-mail: ap@aero.iitkgp.ernet.in

S. Manna et al. (eds.), *Mathematical Modelling and Scientific Computing with Applications*, Springer Proceedings in Mathematics & Statistics 308,
https://doi.org/10.1007/978-981-15-1338-1_18

implementation, high stability and efficient parallelizable nature [4]. LBM has wide range of applications in multiphase flows, porus media, and turbulents flows. The interested reader is referred to Succi [5], Chen and Doolen [2], and Aidun et al. [6] for more information on the LBM.

Boundary treatment plays an important role in the lattice Boltzmann method [7]. Most commonly used boundary condition for the implementation of no-slip are standard bounce-back scheme, half-way bounce back scheme and modified bounce-back scheme [8]. Later, in 1997 Zuo and He [1] proposed velocity wall boundary condition and pressure (density) flow boundary condition. This study is a comparison of pressure boundary condition proposed by [1] with the various no-slip boundary condition on 2D steady plane Poiseuille's flow. The results with these schemes of boundary conditions gets validated with the analytical solution and comparison amongst these boundary conditions are made for accuracy.

The remainder of the paper is organized as follows: Sect. 2 provides a succinct introduction of the single-relaxation time-lattice Boltzmann method (SRT-LBM), and describes different boundary conditions. Section 3, explains the flow domain and fluid properties used during the study. Results and discussion are included in Sect. 4 that particularly focuses on the validation, and comparison of different boundary conditions in terms of error accuracy. Finally, in Sect. 5, conclusion is addressed.

2 Simulation Procedure

2.1 Single Relaxation Time (SRT) LBM Model

The single relaxation time (SRT) is the most commonly used model of LBM for the complex fluid flow problems. The lattice Boltzmann equation (LBE) for the SRT-LBM model can be written as:

$$\underbrace{f_k(x, \mathbf{e}_k \Delta t, t + \Delta t)}_{streaming-step} = \underbrace{f_k(x, t) - \frac{1}{\tau}\left[f_k(x, t) - f_k^{eq}(x, t)\right]}_{collision-step} \tag{1}$$

where f is the probability density distribution function (i.e. PDDF) describing the probability of particle at position vector $\mathbf{x} = (x, y)$ at time t for the discrete velocity, $\mathbf{e}_k$ $(k = 1, \ldots, Q)$ in D-dimensional space, and τ is the non-dimensional relaxation time parameter due to fluid-particle collision, describes the time taken by the fluid-particle to reach to equilibrium [9]. The equilibrium probability density distribution function (i.e. EPDDF) f_k^{eq} can be expressed as:

$$f_k^{eq} = \rho w_k \left[1 + \frac{(\mathbf{e}_k \,.\, \mathbf{u})}{c_s^2} + \frac{(\mathbf{e}_k \,.\, \mathbf{u})^2}{2c_s^4} - \frac{\mathbf{u}^2}{2c_s^2}\right] \tag{2}$$

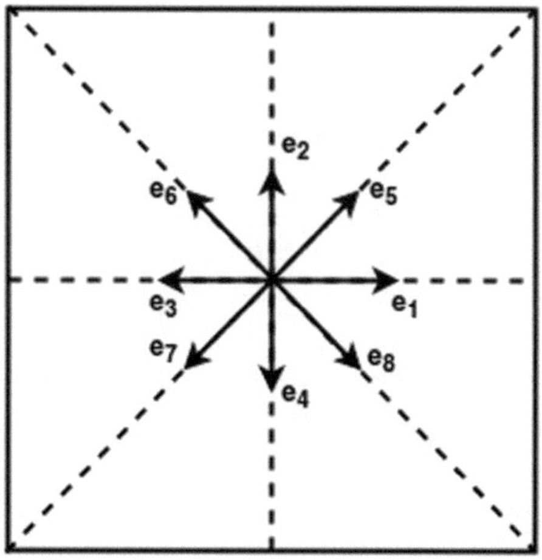

Fig. 1 Schematic diagram of D2Q9 square lattice

where $\mathbf{u} = (u, v)$ is the liquid velocity vector and c_s is the speed of sound. It is related to the lattice speed c as:

$$c_s = \frac{c}{\sqrt{3}}$$
$$c = \frac{\Delta x}{\Delta t} \tag{3}$$

Δx and Δt are the size of lattice and time step size respectively. In this study, we use the twodimensional nine velocity (D_2Q_9) square lattice as shown in Fig. 1. The discrete velocity vector $\mathbf{e}_k$ and the corresponding weighted function $\mathbf{w}_k$ for the (D_2Q_9) square lattice are [7]:

$$\mathbf{e}_k = \begin{cases} [\cos\{(k-1)\pi/2\}, \sin\{(k-1)\pi/2\}], \quad k = 1\text{–}4 \\ [\cos\{(k-1)\pi/2 + \pi/4\}, \sin\{(k-1)\pi/2 + \pi/4\}], \quad k = 5\text{–}8 \\ (0, 0), \quad k = 9 \end{cases} \tag{4}$$

$$\mathbf{w}_k = \begin{cases} 1/9, \quad k = 1\text{–}4 \\ 1/36, k = 5\text{–}8 \\ 4/9, \quad k = 9 \end{cases} \tag{5}$$

The relaxation time parameter τ is related to the kinematic viscosity that fixes the rate of approach to equilibrium given by [10]:

$$\nu = \frac{2\tau - 1}{6} \frac{(\Delta x)^2}{\Delta t} \tag{6}$$

The macroscopic variables such as density and momentum density can be evaluated directly from the real-valued probability density distribution function (PDDF) respectively [9]:

$$\rho = \sum_k f_k = \sum_k f_k^{eq}$$
$$\rho \mathbf{u} = \sum_k \mathbf{e}_k f_k = \sum_k \mathbf{e}_k f_k^{eq} \tag{7}$$

The above expression of density, and momentum density satisfy the traditional pressure-based solver (i.e. Navier-Stokes solver) for incompressible flow explained by He and Luo, 1997 using the Chapman-Enskog (CE) expansion [11].

2.2 *Boundary Conditions*

2.2.1 Pressure Boundary Condition

Zuo and He [1] proposed pressure (density) flow boundary condition in the year 1997. For the case of steady plane Poiseuille's plane, we applied pressure flow boundary conditions at the inlet and outlet nodes. Consider a node at the inlet, after the streaming process, distribution function (f_2, f_3, f_4, f_6, f_7) are known (see Fig. 1). To evaluate unknown fluid populations (f_1, f_5, f_8) at inlet nodes formulation is given below:

$$f_1 + f_5 + f_8 = \rho - (f_2 + f_3 + f_4 + f_6 + f_7 + f_9) \tag{8}$$

$$f_1 + f_5 + f_8 = \rho u - (f_3 + f_6 + f_7) \tag{9}$$

$$f_5 - f_8 = \rho v - f_2 + f_4 - f_6 + f_7 \tag{10}$$

for the present case $v = 0$, thus Eq. (10) gives:

$$f_5 - f_8 = -f_2 + f_4 - f_6 + f_7 \tag{11}$$

Substituting Eq. (9) in Eq. (8) gives:

$$u = 1 - \frac{[f_2 + f_4 + f_9 + 2 * (f_3 + f_6 + f_7)]}{\rho} \tag{12}$$

The bounce-back rule was used to obtained the population f_1 (normal to the inlet). It gives:

$$f_1 - f_1^{eq} = f_3 - f_3^{eq} \tag{13}$$

From Eq. (2)

$$f_1 = f_3 + \frac{2}{3}u \tag{14}$$

Using f_1, the other unknown fluid populations f_5 and f_8 can be evaluated from expression in Eqs. (9)–(10)

$$\begin{aligned} f_5 &= f_7 - \frac{1}{2}(f_2 - f_4) + \frac{1}{6}\rho u \\ f_8 &= f_8 + \frac{1}{2}(f_2 - f_4) + \frac{1}{6}\rho u \end{aligned} \tag{15}$$

Similar procedure can be used to evaluate the fluid population (f_3, f_6, f_7) at the outlet nodes.

2.2.2 Wall Boundary Condition

Full-Way Bounce Back

The full-way bounce back scheme implements no-slip boundary condition at the wall, also referred as to the standard bounce back method [12]. In this method lattices are located directly at the solid surface (i.e. wall). The method sees to it that no collision operation is performed, on the fluid node, at the wall, during its implementation. After the streaming process, the unknown fluid population at the wall is obtained using the expression below:

$$f_i(x, y, t + \Delta t) = f_{\bar{i}}(x, y, t + \Delta t) \tag{16}$$

where, f is the post-streaming distribution function, and $(i, \bar{i})$ are the distribution functions (fluid population) in the direction opposite to each other.

Half-Way Bounce Back

In this approach, wall is located at a half distance between two lattice sites. The collision operation is performed on all lattice sites. Consider D_2Q_9 (see Fig. 1) lattice structure at the upper wall node. After, the streaming process, fluid population (f_2, f_5, and f_6), the unknown fluid population f_4, f_7, and f_8 can be computed as:

$$\begin{aligned} f_4(x, l_y, t + \Delta t) &= f_2(x, l_y - 1, t + \Delta t) \\ f_7(x, l_y, t + \Delta t) &= f_5(x - 1, l_y - 1, t + \Delta t) \\ f_8(x, l_y, t + \Delta t) &= f_6(x + 1, l_y - 1, t + \Delta t) \end{aligned} \tag{17}$$

where l_y are the upper wall boundary nodes. Similar procedure can be applied to lower wall boundary nodes for 2D steady plane Poiseuille's flow.

Modified Bounce-Back

The implementation of modified bounce back boundary condition is similar to the full-way bounce back method. It differs only in the sense that it includes collision operation at the wall nodes during the process.

3 Numerical Setup and Flow System

The simulation were carried out for 2D steady plane Poiseuille's flow using D_2Q_9 square lattice for SRT-LBM model. The fluid properties and geometric parameters are given in Table 1. Figure 2. shows the schematic representation of flow between two stationary plates. The relaxation parameter for the collision step in Eq. (1) is choosen to be $\tau = 0.6$.

The convergence criterion for the steady state solution is set to be:

$$\frac{\sum_{i,j}\left|u_{i,j}^{m+1} - u_{i,j}^{m}\right|}{\sum_{i,j}\left|u_{i,j}^{m+1}\right|} \leq 1.0 \times 10^{-8} \tag{18}$$

Table 1 Fluid properties and geometric parameters

Density, ρ (kg/m^3)	1000
Dynamic Viscosity, μ (kg/m s)	0.001
Separation between plates, W (m)	0.1
Inlet velocity, u (m/s)	7.5×10^{-4}
Length of plates, L (m)	1
Reynolds number, Re	75

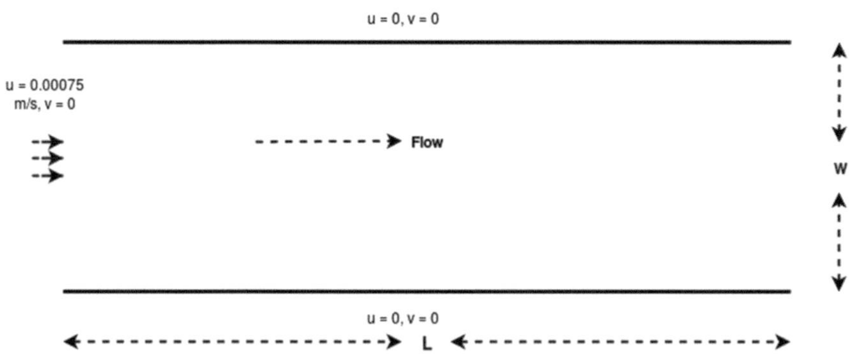

Fig. 2 Schematic diagram for flow between two stationary parallel plates

where $u_{i,j}^{m} = u(x, y, m\Delta t)$ is the axial component of velocity vector $\mathbf{u}$ at different grid point and at time instant m. The results are compared with the analytical solution of plane Poiseuille's flow governed by the expression:

$$u(x, y, t) = \frac{1}{2\mu}\frac{dp}{dx}y(y - W)$$
$$v(x, y, t) = 0 \tag{19}$$

4 Results and Discussion

In this section, simulation results of 2D plane Poiseuille's flow with different boundary condition methods are compared with the analytical solution as given in Eq. (19). Further, grid independence test was performed to compare different approaches of

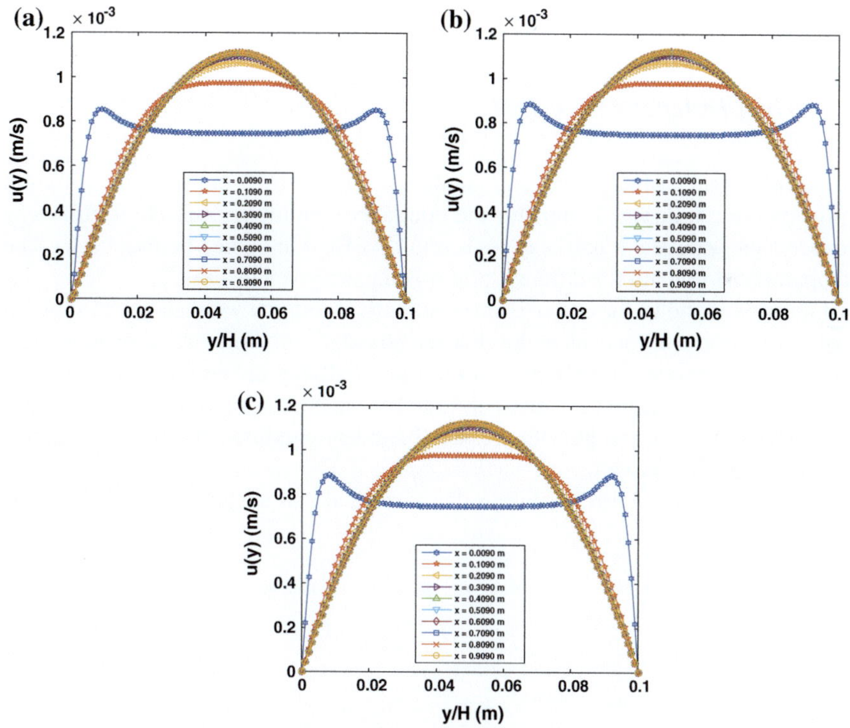

Fig. 3 Axial velocity along vertical direction at different position along x-axis with **a** Zuo-He + full way bounce back, **b** Zuo-He + half way bounce back, **c** Zuo-He + modified bounceback boundary condition

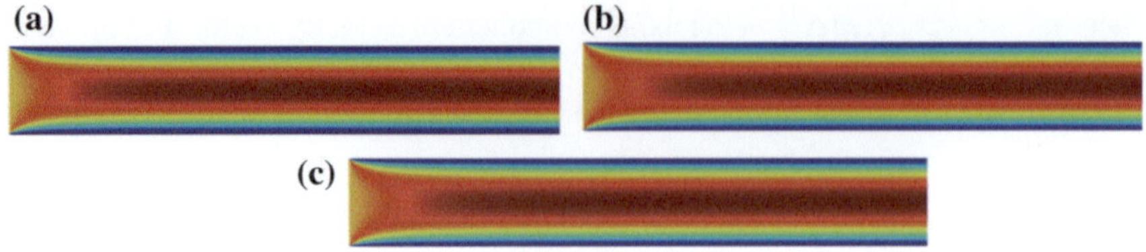

Fig. 4 Velocity distribution of steady plane Poiseuille's flow with **a** Zuo-He + full way bounce back, **b** Zuo-He + half-way bounce back, and **c** Zuo-He + modified bounce back boundary conditions

boundary conditions. Figure 3(a–c) shows the axial velocity along vertical direction [i.e. $u(y)$] at different position along x-axis for various boundary conditions. Figure 3a is the results obtained after implementing pressure boundary condition at the inlet/outlet nodes with the full-way bounce back boundary condition method for the no-slip boundary at the top/bottom wall nodes, Fig. 3b, c are the results from pressure boundary condition at the inlet/outlet nodes with the half-way bounce-back and modified bounce back boundary condition method for the no-slip boundary respectively. Figure 4 shows velocity distribution of steady plane Poiseuille's flow.

4.1 Grid Independence Test

To perform grid convergence test, simulation were carried out on three different grid size (i.e., 4, 2, and 1 mm) for the boundary conditions (Zuo-He + full- way bounce back, Zuo-He + half-way back, and Zuo-He + modified bounce back). The computational domain for different grid spacing are given in Table 2.

Figure 5a–c shows the comparison of numerical solution with analytical solution on different grid size for each of the choosen boundary condition. As seen, numerical scheme with pressure boundary condition and halfway or modified bounce back shown similar results and are more close to the analytical solution for the fine grid as comparison to the results with pressure boundary condition and full way bounce back condition at the wall.

The relative global error between the numerical and analytical solution can be calculated using:

Table 2 Simulation parameters and cases

	Case 1	Case 2	Case 3
Grid spacing, Δx (mm)	4	2	1
Computational domain $l_x \times l_y$	250 × 25	500 × 50	1000 × 100

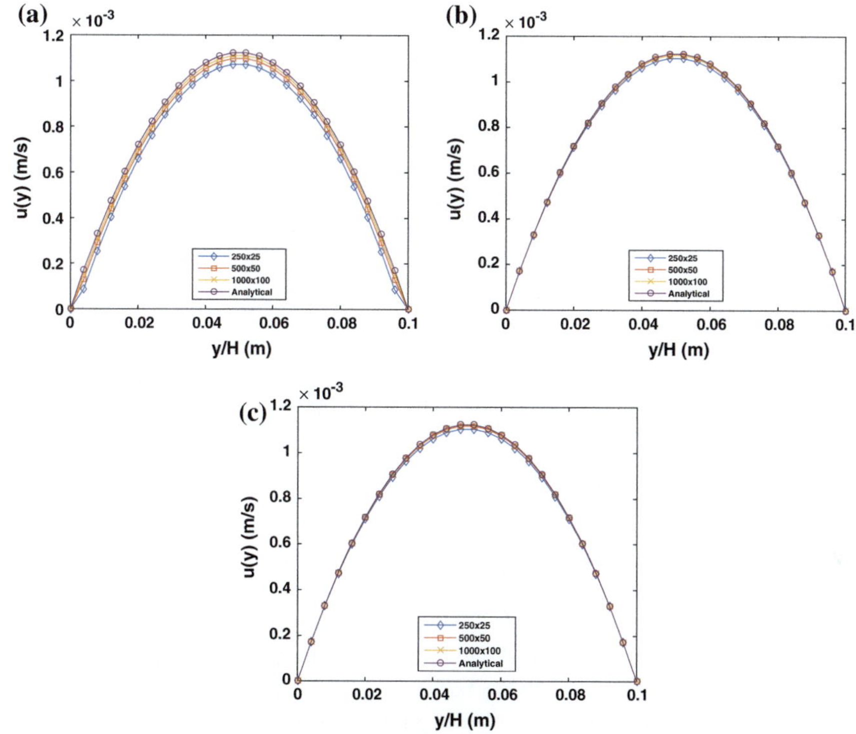

Fig. 5 Comparison of numerical solution with analytical solution on different grid size for **a** Zuo-He + full way bounce back, **b** Zuo-He + half way bounce back, **c** Zuo-He + modified bounce back boundary condition

$$\text{Relative global error} = \frac{\left\| u_{LBM} - u_{analytical} \right\|}{\left\| u_{analytical} \right\|_2} \tag{20}$$

The corresponding error obtained on different grid size with various boundary conditions are presented in Table 3. As seen, SRT-LBM scheme with full way bounce back is approximately 1st order accurate whereas scheme with other two boundary conditions (i.e. half way bounce back and modified bounce back) shown approximately 2nd order accuracy for incompressible laminar flow.

5 Conclusion

In this study, various boundary conditions in SRT-LBM were assessed for incompressible laminar flow through benchmark fluid flow problem—steady plane Poiseuille's flow. The results are compared with the analytical solution, and grid

Table 3 Relative global error and order of accuracy with different boundary conditions

Boundary Condition	Case	Δx (mm)	Domain ($l_x \times l_y$)	Relative global error	Order of accuracy
Zuo-He + full-way bounce back	1	4	250 × 25	0.0800	
	2	2	500 × 50	0.0400	0.9988
	3	1	1000 × 100	0.0200	0.9998
Zuo-He + half-way bounce back	1	4	250 × 25	0.0180	
	2	2	500 × 50	0.0051	1.8108
	3	1	1000 × 100	0.0014	1.8254
Zuo-He + modified bounce back	1	4	250 × 25	0.0180	
	2	2	500 × 50	0.0051	1.8108
	3	1	1000 × 100	0.0014	1.8254

convergence study was performed to compare various boundary conditions. As presented in results and discussion section, half-way bounce back and modified bounce back with the pressure boundary condition of Zuo-He [1] show approximately second order accuracy comparative to full-way bounce back boundary method that results with the accuracy of first order. This study is a part of development of solver for incompressible flow using LBM and further study regarding the stability issue in boundary conditions is required. Also, based on the work presented here, we continue to develop the algorithm for other challenging problem of fluid flow, that includes turbulent flow simulation, multiphase flows, etc.

References

1. Zou, Q., He, X.: On pressure and velocity boundary conditions for the lattice Boltzmann BGK model. Phys. Fluids **9**(6), 1591–1598 (1997)
2. Chen, S., Doolen, G.D.: Lattice Boltzmann method for fluid flows. Annu. Rev. Fluid Mech. **30**(1), 329–364 (1998)
3. Gunstensen, A.K., Rothman, D.H.: Microscopic modeling of immiscible fluids in three dimensions by a lattice Boltzmann method. EPL (Europhys. Lett.) **18**(2), 157 (1992)
4. McClure, J.E., Prins, J.F., Miller, C.T.: Comparison of CPU and GPU implementations of the lattice Boltzmann method. In: XVIII International Conference on Water Resources, CMWR (2010)
5. Succi, S.: The Lattice Boltzmann Equation: for Fluid Dynamics and Beyond. Oxford University Press (2001)
6. Aidun, C.K., Clausen, J.R.: Lattice-Boltzmann method for complex flows. Annu. Rev. Fluid Mech. **42**, 439–472 (2010)
7. Perumal, D.A., Dass, A.K.: A review on the development of lattice Boltzmann computation of macro fluid flows and heat transfer. Alexandria Eng. J. **54**(4), 955–971 (2015)
8. Mohammadipour, O.R., Niazmand, H., Succi, S.: General velocity, pressure, and initial condition for two-dimensional and three-dimensional lattice Boltzmann simulations. Phys. Rev. E **95**(3), 033301 (2017)

9. Kang, S.K., Hassan, Y.A.: A comparative study of direct-forcing immersed boundary-lattice Boltzmann methods for stationary complex boundaries. Int. J. Numer. Meth. Fluids **66**(9), 1132–1158 (2011)
10. Zhang, R., Chen, H.: Lattice Boltzmann method for simulations of liquid-vapor thermal flows. Phys. Rev. E **67**(6), 066711 (2003)
11. He, X., Luo, L.-S.: Lattice Boltzmann model for the incompressible Navier-Stokes equation. J. Stat. Phys. **88**(3), 927–944 (1997)
12. Guo, Z., Shu, C.: Lattice Boltzmann Method and Its Applications in Engineering, vol. 3. World Scientific, Singapore (2013)

Stability Analysis of a Film Flow Down an Incline in the Presence of a Floating Flexible Membrane

M. Sani, H. Behera and S. Ghosh

Abstract The present study deals with the effects of floating flexible membrane on the instability of a gravity-driven flow down an incline. Linear stability of the flow system is explored using normal-mode analysis. Free surface gravity-driven flow is unstable at much lower Reynolds numbers. Instability of such a flow can be controlled either by changing behavioral of the lower wall or by altering the surface waves at the free surface which is done here by including a floating flexible membrane at the top of the liquid layer. Influence of membrane tension is taken into account in terms of stress jump at the free surface. The Orr-Sommerfeld system of the flow is solved numerically using spectral collocation method. The results displays a destabilizing role of membrane tension for a wide range of parameters. The growth rate of the perturbation waves increases with an increase of membrane tension and the critical Reynolds number becomes smaller. Therefore, it is possible to enhance the instability of the flow system with help of membrane properties, which may be useful in Ocean engineering and coating industries.

Keywords Hydrodynamic stability · Film flow · Floating membrane · Orr-Sommerfeld analysis

M. Sani · H. Behera
Department of Mathematics, SRM Institute of Science and Technology, Kattankulathur, Tamil Nadu, India
e-mail: msdjunior9@gmail.com

H. Behera
e-mail: hkb.math@gmail.com

S. Ghosh (✉)
Department of Mathematics, Indian Institute of Technology, Jodhpur 342037, India
e-mail: sukhendu.math@gmail.com

S. Manna et al. (eds.), *Mathematical Modelling and Scientific Computing with Applications*, Springer Proceedings in Mathematics & Statistics 308,
https://doi.org/10.1007/978-981-15-1338-1_19

1 Introduction

Hydrodynamic modal stability analysis of wall bounded channel flow or semi-bounded free surface flow is a flourishing area of modern fluid dynamic research. The knowledge of instability for a flow system has wide range of applications [7, 16, 18] in mechanical, civil, biomedical, geophysics, and agricultural engineering etc. This among other reasons is why it attracted and continues to attract the attention of large number of researchers from applied sciences and engineering. In the past semicentennial, a lot of work has been done on the instability of film flow with a free boundary down an incline (see [1–3, 10, 17]). Amaouche et al. [1] studied the long wave instability of an electroconductor fluid film, flowing down an inclined plane under the action of electromagnetic field. They used the lubrication theory and the weighted residual approach to check the instability behaviour of the flow system.

A two-dimensional motion of a thin film flowing down an inclined plane under the influence of the surface tension and the gravity was studied by Ueno and Iguchi [15]. In their work, a mathematical justification of a thin film approximation between the solution of the Navier–Stokes equations and those of approximate equations is elaborated. Earlier, Tshehla [14] investigated the effect of a temperature dependent variable viscosity on a free surface flow down an inclined plane. The full solution of the physical problem was handled using Runge–Kutta numerical method.

In the recent decades, researchers have great interest to find out passive or active control technique to encourage/suppress the free surface flow instabilities depending on the applications. To control instability of a film flow by passive way, several mathematical models have been developed by considering the effects of surface tension or floating insoluble surfactants at the free surface and slippery/hydrophobic/porous wall boundary. The effect of surfactant on the instability of film flow down an inclined plane has been studied by Blyth and Pozrikidis [3]. Anjalaiah et al. [2] investigated the stability analysis for thin film flow over a porous substrate in the presence of an insoluble surfactants and they have shown the competing influence of porous layer (which is modeled by a velocity slip condition) and insoluble surfactants in an effective way. Apart from stability analysis, gravity wave interaction with floating/submerged flexible membranes is another branch of study that draws significant attention of the scientific community. Using mode expansion method and boundary element method, Cho and Kim [5] studied oblique wave interaction with a submerged horizontal membrane. From their study, they found that the horizontal membrane improved the wave blocking efficiency as compared to a rigid submerged plate. Later, Cho and Kim [6] analyzed the effect of circular membrane on wave scattering. Using eigenfunction expansion method, Sahoo et al. [12] investigated wave scattering by a semi-infinite floating horizontal elastic plate by considering different edge conditions which were generalized by Yip et al. [19] for wave scattering by multiple floating membranes. Karmakar and Sahoo [8] developed an analytic method using eigenfunction expansion method along with expansion formulae for gravity wave interaction with floating membrane due to abrupt change in water depth. Karmakar and Soares [9] also studied the interaction of oblique incident wave with a moored floating membrane as a breakwater for both the cases of finite water depth and shallow water approximation based on the linearized water wave theory. Mondal et al. [11] studied gravity wave interaction with floating circular flexible plate and membrane in both single-and two-layer fluid using Fourier-Bessel series along with mode-coupling relation. However, in all the aforementioned studies on floating elastic membrane, the fluid is assumed to be inviscid. To the best knowledge of the authors, there is no study in the literature on the instability of viscous free surface flow in the presence of floating flexible membrane.

In this work, the effects of floating flexible membrane on the instability of a gravity-driven flow down an inclined plane has been discussed. Linear stability of the flow system is explored using normal-mode analysis. The instability of the fluid is passively controlled here using a floating flexible membrane at the top of the liquid layer. Influence of the membrane tension is taken into account in terms of stress jump at the free surface. We have studied a Newtonian thin film flow in two dimensional Cartesian coordinate system. The Orr-Sommerfeld system of the flow is solved numerically using spectral collocation method. Systematic derivation of the Orr-Sommerfeld equation for linear stability and numerical results are presented in the next two sections.

2 Mathematical Formulation

A two-dimensional incompressible Newtonian fluid flowing over an inclined substrate is considered. The top of the flow is bounded by a floating membrane. Interaction of fluid surface with a floating flexible membrane is studied in a Cartesian coordinate system where the x-axis is along the inclined plane and y-axis is being placed vertically positive upward direction on the plane (see Fig. 1). It is assumed that a flexible membrane of uniform thickness d_m floats on the top of the fluid layer with membrane deflection $y = h(x, t)$, and the mean surface (at $y = H$) is parallel to the inclined plane $y = 0$. The fluid motion is incompressible and irrotational. The governing equations for the flow system are

$$u_x + v_y = 0, \tag{1}$$

$$\rho\big[u_t + uu_x + vu_y\big] = -p_x + \mu\big[u_{xx} + u_{yy}\big] + g\sin\theta, \tag{2}$$

$$\rho\big[v_t + uv_x + vv_y\big] = -p_y + \mu\big[v_{xx} + v_{yy}\big] - g\cos\theta, \tag{3}$$

where ρ, μ are the density and viscosity of the fluid, respectively. u, v are the components of velocity in the x and y increasing directions, respectively and p is the dynamic pressure exerted by the fluid on the flexible membrane. The term g represents the acceleration due to gravity and θ is the angle of inclination of the substrate. The set of dimensional boundary conditions at the free surface ($y = h(x, t)$) which

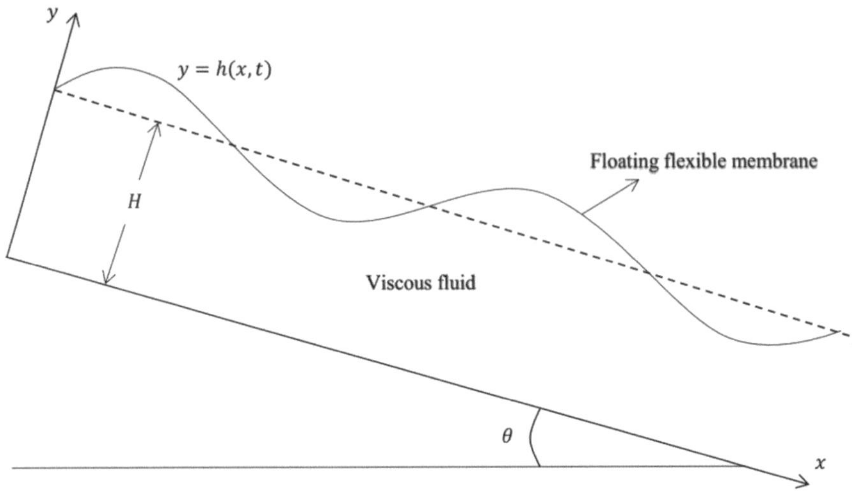

Fig. 1 Schematic diagram for a Newtonian fluid flowing down an inclined plane with a floating flexible membrane at the free surface

is aligned with the flexible membrane, are the kinematic condition, the balance of the tangential and normal stresses given by

$$v = h_t + uh_x \quad \text{at } y = h(x,t), \tag{4}$$

$$\mu\left[-4u_x h_x + \left(u_y + v_x\right)\left(1 - h_x^2\right)\right] = T_x\sqrt{\left(1 + h_x^2\right)} \quad \text{at } y = h(x,t) \tag{5}$$

$$p - p_\infty = \frac{2\mu}{\left(1 + h_x^2\right)}\left[u_x h_x^2 - \left(u_y + v_x\right)h_x + v_y\right] - \frac{Th_{xx}}{(1 + h_x^2)^{\frac{3}{2}}} - m_m h_{tt} \quad \text{at } y = h(x,t), \tag{6}$$

where, T is the membrane tension per length, p_∞ is the atmospheric pressure and $m_m = \rho_m d_m$ is the uniform mass per unit length of the membrane. Note that, when the elastic membrane stretched with a constant tension per length (i.e. $T =$ constant) then $T_x = 0$ and therefore, the right hand side of the Eq. (5) will be zero. The boundary conditions at the inclined wall are

$$u = 0 \text{ and } v = 0 \quad \text{at } y = 0, \tag{7}$$

which confirms no velocity slip at the wall. The set of equations and boundary conditions are made nondimensional using the following dimensionless variables:

$$\bar{x} = \frac{x}{H},\ \bar{y} = \frac{y}{H},\ \bar{u} = \frac{u}{V},\ \bar{v} = \frac{v}{V},\ \bar{t} = \frac{tV}{H},\ \bar{p} = \frac{p}{\rho V^2},\ \bar{h} = \frac{h}{H},\ \bar{T} = \frac{T}{T_0}$$

where H is the mean film thickness of the fluid layer and T_0 is a mean reference tension of the membrane. The characteristic velocity scale V for the fluid layer is chosen as the maximum velocity of a uniform flat Nusselt film over a rigid substrate and given by $V = \frac{gH^2 \sin\theta}{2\nu}$ (ν is kinematic viscosity). The present study is motivated by the investigations of Blyth and Pozrikidis [3], Anjalaiah et al. [2] and hence, in order to compare the results with available results, the formulation is in terms of the Reynolds number based on the Nusselt film free surface velocity for a film without membrane over a rigid substrate has been used. The set of dimensionless equations and conditions look like (after suppressing the over bars):

$$u_x + v_y = 0, \tag{8}$$

$$u_t + uu_x + vu_y = -p_x + \frac{1}{\text{Re}}\left[u_{xx} + u_{yy}\right] + G, \tag{9}$$

$$v_t + uv_x + vv_y = -p_y + \frac{1}{\text{Re}}\left[v_{xx} + v_{yy}\right] - G\cot\theta, \tag{10}$$

$$v = h_t + uh_x \quad \text{at } y = h(x, t), \tag{11}$$

$$\left(1 - h_x^2\right)\left(u_y + v_x\right) - 4u_x h_x = 0 \quad \text{at } \; y = h(x, t), \tag{12}$$

$$Rep = \frac{2}{\left(1 + h_x^2\right)}\left[u_x h_x^2 - \left(u_y + v_x\right) + v_y\right] - \frac{Th_{xx}}{(1 + h_x^2)^{\frac{3}{2}}} - m_m h_{tt} \quad \text{at } y = h(x, t), \tag{13}$$

$$u = 0 \text{ and } v = 0 \quad \text{at } y = 0, \tag{14}$$

where Reynolds number $\text{Re} = VH/\nu$ and G is the dimensionless gravitational force. Further, in above equations T and m_m are dimensionless membrane tension and unit mass, respectively.

2.1 Base Solution and Stability Equation

The base flow solution (for velocity and pressure) corresponding to the considered flow system is obtained from above set of equation and boundary conditions with unidirectional parallel flow assumption and are given by

$$U(y) = -y^2 + 2y$$

$$P(y) = 2\cot\theta(1 - y)$$

Following the work of Anjalaiah et al. [2], the base flow calculation is done by fixing $G\,\mathrm{Re} = 2(G \simeq O(1))$. It is important to note that there will be no change in the behavior of the base flow due to the presence of membrane at the free surface.

Next, the stability of the base state with respect to infinitesimal perturbations is considered and the flow variables are now taken as the sum of basic state and perturbed state solution. Substituting $u(x, y, t) = U(y) + \tilde{u}(x, y, t)$, $v(x, y, t) = \tilde{v}(x, y, t)$, $p(x, y, t) = P(y) + \tilde{p}(x, y, t)$, $h(x, t) = 1 + \tilde{h}(x, t)$ into the equations of motion and boundary conditions and linearizing with respect to the small amplitude perturbations, the equations for the perturbed quantities are obtained as

$$\tilde{u}_x + \tilde{v}_y = 0, \tag{15}$$

$$\mathrm{Re}\left[\tilde{u}_t + U\tilde{u}_x + vU_y\right] = -\mathrm{Re}\tilde{p}_x + \left[\tilde{u}_{xx} + \tilde{u}_{yy}\right], \tag{16}$$

$$\mathrm{Re}\left[\tilde{v}_t + U\tilde{v}_x\right] = -\mathrm{Re}\tilde{p}_y + \left[\tilde{v}_{xx} + \tilde{v}_{yy}\right], \tag{17}$$

$$\tilde{u} = 0 \quad \text{and} \quad \tilde{v} = 0 \quad \text{at} \quad y = 0, \tag{18}$$

$$\tilde{v} = \tilde{h}_t + U\tilde{h}_x \quad \text{at } y = 1, \tag{19}$$

$$\tilde{h}U_{yy} = -\tilde{u}_y - \tilde{v}_x \quad \text{at } y = 1 \tag{20}$$

$$\mathrm{Re}\tilde{p} + \tilde{h}P_y = 2\tilde{v}_y - T\tilde{h}_{xx} - m_m\tilde{h}_{tt} \quad \text{at} \quad y = 1 \tag{21}$$

Using Eq. (15) in Eqs. (20) and (21) one can rewrite these equations after some manipulation as,

$$\tilde{h}_x U_{yy} = \tilde{v}_{yy} - \tilde{v}_{xx} \quad \text{at } y = 1 \tag{22}$$

$$\begin{aligned} &2\cot\theta\tilde{h}_{xx} - T\tilde{h}_{xxxx} - m_m\left(\tilde{v}_{xxt} - U\tilde{h}_{xxtt}\right) + 3\tilde{v}_{xxy} + \tilde{v}_{yyy} \\ &- \mathrm{Re}\tilde{v}_{yt} - \mathrm{Re}U\tilde{v}_{xy} = 0 \quad \text{at} \quad y = 1 \end{aligned} \tag{23}$$

Let $\tilde{\psi}$ is the stream function of the two dimensional flow and in the form of normal mode solution,

$$\tilde{\psi}(x, y, t) = \phi(y)\mathrm{e}^{\mathrm{i}k(x-ct)}$$

$$\tilde{h}(x, y, t) = \eta(y)\mathrm{e}^{\mathrm{i}k(x-ct)}$$

where, k, c are the wave number and the complex wave speed, respectively and $\mathrm{i} \equiv \sqrt{-1}$. Expressing velocity components in terms of stream function and using all in the perturbation equations and boundary conditions we obtained the following Orr-Sommerfeld system,

$$\left(D^2 - k^2\right)^2\phi - \mathrm{Re}\left[ik(U - c)\left(D^2 - k^2\right)\phi - ikU_{yy}\phi\right] = 0, \tag{24}$$

$$D\phi = 0 \quad \text{and} \quad \phi = 0 \quad \text{at} \quad y = 0, \tag{25}$$

$$\phi + \eta(U - c) = 0 \quad \text{at} \quad y = 1, \tag{26}$$

$$U_{yy}\eta + D^2\phi + k^2\phi = 0 \quad \text{at} \quad y = 1, \tag{27}$$

$$\begin{aligned} &- ikD^3\phi + \left[\mathrm{Re}k^2(c - U) + 3ik^3\right]D\phi - \tilde{m}_m k^4 c\phi \\ &- \left[\left(2k^2\cot\theta + k^4 T\right) + \tilde{m}_m k^4 Uc\right]\eta = 0 \quad \text{at} \quad y = 1. \end{aligned} \tag{28}$$

where D presents derivative with respect to y. Equations (25)–(28) describe a generalized eigenvalue problem with c as eigenvalue and we are interested to obtain a non-trivial solution of the system. The parameter $c = c_r + ic_i$ where c_r and c_i are respectively, the wave speed and the growth rate. In this study we have considered $\omega_i = kc_i$ as scaled growth rate of the disturbances. When no membrane is present on the free surface the above system, in the absence of membrane tension at the free surface, reduces to the Orr-Sommerfeld equations represent a Newtonian flow down a rigid inclined substrate given by Yih [17].

3 Numerical Results

The Orr-Sommerfeld system is solved using Spectral collocation method of Canuto et al. [4] based on Chebyshev polynomials and points which are used for discretizing the generalized eigenvalue problem described by Eqs. (25)–(28). The solution of this system yields complex phase velocity $c = c_r + ic_i$, from which the dimensionless scaled growth rate $\omega_i = kc_i$ is obtained for a wide range of wave number k. The accuracy of the eigenvalues is tested by varying the number of collocation points used in the computation and the spurious eigenvalues are eliminated in a special process. The employed 'MATLAB' code for the computations is validated by obtaining the numerical results for the growth rate of a film flow in the absence of the membrane on the free surface over a rigid inclined substrate.

We start our numerical computation with angle of inclination $\theta = 4°$ and $\theta = 90°$ in Fig. 2 to validate our results with the results of Blyth and Pozrikidis [3] and Skotheim et al. [13], respectively in the absence of floating flexible membrane (for $m_m = 0$). It is seen that both the curves are in good agreement. In Fig. 3a, we see existence of two humps in the growth rate curve when the membrane tension is very low ($T = 0.001$). This in turn indicates subsistence of two different unstable

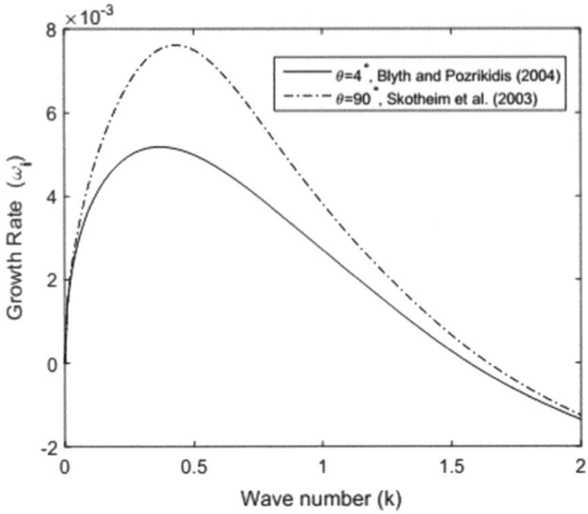

Fig. 2 The growth rate of the dominant disturbance as a function of wave number k for a comparison between present results with the results of Blyth and Pozrikidis [3] ($\theta = 4°$) and Skotheim et al. [13] ($\theta = 90°$) with $m_m = 0$, $T \approx Ca = 0.017716$ and Re $= 2500$

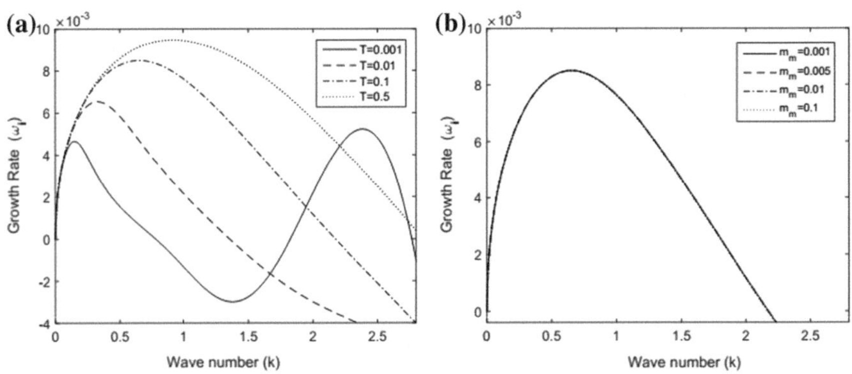

Fig. 3 The growth rate of the dominant disturbance as a function of wave number k for different values of **a** membrane tension T for $m_m = 0.1$ and **b** mass per unit length of the membrane m_m for $T = 0.1$ when $\theta = 10°$ and Re $= 2000$

modes for the perturbed flow with one mode for small wave numbers and other one at moderate wave numbers ($k > O(1)$). The long wave instability is weaker than the instability of shorter waves. An increase in the magnitude of membrane tension enhances the long wave instability by increasing the growth rate of the unstable mode, but simultaneously it suppresses the instabilities at higher wave numbers.

The membrane placed on the free surface is very thin and flexible thus perturbation inside the flow gives rise disturbance to the membrane and the membrane can generate some extra waves at the surface. Quite naturally the membrane waves are of small

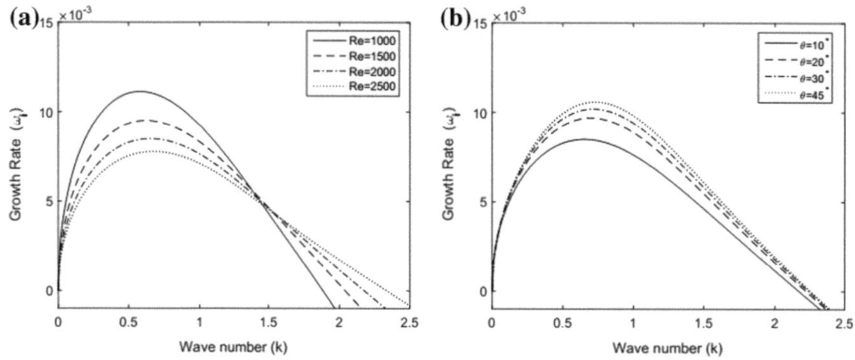

Fig. 4 Influence of **a** Reynolds Number Re with $\theta = 10°$ and **b** inclination angle θ with Re $= 2000$ on the growth rate ω_i for $T = 0.1$ and $m_m = 0.1$

amplitude and long, as a consequence long wave disturbances inside the flow are becoming stronger and therefore, more unstable in the presence of the membrane. Uniform mass per unit length of the membrane, m_m has very weak influence on the growth rate of the two dimensional disturbances (Fig. 3b). As the membrane is thin (d_m is small), the unit mass of the membrane is small and thus, we can not use larger value of m_m.

In Fig. 4, we have shown the effects of inertial/viscous forces (Re) and inclination angle (θ) on the growth rate of the dominant disturbances. Figure 4a displays, for a fixed θ as Re increases, the growth of the long wave instability diminishes but an opposite trend is seen for higher range of wave numbers. This bifurcation of the growth rate is happening around the wave number $k = 1.25$. If we keep the Re fixed and vary θ, a destabilizing effect of θ is found in Fig. 4b, since the growth rate is increasing with an increase of θ. As a consequence, a vertical film flow is unstable for all Reynolds numbers and horizontal flow is always much stable.

Finally, to get a clear idea about the instability behavior and the effects of membrane properties, maximum growth rate ($\omega_{i,max}$) over a range of wave numbers is calculated and plotted as a function of Re in Fig. 5. Figure 5a displays that the flow system is stable at very small Reynolds numbers since $\omega_{i,max} < 0$. But, for each T there exists a critical Re (Re$_c$) and the flow is unstable for all Re $>$ Re$_c$, due to the positive growth rate of the dominant disturbance. It is also clear from the figure that the growth rate of the unstable mode amplifying up to a certain value of Reynolds number and at higher magnitude of inertial forces $\omega_{i,max}$ is almost constant for Re. The membrane tension T is destabilizing the flow by increasing the disturbance growth rate for all Reynolds numbers. In order to facilitate the effects of membrane thickness by means of uniform unit mass m_m over all Reynolds number, we have plotted $\left(\omega_{i,max}\right)$ as a function of Re in Fig. 5b. We see that the influence of m_m is not significant for the considered values of Re, θ and T. The physical reason behind this is not very clear to us in this stage. A detailed energy budget analysis may give some

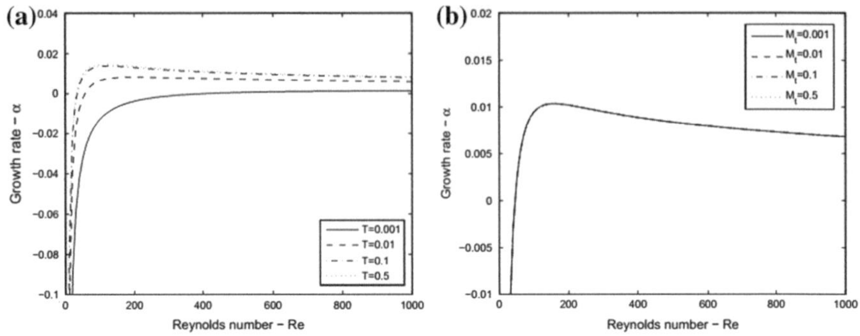

Fig. 5 Maximum growth rate as a function of Re for different values of **a** T with $m_m = 0.1$ and **b** m_m with $T = 0.1$, and for both the cases $\theta = 10°$

knowledge on the mechanisms of the instability and we have planed to investigate the same in future work.

4 Conclusions

The linear stability of a film flow down an inclined substrate in the presence of floating flexible membrane at the free surface is examined. A hybrid Chebyshev spectral method is employed to obtain the growth rate of the disturbances and the critical Reynolds number. The influence of the parameters characterizing the floating flexible membrane and the inclined substrate such as membrane tension T, membrane thickness d_m and inclination angle θ have been shown by properly choosing the various fluid and structural parameters. Numerical results show the existence of both long and short waves instabilities. It is possible to find a $k-\text{Re}$ window for a range of other parameters, where the flow over an inclined substrate can be more destabilized in the presence of a floating membrane at the free surface. The results indicate that the presence of the flexible membrane enlarges the range of unstable wave numbers beyond the onset of instability and the unstable region grows. The growth rate of the perturbation waves increases with an increase of membrane tension and the critical Reynolds number becomes smaller. Therefore, it is possible to enhance the instability of the flow system with help of the flexible floating membrane on the top of the fluid layer.

Acknowledgements HB and SG are thankful to SERB, DST, Government of India (Award No. CRG/2018/004521) for the financial support.

References

1. Amaouche, M., Abderrahmane, H.A., Bourdache, L.: Hydromagnetic thin film flow: linear stability. Phys. Rev. E **88**, 023028 (2013)
2. Anjalaiah, Usha, R., Millet, S.: Thin film flow down a porous substrate in the presence of an insoluble surfactant: stability analysis. Phys. Fluids **25**, 022101 (2013)
3. Blyth, M., Pozrikidis, C.: Effect of surfactant on the stability of film flow down an inclined plane. J. Fluid Mech. **521**, 241–250 (2004)
4. Canuto, C., Hussaini, M.Y., Quarteroni, A., Thomas Jr., A.: Spectral Methods in Fluid Dynamics. Springer Science & Business Media (2012)
5. Cho, I., Kim, M.: Interactions of a horizontal flexible membrane with oblique incident waves. J. Fluid Mech. **367**, 139–161 (1998)
6. Cho, I., Kim, M.: Wave deformation by a submerged flexible circular disk. Appl. Ocean Res. **21**, 263–280 (1999)
7. Criminale, W.O., Jackson, T.L., Joslin, R.D.: Theory and Computation of Hydrodynamic Stability. Cambridge University Press, Cambridge (2003)
8. Karmakar, D., Sahoo, T.: Gravity wave interaction with floating membrane due to abrupt change in water depth. Ocean Eng. **35**, 598–615 (2008)
9. Karmakar, D., Soares, C.G.: Oblique scattering of gravity waves by moored floating membrane with changes in bottom topography. Ocean Eng. **54**, 87–100 (2012)
10. Liu, R., Liu, Q.: Instabilities of a liquid film flowing down an inclined porous plane. Phys. Rev. E **80**, 036316 (2009)
11. Mondal, R., Mandal, S., Sahoo, T.: Surface gravity wave interaction with circular flexible structures. Ocean Eng. **88**, 446–462 (2014)
12. Sahoo, T., Yip, T.L., Chwang, A.T.: Scattering of surface waves by a semi-infinite floating elastic plate. Phys. Fluids **13**, 3215–3222 (2001)
13. Skotheim, J.M., Thiele, U., Scheid, B.: On the instability of a falling film due to localized heating. J. Fluid Mech. **475**, 1–19 (2003)
14. Tshehla, M.: The flow of a variable viscosity fluid down an inclined plane with a free surface. Math. Probl. Eng. **2013** (2013)
15. Ueno, H., Iguchi, T.: A mathematical justification of a thin film approximation for the flow down an inclined plane. J. Math. Anal. Appl. **444**, 804–824 (2016)
16. Wang, C.: Thin film flowing down a curved surface. Z. Angew. Math. Phys. ZAMP **35**, 532–544 (1984)
17. Yih, C.S.: Stability of liquid flow down an inclined plane. Phys. Fluids **6**, 321–334 (1963)
18. Yih, C.S.: Stability of liquid flow down an inclined plane. In: Selected Papers By Chia-Shun Yih (In 2 Volumes). World Scientific, pp. 357–370 (1991)
19. Yip, T., Sahoo, T., Chwang, A.T.: Wave scattering by multiple floating membranes. In: The Eleventh International Offshore and Polar Engineering Conference. International Society of Offshore and Polar Engineers (2001)

On the Dynamics of Transverse Domain Walls in Ferromagnetic Heterostructures with Nonlinear Dissipative Effects

Sharad Dwivedi

Abstract This work delineates the investigation of static and dynamical behavior of transverse domain walls in ferromagnetic heterostructure in which the thin ferromagnetic layer is sandwiched between a heavy metal and a metal oxide layers. We consider the metallic ferromagnet with high perpendicular magnetic anisotropy and exhibits structural inversion asymmetry. The presence of a strong inversion asymmetry in the ferromagnetic material leads to additional spin-orbit coupling torque terms into the dynamics qualitatively different from the standard spin-transfer torques. The evolution of magnetization inside the ferromagnetic layer is governed by the one-dimensional model of generalized Landau-Lifshitz-Gilbert-Slonczewski equation consisting of the standard spin transfer and spin-orbit coupling torque terms along with the nonlinear dissipation factors *viz.* viscous and dry-friction. Under this framework, first, we establish the static magnetization profile in the two faraway domains in the presence of transverse magnetic field once the electric and longitudinal magnetic fields are switched off. Then, by means of regular perturbation expansion method, we derive the zero order traveling wave solutions under the influence of small applied magnetic field and spin-polarized electric currents.

Keywords Domain wall · Micromagnetics · Nonlinear dissipations · Traveling wave

1 Introduction

The ferromagnetic materials have been widely used in the different applications emerging from various disciplines of science and engineering. For instance, it has been used in MRI's, magnetic sensors, nuclear systems, logic devices, and modern storage media etc. (cf. [1–3]). In particular, in modern electronic devices,

S. Dwivedi (✉)
Department of Mathematics, School of Sciences, National Institute of Technology Andhra Pradesh, Tadepalligudem 534102, Andhra Pradesh, India
e-mail: sharadiitm@gmail.com

S. Manna et al. (eds.), *Mathematical Modelling and Scientific Computing with Applications*, Springer Proceedings in Mathematics & Statistics 308,
https://doi.org/10.1007/978-981-15-1338-1_20

ferromagnetic nanowires (nanostrips) have been used to encode the data as a pattern of domains along these wires (strips). *Domains* are the uniformly oriented magnetized regions separated by a thin transition zones referred as domain walls (DW). In literature, various types of domain wall have been reported based on their characteristics such as Bloch, Néel, transverse, chiral, vortex, and spiral domain walls. In this work, we examine the dynamics of transverse domain wall (TDW) (a type of DW in which the magnetization at the center of the DW is oriented along the transverse direction to the wire (strip) axis) in the presence of small applied magnetic and electric fields. For the complete description of domain wall motions from both the theoretical and application viewpoints, we would like to draw the attention of our readers to the excellent monographs [1, 2].

Experimentally, it has been observed that the DW motions experience the phase transition in the presence of external sources. For the small value of applied magnetic (electric) field, DW moves rigidly with a constant velocity along the wire (strip) axis. This dynamic regime is known as the steady-state regime. The minimum and the maximum value of the strength of the external sources for which DW motions remains in the steady-state dynamic regime are referred as the threshold and Walker-breakdown value. As the strength of the external sources surpasses the breakdown value, the DW motion becomes oscillatory due to its internal deformation. This dynamic regime is referred as the precessional dynamic regime. To understand the DW motions in these dynamic regimes, various techniques have been reported in literature such as energy estimate methods and the classical theory of differential equations (cf. [4–8]), approximating the dynamics near the center of DW (cf. [9–14]), and the regular perturbation expansion method (cf. [15–20]) etc. We adopted the latter approach in our analysis to draw the analytical results.

In this work, we analyze the TDW motion in ferromagnetic heterostructure (in which the thin ferromagnetic layer is sandwiched between a heavy metal and a metal oxide layers) with structural inversion asymmetry in the presence of applied magnetic and electric fields. We consider the evolution of magnetization inside the ferromagnetic layer is governed by the one-dimensional model of generalized Landau-Lifshitz-Gilbert-Slonczewski equation with an inclusion of nonlinear dissipative effects. Moreover, in these heterostructures, the current induced domain wall motion is described through the standard spin-transfer torques (STT) which originate due to the transfer of spin angular momentum.

The theoretical evidences (cf. [13, 21–23]) demonstrate that in ferromagnetic material with structural inversion asymmetry, an additional torque acts due to the strong spin-orbit coupling referred as spin-orbit torque (SOT). These SOTs are mainly due to the Rashba and Spin-Hall effects. More precisely, Rashba effect is due to the spin-orbit coupling and structural inversion asymmetry that occurs between the different interfaces namely heavy-metal and ferromagnetic layers and metal-oxide and ferromagnetic layers. Moreover, the Spin-Hall effect (SHE) is due to the Spin-Hall current that induced from the heavy-metal layer and enters in the ferromagnetic layer. In addition, these heterostructures experiences an interfacial anisotropic exchange due to the Dzyaloshinskii-Moriya interaction (DMI) which is an additional torque qualitatively different from the STTs and SOTs.

Structure of the article

This article is organized in the following manner: In the next section, we explain the governing dynamics and the various terms appear in the equation and their physical significance. In section "Problem under consideration", we describe the theoretical model and derive the governing equation under the light of considered assumptions. In section "Characterization of DW motion", we investigate the static and dynamic features of TDW in the presence of small applied magnetic and electric fields. To be precise, first, we derive the static magnetization profile in the two faraway distant domains then, we study the zero order traveling wave solutions in the presence of external sources. Finally, in the last section, we summarize our results and emphasize the significant issues emerging from this work.

2 Governing Equation

The one-dimensional model of generalized Lifshitz-Gilbert-Slonczewski equation which describes the magnetization dynamics in ferromagnetic heterostructures with an inclusion of nonlinear dissipative effects along with the STT and SOT contributions is given as (cf. [11–13, 22, 23]):

$$\frac{\partial \mathbf{m}}{\partial t} = \gamma(\mathbf{H}_{eff} \times \mathbf{m}) + \left[\alpha_G\left(1 + \frac{\alpha_v}{\gamma^2}\left(\frac{\partial \mathbf{m}}{\partial t}\right)^2\right) + \frac{\gamma \alpha_d}{|\partial \mathbf{m}/\partial t|}\right]\left(\mathbf{m} \times \frac{\partial \mathbf{m}}{\partial t}\right)$$
$$+ v_s\left(-1 + \zeta \mathbf{m}\times\right)\frac{\partial \mathbf{m}}{\partial x} + \gamma \frac{v_s}{v_0}\left[\alpha_{RE}\left(\mathbf{e}_2 \times \mathbf{m}\right) + \left(\zeta \alpha_{RE} + H_{SH}\right)\mathbf{m} \times \left(\mathbf{m} \times \mathbf{e}_2\right)\right], \tag{1}$$

where, $\times$ denotes the vectorial cross product. Also, $\mathbf{m} = \mathbf{M}/M_s$, represents the time varying normalized magnetization vector field given by:

$$\mathbf{m} : \mathbb{R}^+ \times \mathbb{R} \rightarrow \mathbb{S}^2.$$

$\mathbb{S}^2$ is the unit sphere in $\mathbb{R}^3$. Also, $\mathbf{M}$ and M_s stand for the magnetization vector field and saturation value of magnetization respectively.

Now, we elaborate the physical significance of each term present on the right-hand side of Eq. (1). The first term depicts the undamped precessional motion of magnetization towards the total effective field $\mathbf{H}_{eff}$ and known as Larmor precession.The positive constant γ is defined in terms Landè factor g, permeability of vacuum μ_0, electron charge e, and electron mass m_e as $\gamma = M_s\mu_0 ge/m_e$. Moreover, The total effective field is connected with the micromagnetic energy $\mathscr{E}$ by the relation $\mathbf{H}_{eff} = -\nabla_{\mathbf{m}}\mathscr{E}$, the first variation of the energy functional. The total effective field constitutes of several energy contributions. In our analysis, we take into the account of exchange, anisotropy, demagnetizing, Zeeman (applied field), and DMI energy contributions in the total effective field which in turn takes the following form:

$$\mathbf{H}_{eff}(\mathbf{m}) = \mathbf{H}_{ex}(\mathbf{m}) + \mathbf{H}_{an}(\mathbf{m}) + \mathbf{H}_{d}(\mathbf{m}) + \mathbf{H}_{a} + \mathbf{H}_{DMI}. \tag{2}$$

Now, we give insight to individual field contribution appears in Eq. (2).

- Exchange field ($\mathbf{H}_{ex}$): This field is responsible for the parallel alignment of magnetization orientation inside the ferromagnetic medium and is given as (cf. [12, 14, 20]):

$$\mathbf{H}_{ex}(\mathbf{m}) = \frac{2A_{ex}}{\mu_0 M_s^2} \frac{\partial^2 \mathbf{m}}{\partial x^2},$$

  where A_{ex} denotes the exchange constant.
- Anisotropy field ($\mathbf{H}_{an}$): In the absence of external stimulus, the orientation of magnetization vector tends to align itself along one or more energetically preferred directions in a ferromagnetic crystal referred as *easy-axes*. In case of uniaxial ferromagnetic nanostrip with high perpendicular magnetocrystalline anisotropy, the anisotropy field is given by (cf. [10, 12, 14, 20]):

$$\mathbf{H}_{an}(\mathbf{m}) = \frac{2K_e}{\mu_0 M_s^2} (\mathbf{m} \cdot \mathbf{e})\, \mathbf{e}$$

  where K_e and $\mathbf{e}$ represent the anisotropy constant and the direction of easy axis respectively.
- Demagnetizing field ($\mathbf{H}_d$): It is due to the magnetic field generated by the medium itself and defined as (cf. [10, 12, 13]):

$$\mathbf{H}_d(\mathbf{m}) = -N_1 (\mathbf{m} \cdot \mathbf{e}_1)\, \mathbf{e}_1 - N_2 (\mathbf{m} \cdot \mathbf{e}_2)\, \mathbf{e}_2 - N_3 (\mathbf{m} \cdot \mathbf{e}_3)\, \mathbf{e}_3$$

  where N_1, N_2 and N_3 denote the demagnetizing factors that depend on the geometry of the material and satisfy the relation $N_1 + N_2 + N_3 = 1$. Also, $\mathbf{e}_1$, $\mathbf{e}_2$, and $\mathbf{e}_3$ represent the canonical basis of $\mathbb{R}^3$. We would like to mention that an explicit expression of $\mathbf{H}_d$ is known only for few favourable geometries namely a uniform magnetized ellipsoid (cf. [7, 8, 24]) and a straight nanowire with the circular cross-section (cf. [5, 6, 11, 14]). However, the above expression of $\mathbf{H}_d$ is often used in literature as a good approximation also for non-ellipsoidal geometries.
- Zeeman field ($\mathbf{H}_a$): This contribution is due to the effect of applied magnetic field and is given as:

$$\mathbf{H}_a = h_1 \mathbf{e}_1 + h_2 \mathbf{e}_2 + h_3 \mathbf{e}_3.$$

  where h_1, h_2, and h_3 corresponds to the longitudinal and transverse components of the applied magnetic field. In addition, we also assume that $\mathbf{H}_a$ is constant in both the time and space variables.
- Dzyaloshinskii-Moriya interaction field ($\mathbf{H}_{DMI}$): It is an additional term in the exchange interaction occurs in the ferromagnetic heterostructures due to the strong spin-orbit coupling and given as (cf. [13, 22, 23]):

$$\mathbf{H}_{DMI} = \frac{2D}{\mu_0 M_s^2}\left(\frac{\partial m_3}{\partial x}\mathbf{e}_1 - \frac{\partial m_1}{\partial x}\mathbf{e}_3\right).$$

where $\mathbf{m} = (m_1, m_2, m_3)$ and the positive constant D stands for the DMI parameter.

The second term appears on the right-hand side of Eq. (1) describes the dissipation of energy in the system comprising of three terms. In which, the first term exhibits the classical viscous Gilbert dissipation torque responsible for the decay of energy in ideal ferromagnets without any crystallographic defects. However, the second and third reflects the nonlinear viscous and *dry-friction* dissipations corresponding to the dissipation of energy due to the large angle precession and impurities present in the ferromagnetic material respectively (cf. [10–12, 25]). The positive dimensionless constants α_G, α_v, and α_d stand for the standard Gilbert damping, nonlinear viscous and *dry-friction* damping coefficients respectively.

The subsequent torque term appearing in Eq. (1) reflects the standard spin-transfer torque term arising from the injection of spin-polarized electric currents into the ferromagnet. The STT comprises of adiabatic and non-adiabatic contributions corresponding to the formation and evolution of the domain wall. More precisely, adiabatic component describes a reactive STT whereas the non-adiabatic term shows the dissipative effect (cf. [11, 13, 21, 23]). The spin-torque velocity $\mathbf{v}_s$ is directed along the current motion direction with the amplitude $v_s = v_0 J$, J being the current density. Also, $v_0 = g\mu_B P/2eM_s$, in which μ_B and P denote the Bohr magneton and polarization factor of current respectively. The positive dimensionless constant ζ stands for the phenomenological non-adiabatic spin-torque parameter.

Finally, the last torque term emerges on the right-hand side of Eq. (1) characterizes the combined effect of Rashba and Spin-Hall effect due to the strong spin-orbit coupling taking place between the heavy metal, ferromagnet and metal oxide interfaces. The first two terms depict the contribution of Rashba effect and the last STT like torque term reflects the effect of Spin-Hall current (cf. [11, 13, 22, 23]). Also, $\alpha_{RE} = \left(\alpha_R P/\mu_0\mu_B M_s^2\right)$ and $H_{SH} = (\mu_B\theta_{SH}/\gamma e\delta M_s)$, where α_R, θ_{SH}, and δ denote the Rashba parameter, Spin-Hall angle and the thickness of the ferromagnetic layer respectively.

In the subsequent section, we propose the considered problem and emphasize on the analytical and methodological boundaries within which the problem has been carried out.

3 Problem Under Consideration

3.1 Theoretical Model

We consider a thin ferromagnetic layer (FML) sandwiched between the heavy metal layer (HML) and a metal oxide layer (MOL). To be precise, we assume the length,

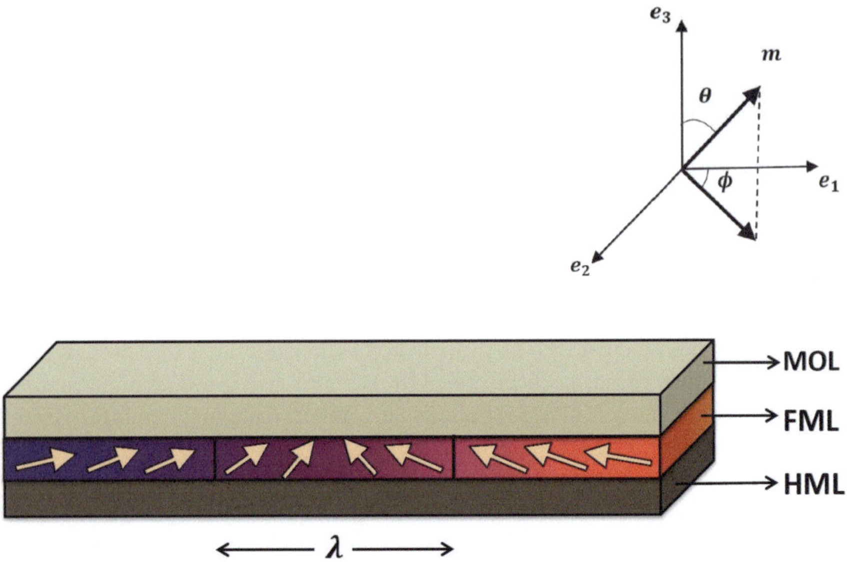

Fig. 1 Schematics of a head-to-head transverse domain wall in ferromagnetic nanostrip sandwiched between the heavy metal and metal oxide layers along with the coordinate axes

width, and thickness of the FML as L, w, δ together with the condition $L \gg w > \delta$. We arrange this multilayer structure (heterostructure) in such a manner that its length, width, and thickness is in the direction of $\mathbf{e}_1$, $\mathbf{e}_2$, and $\mathbf{e}_3$ respectively. In addition, we consider the ferromagnetic material exhibits the high perpendicular magnetocrystalline anisotropy and the easy-axis is oriented along the $\mathbf{e}_3$-direction. We, then subject this heterostructure to a static applied magnetic and electric field which are both constant in time and space variables. We consider the applied magnetic and electric field of the form $\mathbf{H}_a = h_1\mathbf{e}_1 + h_2\mathbf{e}_2 + h_3\mathbf{e}_3$ and $\mathbf{v}_s = v_s\mathbf{e}_1$ respectively. A head-to-head domain wall of width λ is nucleated along the FML that separates the domains with $\mathbf{m}(x,t) \sim \mathbf{m}_L$ as $x \to -\infty$ and $\mathbf{m}(x,t) \sim \mathbf{m}_R$ as $x \to +\infty$ as portrayed in Fig. 1.

3.2 Mathematical Formulation

Under the light of aforementioned assumptions, we examine the following equation:

$$\frac{\partial \mathbf{m}}{\partial t} = \gamma(\mathbf{H}_{eff} \times \mathbf{m}) + \left[\alpha_G\left(1 + \frac{\alpha_v}{\gamma^2}\left(\frac{\partial \mathbf{m}}{\partial t}\right)^2\right) + \frac{\gamma\alpha_d}{|\partial \mathbf{m}/\partial t|}\right]\left(\mathbf{m} \times \frac{\partial \mathbf{m}}{\partial t}\right)$$
$$+ v_s(-1 + \zeta\mathbf{m}\times)\frac{\partial \mathbf{m}}{\partial x} + \gamma\frac{v_s}{v_0}\left[\alpha_{RE}(\mathbf{e}_2 \times \mathbf{m}) + (\zeta\alpha_{RE} + H_{SH})\mathbf{m} \times (\mathbf{m} \times \mathbf{e}_2)\right], \tag{3}$$

with,

$$\mathbf{H}_{eff} = \frac{2A_{ex}}{\mu_0 M_s^2}\frac{\partial^2 \mathbf{m}}{\partial x^2} + \frac{2K_e}{\mu_0 M_s^2} m_3 \mathbf{e}_3 + (-N_1 m_1 \mathbf{e}_1 - N_2 m_2 \mathbf{e}_2 - N_3 m_3 \mathbf{e}_3) + (h_1 \mathbf{e}_1 + h_2 \mathbf{e}_2 + h_3 \mathbf{e}_3) + \left[\frac{2D}{\mu_0 M_s^2}\left(\frac{\partial m_3}{\partial x}\mathbf{e}_1 - \frac{\partial m_1}{\partial x}\mathbf{e}_3\right)\right], \quad (4)$$

along with the physical constraints

$$\frac{\partial \mathbf{m}}{\partial t} \to 0 \text{ and } \frac{\partial \mathbf{m}}{\partial x} \to 0 \text{ as } |x| \to \infty.$$

We introduce the following transformations to convert Eq. (3) in the dimensionless form (cf. [20, 26]):

$$\tilde{x} = x\sqrt{\frac{K_e}{A_{ex}}}, \quad \tilde{v}_s = \frac{v_s}{\gamma}\sqrt{\frac{K_e}{A_{ex}}}, \quad \tilde{D} = \left(\frac{2D}{\mu_0 M_s^2}\right)\sqrt{\frac{K_e}{A_{ex}}}$$
$$\tilde{\alpha}_{RE} = \left(\frac{\gamma \alpha_{RE}}{v_0}\right)\sqrt{\frac{A_{ex}}{K_e}}, \quad \tilde{H}_{SH} = \left(\frac{\gamma H_{SH}}{v_0}\right)\sqrt{\frac{A_{ex}}{K_e}}, \quad \tilde{t} = \gamma t. \quad (5)$$

With the help of Eqs. (5), (3) renders the following dimensionless form:

$$\frac{\partial \mathbf{m}}{\partial \tilde{t}} = (\tilde{\mathbf{H}}_{eff} \times \mathbf{m}) + \left[\alpha_G\left(1 + \alpha_v\left(\frac{\partial \mathbf{m}}{\partial \tilde{t}}\right)^2\right) + \frac{\alpha_d}{|\partial \mathbf{m}/\partial \tilde{t}|}\right]\left(\mathbf{m} \times \frac{\partial \mathbf{m}}{\partial \tilde{t}}\right) + \tilde{v}_s\left(-1 + \zeta \mathbf{m}\times\right)\frac{\partial \mathbf{m}}{\partial \tilde{x}} + \tilde{v}_s\left[\tilde{\alpha}_{RE}\left(\mathbf{e}_2 \times \mathbf{m}\right) + \left(\zeta\tilde{\alpha}_{RE} + \tilde{H}_{SH}\right)\mathbf{m} \times (\mathbf{m} \times \mathbf{e}_2)\right] \quad (6)$$

with,

$$\tilde{\mathbf{H}}_{eff} = \tilde{K}_e\frac{\partial^2 \mathbf{m}}{\partial \tilde{x}^2} + \tilde{K}_e m_3 \mathbf{e}_3 + (-N_1 m_1 \mathbf{e}_1 - N_2 m_2 \mathbf{e}_2 - N_3 m_3 \mathbf{e}_3) + \left(\tilde{h}_1 \mathbf{e}_1 + \tilde{h}_2 \mathbf{e}_2 + \tilde{h}_3 \mathbf{e}_3\right) + \left[\tilde{D}\left(\frac{\partial m_3}{\partial x}\mathbf{e}_1 - \frac{\partial m_1}{\partial x}\mathbf{e}_3\right)\right].$$

where $\tilde{K}_e = 2K_e/(\mu_0 M_s^2)$ represent the dimensionless anisotropy constant. Also, $\tilde{v}_s, \tilde{\alpha}_{RE}, \tilde{H}_{SH}, \tilde{D}, \tilde{h}_1, \tilde{h}_2, \tilde{h}_3$ depict the respective dimensionless entities.

In the next section, we perform the analysis and study the TDW motions using the regular perturbation expansion technique by transforming Eq. (6) into the spherical polar coordinate system.

4 Characterization of DW Motion

We consider the unitary magnetization vector in the framework of spherical coordinate system is expressed as:

$$\mathbf{m}(\tilde{x}, \tilde{t}) = (\sin\theta\cos\phi, \sin\theta\sin\phi, \cos\theta)\,. \tag{7}$$

where $\theta(\tilde{x}, \tilde{t})$ represent the angle between the easy-axis and the magnetization vector known as polar angle and $\phi(\tilde{x}, \tilde{t})$ referred as the azimuthal angle as depicted in Fig. 1. We substitute the Eq. (7) in Eq. (6) which in turn render a couple of partial differential equations of the form:

$$\begin{aligned}
&\sin\theta\left(\frac{\partial\phi}{\partial\tilde{t}}\right)-\left[\alpha_G\left(1+\alpha_v\left(\left(\frac{\partial\theta}{\partial\tilde{t}}\right)^2+\sin^2\theta\left(\frac{\partial\phi}{\partial\tilde{t}}\right)^2\right)\right)\right.\\
&\left.+\alpha_d\left(\left(\frac{\partial\theta}{\partial\tilde{t}}\right)^2+\sin^2\theta\left(\frac{\partial\phi}{\partial\tilde{t}}\right)^2\right)^{-1/2}\right]\left(\frac{\partial\theta}{\partial\tilde{t}}\right)=-\tilde{K}_e\left(\frac{\partial^2\theta}{\partial\tilde{x}^2}\right)\\
&+\tilde{K}_e\sin\theta\cos\theta\left(\frac{\partial\phi}{\partial\tilde{x}}\right)^2+\zeta\tilde{v}_s\left(\frac{\partial\theta}{\partial\tilde{x}}\right)-\tilde{v}_s\sin\theta\left(\frac{\partial\phi}{\partial\tilde{x}}\right)+\tilde{D}\sin^2\theta\sin\phi\left(\frac{\partial\phi}{\partial\tilde{x}}\right)\\
&-\tilde{\alpha}_{RE}\tilde{v}_s\cos\theta\sin\phi-\left(\tilde{\alpha}_{RE}\zeta+\tilde{H}_{SH}\right)\tilde{v}_s\cos\phi-\tilde{h}_1\cos\theta\cos\phi-\tilde{h}_2\cos\theta\sin\phi\\
&+\tilde{h}_3\sin\theta+\left(\tilde{K}_e+N_1\cos^2\phi+N_2\sin^2\phi-N_3\right)\sin\theta\cos\theta,
\end{aligned}\tag{8}$$

$$\begin{aligned}
&\left(\frac{\partial\theta}{\partial\tilde{t}}\right)+\left[\alpha_G\left(1+\alpha_v\left(\left(\frac{\partial\theta}{\partial\tilde{t}}\right)^2+\sin^2\theta\left(\frac{\partial\phi}{\partial\tilde{t}}\right)^2\right)\right)\right.\\
&\left.+\alpha_d\left(\left(\frac{\partial\theta}{\partial\tilde{t}}\right)^2+\sin^2\theta\left(\frac{\partial\phi}{\partial\tilde{t}}\right)^2\right)^{-1/2}\right]\sin\theta\left(\frac{\partial\phi}{\partial\tilde{t}}\right)=\tilde{K}_e\sin\theta\left(\frac{\partial^2\phi}{\partial\tilde{x}^2}\right)\\
&+2\tilde{K}_e\cos\theta\left(\frac{\partial\theta}{\partial\tilde{x}}\right)\left(\frac{\partial\phi}{\partial\tilde{x}}\right)-\zeta\tilde{v}_s\sin\theta\left(\frac{\partial\phi}{\partial\tilde{x}}\right)-\tilde{v}_s\left(\frac{\partial\theta}{\partial\tilde{x}}\right)+\tilde{D}\sin\theta\sin\phi\left(\frac{\partial\theta}{\partial\tilde{x}}\right)\\
&+\tilde{\alpha}_{RE}\tilde{v}_s\cos\phi-\left(\zeta\tilde{\alpha}_{RE}+\tilde{H}_{SH}\right)\tilde{v}_s\cos\theta\sin\phi-\tilde{h}_1\sin\phi+\tilde{h}_2\cos\phi\\
&+(N_1-N_2)\sin\theta\sin\phi\cos\phi.
\end{aligned}\tag{9}$$

Static magnetization profile in two faraway domains:

In order to understand the dynamics of TDW profile in the presence of applied magnetic and electric fields, first, we obtain the static magnetization orientations in the two distant domains separated by a DW width λ. The static magnetization profile can be evaluated in the presence of transverse applied magnetic field once the electric and longitudinal magnetic fields are switched off. To perform the analysis, we assume that the polar and azimuthal angles in the left domain ($\tilde{x} \to -\infty$) as θ_L and ϕ_L respectively. Also, by the symmetrical property of TDW about the center of the DW, the polar and azimuthal angles in the right domain ($\tilde{x} \to +\infty$) is given as $\theta_R = \pi - \theta_L$ and $\phi_R = \phi_L$ respectively. Therefore, as $\tilde{x} \to -\infty$, the magnetization profile in the left domain is given as $\mathbf{m}_L = (\sin\theta_L \cos\phi_L, \sin\theta_L \sin\phi_L, \cos\theta_L)$ whereas in the right domain it takes the form as $\mathbf{m}_R = (\sin\theta_L \cos\phi_L, \sin\theta_L \sin\phi_L, -\cos\theta_L)$.

Next, to evaluate the closed form solution of magnetization profile in the two faraway domains, we apply the following physical constraints:
The static condition:

$$\frac{\partial \theta_L}{\partial \tilde{t}} = 0, \quad \frac{\partial \phi_L}{\partial \tilde{t}} = 0, \tag{10}$$

and the domain condition:

$$\frac{\partial \theta_L}{\partial \tilde{x}} = \frac{\partial^2 \theta_L}{\partial \tilde{x}^2} = 0; \quad \text{and} \quad \frac{\partial \phi_L}{\partial \tilde{x}} = \frac{\partial^2 \phi_L}{\partial \tilde{x}^2} = 0. \tag{11}$$

By substituting Eqs. (10) and (11) in Eqs. (8) and (9), we obtain:

$$\begin{aligned} &-\tilde{h}_1 \cos\theta_L \cos\phi_L - \tilde{h}_2 \cos\theta_L \sin\phi_L + \tilde{h}_3 \sin\theta_L \\ &-\tilde{\alpha}_{RE} \tilde{v}_s \cos\theta_L \sin\phi_L - \left(\zeta \alpha_{RE} + \tilde{H}_{SH}\right) \tilde{v}_s \cos\phi_L \\ &+\left(\tilde{K}_e + N_1 \cos^2\phi_L + N_2 \sin^2\phi_L - N_3\right) \sin\theta_L \cos\phi_L = 0, \end{aligned} \tag{12}$$

$$\begin{aligned} &-\tilde{h}_1 \sin\phi_L + \tilde{h}_2 \cos\phi_L - \left(\zeta \alpha_{RE} + \tilde{H}_{SH}\right) \tilde{v}_s \cos\theta_L \sin\phi_L + \tilde{\alpha}_{RE} \tilde{v}_s \cos\phi_L \\ &+ (N_1 - N_2) \sin\theta_L \sin\phi_L \cos\phi_L = 0. \end{aligned} \tag{13}$$

On solving Eqs. (12) and (13), a straightforward rather a lengthy algebraic computation yields the following expression:

$$\theta_L = \arccos\left(\frac{-\tilde{h}_3}{\tilde{K}_e + N_1 - N_3}\right), \tag{14}$$

$$\phi_L = \arcsin\left(\frac{-\tilde{h}_2\left(\tilde{K}_e + N_1 - N_3\right)}{(N_1 - N_2)\left(\left(\tilde{K}_e + N_1 - N_3\right)^2 - \left(\tilde{h}_3\right)^2\right)^{1/2}}\right). \tag{15}$$

With the information from Eqs. (14) and (15), we obtain the magnetization in the left and right domains. However, it is evident from Eq. (15) that the TDW structure prevails as long as $\tilde{h}_3$ remains smaller than the quantity $\left(\tilde{K}_e + N_1 - N_3\right)$.
Dynamics of TDW under small applied magnetic and electric field:

In the presence of external sources, we study the dynamical behavior of magnetization profile under the influence of various parameters involved in the model. We carry out the analysis using the regular perturbation expansion method. To begin with, first we introduce an infinitesimal dimensionless parameter ε and set $\tilde{h}_1 = \varepsilon\bar{h}_1, \tilde{h}_2 = \varepsilon\bar{h}_2, \tilde{h}_3 = \varepsilon\bar{h}_3$, and $\tilde{v}_s = \varepsilon\bar{v}_s$. Also, to understand the long-time behavior of traveling wave solutions, we scale the time variable as $\tilde{t} = \varepsilon\tau$.

In view of the above scaling parameters, we seek the solutions of Eqs. (8) and (9) in the following form:

$$\theta\left(\tilde{x}, \tilde{t}\right) = \theta_0\left(\tilde{x}, \tau\right) + \varepsilon\theta_1\left(\tilde{x}, \tau\right) + \cdots, \tag{16}$$

$$\phi\left(\tilde{x}, \tilde{t}\right) = \phi_0\left(\tilde{x}, \tau\right) + \varepsilon\phi_1\left(\tilde{x}, \tau\right) + \cdots. \tag{17}$$

together with the physical constraint:

$$\mathbf{m}\left(\pm\infty, \tau\right) = \left(\varepsilon\bar{h}_1, \varepsilon\bar{h}_2, \pm 1\right) + O\left(\varepsilon^2\right). \tag{18}$$

By inserting Eqs. (16) and (17) in Eqs. (8) and (9), we achieve the following couple of partial differential equations in the leading order of ε:

$$\frac{\alpha_d}{\left(\left(\frac{\partial\theta_0}{\partial\tau}\right)^2 + \sin^2\theta_0\left(\frac{\partial\phi_0}{\partial\tau}\right)^2\right)^{1/2}}\left(\frac{\partial\theta_0}{\partial\tau}\right) = \tilde{K}_e\left(\frac{\partial^2\theta_0}{\partial\tilde{x}^2}\right) - \tilde{K}_e\sin\theta_0\cos\theta_0\left(\frac{\partial\phi_0}{\partial\tilde{x}}\right)^2$$
$$-\tilde{D}\sin^2\theta_0\sin\phi_0\left(\frac{\partial\phi}{\partial\tilde{x}}\right) - \left(\tilde{K}_e + N_1\cos^2\phi_0 + N_2\sin^2\phi_0 - N_3\right)\sin\theta_0\cos\theta_0, \tag{19}$$

$$\frac{\alpha_d}{\left(\left(\frac{\partial\theta_0}{\partial\tau}\right)^2 + \sin^2\theta_0\left(\frac{\partial\phi_0}{\partial\tau}\right)^2\right)^{1/2}}\sin\theta_0\left(\frac{\partial\phi_0}{\partial\tau}\right) = \tilde{K}_e\sin\theta_0\left(\frac{\partial^2\phi_0}{\partial\tilde{x}^2}\right)$$
$$+2\tilde{K}_e\cos\theta_0\left(\frac{\partial\theta_0}{\partial\tilde{x}}\right)\left(\frac{\partial\phi_0}{\partial\tilde{x}}\right) + \tilde{D}\sin\theta_0\sin\phi_0\left(\frac{\partial\theta_0}{\partial\tilde{x}}\right)$$
$$+\left(N_1 - N_2\right)\sin\theta_0\sin\phi_0\cos\phi_0. \tag{20}$$

The desired solution of Eq. (20) that satisfies the physical constraint take the form $\phi_0\left(\tilde{x}, \tau\right) = (2n+1)\frac{\pi}{2}$ which further reduces the Eq. (19) into:

$$\left(\frac{\partial^2\theta_0}{\partial\tilde{x}^2}\right) - \left(\frac{\tilde{K}_e + N_1 - N_3}{2\tilde{K}_e}\right)\sin 2\theta_0 = \left(\frac{\alpha_d}{\tilde{K}_e}\right). \tag{21}$$

We remark that Eq. (21) is difficult to solve, however, an approximate solution assuming a small angle deviation of magnetization ($\sin 2\theta_0 \approx 2\theta_0$) can be obtained. In view of the assumption of small angle deviation of magnetization and the prescribed physical constraint $(\partial\theta_0/\partial\tilde{x}) \to 0$ as $|\tilde{x}| \to \infty$, solution to Eq. (21) yields the following form:

$$\left[\theta_0 + \left(\frac{2\alpha_d}{\tilde{K}_e + N_1 - N_3}\right)\right] + \left[\left(\theta_0 + \left(\frac{2\alpha_d}{\tilde{K}_e + N_1 - N_3}\right)\right)^2 - \left(\left(\frac{2\alpha_d}{\tilde{K}_e + N_1 - N_3}\right)^2 - \frac{C^\sharp}{\eta}\right)\right]^{1/2} = e^{\eta\tilde{x}+C^*}, \tag{22}$$

where, the constants $C^\sharp$ and η are given as $C^\sharp = -\left(\left(2\alpha_d/\tilde{K}_e\right)\theta_L + \eta\left(\theta_L\right)^2\right)$ and $\eta = \left(\left(\tilde{K}_e + N_1 - N_3\right)/2\tilde{K}_e\right)$. Moreover, C^* represents the integration constant and calculated at the center of the DW. To be precise, we impose the physical constraint that $\theta_0 = (\pi/2), \tilde{x} = \bar{x}(\tau)$, where $\bar{x}(\tau)$ denote the time-varying position of the center of the DW and $(d\bar{x}/d\tau)$ represent the DW velocity. Therefore, the desired solution of Eq. (21) takes the following form:

$$\left[\theta_0 + \left(\frac{2\alpha_d}{\tilde{K}_e + N_1 - N_3}\right)\right] + \left[\left(\theta_0 + \left(\frac{2\alpha_d}{\tilde{K}_e + N_1 - N_3}\right)\right)^2 - \left(\left(\frac{2\alpha_d}{\tilde{K}_e + N_1 - N_3}\right)^2 - \frac{C^\sharp}{\eta}\right)\right]^{1/2} = \Gamma e^{\eta\xi}, \tag{23}$$

where, ξ and Γ are given as:

$$\xi = (\tilde{x} - \bar{x}(\tau));\ \Gamma = \left[\left(\frac{\pi}{2}\right) + \left(\frac{2\alpha_d}{\tilde{K}_e + N_1 - N_3}\right)\right] + \left[\left(\left(\frac{\pi}{2}\right) + \left(\frac{2\alpha_d}{\tilde{K}_e + N_1 - N_3}\right)\right)^2 - \left(\left(\frac{2\alpha_d}{\tilde{K}_e + N_1 - N_3}\right)^2 - \frac{C^\sharp}{\eta}\right)\right]^{1/2}.$$

It is apparent from Eq. (23) that the zero order traveling wave solution $\theta_0(\tilde{x}, \tau)$ depends on space and time variables $\tilde{x}$ and τ through the traveling wave variable ξ. It is worth mentioning that the obtained analytical zero order traveling wave solution exist locally and valid only in the vicinity of the center of DW. However, in the absence of nonlinear dissipation (*dry-friction*) factor α_d, the solution (23) agrees with the classical Walker-type form (cf. [15, 16, 20, 27]):

$$\theta_0(\tilde{x}, \tau) = 2\arctan\left(e^{-\sqrt{\eta}(\tilde{x} - \bar{x}(\tau))}\right).$$

In the next section, we summarize the obtained results and discuss the further problems which branch out from this investigation and are beyond the scope of the present study.

5 Conclusion

In this work, we study the static and dynamic feature of TDW in a ferromagnetic heterostructure in the presence of small applied magnetic field and electric currents under the nonlinear dissipative effects. We obtain a sufficient condition on the applied magnetic field which ensures the existence of TDW structure. Also, the zero order traveling solution reflects the magnetization dynamics in the vicinity of the center of the DW under the influence of small applied magnetic and electric fields. We noticed that the zero order traveling wave solution yields the magnetization profile in the ferromagnetic medium. In addition, the higher order traveling wave solutions not only delineate the various dynamic feature such as the DW velocity in the steady-state regime, bifurcation value of the external sources for which the motion remains in the steady-state regime, and the average DW velocity in the precessional dynamic regime but also incorporate the impact of Rashba, Spin-Hall, and DMI effects into the DW motion. We intend to address these significant issues in our future work.

Acknowledgements This work has been supported by Science and Engineering Research Board (SERB), Department of Science and Technology, Government of India through Project CRG/2019/003101, "Investigation of Ultrafast Magnetization Dynamics in a Network of Ferromagnetic Nanostructures".

References

1. Hubert, A., Schäfer, R.: Magnetic Domains: The Analysis of Magnetic Microstructures. Springer Science & Business Media, Berlin (2008)
2. Mayergoyz, I.D., Bertotti, G., Serpico, C.: Nonlinear Magnetization Dynamics in Nanosystems. Elsevier, Amsterdam (2009)
3. Parkin, S.S., Hayashi, M., Thomas, L.: Magnetic domain-wall racetrack memory. Science **320**(5873), 190–194 (2008)
4. Carbou, G.: Stability of static walls for a three-dimensional model of ferromagnetic material. J. Math. Appl. **93**(2), 183–203 (2010)
5. Dwivedi, S., Dubey, S.: On the stability of steady-states of a two-dimensional system of ferromagnetic nanowires. J. Appl. Anal. **23**(2), 89–100 (2017)
6. Dwivedi, S., Dubey, S.: On stability of steady-states for a two-dimensional network model of ferromagnetic nanowires. In: Mathematical Analysis and its Applications, pp. 405–416. Springer, New Delhi (2015)
7. Agarwal, S., Carbou, G., Labbè, S., Prieur, C.: Control of a network of magnetic ellipsoidal samples. Math. Control Relat. Fields **1**(2), 129–147 (2011)
8. Dubey, S., Dwivedi, S.: On controllability of a two-dimensional network of ferromagnetic ellipsoidal samples. Differ. Equ. Dyn. Syst. **1–21** (2018)

9. Mougin, A., Cormier, M., Adam, J.P., Metaxas, P.J., Ferré, J.: Domain wall mobility, stability and Walker breakdown in magnetic nanowires. EPL (Europhys. Lett.) **78**(5), 57007 (2007)
10. Consolo, G., Currò, C., Martinez, E., Valenti, G.: Mathematical modeling and numerical simulation of domain wall motion in magnetic nanostrips with crystallographic defects. Appl. Math. Model. **36**(10), 4876–4886 (2012)
11. Dwivedi, S., Dubey, S.: On dynamics of current-induced static wall profiles in ferromagnetic nanowires governed by the Rashba field. Int. J. Appl. Comput. Math. **3**(1), 27–42 (2017)
12. Consolo, G., Valenti, G.: Traveling wave solutions of the one-dimensional extended Landau-Lifshitz-Gilbert equation with nonlinear dry and viscous dissipations. Acta Appl. Math. **122**(1), 141–152 (2012)
13. Consolo, G.: Modeling magnetic domain-wall evolution in trilayers with structural inversion asymmetry. Ricerche di Matematica **1–15** (2018)
14. Dwivedi, S., Dubey, S.: Field-driven motion of ferrofluids in ferromagnetic nanowire under the influence of inertial effects. Proc. Eng. **127**, 3–9 (2015)
15. Goussev, A., Lund, R.G., Robbins, J.M., Slastikov, V., Sonnenberg, C.: Domain wall motion in magnetic nanowires: an asymptotic approach. Proc. R. Soc. A **469**(2160), 20130308 (2013)
16. Goussev, A., Lund, R.G., Robbins, J.M., Slastikov, V., Sonnenberg, C.: Fast domain-wall propagation in uniaxial nanowires with transverse fields. Phys. Rev. B **88**(2), 024425 (2013)
17. Lu, J.: Statics and field-driven dynamics of transverse domain walls in biaxial nanowires under uniform transverse magnetic fields. Phys. Rev. B **93**(22), 224406 (2016)
18. Lu, J., Wang, X.R.: Motion of transverse domain walls in thin magnetic nanostripes under transverse magnetic fields. J. Appl. Phys. **107**(8), 083915 (2010)
19. Li, M., Wang, J., Lu, J.: General planar transverse domain walls realized by optimized transverse magnetic field pulses in magnetic biaxial nanowires. Sci. Rep. **7**, 43065 (2017)
20. Dwivedi, S., Dubey, S.: On the evolution of transverse domain walls in biaxial magnetic nanowires. Mater. Today: Proc. **4**(9), 10555–10559 (2017)
21. Slonczewski, J.C.: Current-driven excitation of magnetic multilayers. J. Magn. Magn. Mater. **159**(1–2), L1–L7 (1996)
22. Vandermeulen, J., Nasseri, S.A., Van de Wiele, B., Durin, G., Van Waeyenberge, B., Duprè, L.: The effect of Dzyaloshinskii-Moriya interaction on field-driven domain wall dynamics analysed by a semi-analytical approach. J. Phys. D: Appl. Phys. **49**(46), 465003 (2016)
23. Martinez, E., Emori, S., Beach, G.S.: Current-driven domain wall motion along high perpendicular anisotropy multilayers: the role of the Rashba field, the spin Hall effect, and the Dzyaloshinskii-Moriya interaction. Appl. Phys. Lett. **103**(7), 072406 (2013)
24. Osborn, J.A.: Demagnetizing factors of the general ellipsoid. Phys. Rev. **67**(11–12), 351 (1945)
25. Tiberkevich, V., Slavin, A.: Nonlinear phenomenological model of magnetic dissipation for large precession angles: generalization of the Gilbert model. Phys. Rev. B **75**(1), 014440 (2007)
26. Depassier, M.C.: Speed of field-driven domain walls in nanowires with large transverse magnetic anisotropy. EPL (Europhys. Lett.) **111**(2), 27005 (2015)
27. Schryer, N.L., Walker, L.R.: The motion of 180° domain walls in uniform DC magnetic fields. J. Appl. Phys. **45**(12), 5406–5421 (1974)

Heat Transfer Analysis of CNT-Nanofluid Between Two Rotating Plates in the Presence of Viscous Dissipation Effect

A. Kumar, R. Singh and R. Tripathi

Abstract In this research study, an investigation of three-dimensional (3D) CNT based nanofluid flow through a horizontal rotating channel under the influence of viscous dissipation, is carried out. We have considered that the upper sheet of the channel is permeable and fixed while the lower sheet is impermeable and the sole reason for fluid flow initiation is the stretching of the lower sheet. The mathematical model of the problem is developed and is presented in the form of a system of nonlinear partial differential equations. Suitable similarity technique is employed to transform these governing partial differential equations into the set of ordinary differential equations which are nonlinear. The transformed equations are then solved numerically by the bvp4c routine of MATLAB. Computations for the nanofluid velocity and nanofluid temperature along with skin friction coefficient and Nusselt number are, carried out for relevant flow parameters. A comparative analysis of single-wall carbon nanotubes as well as multiwall carbon nanotubes on temperature and velocity distribution is carried out. Three dimensional flow of CNT-based nanofluid inside a horizontal channel whose one wall is permeable and the other is not, has not been considered before. Although up to some extent, such an analysis has practical bearings in the industries related with lubrication under the influence of magnetic field. The temperature of the fluid is getting increased with growing values of Eckert number and rotation parameter while a completely opposite trend is found for suction/injection and Reynolds number.

Keywords SWCNT · MWCNT · Rotating channel · Viscous dissipation

A. Kumar (✉) · R. Singh · R. Tripathi
Department of Mathematics, National Institute of Technology, Jamshedpur, Jharkhand 831004, India
e-mail: amit65chauhan@gmail.com

R. Singh
e-mail: rsingh.math@nitjsr.ac.in

R. Tripathi
e-mail: rajat17mnnit@gmail.com

S. Manna et al. (eds.), *Mathematical Modelling and Scientific Computing with Applications*, Springer Proceedings in Mathematics & Statistics 308,
https://doi.org/10.1007/978-981-15-1338-1_21

1 Introduction

In recent years, a considerable amount of research has been carried out on nanotechnology and nano-composites. The carbon nanotube (CNT) embodies one of the most significant developments in the area of nanotechnology. Since the last two or three decades, CNTs have received a great amount of attention by many scientists and researchers around the world due to their significant applications in different fields of science. These include applications in materials applications, optics and nano-electronics. CNTs are a geometrical arrangements of carbon atoms which are rolled into tubes. Such tubes have a diameter less than 100 nm and are as thin as 1 or 2 nm. We have discussed the flow of nanofluids with two kinds of CNTs, CNTs with single wall (SWCNTs) and CNTs with multiple walls (MWCNTs) as they pave the way for a farfetched range of nanotube applications in science, engineering and in many industries due to their exclusive construction and extraordinary physical properties. Considering all the important applications, researchers carried out a comprehensive amount of research work on the flow of nanofluids suspended with carbon nanotubes. Wen and Ding [1] analyzed the concentration effects of CNTs on thermal conductivity. In the existence of MWCNTs, the thermal conductivity improvements in base fluid synthetic engine oil and ethylene glycol scrutinized by Liu et al. [2]. In his investigation author conclude that the ratio of thermal conductivity for CNT–synthetic engine oil suspension higher as compared to CNT–ethylene glycol suspension. Xue [3] presented a new model for effective thermal conductivity of composites based on CNTs. It is noticed that theoretical outcomes on thermal conductivity of suspension for CNTs in oil and CNTs in decene are in fine correspondence with the experimental results. Kamali and Binesh [4] achieved numerical examination for heat transfer improvement in the presence of carbon nanotube. Sheikholeslami and Ganji [5] used the OHAM technique to discuss the laminar hydromagnetic nanofluids flow, passing through a channel filled with porous materials and observed that increasing the nanoparticle volume fraction results into an improved momentum boundary layer. Kumaresan et al. [6] in their article, study on the convective heat transfer characteristics of a secondary refrigerant based CNT nanofluids in a tubular heat exchanger. Hussain et al. [7] presented an analysis of flow and heat transfer effects for single wall (SW) as well as multiwall (MW) carbon nanotubes (CNTs) between two rotating plates, taking water as the base fluid. They noticed that SWCNTs produce fewer drag and high heat transfer rate as compared to the MWCNTs. Recently, Hayat et al. [8] discussed the water-based carbon nanotube flow in the Darcy-Forchheimer environment due to a rotating disk together with the no-slip condition.

The main aim of all the above conferred research studies made on nanofluids, was to understand and exploit the heat transfer mechanism, offered by nanofluids suspended with carbon nanotubes (CNTs). Viscous dissipation phenomena are something which affects the heat transfer mechanism considerably. Distribution of temperature in the boundary layer is seriously affected by viscous dissipation and this works as an internal heat source. Brinkman [9] is considered to be the first person, depicting the generation of heat owing to viscous dissipation. One thing which makes the study of consideration of viscous dissipation in the nanofluid flow dynamics worthwhile is that the solid particles present in the fluid brings out supplementary disturbance in the stream, and consequently, the flow field gets altered which gives rise to an enhancement in the dissipation of heat [10]. Inspired by the importance of viscous dissipation effects, Mah et al. [11] studied the forced convection flow of water-based nanofluid in which nanoparticles of Al_2O_3 dispersed uniformly, through a microchannel, taking the influence of viscous dissipation on entropy generation into consideration. They noticed that viscous dissipation effect has a contrary effect on the thermal performance for forced convection of nanofluids in the channel. Sheikholeslami and Ganji [12] scrutinized analytically, heat transfer of a water-based nanofluid in which Cu nanoparticles dispersed uniformly in the base fluid, which is squeezed between parallel flat walls. The obtained results show that rate of heat transfer has direct relationship with the squeeze number as well as Eckert number when two walls are separated but it has completely reverse association with the squeeze number when both the walls are squeezed. A numerically study to describe the viscous dissipation effect on natural and forced convection flow of water-Al_2O_3 nanofluid in a channel, when the walls of the channel were heated unequally presented by Chen et al. [13]. Khan et al. [14] presented numerical as well as analytical study for two types of squeezing flows viz. two-dimensional and axisymmetric flows of Cu-water and Cu-kerosene nanofluids between two parallel walls, considering viscous dissipation effect into account. Hayat et al. [15] studied the free and forced convection for Casson fluid with temperature-dependent thermal conductivity in the presence of viscous dissipation effect. Using homotopy analysis method, Nayak [16] presented a heat transfer analysis related with thermal radiation as well as viscous dissipation of nanofluid over a shrinking surface. Kumar et al. [17] studied the viscous dissipation effect on the three dimensional (3D) flow of ferro-nanoliquid over a stretchable surface. Sithole et al. [18] studied the hydromagnetic second grade nanofluid flow over a convectively heated stretching sheet under the influence of viscous dissipation. Hussain et al. [19] discussed the magnetohydrodynamics Sisko nanofluid with cumulative effects of viscous dissipation and Joule heating in the presence of nanoparticles.

As far as our knowledge is concerned, in all the previous research works, the heat transfer analysis for the flow of nanofluids suspended with nano carbon tubes through a channel is carried out without the consideration of viscous dissipation effects, although such an analysis may have impactful applications in the industries dealing with heat transfer problems. Our focus in carrying out this research work is to scrutinize the effect of viscous dissipation on the hydromagnetic flow of nanofluids suspended with carbon nanotubes (CNTs) in a rotating horizontal channel consisting of two sheets, one is being stretched (the lower sheet) and other sheet is kept at rest (the upper sheet) and the fluid suction is allowed to take place on this sheet.

2 Mathematical Formulation of the Problem

Consider an incompressible, steady three dimensional (3D) laminar flow of water-based carbon nanotubes flow, between two horizontal parallel sheets in a rotating channel, taking into consideration the viscous dissipation effect. The mathematical model of the fluid flow problem is presented in Fig. 1. The Cartesian coordinate system is chosen in such a way that, sheets are in xz-plane and y-axis is taken in the direction normal to sheets. The upper sheet, which is considered as porous, assumed to be situated at $y = h$ while the lower sheet is located at $y = 0$ and being stretched by two opposite but equal forces, applied at either end which varies linearly with the distance; ax $(a > 0)$. It is considered that both the sheets and fluid is rotating anticlockwise with an angular velocity Ω about y-axis. The temperature of the fluid at the upper and lower plates are maintained at T_h and T_0 respectively. Since the upper sheet is taken porous, fluid suction at this sheet is also taken into consideration.

Under the assumption made above, governing equations for the nanofluid flow, are given as:

$$\nabla.q = 0 \tag{1}$$

$$\rho_{nf}[(q.\nabla)q] + 2\Omega \times q = -\nabla p^* + \mu_{nf}\nabla^2 q \tag{2}$$

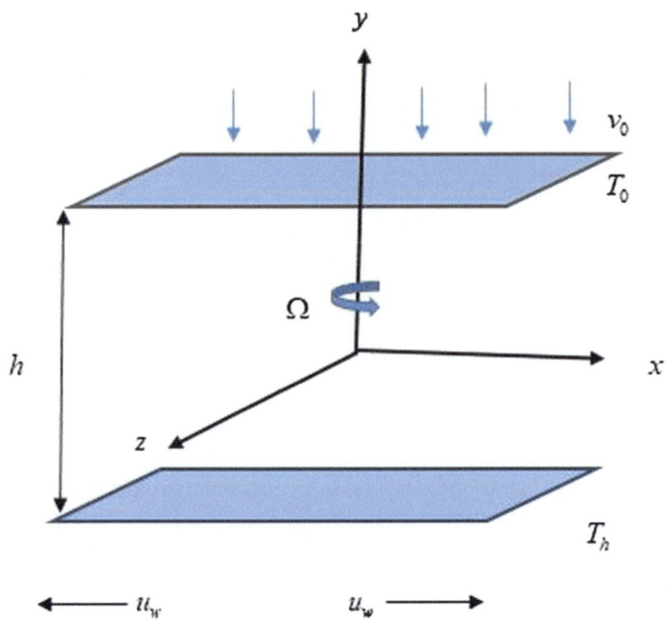

Fig. 1 Geometry of the problem

$$(\rho c_p)_{nf}[(q.\nabla)T] = k_{nf}\nabla^2 T + \mu_{nf}(\nabla q)^2 \tag{3}$$

where $q(u, v, w)$ is the velocity of nanofluid. where u, v and w represent the fluid velocity components in the direction of x, y, z-direction, respectively. T is the temperature of nanofluid. ρ_{nf}, $(\rho c_p)_{nf}$, α_{nf}, μ_{nf} and k_{nf} are, respectively, expression for the nanofluid density, heat capacity, thermal diffusivity, dynamic viscosity and which can be expressed as follows:

$$\mu_{nf} = \frac{\mu_f}{(1-\phi)^{2.5}}, \quad \rho_{nf} = \rho_f(1-\phi) + \rho_{CNT}\phi,$$
$$(\rho c_p)_{nf} = (\rho c_p)_f(1-\phi) + \phi(\rho c_p)_{CNT},$$

$$\alpha_{nf} = \frac{k_{nf}}{(\rho c_p)_{nf}}, \quad \frac{k_{nf}}{k_f} = \frac{(1-\phi) + \frac{k_{CNT}}{k_{CNT}-k_f} 2\phi \ln\left(\frac{k_{CNT}+k_f}{2k_f}\right)}{(1-\phi) + \frac{k_f}{k_{CNT}-k_f} 2\phi \ln\left(\frac{k_{CNT}+k_f}{2k_f}\right)}.$$

where ϕ is the nanoparticles volume fraction, k_f is the thermal conductivity of the base fluid, k_{CNT} is the thermal conductivity of CNTs, ρ_f is the density of base fluid, ρ_{CNT} is the density of CNTs, μ_f is the dynamic viscosity of the base fluid. The upper plate, which is considered as porous, hence at the upper plate, y component of the fluid velocity is not zero.

Keeping in view, the Prandtl's boundary layer theory, the governing equations for the problem under consideration, are given as:

$$\frac{\partial u}{\partial x} + \frac{\partial v}{\partial y} + \frac{\partial w}{\partial z} = 0 \tag{4}$$

$$\rho_{nf}\left(u\frac{\partial u}{\partial x} + v\frac{\partial v}{\partial y} + 2\Omega_0 w\right) = -\frac{\partial p}{\partial x} + \mu_{nf}\left(\frac{\partial^2 u}{\partial x^2} + \frac{\partial^2 u}{\partial y^2}\right) \tag{5}$$

$$\rho_{nf}\left(u\frac{\partial v}{\partial x} + v\frac{\partial v}{\partial y}\right) = -\frac{\partial p}{\partial y} + \mu_{nf}\left(\frac{\partial^2 v}{\partial x^2} + \frac{\partial^2 v}{\partial y^2}\right) \tag{6}$$

$$\rho_{nf}\left(u\frac{\partial w}{\partial x} + v\frac{\partial w}{\partial y} - 2\Omega_0 u\right) = \mu_{nf}\left(\frac{\partial^2 w}{\partial x^2} + \frac{\partial^2 w}{\partial y^2}\right) \tag{7}$$

$$\left(u\frac{\partial T}{\partial x} + v\frac{\partial T}{\partial y} + w\frac{\partial T}{\partial z}\right) = \frac{k_{nf}}{(\rho c_p)_{nf}}\left(\frac{\partial^2 T}{\partial x^2} + \frac{\partial^2 T}{\partial y^2} + \frac{\partial^2 T}{\partial z^2}\right)$$
$$+ \frac{\mu_{nf}}{(\rho c_p)_{nf}}\left[2\left\{\left(\frac{\partial u}{\partial x}\right)^2 + \left(\frac{\partial v}{\partial y}\right)^2 + \left(\frac{\partial w}{\partial z}\right)^2\right\}\right.$$
$$\left. + \left(\frac{\partial v}{\partial x}\right)^2 + \left(\frac{\partial w}{\partial z}\right)^2 + \left(\frac{\partial w}{\partial x} + \frac{\partial u}{\partial z}\right)^2\right] \tag{8}$$

Subject to the relevant conditions on the boundary:

$$v = 0, u = u_w = ax, w = 0, T = T_h \quad at\ y = 0, \tag{9}$$

$$v = -v_0, u = 0, w = 0, T = T_0 \quad at\ y = h. \tag{10}$$

In the last Eq. (10), v_0 represents the uniform suction ($v_0 > 0$)/injection ($v_0 < 0$).

Similarity variables are introduced to translate the governing Eqs. (4)–(8) along with the boundary conditions (9) and (10) into ordinary differential equations (ODEs):

$$u = axf'(\eta), v = -ahf(\eta), w = axg(\eta), \theta(\eta) = \frac{T - T_0}{T_h - T_0}, \eta = \frac{y}{h}. \tag{11}$$

Making use of similarity variables given in Eq. (11), the governing Eqs. (4)–(8) are converted into non-linear ODEs which are presented as follows:

$$f^{iv} - \mathrm{Re}\, E_1\left(f'f'' - ff'''\right) - 2\mathrm{Kr}\, E_1 g' = 0, \tag{12}$$

$$g'' - \mathrm{Re}\, E_1\left(gf' - fg'\right) + 2\mathrm{Kr}\, E_1 f' = 0, \tag{13}$$

$$\theta'' + \mathrm{Pr}\,\mathrm{Re}\frac{E_2}{E_3} f\theta' + \mathrm{Pr}\,\mathrm{Ec}\frac{E_4}{E_3}\left(4f'^2 + g^2\right) = 0, \tag{14}$$

The transformed conditions at the boundary are obtained as:

$$f' = 0, f = 0, g = 0, \theta = 1 \quad at\ \eta = 0, \tag{15}$$

$$f' = 0,\ f = S,\ g = 0,\ \theta = 0 \quad at\ \eta = 1. \tag{16}$$

where $\mathrm{Re} = \frac{ah^2}{\nu_f}$ is Reynolds number, $\mathrm{Kr} = \frac{\Omega h^2}{\nu_f}$ is rotation parameter, $\mathrm{Ec} = \frac{\rho_{nf} a^2 h^2}{(\rho c_p)_{nf}(\theta_0 - \theta_h)}$ is Eckert number, $S = \frac{v_0}{ah}$. Other dimensionless quantities E_1, E_2, E_3 and E_4, appearing in (12)–(14) are defined as:

$$E_1 = \frac{\rho_{nf}}{\rho_f}\frac{\mu_f}{\mu_{nf}}, E_2 = \frac{(\rho c_p)_{nf}}{(\rho c_p)_f}, E_3 = \frac{k_{nf}}{k_f} \text{ and } E_4 = \frac{\mu_{nf}}{\mu_f}.$$

3 Skin Friction Coefficient and Nusselt Number

The important physical quantities are skin friction coefficient together with Nusselt number, defined by:

$$\mathrm{C_f} = \frac{\tau_{\mathrm{w}}}{0.5\rho_{\mathrm{nf}}\mathrm{u}_{\mathrm{w}}^2} \tag{17}$$

$$\mathrm{Nu} = \frac{\mathrm{hq_w}}{\mathrm{k_{nf}(T_0 - T_h)}}, \tag{18}$$

where τ_w denotes shear wall stress and q_w is heat flux at wall

$$\tau_w = \mu_{nf}\left.\frac{\partial u}{\partial y}\right|_{y=0} \tag{19}$$

$$q_w = -k_{nf}\left.\frac{\partial T}{\partial y}\right|_{y=0} \tag{20}$$

Using the similarity variables (11), skin friction coefficient together with Nusselt number takes the following form,

$$\mathrm{Re}_x^{1/2} Cf_x = E_1 f''(0) \tag{21}$$

$$\mathrm{Re}_x^{-1/2} Nu_x = -E_3\theta'(0) \tag{22}$$

in which $\mathrm{Re}_x = \frac{xU_w}{\nu_f}$ denotes the local Reynolds number.

4 Numerical Method for Solution

Equations (12)–(14), along with boundary conditions (15) and (16) are solved numerically using the solver bvp4c of MATLAB. This is based on a relaxation method which makes use of polynomial interpolation on a systematically defined grid. This method has an accuracy of fourth order.

5 Result and Discussion

Extensive numerical computations are performed for velocity profile i.e. for both primary and secondary velocities and temperature profile together with the skin friction coefficient and Nusselt number, to get an insight of the physics involved in the flow regime for various value of flow parameters which characterize the feature of the flow. Also, the effect of the two different types of nanotubes viz. single wall (SW) and multiwall (MW) carbon nanotubes (CNTs), on the velocity profile and temperature profile together with the skin friction coefficient and the Nusselt number for the various value of emerging parameter are shown.

The effect of CNTs with single wall as well as multiple walls, on the primary velocity f' as well as secondary velocity g and temperature θ against the various value of nanoparticle volume fraction φ are presented in Figs. 2, 3 and 4. It is obvious from Fig. 2 that, on increasing nanoparticle volume fraction gradually, the changes in the primary velocity f' is not very appreciable. It is found that nanofluids with SWCNTs are relatively slower as compared to nanofluids with MWCNTs, while a completely opposite behavior is observed in Fig. 3 for secondary velocity profile i.e. nanofluids with SWCNTs have somewhat advanced velocity profile as compared to the nanofluids with MWCNTs. Figure 4 is drawn to observe the behavior of temperature profile against the nanoparticle volume fraction. It is detected that temperature profile declines as the nanoparticle volume fraction is increased, though the temperature profiles for nanofluids with SWCNTs as well as MWCNTs are nearly the same.

Figures 5, 6, 7, 8, 9 and 10 portray the behavior of primary fluid velocity f' and secondary fluid velocity g, under the influence of rotation parameter Kr, Reynolds number Re, and suction/injection parameter S respectively. Figures 5 and 6 are drawn to analyze the behavior of both the primary and secondary velocity profiles against the important factor of the existence of the secondary flow, which is a rotation. It is seen from Fig. 5 that initially, primary velocity decreases near the lower sheet and as we move away from lower sheet towards the upper sheet, the behavior is quite opposite. Since the Coriolis force (which is induced due to rotation) has a tendency

Fig. 2 Primary velocity profile when Re = Kr = 2, Ec = 0.2 and S = 1

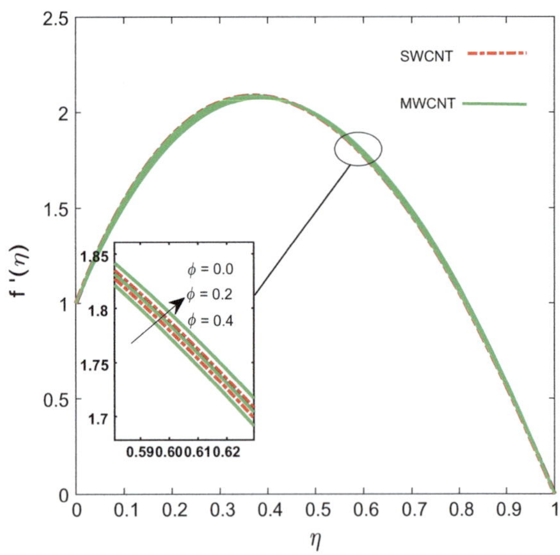

Fig. 3 Secondary velocity profile when Re = Kr = 2, Ec = 0.2 and S = 1

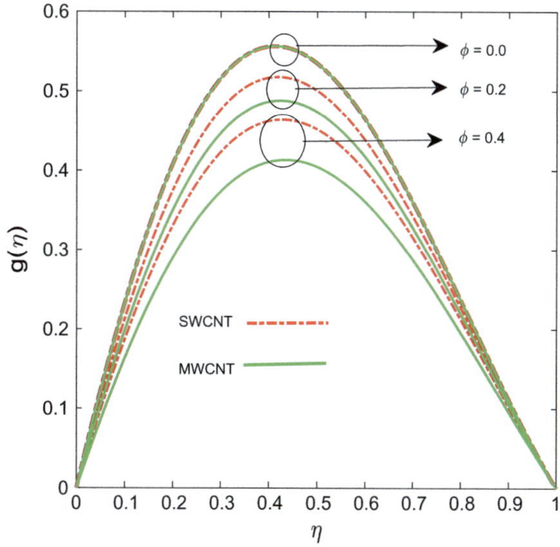

Fig. 4 Temperature profile when Re = Kr = 2, Ec = 0.2 and S = 1

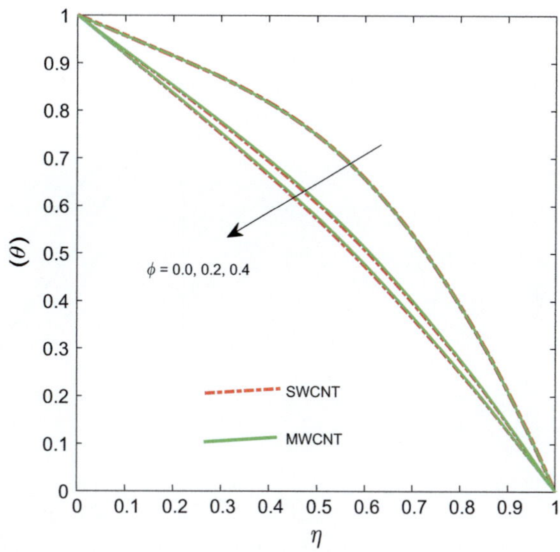

to suppress the main flow i.e. the primary flow to induce a secondary flow in the flow field. Therefore, we observe an acceleration in the secondary flow with an increase in rotation parameter near the lower sheet. Figure 6 elucidates that secondary velocity profile decelerates with the increasing values of rotation parameter Kr. Figures 7 and 8 have been plotted to analyzed the behaviour fluid velocity against the variations in Reynolds number Re. It is noticed from these two figures that primary as well as secondary fluid velocities both are getting diminished on increasing the Re. Since

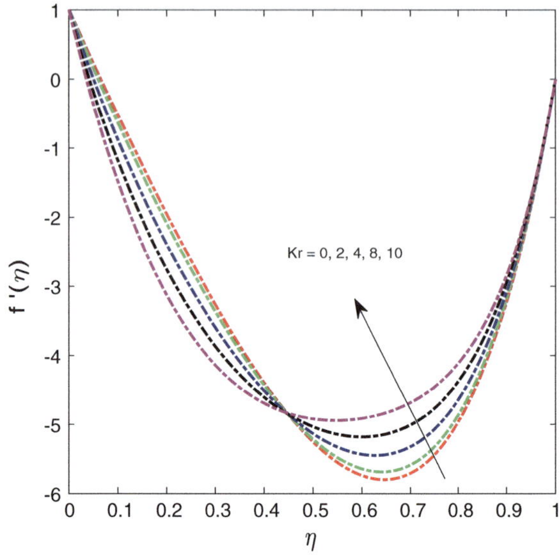

Fig. 5 Primary velocity profile when Re = 2, Ec = 0.2, $\phi = 0.4$ and S = −3

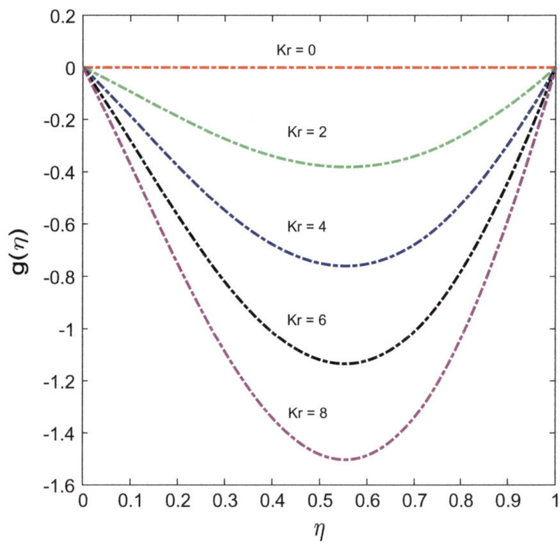

Fig. 6 Secondary velocity profile when Re = 2, Ec = 0.2, $\phi = 0.4$ and S = −3

Re expresses the ratio of inertial force (resistance to the change in motion) to the viscous force, consequently, a rise in Re causes such behavior of fluid velocities. Figures 9 and 10 are drawn to see how the fluid flow behaves against the variation in suction/injection parameter S. it is evident from these two figures that for each suction and injection situation, the velocity in the x-direction as well the velocity in the w-direction i.e. primary and secondary velocities increase, this is happening because of an increase in the injecting fluid from the upper plate which results in

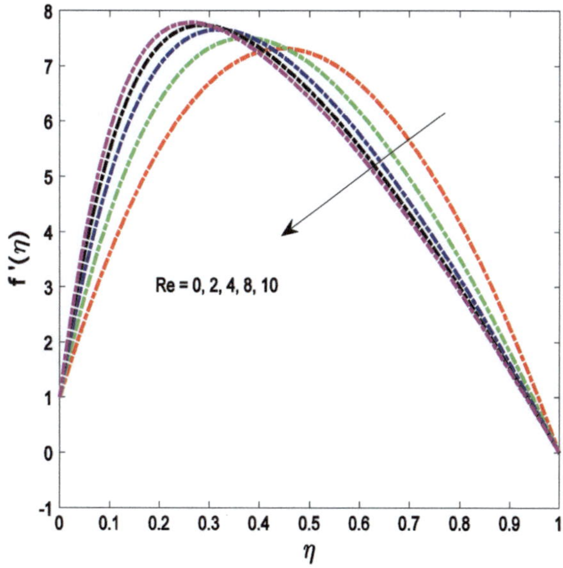

Fig. 7 Primary velocity profile when Kr = 2, Ec = 0.2, $\phi = 0.4$ and S = 3

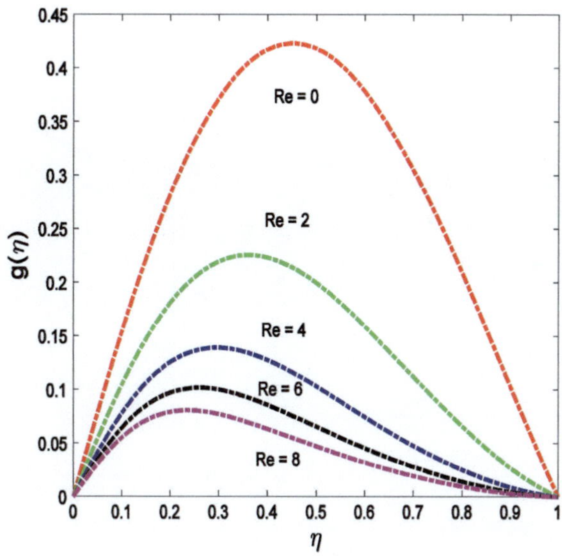

Fig. 8 Secondary velocity profile when Kr = 2, Ec = 0.2, $\phi = 0.4$ and S = 1

enhancing the fluid velocities. Also, it is witnessed that, near the central region of the channel, change in the primary and secondary velocities is much important as compared to a velocity near the neighborhood of the lower and upper plate of the channel. These two figures also portray that the peak value of the velocity gradually shifts toward the lower sheet for the positive value of *S* while for negative values of S, it slightly shifts toward the upper sheet.

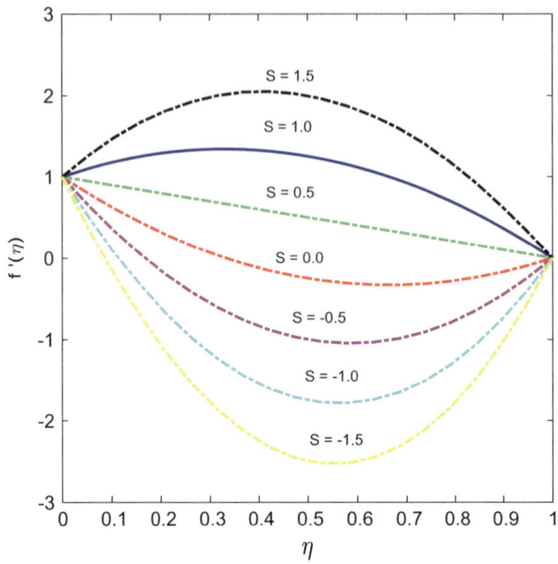

Fig. 9 Primary velocity profile when Re = Kr = 2, Ec = 0.2 and $\phi = 0.4$

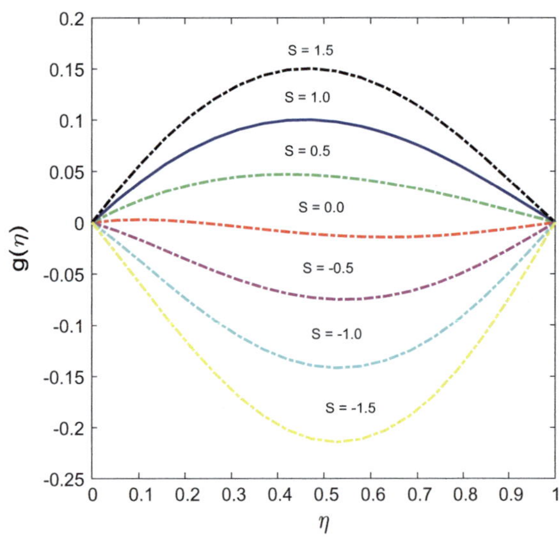

Fig. 10 Secondary velocity profile when Re = Kr = 2, Ec = 0.2 and $\phi = 0.4$

Figures 11, 12, 13 and 14 have been plotted to analyze the behavior of nanofluid temperature θ against Eckert number Ec that symbolizes the viscous dissipation effect, rotation parameter Kr and suction/injection parameter S respectively. As observed from Fig. 11, one can one can make a statement that nanofluid temperature is getting increased on increasing the Eckert number Ec; Such a behaviour is observed due to the fact that Eckert number signifies the quantitative relation of kinetic energy and enthalpy, an increase in Ec means the dissipated heat is stored in

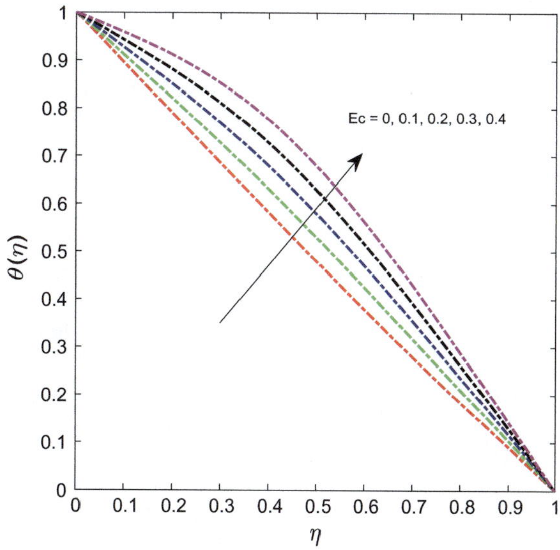

Fig. 11 Temperature profile when Re = Kr = 2, $\phi = 0.4$ and S = 1

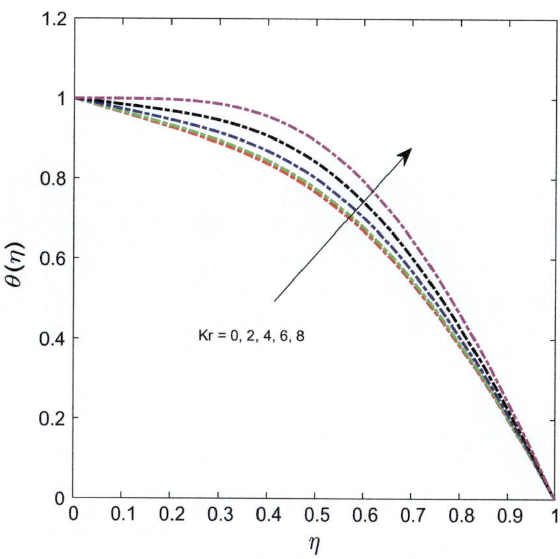

Fig. 12 Temperature profile when Re = 2, Ec = 0.2, $\phi =$ 0.4 and S = −3

the liquid, which raises the fluid temperature. The effect of rotation parameter Kr on the nanofluid temperature is portrayed in the Fig. 12. It is fascinating to see from this figure that temperature profile is getting enhanced on increasing values of Kr, while the completely opposite trend is found in Figs. 13 and 14 for the suction/injection parameter *S* i.e. temperature profile is getting reduced on increasing the value of *S* because more fluid is injecting from the upper plate results in a decrease in fluid temperature within the plates.

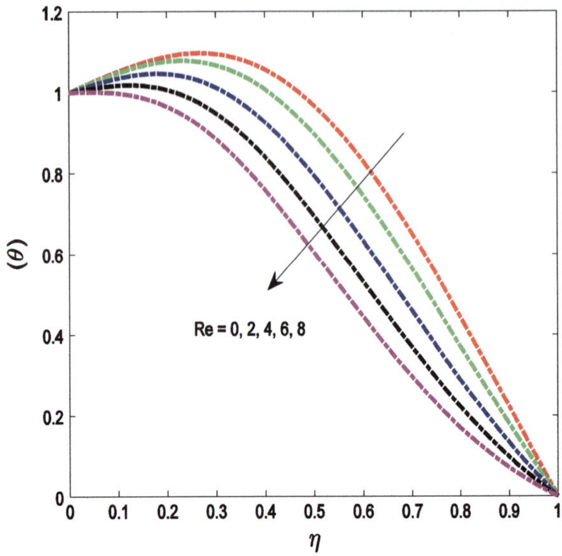

Fig. 13 Temperature profile when Re = Kr = 2, Ec = 0.2 and $\phi = 0.4$

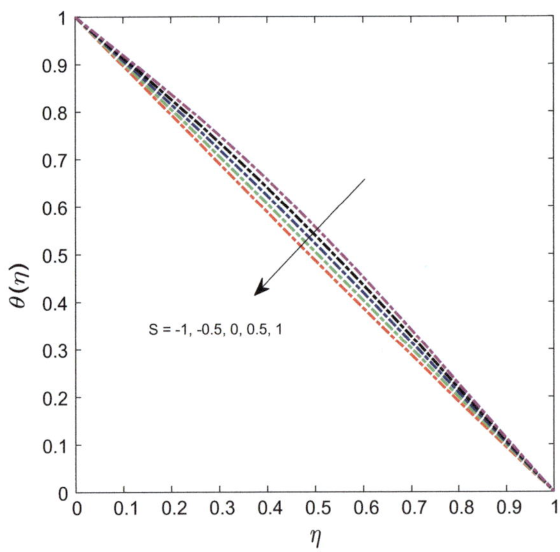

Fig. 14 Temperature profile when Kr = 2, Ec = 0.2, $\phi = 0.4$ and S = −3

The variation in behavior of skin friction coefficient which is a measure of wall velocity gradient and Nusselt number, which accounts for heat transfer rate at the sheet, under the impact of various pertinent flow parameter viz. suction/injection parameter S, rotation parameter Kr, Reynolds number Re, nanoparticle volume fraction φ and Eckert number Ec are provided in tabular form in Tables 2 and 3 for both SWCNTs and MWCNTS respectively. From the Table 2 it can be noticed that

Table 1 Thermophysical characteristic of carbon nanotubes and base fluid [20]

Physical properties	Nanoparticles		H_2O
	MWCNTs	SWCNTs	
ρ (kg/m^3)	1600	2600	997.1
cp (J/kg K)	796	425	4179
k (W/mK)	3000	6600	0.613

Table 2 Skin friction coefficient for SWCNT and MWCNT

S	Kr	Re	φ	$\mathrm{Re}_x^{1/2} Cf_x$	
				SWCNT	MWCNT
0.5	2	2	0.2	1.5965	1.5860
1				3.5699	3.4835
1.5				9.2229	8.9717
1	0	2	0.2	3.5065	3.1019
	2			3.5699	3.4835
	4			3.7615	3.8657
1	2	0	0.2	3.1334	3.4336
		2		3.5699	3.4835
		4		4.0105	3.6338
1	2	2	0.0	3.7513	3.7513
			0.1	3.6713	3.6138
			0.2	3.5699	3.4835

skin friction coefficient is the increasing function for S, Kr, and Re while on the other hand Cf_x tends to decrease with increasing value of φ. Table 3 reveals that a rise, either in the Reynolds number Re or in the value of volume fraction, results in the enhancement of Nusselt number while it is decreasing on increasing either of suction/injection parameter, rotation parameter and Eckert number (Table 3).

6 Conclusions

A mathematical model is established for the three-dimensional flow of water-based nanofluid suspended with carbon nanotubes (CNTs) through a horizontal channel, taking suction/injection and viscous dissipation effects into consideration. Noteworthy outcomes of this study can be summarized in the following points:

- More enhancement in primary as well as secondary velocity is observed for SWCNTs when volume fraction of nanotubes is improved as compared to MWCNTs.

Table 3 Nusselt number for SWCNT and MWCNT

S	Kr	R	Ec	φ	$-\mathrm{Re}_x^{-1/2} Nu_x$	
					SWCNT	MWCNT
0.5	2	2	0.2	0.2	1.4803	1.4782
1					1.3636	1.3524
1.5					1.1460	1.1179
1	0	2	0.2	0.2	1.3655	1.4376
	2				1.3636	1.3524
	4				1.3578	1.2699
1	2	0	0.2	0.2	1.2887	1.3510
		2			1.3636	1.3524
		4			1.4410	1.3674
1	2	2	0.1	0.2	1.4758	1.4740
			0.2		1.3636	1.3524
			0.3		1.2514	1.2308
1	2	2	0.2	0.0	1.3420	1.3420
				0.1	1.3624	1.3539
				0.2	1.3636	1.3524

- Both primary and secondary velocities are increased as we increase the value of suction/injection parameter.
- The temperature of the fluid is getting increased with growing values of Eckert number and rotation parameter while a completely opposite trend is found for suction/injection and Reynolds number.
- Coefficient of skin friction is getting boosted on increasing either of suction/injection or Reynolds number whereas the nanoparticle volume fraction does the vice versa. On the other hand, Nusselt number is perceived to rise on increasing Reynolds number while rotation parameter and Eckert number have adverse effect on this physical quantity.

References

1. Wen, D., Ding, Y.: Effective thermal conductivity of aqueous suspensions of carbon nanotubes. J. Thermophys. Heat Transfer **18**(4), 481–485 (2004)
2. Liu, M., Lin, M.C.-C., Huang, I.-T., Wang, C.-C.: Enhancement of thermal conductivity with carbon nanotube for nanofluids. Int. Commun. Heat Mass Transfer **32**, 1202–1210 (2005)
3. Xue, Q.Z.: Model for thermal conductivity of carbon nanotube-based composites. Phys. B **368**, 302–307 (2005)
4. Kamali, R., Binesh, A.: Numerical investigation of heat transfer enhancement using carbon nanotube based non-Newtonian nanofluids. Int. Commun. Heat Mass Transfer **37**, 1153–1157 (2010)

5. Sheikholeslami, M., Ganji, D.D.: Magnetohydrodynamic flow in a permeable channel filled with nanofluid. Sci. Iranica B **21**(1), 203–212 (2014)
6. Kumaresan, V., Velraj, R., Das, S.K.: Convective heat transfer characteristics of secondary refrigerant based CNT nanofluids in a tubular heat exchanger. Int. J. Refrig **35**(2012), 2287–2296 (2012)
7. Hussain, S.T., Ul-Haq, R., Khan, Z.H., Nadeem, S.: Water driven flow of carbon nanotubes in a rotating channel. J. Mol. Liq. **214**, 136–144 (2016)
8. Hayat, T., Haider, F., Muhammad, T., Alsaedi, A.: On Darcy-Forchheimer flow of carbon nanotubes due to a rotating disk. Int. J. Heat Mass Transfer. **112**, 248–254 (2017)
9. Brinkman, H.C.: Heat effects in capillary flow I. Appl. Sci. Res. **2**, 120–124 (1951). https://doi.org/10.1007/BF00411976
10. Anoop, K.B., Kabelac, S., Sundrarajan, T., Das, S.K.: Rheological and flow characteristics of nanofluids: influence of electro-viscous effects and particle agglomeration. J. Appl. Phys. **106**, Article id: 034909 (2009)
11. Mah, W.H., Hung, M., Guo, N.: Entropy generation of viscous dissipative nanofluid flow in microchannels. Int. J. Heat Mass Transf. **55**, 4169–4182 (2012)
12. Sheikholeslami, M., Ganji, D.D.: Heat transfer of Cu-water nanofluid flow between parallel plates. Powder Technol. **235**, 873–879 (2013)
13. Chen, C.K., Chen, B.S., Liu, C.C.: Heat transfer and entropy generation in fully-developed mixed convection nanofluid flow in vertical channel. Int. J. Heat Mass Transf. **79**, 750–758 (2014)
14. Khan, U., Ahmed, N., Asadullah, M., Mohyud-din, S.T.: Effects of viscous dissipation and slip velocity on two-dimensional and axisymmetric squeezing flow of Cu-water and Cu-kerosene nanofluids. Propul. Power Res. **4**(1), 40–49 (2015)
15. Hayat, T., Ijaz Khan, M., Waqas, M., Yasmeen, T., Alsaedi, A.: Viscous dissipation effect in flow of magnetonanofluid with variable properties. J. Mol. Liq. (2016). https://doi.org/10.1016/j.molliq.2016.06.096
16. Nayak, M.K.: MHD 3D flow and heat transfer analysis of nanofluid by shrinking surface inspired by thermal radiation and viscous dissipation. Int. J. Mech. Sci. http://dx.doi.org/10.1016/j.ijmecsci.2017.03.014
17. Kumar, R., Kumar, R., Shehzad, S.A., Sheikholeslami, M.: Rotating frame analysis of radiating and reacting ferro-nanofluid considering Joule heating and viscous dissipation. Int. J. Heat Mass Transf. **120**, 540–551 (2018)
18. Sithole, H., Mondal, H., Sibanda, P.: Entropy generation in a second grade magnetohydrodynamic nanofluid flow over a convectively heated stretching sheet with nonlinear thermal radiation and viscous dissipation. Res. Phys. **9**, 1077–1085 (2018)
19. Hussain, A., Malik, M.Y., Salahuddin, T., Bilal, S., Awais, M.: Combined effects of viscous dissipation and Joule heating on MHD Sisko nanofluid over a stretching cylinder. J. Mol. Liq. **231**, 341–352 (2017)
20. Khan, W.A., Khan, Z.H., Rahi, M.: Fluid flow and heat transfer of carbon nanotubes along a plate with Navier slip boundary. Appl. Nanosci. **4**, 633–641 (2014)

Modeling of Viscoelastic Fluid Flow Past a Non-linearly Stretching Surface with Convective Heat Transfer: OHAM Analysis

A. Bhattacharyya, G. S. Seth and R. Kumar

Abstract The intention behind carrying out this research problem is to understand the nature of two-dimensional hydromagnetic flow of an electro-conductive and thermally radiating viscoelastic fluid past a non-linear stretching surface, considering viscous and Joule dissipation. With a perspective of converting the governing PDEs into a system of ODEs, appropriate similarity transformation has been incorporated. Optimal Homotopy Analysis Method (OHAM) is used for solving the converted non-linear and coupled equations. Impact of several regulatory flow parameters on temperature, velocity and species concentration are explained via graphs while the variation of some engineering quantities such as Nusselt number, skin friction coefficient and Sherwood number are shown by tables. One of the major outcomes of this investigation is that velocity is getting decreased by viscoelastic parameter whereas the convective heat transfer condition enhances the fluid temperature.

Keywords Viscoelastic fluid · Stretching sheet · Thermal radiation · Heat transfer · OHAM

1 Introduction

Because of plentiful noteworthy applications in various industrial and scientific fields, the problems concerning the magnetohydrodynamic flow of incompressible and viscous fluid over a stretching surface have earned a huge significance by the fluid dynamics fraternity. The pioneering attempt in this regard was made by Sakiadis [1] who investigated the hydromagnetic flow past a solid surface that moves with a uniform velocity. Crane [2] inspected two-dimensional MHD flow induced by a continuously deforming surface. He assumed that the surface deforms linearly with distance from the origin. Later, extending Crane's [2] work, Gupta and Gupta [3]

A. Bhattacharyya (✉) · G. S. Seth · R. Kumar
Department of Mathematics and Computing, Indian Institute of Technology (Indian School of Mines), Dhanbad 826004, India
e-mail: arnab.ism1611@gmail.com

S. Manna et al. (eds.), *Mathematical Modelling and Scientific Computing with Applications*, Springer Proceedings in Mathematics & Statistics 308,
https://doi.org/10.1007/978-981-15-1338-1_22

made an effort to analyze the mass and heat transfer features for magnetohydrodynamic flow past a stretching surface subject to blowing or suction. Some other remarkable research works in this background are due to [4–8].

It is a well-known fact that almost all the liquids we encounters in our life or the liquids used in various industries do not follow Newton's law of viscosity, where the shear stress between two adjacent fluid layers is linearly proportional to the velocity gradient. Thus it is important to understand the effect of rheological behavior of such kind of fluids over flow characteristics. Beard and Walters [9] devised a model to investigate the stagnation point flow of elastic-viscous fluid past a stretching surface. Siddappa and Abel [10] extended the problem of Crane [2] for Walters liquid-B model of viscoelastic fluid. Andersson [11] analyzed the problem of Siddappa and Abel [10] with the existence of a transverse magnetic field. Liu [12] found the solution of MHD viscoelastic fluid past a stretching surface along with the heat transfer analytically. Many authors have contributed in the literature of viscoelastic fluid flows, some of them are the worth mentioning [13–15].

When the considered fluid is sufficiently viscous, it becomes absolutely worthwhile to discuss the consequence of viscous dissipation in heat transfer analysis. Significant applications of this effect can be found in food processing, instrumentation, tribology, lubrication, polymer manufacturing etc. Thus, keeping in mind such a fruitful impact of viscous dissipation, several leading research works have been carried out. Cortell [16] inspected the consequence of viscous dissipation on the thermal boundary layer past a stretching surface having the property of non-linearity. Recently, viscous dissipation effect on a chemically reactive mixed convection MHD Casson nanofluid flow over a permeable stretching surface was explored by Ibrahim et al. [17]. Apart from viscous dissipation, there also acts a volumetric heat source in hydromagnetic flows, known as Joule heating. Thus, a lot of profound research attempts have been made to investigate the process of heat transfer taking into account the simultaneous actions of viscous dissipation as well as Joule heating. Some of the relevant research investigations are done by Yih [18], Abo-Eldahab and El Aziz [19], Alam et al. [20] and Jaber [21].

However, in all the above-cited literatures, the consequence of thermal diffusion (Soret effect) is not taken into consideration. This supposition is justified when the level of concentration is very low. In general, this effect is neglected in the process of mass transfer under the postulation that it is of very small order of magnitude as compared to the effect explained by Fick's law. But, when a flux in the species concentration is induced due to temperature gradient, this effect becomes more significant. The study of Soret effect finds several important applications in metal casting industry, refinement of crude oil etc. In the space science, NASA have created a laser spectroscopy instrument called the SCOF/FACET interferometer that can measure Soret phenomena which can be used to study thermodynamic effects in the space. Eckert and Drake Jr [22] have emphasized the importance of the Soret effect in the cases concerning separation of isotope and in the mixing of gases having medium molecular weight (N_2, air) and very low molecular weight (H_e, H_2). Kumar and Singh [23] examined Soret effect on magnetohydrodynamic natural convection flow over a vertical infinite impulsively started plate. Some other productive research studies in this context are due to Pal et al. [24], Reddy and Chamkha [25] and Kataria and Patel [26].

The present article intents to analyze the MHD two-dimensional viscoelastic fluid flow past a stretching surface which is non-linear, reflecting the effects of thermal diffusion, viscous dissipation and Joule heating. The physical model considered in the present investigation finds significant applications in melting metal industries, polymer etc.

2 Development of the Flow

A two-dimensional steady laminar flow of an electro-conductive, incompressible and thermally radiating viscoelastic fluid past a non-linearly stretching surface has been taken under consideration. The orientation of Cartesian coordinate axes is taken in such a fashion that x-axis is along the stretching surface in the direction of motion and y-axis is perpendicular to it. The fluid flow takes place in the region $y \geq 0$ under the action of a uniform transverse magnetic field of intensity B_0, applied normal to the stretching surface. The configuration of the model is presented in Fig. 1.

The following assumptions are made in the present study:

- The surface is being stretched with a velocity $u_w(x) = cx^n$, where n is stretching index c being a constant.

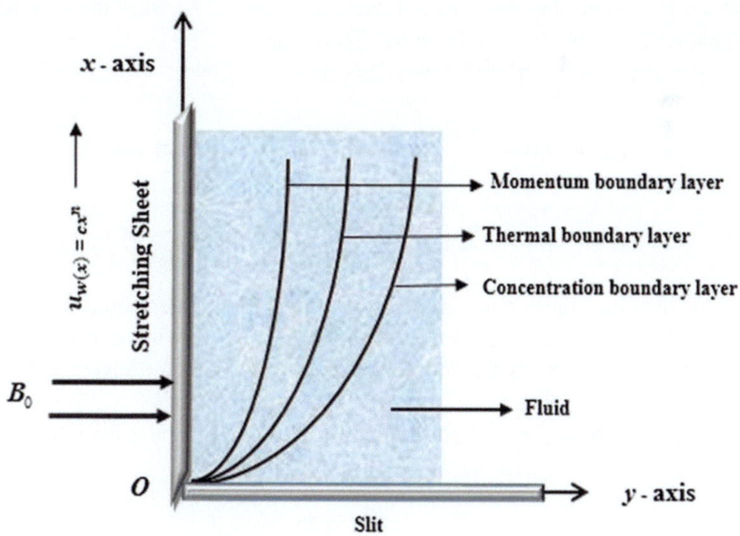

Fig. 1 Physical sketch of the problem

- The effect of polarization is ignored due to the absence of externally applied electric field.
- Because of the very low value of magnetic Reynolds number, the effect of magnetic field which is induced by the fluid flow is neglected [27].

Keeping in mind the assumptions made above and employing Prandtl boundary layer approximation along with the approximation of Boussinesq, the governing set of equations representing the flow regime, are given as

$$\frac{\partial u}{\partial x}+\frac{\partial v}{\partial y}=0, \tag{1}$$

$$u\frac{\partial u}{\partial x}+v\frac{\partial u}{\partial y}=\nu\frac{\partial^2 u}{\partial y^2}-\frac{k_0}{\rho}\left(u\frac{\partial^3 u}{\partial x\partial y^2}+\frac{\partial u}{\partial x}\frac{\partial^2 u}{\partial y^2}+\frac{\partial u}{\partial y}\frac{\partial^2 v}{\partial y^2}+v\frac{\partial^3 u}{\partial y^3}\right)-\frac{\sigma B_0^2}{\rho}u, \tag{2}$$

$$u\frac{\partial T}{\partial x}+v\frac{\partial T}{\partial y}=\alpha_m\frac{\partial^2 T}{\partial y^2}-\frac{1}{\rho c_p}\frac{\partial q_r}{\partial y}+\frac{\nu}{c_p}\left(\frac{\partial u}{\partial y}\right)^2+\frac{\sigma B_0^2}{\rho c_p}u^2, \tag{3}$$

$$u\frac{\partial C}{\partial x}+v\frac{\partial C}{\partial y}=D_e\frac{\partial^2 C}{\partial y^2}+D_{CT}\frac{\partial^2 T}{\partial y^2}, \tag{4}$$

where u and v denote the components of velocity along x and y-axes, respectively, T is the fluid temperature, ν is the kinematic coefficient of viscosity, C is the species concentration, k_0 is the viscoelastic parameter, σ is the electrical conductivity, ρ is the fluid density, q_r is the radiative heat flux, α_m denotes the thermal diffusivity of the fluid, D_e represents the coefficient of mass diffusion, c_p signifies the specific heat at constant pressure and D_{CT} is the Soret diffusivity.

The associated conditions at the boundary are

$$u=u_w,\ v=0,\ -k\frac{\partial T}{\partial y}=h_f(T_f-T),\ C=C_w \text{ at } y=0, \tag{5}$$

$$u\to 0,\ \frac{\partial u}{\partial y}\to 0,\ T\to T_\infty,\ C\to C_\infty \text{ as } y\to\infty. \tag{6}$$

where T_f is the convective fluid temperature, T_∞ is the ambient fluid temperature, C_w is the concentration of the species at the surface, C_∞ is the concentration far away from the surface, h_f is the coefficient of convective heat transfer and k is the thermal conductivity.

With the help of Rosseland's approximation, q_r for an optically thick fluid can be given by

$$q_r=-\frac{4\sigma^*}{3K^*}\frac{\partial T^4}{\partial y}, \tag{7}$$

where K^* stands for the coefficient of mean absorption and σ^* denotes the Stefan-Boltzmann constant.

Making an assumption that the difference of temperature within the fluid inside the boundary layer region and free stream is sufficiently small, we can present T^4 by Taylor series about T_∞. Thus expanding it and avoiding the terms of higher orders, we get

$$T^4 \cong 4T_\infty^3 T - 3T_\infty^4. \tag{8}$$

Incorporating Eqs. (7) and (8) in Eq. (3), we attain

$$u\frac{\partial T}{\partial x} + v\frac{\partial T}{\partial y} = \left(\alpha_m + \frac{1}{\rho c_p}\frac{16\sigma^* T_\infty^3}{3K^*}\right)\frac{\partial^2 T}{\partial y^2} + \frac{\nu}{c_p}\left(\frac{\partial u}{\partial y}\right)^2 + \frac{\sigma B_0^2}{\rho c_p}u^2. \tag{9}$$

2.1 Similarity Transform

For the purpose of converting the governing flow equations as well as the boundary conditions into a system of ODE, the following similarity variables have been introduced.

$$\eta = y\sqrt{\frac{(n+1)u_w}{2\nu x}},\ \psi = \sqrt{\frac{2u_w \nu x}{(n+1)}}f(\eta),\ \theta(\eta) = \frac{T - T_\infty}{T_f - T_\infty},\ \phi(\eta) = \frac{C - C_\infty}{C_w - C_\infty}, \tag{10}$$

where η, ψ, θ and ϕ are, respectively, similarity variable, stream function, dimensionless temperature and dimensionless concentration. With these assumptions, taking $u = \frac{\partial \psi}{\partial y}$ and $v = -\frac{\partial \psi}{\partial x}$, Eqs. (2), (9) and (4) yield the form

$$f''' + ff'' - \left(\frac{2n}{n+1}\right)f'^2 + \alpha\left[\frac{n+1}{2}ff^{iv} - (3n-1)f'f''' + \frac{3n-1}{2}f''^2\right] - \left(\frac{2}{n+1}\right)Mf' = 0, \tag{11}$$

$$\frac{1}{\mathrm{Pr}_{eff}}\theta'' + f\theta' + \mathrm{Ec}\left[f''^2 + \left(\frac{2}{n+1}\right)Mf'^2\right] = 0, \tag{12}$$

$$\phi'' + \mathrm{PrLe}f\phi' + \mathrm{SrLe}\theta'' = 0. \tag{13}$$

The corresponding boundary conditions are given by

$$f = 0,\ f' = 1,\ \theta' = -\mathrm{Bi}(1-\theta),\ \phi = 1 \text{ at } \eta = 1, \tag{14}$$

$$f' \to 0,\ f'' \to 0,\ \theta \to 0,\ \phi \to 0 \text{ as } \eta \to \infty, \tag{15}$$

where $M = \frac{x\sigma B_0^2}{\rho u_w}$ is the local magnetic parameter, $\alpha = \frac{k_0 u_w}{\mu x}$ is the local viscoelastic parameter, $\mathrm{Pr}_{eff} = \frac{\mathrm{Pr}}{1+N_r}$ is the effective Prandtl number [28] where $\mathrm{Pr} = \frac{\nu}{\alpha_m}$ denotes the Prandtl number and $N_r = \frac{16\sigma^* T_\infty^3}{3kK^*}$ stands for the radiation parameter, $\mathrm{Ec} = \frac{u_w^2}{c_p(T_w - T_\infty)}$ indicates the Eckert number, $\mathrm{Bi} = \frac{h_f}{k}\sqrt{\frac{2\nu x}{(n+1)u_w}}$ is the Biot number, $\mathrm{Sr} = \frac{D_{CT}}{\alpha_m}\frac{(T_w - T_\infty)}{(C_w - C_\infty)}$ signifies the Soret number and $\mathrm{Le} = \frac{\alpha_m}{D_e}$ is the Lewis number.

2.2 Quantities of Engineering Interests

Physical quantities of practical importance such as local skin friction coefficient Cf_x, local Nusselt number Nu_x and local Sherwood number Sh_x are presented as:

$$Cf_x = \frac{\tau_w}{\rho u_w^2},\quad Nu_x = \frac{x q_w}{k(T_w - T_\infty)} \text{ and } Sh_x = \frac{x q_m}{D_e(C_w - C_\infty)}, \tag{16}$$

where τ_w, q_w and q_m are, respectively, the wall velocity gradient, the wall heat flux and the mass flux from the surface, given by

$$\begin{aligned}\tau_w &= \mu\left(\frac{\partial u}{\partial y}\right)_{y=0} - k_0\left(u\frac{\partial^2 u}{\partial x \partial y} + 2\frac{\partial u}{\partial x}\frac{\partial u}{\partial y} + v\frac{\partial^2 u}{\partial y^2}\right)_{y=0},\\ q_w &= -k\left(\frac{\partial T}{\partial y}\right)_{y=0} \text{ and } q_m = -D_e\left(\frac{\partial C}{\partial y}\right)_{y=0}.\end{aligned} \tag{17}$$

Transforming into dimensionless form, we obtain

$$Cf_x \mathrm{Re}_x^{1/2} = \sqrt{\frac{n+1}{2}}\left[1 - \frac{7n-1}{2}\alpha f'(0)\right] f''(0), \tag{18}$$

$$Nu_x \mathrm{Re}_x^{-1/2} = -\sqrt{\frac{n+1}{2}}\theta'(0) \tag{19}$$

$$Sh_x \mathrm{Re}_x^{-1/2} = -\sqrt{\frac{n+1}{2}}\phi'(0), \tag{20}$$

where $\mathrm{Re}_x = \frac{u_w x}{\nu}$ is the local Reynolds number.

3 Solution Methodology

The system of Eqs. (11)–(13), being coupled and non-linear in nature, restricts us to obtain solution in closed form. So with the purpose of acquiring solution analytically, optimal homotopy analysis method (OHAM) has been implemented. In OHAM, the main goal is to express the solutions in terms of series.

Bearing in mind the problem considered by us, the solutions $f(\eta)$, $\theta(\eta)$ and $\phi(\eta)$ in terms of series can be expressed as

$$f(\eta) = f_0(\eta) + \sum_{m=1}^{\infty} f_m(\eta),\ \theta(\eta) = \theta_0(\eta) + \sum_{m=1}^{\infty} \theta_m(\eta) \text{ and}$$

$$\phi(\eta) = \phi_0(\eta) + \sum_{m=1}^{\infty} \phi_m(\eta), \tag{21}$$

where the initial guesses $f_0(\eta)$, $\theta_0(\eta)$ and $\phi_0(\eta)$ chosen from the boundary conditions, are like below

$$f_0(\eta) = 1 - \exp(-\eta),\ \theta_0(\eta) = \frac{\text{Bi}}{1+\text{Bi}} \exp(-\eta) \text{ and } \phi_0(\eta) = \exp(-\eta) \tag{22}$$

and the expressions for $f_m(\eta)$, $\theta_m(\eta)$ and $\phi_m(\eta)$ are attained from higher order HAM deformation systems, given by

$$L_f\left[f_m(\eta) - \chi_m f_{m-1}(\eta)\right] = \hbar_f \Re_m^f(\eta), \tag{23}$$

$$L_\theta\left[\theta_m(\eta) - \chi_m \theta_{m-1}(\eta)\right] = \hbar_\theta \Re_m^\theta(\eta), \tag{24}$$

$$L_\phi\left[\phi_m(\eta) - \chi_m \phi_{m-1}(\eta)\right] = \hbar_\phi \Re_m^\phi(\eta), \tag{25}$$

along with the boundary conditions

$$f_m(0) = 0,\ f_m'(0) = 0,\ f_m'(\eta) \to 0 \text{ and } f_m''(\eta) \to 0 \text{ as } \eta \to \infty, \tag{26}$$

$$\theta_m'(0) = \text{Bi}\theta_m(0) \text{ and } \theta_m(\eta) \to 0 \text{ as } \eta \to \infty, \tag{27}$$

$$\phi_m(0) = 0 \text{ and } \phi_m(\eta) \to 0 \text{ as } \eta \to \infty, \tag{28}$$

where $(\hbar_f, \hbar_\theta, \hbar_\phi)$ are convergence control parameters and $L_f(f)$, $L_\theta(\theta)$ and $L_\phi(\phi)$ denote the auxiliary linear operators given by

$$L_f(f) = \frac{d^4 f}{d\eta^4} - \frac{d^2 f}{d\eta^2},\ L_\theta(\theta) = \frac{d^2\theta}{d\eta^2} - \theta \text{ and } L_\phi(\phi) = \frac{d^2\phi}{d\eta^2} - \phi \tag{29}$$

and

$$\Re_m^f(\eta) = f'''_{m-1} + \sum_{k=0}^{m-1} f_{m-1-k} f''_k - \left(\frac{2n}{n+1}\right) \sum_{k=0}^{m-1} f'_{m-1-k} f'_k - \left(\frac{2M}{n+1}\right) f'_{m-1}$$
$$+ \alpha\left(\frac{n+1}{2}\right) \sum_{k=0}^{m-1} f_{m-1-k} f_k^{iv} - \alpha(3n-1) \sum_{k=0}^{m-1} f'_{m-1-k} f'''_k + \alpha\left(\frac{3n-1}{2}\right) \sum_{k=0}^{m-1} f''_{m-1-k} f''_k, \quad (30)$$

$$\Re_m^\theta(\eta) = \frac{1}{\Pr_{eff}} \theta''_{m-1} + \sum_{k=0}^{m-1} f_{m-1-k} \theta'_k + \mathrm{Ec}\left[\sum_{k=0}^{m-1} f''_{m-1-k} f''_k + \left(\frac{2M}{n+1}\right) \sum_{k=0}^{m-1} f'_{m-1-k} f'_k\right], \quad (31)$$

$$\Re_m^\phi(\eta) = \phi''_{m-1} + \Pr \mathrm{Le} \sum_{k=0}^{m-1} f_{m-1-k} \phi'_k + Sr \mathrm{Le} \theta''_{m-1}, \quad (32)$$

where

$$\chi_m = \begin{cases} 0, & m \le 1 \\ 1, & m > 1 \end{cases}. \quad (33)$$

Now the mth order averaged squared residual errors are defined as:

$$\varepsilon_m^f = \frac{1}{L+1} \sum_{j=0}^{K} \left[N_f\left(\sum_{i=0}^{m} f_i(\eta_j) \right) \right]^2, \quad (34)$$

$$\varepsilon_m^\theta = \frac{1}{L+1} \sum_{j=0}^{K} \left[N_\theta\left(\sum_{i=0}^{m} f_i(\eta_j), \sum_{i=0}^{m} \theta_i(\eta_j), \sum_{i=0}^{m} \phi_i(\eta_j) \right) \right]^2, \quad (35)$$

$$\varepsilon_m^\phi = \frac{1}{L+1} \sum_{j=0}^{K} \left[N_\phi\left(\sum_{i=0}^{m} f_i(\eta_j), \sum_{i=0}^{m} \theta_i(\eta_j), \sum_{i=0}^{m} \phi_i(\eta_j) \right) \right]^2, \quad (36)$$

where L represents the number of discrete points of dimensionless coordinate axis η, $\eta_j = j\delta\eta$ and $\delta\eta = \frac{\eta_{\max}}{L+1}$. In our problem we have selected $\eta_{\max} = 7$. The total averaged squared residual error ε_m^t at mth order of approximation is given by

$$\varepsilon_m^t = \varepsilon_m^f + \varepsilon_m^\theta + \varepsilon_m^\phi. \quad (37)$$

Table 1 shows the values of total averaged squared residual errors with respect to optimal convergence control parameters at different order of approximations. Also the individual averaged square residual errors at different order of approximations have been presented in Table 2. From the tables one can observe that both the total and individual averaged squared residual errors reduce gradually as we keep on increasing the order of approximations.

4 Results and Discussion

This segment reveals the physics involved in the flow regime as extensive computations have been done for the velocity field, temperature distribution and species concentration together with the entities having physical importance under the influence of several regulatory flow parameters. The results are well demonstrated either by graphically or by tabular form. The default values of the parameters are chosen as $M = 1.5, \alpha = 0.2, n = 1.5, \mathrm{Pr}_{eff} = 1$, $\mathrm{Ec} = 0.1$, $\mathrm{Bi} = 0.2$, $\mathrm{Sr} = 0.2$ and $\mathrm{Le} = 10$, until otherwise specified particularly.

4.1 *Velocity Profiles*

Influence of magnetic parameter M, stretching index n and viscoelastic parameter α upon the velocity field $f'(\eta)$ have been displayed in Figs. 2, 3 and 4. From Fig. 2 one can observe that fluid velocity is getting reduced as we increase the intensity of magnetic field. This is due to the effect of transverse magnetic field induces a retarding body force, namely, Lorentz force, which resist the motion of the flow. As a result of which the velocity is getting lowered. Figure 3 elucidates the influence of viscoelastic parameter on the fluid flow. We perceive from the curves of this figure that viscoelastic parameter slows down the fluid velocity. The viscoelastic parameter signifies the influence of normal stress coefficient on the flow. So in case of elastic liquid, strain is responsible for the recovery to the original state. So we can't neglect it,

Table 1 Total averaged squared residual errors with total CPU time at different order of approximations

m	$\hbar_f$	$\hbar_\theta$	$\hbar_\phi$	ε_m^t	CPU time (s)
2	−0.456965	−1.53195	−1.03710	0.00577122	11.1464
4	−0.412080	−1.49244	−1.02869	0.000444044	46.2204
6	−0.411726	−1.45456	−1.04551	0.000113409	263.811
8	−0.412192	−1.39185	−1.03182	0.0000453433	1014.35

Table 2 Individual averaged squared residual errors with total CPU time for optimal convergence control parameters at $m = 8$ from Table 1

m	ε_m^f	ε_m^θ	ε_m^ϕ	CPU Time (s)
2	0.00425313	0.000279281	0.00181989	2.68763
4	0.000191104	0.0000625058	0.000194724	10.2349
6	0.0000177173	0.0000234863	0.0000732957	25.4046
8	3.13846×10^{-6}	9.92568×10^{-6}	0.0000322758	51.4275
10	7.71555×10^{-7}	4.67331×10^{-6}	0.0000142378	79.4893

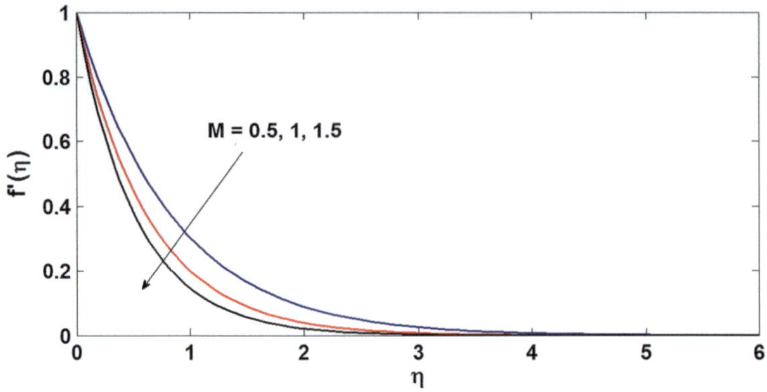

Fig. 2 Velocity distribution for M

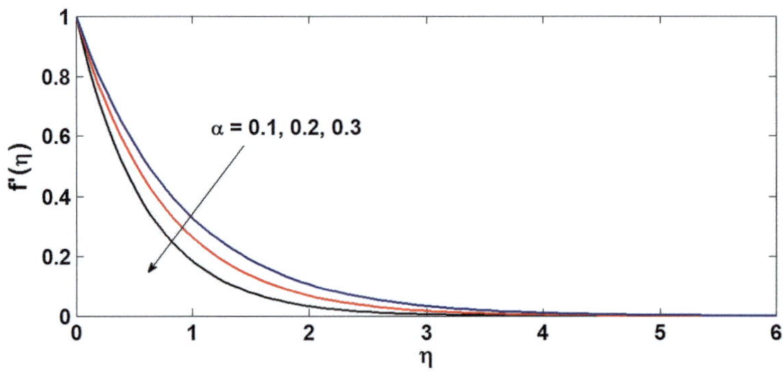

Fig. 3 Velocity distribution for α

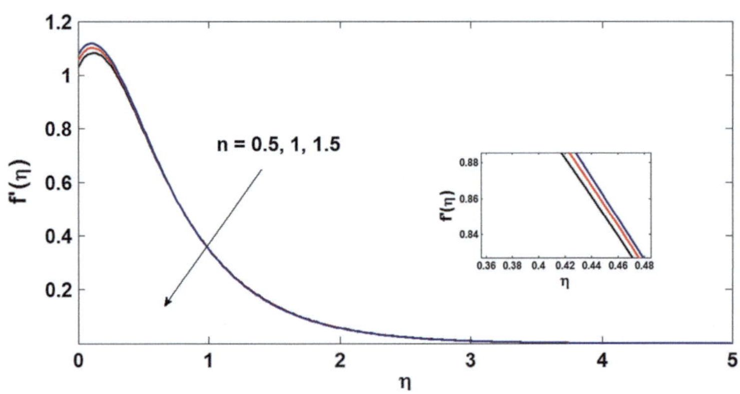

Fig. 4 Velocity distribution for n

however small it may be. Hence, in viscoelastic liquid, the greater degree of recovery leads to a stronger retardation in the velocity profile due to enhancement of elastic parameter. Figure 4 shows that a gradual elevation in the stretching index results in a reducing velocity profile i.e. fluid velocity is getting opposed as we increase the non-linearity of the stretching sheet.

4.2 Temperature Distribution

Figures 5, 6 and 7 are, respectively, depicted to examine the behavior of temperature field $\theta(\eta)$ with respect to Pr_{eff}, Ec and Bi. Fig. 5 portrays the nature of temperature profiles under the variation of effective Prandtl number i.e. the combined outcome of the radiation parameter and Prandtl number for heat radiating optically thick fluid. It

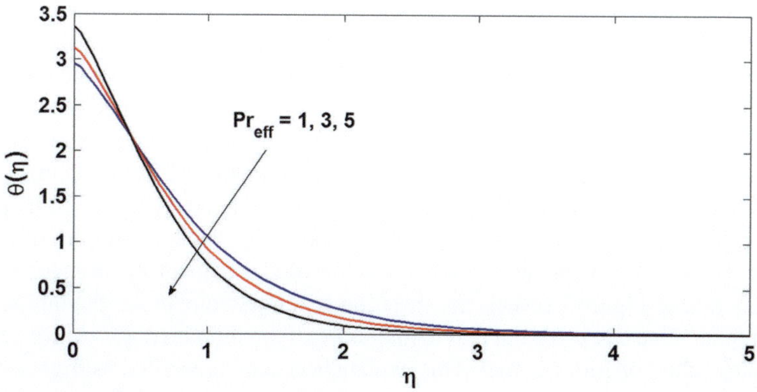

Fig. 5 Temperature distribution for Pr_{eff}

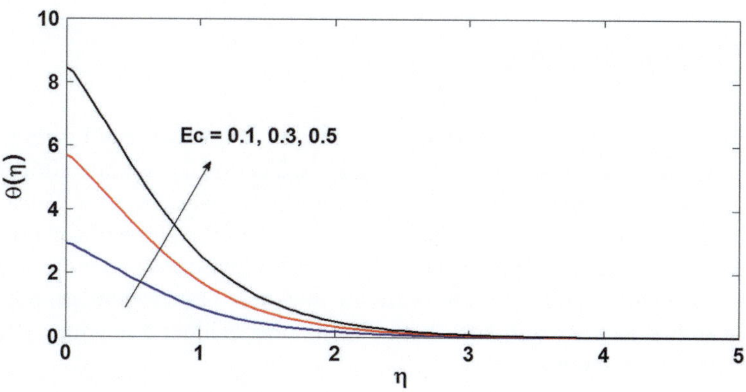

Fig. 6 Temperature distribution for Ec

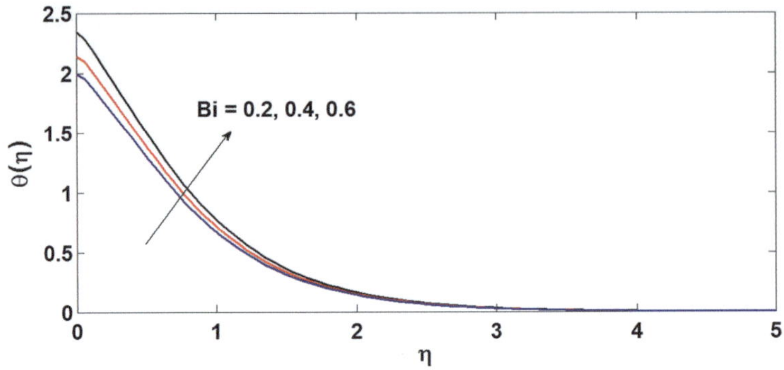

Fig. 7 Temperature distribution for Bi

is witnessed from Fig. 5 that the thickness of thermal boundary layer as well as temperature diminishes with the growing value of Pr_{eff} . From the expression of Pr_{eff} it is clear that the effect of Pr_{eff} is inversely proportional to the effect of the radiation parameter N_r. N_r has a tendency to escalate the effect of conduction and due to this reason thermal boundary layer becomes thicker. Naturally, an augmentation in Pr_{eff} results in a downfall in the fluid temperature. An enhancement in the temperature of fluid is being noticed from Fig. 6 with the increment of Eckert number. Actually, Eckert number signifies the relative measure of kinetic energy to enthalpy. So for larger Ec the material particles are more active due to large amount of energy storage. Hence, enhancement of temperature profile takes place by stronger viscous dissipative heat. Figure 7 reveals the variation of temperature under the influence of Biot number. One can point out that temperature of the fluid increases as we keep on increasing value of Biot number. This explanation can be justified as higher values of Biot number reveals the greater convective heating at the sheet. Consequently, the thickness of thermal boundary layer increases.

4.3 Concentration Profiles

The behavior of species concentration $\phi(\eta)$ under the consequence of Lewis number Le and Soret number Sr are demonstrated in Figs. 8 and 9. It is revealed from Fig. 8 that Soret number enhances the concentration of the species. Physical clarification behind this fact is that a mass flux from higher to lower concentration of species driven by temperature gradient is produced by Soret effect and that is why the diffusive species with higher Soret values increase the thickness of concentration boundary layer. From Fig. 9 one can observe that Lewis number exhibits a retarding effect on the species concentration.

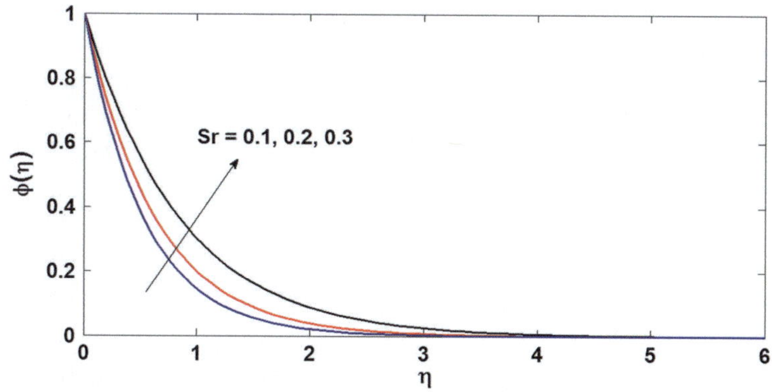

Fig. 8 Variation of species concentration for Sr

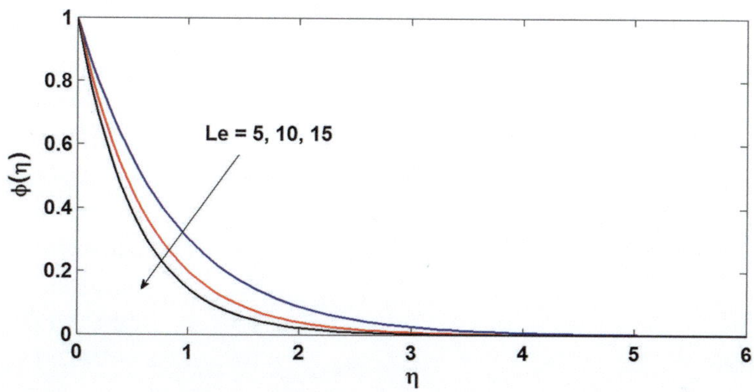

Fig. 9 Variation of species concentration for Le

4.4 Variation of the Physical Quantities

Variation of physical quantities i.e. coefficient of skin friction, Nusselt number and Sherwood number have been displayed in tabular form in Tables 3, 4 and 5. From Table 3 it is clear that skin friction coefficient gets enhanced with the intensification in the values of magnetic parameter, viscoelastic parameter and the stretching index. From Table 4 it is perceived that effective Prandtl number results in an increment in the Nusselt number whereas a reverse trend is noticed in case of Eckert number and Biot number. Table 5 shows that Sherwood number decreases as we increase the value of Soret number whereas an opposite pattern in the values of this physical entity is seen for Lewis number.

Table 3 Variation of local skin friction coefficient

M	α	n	$Cf_x \mathrm{Re}_x^{1/2}$
0.5	0.1	0.5	0.07543526
1.0			0.09453217
1.5			0.12314576
0.5	**0.1**	0.5	0.07543526
	0.2		0.07658472
	0.3		0.07711354
0.5	0.1	**0.5**	0.07543526
		1.0	0.07564582
		1.5	0.07602134

The bold values indicate that the values of that particular parameter is being increased gradually

Table 4 Variation of local Nusselt number

Pr_{eff}	Ec	Bi	$-Nu_x \mathrm{Re}_x^{-1/2}$
1	0.1	0.2	0.12622884
3			0.14092449
5			0.14294935
1	**0.1**	0.2	0.12622884
	0.2		0.10557320
	0.3		0.08484309
1	0.1	**0.2**	0.12622884
		0.4	0.09654213
		0.6	0.07541285

The bold values indicate that the values of that particular parameter is being increased gradually

Table 5 Variation of local Sherwood number

Sr	Le	$-Sh_x \mathrm{Re}_x^{-1/2}$
0.1	5	0.30298226
0.2		0.26257384
0.3		0.22180850
0.1	**5**	0.30298226
	10	0.48288903
	15	0.63754944

The bold values indicate that the values of that particular parameter is being increased gradually

5 Conclusions

Significant outcomes may be summarized as follows:

- The fluid velocity goes down with the intensification in magnetic field. The non-linearity of the stretching index and the viscoelastic parameter also tend to opposes the flowing fluid.
- Fluid temperature gets reduced as we increase the value of effective Prandtl number whereas Eckert number and Biot number do the vice-versa.
- Species concentration increases with the intensification of Soret number while an adverse effect is noticed for Lewis number.
- Coefficient of skin friction gradually takes higher values with the enhancement of magnetic parameter, viscoelastic parameter and stretching index. Local Nusselt number is seen to be increasing as we keep on rising the value of effective Prandtl number. On the other hand, a downfall in this physical quantity is observed when we gradually increase the values of Eckert and Biot number. Sherwood number acts as a decreasing function of Soret number whereas an opposite scenario takes place for Lewis number.

Acknowledgements Mr. Arnab Bhattacharyya is grateful to IIT (ISM), Dhanbad, India, for providing fellowship for successfully carrying out this research problem. The fertility of thoughts and the phenomena explained in the present research work can be expected to lead to extremely productive interactions across disciplines.

References

1. Sakiadis, B.C.: Boundary-layer behavior on continuous solid surfaces: I. boundary-layer equations for two-dimensional and axisymmetric flow. AIChE J. **7**(1), 26–28 (1961)
2. Crane, L.J.: Flow past a stretching plate. Z. Angew. Math. Phys. (ZAMP) **21**(4), 645–647 (1970)
3. Gupta, P.S., Gupta, A.S.: Heat and mass transfer on a stretching sheet with suction or blowing. Can. J. Chem. Eng. **55**(6), 744–746 (1977)
4. Mahabaleshwar, U.S., Sarris, I.E., Lorenzini, G.: Effect of radiation and Navier slip boundary of Walter's liquid B flow over a stretching sheet in a porous media. Int. J. Heat Mass Transf. **127**, 1327–1337 (2018)
5. Hamid, A., Khan, M., Hafeez, A.: Unsteady stagnation-point flow of Williamson fluid generated by stretching/shrinking sheet with Ohmic heating. Int. J. Heat Mass Transf. **126**, 933–940 (2018)
6. Jahan, S., Sakidin, H., Nazar, R., Pop, I.: Unsteady flow and heat transfer past a permeable stretching/shrinking sheet in a nanofluid: a revised model with stability and regression analyses. J. Mol. Liq. **261**, 550–564 (2018)
7. Ghosh, S.K.: Unsteady magnetized flow and heat transfer of a viscoelastic fluid over a stretching surface. J. Magn. Magn. Mater. **443**, 309–318 (2017)
8. Vajravelu, K.: Viscous flow over a nonlinearly stretching sheet. Appl. Math. Comput. **124**(3), 281–288 (2001)
9. Beard, D.W., Walters, K.: Elastico-viscous boundary-layer flows I. Two-dimensional flow near a stagnation point. In: Mathematical Proceedings of the Cambridge Philosophical Society, vol. 60, pp. 667–674 (1964)

10. Siddappa, B., Abel, S.: Non-Newtonian flow past a stretching plate. Z. Angew. Math. Phys. (ZAMP) **36**(6), 890–892 (1985)
11. Andersson, H.I.: MHD flow of a viscoelastic fluid past a stretching surface. Acta Mech. **95**(1), 227–230 (1992)
12. Liu, I.C.: Flow and heat transfer of an electrically conducting fluid of second grade in a porous medium over a stretching sheet subject to a transverse magnetic field. Int. J. Non-Linear Mech. **40**(4), 465–474 (2005)
13. Chen, X., Ye, Y., Zhang, X., Zheng, L.: Lie-group similarity solution and analysis for fractional viscoelastic MHD fluid over a stretching sheet. Comput. Math Appl. **75**(8), 3002–3011 (2018)
14. Li, J., Zheng, L., Liu, L.: MHD viscoelastic flow and heat transfer over a vertical stretching sheet with Cattaneo-Christov heat flux effects. J. Mol. Liq. **221**, 19–25 (2016)
15. Madhu, M., Kishan, N.: Finite element analysis of MHD viscoelastic nanofluid flow over a stretching sheet with radiation. Procedia Eng. **127**, 432–439 (2015)
16. Cortell, R.: Effects of viscous dissipation and radiation on the thermal boundary layer over a nonlinearly stretching sheet. Phys. Lett. A **372**(5), 631–636 (2008)
17. Ibrahim, S.M., Lorenzini, G., Kumar, P.V., Raju, C.S.K.: Influence of chemical reaction and heat source on dissipative MHD mixed convection flow of a Casson nanofluid over a nonlinear permeable stretching sheet. Int. J. Heat Mass Transf. **111**, 346–355 (2017)
18. Yih, K.A.: Viscous and Joule heating effects on non-Darcy MHD natural convection flow over a permeable sphere in porous media with internal heat generation. Int. Commun. Heat Mass Transf. **27**(4), 591–600 (2000)
19. Abo-Eldahab, E.M., El Aziz, M.A.: Viscous dissipation and Joule heating effects on MHD free convection from a vertical plate with power-law variation in surface temperature in the presence of Hall and ion-slip currents. Appl. Math. Model. **29**(6), 579–595 (2005)
20. Alam, M.S., Rahman, M.M., Sattar, M.A.: Effects of chemical reaction and thermophoresis on magneto-hydrodynamic mixed convective heat and mass transfer flow along an inclined plate in the presence of heat generation and (or) absorption with viscous dissipation and Joule heating. Can. J. Phys. **86**(9), 1057–1066 (2008)
21. Jaber, K.K.: Effects of viscous dissipation and Joule heating on MHD flow of a fluid with variable properties past a stretching vertical plate. Eur. Sci. J. **10**(33), 383–393 (2014)
22. Eckert, E.R.G., Drake Jr., R.M.: Analysis of Heat and Mass Transfer (1987)
23. Kumar, J.B., Singh, A.K.: Soret effects on free-convection and mass transfer flow in the stokes problem for a infinite vertical plate. Astrophys. Space Sci. **173**(2), 251–255 (1990)
24. Pal, D., Mandal, G., Vajravalu, K.: Soret and Dufour effects on MHD convective-radiative heat and mass transfer of nanofluids over a vertical non-linear stretching/shrinking sheet. Appl. Math. Comput. **287–288**, 184–200 (2016)
25. Reddy, P.S., Chamkha, A.J.: Soret and Dufour effects on MHD convective flow of Al_2O_3-water and TiO_2-water nanofluids past a stretching sheet in porous media with heat generation/absorption. Adv. Powder Technol. **27**(4), 1207–1218 (2016)
26. Kataria, H.R., Patel, H.R.: Soret and heat generation effects on MHD Casson fluid flow past an oscillating vertical plate embedded through porous medium. Alexandria Eng. J. **55**(3), 2125–2137 (2016)
27. Cramer, K.R., Pai, S.I.: Magnetofluid dynamics for engineers and applied physicist (1973)
28. Magyari, E., Pantokratoras, A.: Note on the effect of thermal radiation in the linearized Rosseland approximation on the heat transfer characteristics of various boundary layer flows. Int. Commun. Heat Mass Transfer **38**, 554–556 (2011)

Scientific Computing and Neural Network

Probabilistic Intuitionistic Fuzzy Set Based Intuitionistic Fuzzy Time Series Forecasting Method

Krishna Kumar Gupta and Sanjay Kumar

Abstract IFS can handle non-stochastic non-determinism that arises due to single membership function for both membership and non-membership grade. PIFS may handle non-determinacy because of both stochastic and non-stochastic reasons. In this study, we propose PIFS based FTS forecasting model to control the both kind of non-determinism along with non-stochastic uncertainty in TS forecasting. The developed model describes issue of non-determinism which rises due to both randomness as well as linguistic representation of TS data. An aggregation operator to aggregate the PIFS into IFS is also used in this study. The presented method has been simulated using financial TS data of TAIEX to confirm it's outperformance using RMSE.

Keywords Probabilistic intuitionistic fuzzy set · Intuitionistic fuzzy logical relation · Stochastic non-determinism · Non-stochastic non-determinism · Forecasting

1 Introduction

Time series (TS) forecasting performs the prominent prelude in the field of finance, medical sciences, meteorology and engineering etc. various statistical techniques e.g. ARIMA, ARMA, moving average and regression analysis are used for TS forecasting. Major drawbacks of TS forecasting methods based on these techniques are their incapability to handling the linguistic values and non-deteminism caused by non-probabilistic reasons. Song and Chissom [25–27] proposed fuzzy time series (FTS) forecasting methods by implementation of fuzzy set [34] to deal with uncertainty caused by the reasons other than randomness. Further Hung and Lin [17],

K. K. Gupta (✉) · S. Kumar
Department of Mathematics, Statistics and Computer Science, G. B. Pant University of Agriculture and Technology, Pantnagar, Uttarakhand 263145, India
e-mail: guptakrishna.gupta@gmail.com

S. Kumar
e-mail: skruhela@hotmail.com

S. Manna et al. (eds.), *Mathematical Modelling and Scientific Computing with Applications*, Springer Proceedings in Mathematics & Statistics 308,
https://doi.org/10.1007/978-981-15-1338-1_23

Singh and Borah [24], Torbat et al. [29] proposed forecasting methods to improve accurateness in forecasted values. Chen and Chen [9], Yolcu [33], Chen and Phuong [10], Xian et al. [31] proposed Granular computing, swarm optimization and other machine learning based FTS forecasting methods. However, these forecasting methods are not capable to handle the both stochastic uncertainty and non-deteminism in TS forecasting simultaneously.

Atanassov [3] introduced intuitionistic fuzzy set (IFS) that contains separate functions for membership and non-membership grade to manage the non-determinacy in system due to non-stochastic factors. Joshi and Kumar [18] proposed the first intuitionistic fuzzy time series (IFTS) forecasting method to include non-determinacy. Later on, Fan et al. [12], Wang et al. [30], Zheng et al. [35, 36] also proposed IFTS forecasting methods based on intuitionistic fuzzy reasoning, intuitionistic fuzzy c-clustering and order decision and adaptive partition algorithm. Gangwar and Kumar [14], Joshi and Kumar [18], Kumar and Gangwar [21] used IFSs with procedure of FTS forecasting to include the non-determinism and degree of hesitation.

Probabilistic fuzzy set (PFS) was introduced by Meghdadi [23] to address issue of both stochastic and non-stochastic uncertainties simultaneously. Liu and Li [22] proposed probabilistic fuzzy logic system to model both types of uncertainties in control problems. Various applications of PFS were explored by Fialho et al. [13], Almeida et al. [2], Hinojosa et al. [15], Tak et al. [28], Huang et al. [16] in different domains. Agarwal et al. [1] proposed an approach to compute the net conditional possibility using probabilistic intuitionistic fuzzy rule.

In this study, we propose probabilistic intuitionistic fuzzy set (PIFS) based IFTS forecasting method that includes non-determinism in the presence of both types of uncertainties. In this study, PIFS are constructed using a probability distribution function that assigns probability to membership and non-membership grades. We also present an aggregation operator to aggregate the probabilistic intuitionistic fuzzy element (PIFE) into intuitionistic fuzzy element (IFE) using corresponding probabilities. Proposed method is simulated using TS data of TAIEX and its performance is compare with other existing methods in terms of RMSE.

2 Preliminaries

In this section we review and present definitions of fuzzy set, PFS, IFS and PIFS.

Definition 1 [34] Let $X = \{x_1, x_2, \ldots, x_n\}$ be a discrete set. A fuzzy set, A on X is defined as $A = \{\langle x, \mu_A(x)\rangle | \forall x \in X\}$. Here, $\mu_A : X \rightarrow [0, 1]$ and $\mu_A(x)$ is membership grade of x in A.

Definition 2 [22] Let $X = \{x_1, x_2, \ldots, x_n\}$ be a discrete set. A PFS $\widetilde{A}$ is defined by probability space $(U_x, \Im, P)$. Here, $x \in X$ with membership grade $\mu \in [0, 1]$, U_x is collection of every possible events ($\mu \in [0, 1]$) and P is probability assigned on $\Im(\sigma - \text{field})$. For some $B_i \in U_x$, following properties are satisfied.$P(\mathrm{B}_i) \geq$

0, $P\left(\sum \mathrm{B}_i\right) = \sum P(\mathrm{B}_i)$, $P(U_x) = 1$. Here, $P(B_i)$ is probability of B_i and $B_i \subseteq [0, 1]$ is a membership grade μ_i. PFS $\widetilde{A}$ can be expressed in terms of union of confined sub-probability space defined as $\widetilde{A} \equiv \bigcup_{x \in X} (U_x, \Im, P)$.

Definition 3 [3] Let $X = \{x_1, x_2, \ldots, x_n\}$ be a fixed discrete set. An IFS I on X is demonstrate as $I = \{\langle x, \mu_I(x), \nu_I(x)\rangle : \forall x \in X\}$, here $\mu_I : X \to [0, 1]$, $\nu_I : X \to [0, 1]$ represents membership, non-membership functions and $\mu_I(x) \in [0, 1]$, $\nu_I(x) \in [0, 1]$ are membership, non-membership grades of x on X with the condition $0 \le \mu_I(x) + \nu_I(x) \le 1, \forall x \in X$ such that $\pi_I(x) = 1 - \mu_I(x) - \nu_I(x)$; $\pi_I(x) \in (0, 1)$ gives the degree of hesitancy.

Definition 4 [1] Let $X = \{x_1, x_2, \ldots, x_n\}$ be a fixed discrete set. A PIFS I' is defined by probability space $((U_x, V_x), \Im, P)$. Here, $x \in X$ with membership and non-membership grade ($\mu \in [0, 1]$ and $\nu \in [0, 1]$). U_x, V_x are collection of every possible events ($\mu \in [0, 1]$, $\nu \in [0, 1]$) and P is probability assigned on $\Im(\sigma - \text{field})$. For some $B_i \in U_x$ and $C_i \in V_x$ following properties are satisfied. $P(\mathrm{B}_i) \ge 0$, $P\left(\sum \mathrm{B}_i\right) = \sum P(\mathrm{B}_i)$, $P(U_x) = 1$ and $P(\mathrm{C}_i) \ge 0$, $P\left(\sum \mathrm{C}_i\right) = \sum P(\mathrm{C}_i)$, $P(V_x) = 1$. Here, $P(B_i)$ is probability of B_i and $P(C_i)$ is probability of C_i. $B_i \subseteq [0, 1]$ is a membership grade μ_i and $C_i \subseteq [0, 1]$ is a membership grade ν_i. PIFS I' can be written in terms of the union of finite sub-probability space defined as $I' \equiv \bigcup_{x \in X} ((U_x, V_x), \Im, P)$.

3 Proposed Method and Experimental Study

Procedure of proposed PIFS based IFTS forecasting method is explained in this section. Developed method is implemented to forecast TAIEX data (Table 1) from 01-11-2004 to 31-12-2004 and the experimental result are presented. Following steps describe the procedure of proposed method and its execution to forecast TAIEX data.

Step 1: Let $X = \left[D_{\text{min}} - \sigma, D_{\text{max}} + \sigma\right]$ be the universe of discourse. Here D_{max}, D_{min} and σ are maximum, minimum value and standard deviation of TAIEX data respectively. Since standard deviation of time series data taken in this study is $\sigma = 94.34$ therefore, universe of discourse is defined as $X = [5561.83, 6234.03]$.
Step 2: X is partitioned into n equal length intervals e_i. TS data are fuzzified using triangular fuzzy sets A_i which are constructed using corresponding intervals $e_i (i = 1, 2, \ldots, n)$.
Following fuzzy sets $A_1, A_2, \ldots, A_{14}$ are constructed with equal length intervals $e_i (i = 1, 2, \ldots, 14)$.

$$A_1 = [5561.83, 5609.84, 5657.86],\ A_2 = [5609.84, 5657.86, 5705.87],$$
$$A_3 = [5657.86, 5705.87, 5753.89],\ A_4 = [5705.87, 5753.89, 5801.9],$$
$$A_5 = [5753.89, 5801.9, 5849.92],\ A_6 = [5801.9, 5849.92, 5897.93],$$

Table 1 Actual TAIEX data

Date	TAIEX data	Date	TAIEX data	Date	TAIEX data
01-11	5656.17	22-11	5838.42	13-12	5878.89
02-11	5759.61	23-11	5851.1	14-12	5909.65
03-11	5862.85	24-11	5911.31	15-12	6002.58
04-11	5860.73	25-11	5855.24	16-12	6019.23
05-11	5931.31	26-11	5778.65	17-12	6009.32
08-11	5937.46	29-11	5785.26	20-12	5985.94
09-11	5945.2	30-11	5844.76	21-12	5987.85
10-11	5948.49	01-12	5798.62	22-12	6001.52
11-11	5874.52	02-12	5867.95	23-12	5997.67
12-11	5917.16	03-12	5893.27	24-12	6019.42
15-11	5906.69	06-12	5919.17	27-12	5985.94
16-11	5910.85	07-12	5925.28	28-12	6000.57
17-11	6028.68	08-12	5892.51	29-12	6088.49
18-11	6049.49	09-12	5913.97	30-12	6100.86
19-11	6026.55	10-12	5911.63	31-12	6139.69

$$A_7 = [5849.92, 5897.93, 5945.94], A_8 = [5897.93, 5945.94, 5993.96],$$
$$A_9 = [5945.94, 5993.96, 6041.97], A_{10} = [5993.96, 6041.97, 6089.99],$$
$$A_{11} = [6041.97, 6089.99, 6138], A_{12} = [6089.99, 6138, 6186.02],$$
$$A_{13} = [6138, 6186.02, 6234.03], A_{14} = [6186.02, 6234.03, 6234.03].$$

Step 3: Using construction method [19], fourteen IFSs $I_i (i = 1, 2, \ldots, 14)$ are constructed corresponding fuzzy sets $A_i (i = 1, 2, \ldots, 14)$ on X.
Probabilities to membership grades are assigned using following Gaussian probability distribution function.

$$p_{\mu_j} = \begin{cases} \frac{l}{\sqrt{2\pi}\zeta_i}\left(e^{-\frac{\left(x_j-\left(\mu_j-1\right)l_i-m_i\right)^2}{2\zeta_i^2}} + e^{-\frac{\left(x_j-\left(1-\mu_j\right)l_i-m_i\right)^2}{2\zeta_i^2}}\right); \mu_j \in [0,1] \\ 0; \quad otherwise \end{cases} : \quad p_{\mu_j} \in [0,1]$$

where, μ_j is membership grade corresponding to crisp input x_j. l_i represents length of interval. ζ_i and m_i are standard deviation and mean of time series data that lie in interval. Probabilities of non-membership grades are computed by complimenting probability of membership grades to 1, i.e. $\left(p_{v_i} = 1 - p_{\mu_i}; p_{v_i} \in [0, 1]\right)$ and PIFSs are constructed. A sample of probabilistic intuitionistic fuzzy elements (PIFEs) of first three values is shown in Table 2.

Table 2 PIFEs of initial three values

<table>
<tr><th>PIFSs/crisp input</th><th>5656.17</th><th>5759.61</th><th>5862.85</th></tr>
<tr><td>I'1</td><td><0.0351 | 0.5187, 0.9636 | 0.4813></td><td><0 | 0, 0 | 0></td><td><0 | 0, 0 | 0></td></tr>
<tr><td>I'2</td><td><0.0667 | 0.5087, 0.0024 | 0.4913></td><td><0 | 0, 0 | 0></td><td><0 | 0, 0 | 0></td></tr>
<tr><td>I'3</td><td><0 | 0, 0 | 0></td><td><0 | 0, 0 | 0></td><td><0 | 0, 0 | 0></td></tr>
<tr><td>I'4</td><td><0 | 0, 0 | 0></td><td><0.8278 | 0.9463, 0.112 | 0.0537></td><td><0 | 0, 0 | 0></td></tr>
<tr><td>I'5</td><td><0 | 0, 0 | 0></td><td><0.1073 | 0.5349, 0.7927 | 0.4651></td><td><0 | 0, 0 | 0></td></tr>
<tr><td>I'6</td><td><0 | 0, 0 | 0></td><td><0 | 0, 0 | 0></td><td><0.6615 | 0.8574, 0.2439 | 0.1426></td></tr>
<tr><td>I'7</td><td><0 | 0, 0 | 0></td><td><0 | 0, 0 | 0></td><td><0.2656 | 0.5864, 0.7204 | 0.4136></td></tr>
<tr><td>I'8</td><td><0 | 0, 0 | 0></td><td><0 | 0, 0 | 0></td><td><0 | 0, 0 | 0></td></tr>
<tr><td>I'9</td><td><0 | 0, 0 | 0></td><td><0 | 0, 0 | 0></td><td><0 | 0, 0 | 0></td></tr>
<tr><td>I'10</td><td><0 | 0, 0 | 0></td><td><0 | 0, 0 | 0></td><td><0 | 0, 0 | 0></td></tr>
<tr><td>I'11</td><td><0 | 0, 0 | 0></td><td><0 | 0, 0 | 0></td><td><0 | 0, 0 | 0></td></tr>
<tr><td>I'12</td><td><0 | 0, 0 | 0></td><td><0 | 0, 0 | 0></td><td><0 | 0, 0 | 0></td></tr>
<tr><td>I'13</td><td><0 | 0, 0 |0></td><td><0 | 0, 0 | 0></td><td><0 | 0, 0 | 0></td></tr>
<tr><td>I'14</td><td><0 | 0, 0 | 0></td><td><0 | 0, 0 | 0></td><td><0 | 0, 0 | 0></td></tr>
</table>

Step 4: We use an aggregation operator [32] to aggregate PIFSs into IFS by using following expression:

$$I_i = \left(1 - (1 - \mu_i)^{p_{\mu_i}}, 1 - \nu_i^{p_{\nu_i}}\right) = (u_i, v_i);\ u_i, v_i \in [0, 1]$$

An IFS corresponding to maximum membership grade of IFE is chosen for the fuzzification and is assigned to TS data.

Table 3 shows the aggregated intuitionistic fuzzy elements of three initial values.

Step 5: IFLRs are constructed using rule as $I_i \rightarrow I_j$. Here, I_i is intuitionistic fuzzy production of year n as present state and I_j is fuzzy production of year $n + 1$ as next state. TAIEX data is fuzzify and applying the procedure for fuzzification which is given in step 4. IFLRs are determined and IFLRGs are computed (Table 4).

Step 6: Intuitionistic fuzzy relation R_i of first order is described as follows:

$$R_i = R_{i1} \cup R_{i2} \cup \cdots \cup R_{ik};\quad 1 \le k \le i$$

$$R_i = \bigcup_j R_{ij}$$

Table 3 IFEs of initial three values

IFSs/crisp input	5656.17	5759.61	5862.85
I_1	<0.0184, 0.0177>	<0, 0>	<0, 0>
I_2	<0.0345, 0.9481>	<0, 0>	<0, 0>
I_3	<0, 0>	<0, 0>	<0, 0>
I_4	<0, 0>	<0.8107, 0.111>	<0, 0>
I_5	<0, 0>	<0.0589, 0.1024>	<0, 0>
I_6	<0, 0>	<0, 0>	<0.6049, 0.1823>
I_7	<0, 0>	<0, 0>	<0.1656, 0.1268>
I_8	<0, 0>	<0, 0>	<0, 0>
I_9	<0, 0>	<0, 0>	<0, 0>
I_{10}	<0, 0>	<0, 0>	<0, 0>
I_{11}	<0, 0>	<0, 0>	<0, 0>
I_{12}	<0, 0>	<0, 0>	<0, 0>
I_{13}	<0, 0>	<0, 0>	<0, 0>
I_{14}	<0, 0>	<0, 0>	<0, 0>
Fuzzified data	I_2	I_4	I_6

Table 4 IFLRs and IFLRGs for the TAIEX data

IFLRs			IFLRGs
$I_2 \rightarrow I_4$	$I_6 \rightarrow I_6$	$I_7 \rightarrow I_7$	$I_2 \rightarrow I_4$
$I_4 \rightarrow I_6$	$I_6 \rightarrow I_7$	$I_7 \rightarrow I_9$	$I_4 \rightarrow I_6$
$I_6 \rightarrow I_6$	$I_7 \rightarrow I_6$	$I_9 \rightarrow I_{10}$	$I_5 \rightarrow I_5, I_6$
$I_6 \rightarrow I_8$	$I_6 \rightarrow I_5$	$I_{10} \rightarrow I_9$	$I_6 \rightarrow I_5, I_6, I_7, I_8$
$I_8 \rightarrow I_8$	$I_5 \rightarrow I_5$	$I_9 \rightarrow I_9$	$I_7 \rightarrow I_6, I_7, I_8, I_9, I_{10}$
$I_8 \rightarrow I_8$	$I_5 \rightarrow I_6$	$I_9 \rightarrow I_9$	$I_8 \rightarrow I_7, I_8$
$I_8 \rightarrow I_8$	$I_6 \rightarrow I_5$	$I_9 \rightarrow I_9$	$I_9 \rightarrow I_9, I_{10}, I_{11}$
$I_8 \rightarrow I_7$	$I_5 \rightarrow I_6$	$I_9 \rightarrow I_9$	$I_{10} \rightarrow I_6, I_9, I_{10}$
$I_7 \rightarrow I_7$	$I_6 \rightarrow I_7$	$I_9 \rightarrow I_{10}$	$I_{11} \rightarrow I_{11}, I_{12}$
$I_7 \rightarrow I_7$	$I_7 \rightarrow I_7$	$I_{10} \rightarrow I_9$	
$I_7 \rightarrow I_7$	$I_7 \rightarrow I_8$	$I_9 \rightarrow I_9$	
$I_7 \rightarrow I_{10}$	$I_8 \rightarrow I_7$	$I_9 \rightarrow I_{11}$	
$I_{10} \rightarrow I_{10}$	$I_7 \rightarrow I_7$	$I_{11} \rightarrow I_{11}$	
$I_{10} \rightarrow I_{10}$	$I_7 \rightarrow I_7$	$I_{11} \rightarrow I_{12}$	
$I_{10} \rightarrow I_6$	$I_7 \rightarrow I_7$		

Intuitionistic fuzzified output is carried out as $I_i = I_{i-1} \circ R_i$. Here, I_i and I_{i-1} are current and previous states, '∘' is fuzzy max-min composition operator. Crisp outputs are done by the following expression:

$$\text{Crisp output} = \frac{\sum_{i=1}^{n} |u_i - v_i| a_i}{\sum_{i=1}^{n} |u_i - v_i|}$$

Here, u_i, v_i are membership, non-membership grades of row vectors respectively and a_i is average of equal intervals.

4 Performance Analysis

Proposed PIFS based IFTS forecasting method is compared with fuzzy set and IFS based forecasting methods suggested by the various researchers [4–9, 11, 18, 20, 26, 30] in terms of RMSE.

$$RMSE = \sqrt{\frac{\sum_{i=1}^{n} (crisp\ input_i - crisp\ output_i)^2}{n}}$$

where, n is index of data points.

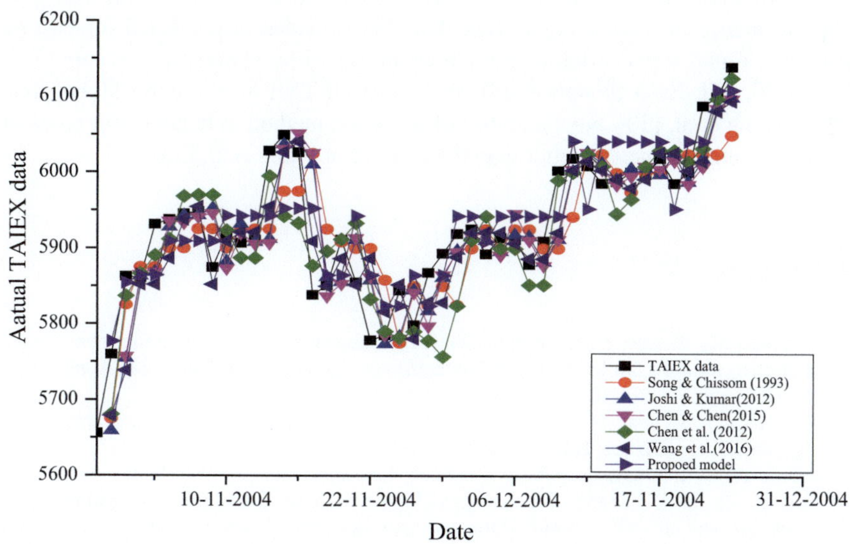

Fig. 1 Crisp outputs of TAIEX by actual, proposed and some other existing methods

Table 5 Comparison in terms of RMSE

Methods	RMSE	Methods	RMSE
Cai et al. [4]	50.33	Cheng et al. [11]	54.25
Chen [5]	84.28	Joshi and Kumar [18]	52.63
Chen and Chang [6]	60.48	Kocak [20]	50.12
Chen and Chen [7]	57.73	Song and Chissom [26]	61.17
Chen et al. [8]	52.27	Wang et al. [30]	43.23
Chen and Chen [9]	53.63	Proposed method	49.03

Figure 1 confirms the close association between actual and forecasted TAIEX. Table 5 shows RMSE of forecasted TAIEX using proposed and other existing fuzzy and intuitionistic fuzzy TS forecasting methods.

5 Conclusion

Occurrence of stochastic and non-stochastic non-determinacy concurrently in the system has been an issue in TS forecasting for long time. In this investigation, we have presented PIFS based forecasting method to address the issue of both types of the non-determinism. In order to show conveniences of developed method and it is applied on financial TS data of TAIEX. Reduced amount of RMSE ratify its outperformance in forecasting TAIEX data. Performance of presented forecasting method is found superior than the methods proposed by various researchers [4–9, 11, 18, 20, 26]. Even though RMSE in forecasted TAIEX is slightly higher than that of Wang et al. [30], but capability of proposed method to handle both types of non-determinism makes it better than the method of Wang et al. [30].

References

1. Agarwal, M., Biswas, K. K., Hanmandlu, M.: Probabilistic intuitionistic fuzzy rule based controller. In: 2011 5th International Conference on Automation, Robotics and Applications (ICARA), pp. 214-219. IEEE, New York (2011 Dec)
2. Almeida, R.J., Kaymak, U.: Probabilistic fuzzy systems in value-at-risk estimation. Intell. Syst. Account. Financ. Manag. **16**(1–2), 49–70 (2009)
3. Atanassov, K.T.: Intuitionistic fuzzy sets. Fuzzy Sets Syst. **20**(1), 87–96 (1986)
4. Cai, Q., Zhang, D., Zheng, W., Leung, S.C.: A new fuzzy time series forecasting model combined with ant colony optimization and auto-regression. Knowl.-Based Syst. **74**, 61–68 (2015)
5. Chen, S.M.: Forecasting enrolments based on fuzzy time series. Fuzzy Sets Syst. **81**(3), 311–319 (1996)

6. Chen, S.M., Chang, Y.C.: Multi-variable fuzzy forecasting based on fuzzy clustering and fuzzy rule interpolation techniques. Inf. Sci. **180**(24), 4772–4783 (2010)
7. Chen, S.M., Chen, C.D.: TAIEX forecasting based on fuzzy time series and fuzzy variation groups. IEEE Trans. Fuzzy Syst. **19**(1), 1–12 (2011)
8. Chen, S.M., Chu, H.P., Sheu, T.W.: TAIEX forecasting using fuzzy time series and automatically generated weights of multiple factors. IEEE Trans. Syst. Man Cybern. Part A Syst. Hum. **42**(6), 1485–1495 (2012)
9. Chen, M.Y., Chen, B.T.: A hybrid fuzzy time series model based on granular computing for stock price forecasting. Inf. Sci. **294**, 227–241 (2015)
10. Chen, S.M., Phuong, B.D.H.: Fuzzy time series forecasting based on optimal partitions of intervals and optimal weighting vectors. Knowl.-Based Syst. **118**, 204–216 (2017)
11. Cheng, S.H., Chen, S.M., Jian, W.S.: Fuzzy time series forecasting based on fuzzy logical relationships and similarity measures. Inf. Sci. **327**, 272–287 (2016)
12. Fan, X., Lei, Y., Wang, Y.: Adaptive partition intuitionistic fuzzy time series forecasting model. J. Syst. Eng. Electron. **28**(3), 585–596 (2017)
13. Fialho, A.S., Vieira, S.M., Kaymak, U., Almeida, R.J., Cismondi, F., Reti, S.R., Sousa, J.M.: Mortality prediction of septic shock patients using probabilistic fuzzy systems. Appl. Soft Comput. **42**, 194–203 (2016)
14. Gangwar, S.S., Kumar, S.: Probabilistic and intuitionistic fuzzy sets–based method for fuzzy time series forecasting. Cybern. Syst. **45**(4), 349–361 (2014)
15. Hinojosa, W.M., Nefti, S., Kaymak, U.: Systems control with generalized probabilistic fuzzy-reinforcement learning. IEEE Trans. Fuzzy Syst. **19**(1), 51–64 (2011)
16. Huang, W.J., Zhang, G., Li, H.X.: A novel probabilistic fuzzy set for uncertainties-based integration inference. In: 2012 IEEE International Conference on Computational Intelligence for Measurement Systems and Applications (CIMSA), pp. 58–62. IEEE, New York (2012 July)
17. Hung, K.C., Lin, K.P.: Long-term business cycle forecasting through a potential intuitionistic fuzzy least-squares support vector regression approach. Inf. Sci. **224**, 37–48 (2013)
18. Joshi, B.P., Kumar, S.: Intuitionistic fuzzy sets based method for fuzzy time series forecasting. Cybern. Syst. **43**(1), 34–47 (2012)
19. Jurio, A., Paternain, D., Bustince, H., Guerra, C., Beliakov, G.: A construction method of Atanassov's intuitionistic fuzzy sets for image processing. In: 2010 5th IEEE International Conference on Intelligent Systems (IS), pp. 337–342. IEEE, New York (2010 July)
20. Kocak, C.: ARMA (p, q) type high order fuzzy time series forecast method based on fuzzy logic relations. Appl. Soft Comput. **58**, 92–103 (2017)
21. Kumar, S., Gangwar, S.S.: Intuitionistic fuzzy time series: an approach for handling non-determinism in time series forecasting. IEEE Trans. Fuzzy Syst. **24**(6), 1270–1281 (2016)
22. Liu, Z., Li, H.X.: A probabilistic fuzzy logic system for modeling and control. IEEE Trans. Fuzzy Syst. **13**(6), 848–859 (2005)
23. Meghdadi, A.H., Akbarzadeh-T, M.R.: Probabilistic fuzzy logic and probabilistic fuzzy systems. In: The 10th IEEE International Conference on Fuzzy Systems, 2001, vol. 3, pp. 1127–1130. IEEE, New York (2001)
24. Singh, P., Borah, B.: Forecasting stock index price based on M-factors fuzzy time series and particle swarm optimization. Int. J. Approx. Reason. **55**(3), 812–833 (2014)
25. Song, Q., Chissom, B.S.: Fuzzy time series and its models. Fuzzy Sets Syst. **54**(3), 269–277 (1993)
26. Song, Q., Chissom, B.S.: Forecasting enrolments with fuzzy time series—part I. Fuzzy Sets Syst. **54**(1), 1–9 (1993)
27. Song, Q., Chissom, B.S.: Forecasting enrolments with fuzzy time series—part II. Fuzzy Sets Syst. **62**(1), 1–8 (1994)
28. Tak, N., Evren, A.A., Tez, M., Egrioglu, E.: Recurrent type-1 fuzzy functions approach for time series forecasting. Appl. Intell. **48**(1), 68–77 (2018)
29. Torbat, S., Khashei, M., Bijari, M.: A hybrid probabilistic fuzzy ARIMA model for consumption forecasting in commodity markets. Econ. Anal. Policy (2018)

30. Wang, Y.N., Lei, Y., Fan, X., Wang, Y.: Intuitionistic fuzzy time series forecasting model based on intuitionistic fuzzy reasoning. Math. Probl. Eng. (2016)
31. Xian, S., Zhang, J., Xiao, Y., Pang, J.: A novel fuzzy time series forecasting method based on the improved artificial fish swarm optimization algorithm. Soft Comput., 1–11 (2017)
32. Xu, Z.: Intuitionistic fuzzy aggregation operators. IEEE Trans. Fuzzy Syst. **15**(6), 1179–1187 (2007)
33. Yolcu, O.C., Yolcu, U., Egrioglu, E., Aladag, C.H.: High order fuzzy time series forecasting method based on an intersection operation. Appl. Math. Model. **40**(19), 8750–8765 (2016)
34. Zadeh, L.A.: Fuzzy sets. Inf. Control. **8**(3), 338–353 (1965)
35. Zheng, K.Q., Lei, Y.J., Wang, R., Wang, Y.: Prediction of IFTS based on deterministic transition. J. Appl. Sci. Electron. Inf. Eng. **31**(2), 204–211 (2013)
36. Zheng, K.Q., Lei, Y.J., Wang, R., Xing, Y.Q.: Method of long-term IFTS forecasting based on parameter adaptation. Syst. Eng. Electron. **36**(1), 99–104 (2014)

Investigation of Transient Stability in IEEE 9-Bus System Using Power World Simulator

Ravishankar Tiwari, Mayank Goyal and Prashant Prakash

Abstract With the day-by-day increase in power consumption, the transmission and distribution networks are operating on the verge of stability. The transient stability studies is one of the important study that helps to maintain the more reliable power supply with increasing demand in complex network configuration. This paper helps to understand the problem of transient stability, its effect on the power system. In this paper IEEE 9-bus system is considered to analyze a list of contingencies and its effect on the system. The contingency analysis is done using a powerful simulation tool for different generator bus.

Keywords Transient stability · Rotor angle · Power world simulator

1 Introduction

The degree of stability of power system is an important factor in planning of new facilities. The sudden major disturbances such as change in load, momentarily fault, and loss of generation or transmission may create stability problem. The information related to ability of power system to remain in synchronism under major disturbances is obtained by performing transient stability studies. These studies provide the change in voltages, currents, powers, machine torques, angles and speeds as well as changes in system voltage and power flows during and immediately after the faults. If the power flow solution after clearing the faults converges then it will be indicated that the system is stable otherwise not. This paper suggested the sequence of corrective actions could be taken to make the system stable, also calculate the degree of stability so that system can plan to operate safely. In this paper the IEEE 9 bus system

R. Tiwari (✉) · M. Goyal · P. Prakash
Department of Electrical Engineering GLA University, Mathura, Uttar Pradesh, India
e-mail: Ravishankar.tiwari@gla.ac.in

M. Goyal
e-mail: Mayank.goyal@gla.ac.in

P. Prakash
e-mail: Prashant.prakash@gla.ac.in

S. Manna et al. (eds.), *Mathematical Modelling and Scientific Computing with Applications*, Springer Proceedings in Mathematics & Statistics 308,
https://doi.org/10.1007/978-981-15-1338-1_24

having three synchronous machines are used. The performances of the system during transient period are obtained by numerical techniques with the use of performance equation (Impedance/admittance form) [1].

2 Mathematical Model Development for Transient Stability Analysis and Assumptions

For transient analysis it is assumed that—(1) The mechanical input is constant during the whole rotor swing, (2) Damping D is proportional to relative speed, (3) Steady state network solution gives the synchronous power, (4) The synchronous machines are represented by series internal voltage connected behind the series direct axis reactance, (5) The mechanical rotor angle δ(i) is in phase with electrical angle for each machine, and (6) The static and induction load are represented by shunt admittance connected to the bus [2].

The system is having 3 states prefault, during fault and post fault. (1) Prefault state give all the initial conditions of angles δ(i). (2) Fault state from $t = 0$ to $t = t_{clr}$ (fault clearing time). (3) Post fault state for, $t > t_{clr}$.

3 Rotor Swing Modeling During Transient

From the law of mechanism related to the rotating bodies, the net torque acting on the rotating bodies is

$$T = \frac{WR^2}{g}\alpha \tag{1}$$

where

T = algebraic sum of all torques, N-m
WR^2 = moment of inertia, kgm^2
g = acceleration due to gravity, equal to 9.8 m/s^2
α = mechanical angular acceleration, rad/s^2

The electrical angle

$$\theta_e = \frac{p}{2}\theta_m \tag{2}$$

Frequency f in cycle per second will be

$$f = \frac{P}{2} * \frac{N}{60} \tag{3}$$

where N is speed in rpm.

Then from Eqs. (2) and (3), the electrical angle in radian is

$$\theta_e = \frac{60f}{N}\theta_m \tag{4}$$

The electrical angular position δ in radians of the rotor with respect to synchronous reference axis

$$\delta = \theta_r + \omega_0 t \tag{5}$$

where ω_0 is rated synchronous speed in radian/second and 't' is time in seconds.

Then angular velocity or slip with respect to reference axis is

$$\frac{d\delta}{dt} = \frac{d\theta_e}{dt} - \omega_0 \tag{6}$$

And the angular acceleration is

$$\frac{d^2\delta}{dt^2} = \frac{d^2\theta_e}{dt^2} \tag{7}$$

On substituting θ_e from Eq. (4)

$$\frac{d^2\delta}{dt^2} = \frac{60f}{N}\frac{d^2\theta_m}{dt^2} \tag{8}$$

where $\frac{d^2\theta_m}{dt^2} = \propto$.

So the net torque will be

$$T = \frac{WR^2}{g}\frac{N}{60f}\frac{d^2\delta}{dt^2} \tag{9}$$

The base torque is torque required to develop at rated power at rated rpm i.e.

$$Base\,Torque = \frac{base\,kva\frac{550}{0.746}}{2\pi\frac{N}{60}} \text{ in foot - pound.}$$

Therefore the torque in per unit

$$T = \frac{\left(WR^2 2\pi/gf\right)\left(\frac{N}{60}\right)^2\frac{0.746}{550}}{Base\,kva}\frac{d^2\delta}{dt^2} \tag{10}$$

The inertia constant H of a machine is kinetic energy at rated speed in kilowatt second/kilo-volt ampere and the kinetic energy in foot pound is

$$kinetic\,energy = \frac{1}{2}\frac{WR^2}{g}\omega_0^2 \quad (11)$$

where $\omega_0 = \frac{2\pi N}{60}$.

Therefore

$$H = \frac{\left(WR^2/g\right)(2\pi)^2\left(\frac{N}{60}\right)^2 \frac{0.746}{550}}{Base\,kva} \quad (12)$$

Substituting in Eq. (10), we get

$$T = \frac{H}{\pi f}\frac{d^2\delta}{dt^2} \quad (13)$$

The total torque acting on an alternator rotor are mechanical input torque, torque due to rotational losses such as frictional, windage losses and core losses, electrical output torque and damping torque due to prime mover and power system since electrical and mechanical losses acting on the rotor are in opposite direction. On neglecting the frictional, windage, core losses and damping, the accelerating torque is given by

$$T_a = T_m - T_e \quad (14)$$

where

T_m = *mechanical torque and*
T_e = *electrical air gap torque*

Thus Eq. (13) becomes

$$\frac{H}{\pi f}\frac{d^2\delta}{dt^2} = T_m - T_e \quad (15)$$

Since the torque and power are equal in per unit so

$$\frac{d^2\delta}{dt^2} = \frac{\pi f}{H}(P_m - P_e) \quad (16)$$

where,

P_m = *mechanical input power*
P_e = *electrical air gap power*

The second order differential equation (16) can be break in two simultaneous first order equation, combined with system performance equation and the solution can be obtain from Runga Kutta Or modified Euler technique. In this paper the Runga Kutta method is used.

Synchronous machine model during transient analysis—Since the transient stability analysis performed for a very small period of time (in seconds or less), the synchronous machines can be represented by a constant voltage source behind a transient reactance with variable angular displacement. Neglect the effect of saliency and assume constant flux linkage and a small change in speed. The voltage behind the transient reactance are determined by

$$E' = E_t + r_a I_t + jx'_d I_t$$

Induction machine representation—Induction motors loads can be represented by shunt impedances whereas in case of larger I.M. have significant effects on transient response of power system so represented in detailed manner. A simple linear representation has been use to represent transient behavior of an induction motor including the effect of mechanical and electrical transients with single time constant. The saliency and the changes of flux can be taken into account by representing the effect of three phase ac quantities of synchronous machine acting along d-axis and q-axis.

Load representation—during the transient period motor loads are represented by their equivalent circuit and other load representation are—static impedance/admittance between a point of connection and ground or a load of fixed power factor at constant current, constant real and reactive power demand or in combination of these [3, 4].

Constant current representation $I_{p0} = \frac{P_L - jQ_L}{E_P^*}$

Static admittance representation $y_p = \frac{I_{p0}}{E_p}$

Connected at any bus p where E_p is the bus voltage with respect to ground and I_{p0} is the current flowing from bus to ground.

4 Network Equation Formulation

The power system network model can be either with in node/bus or mesh or branch reference frame. The mathematical model in bus reference frame of n bus system and taking one bus as a reference gives n − 1 number of linear independent equations. The equation in admittance from can write as [2]:

$$\begin{bmatrix} \text{Vector of injected} \\ \text{bus currents } I_{BUS} \end{bmatrix} = \begin{bmatrix} \text{BUS admittance matrix} \\ \text{of Power system } Y_{BUS} \end{bmatrix} \begin{bmatrix} \text{Vector of node} \\ \text{voltage w.r.t. ground } E_{BUS}^* \end{bmatrix}$$

$$\begin{bmatrix} I_1 \\ I_2 \\ \vdots \\ I_{n-1} \end{bmatrix} = \begin{bmatrix} Y_{11} & Y_{12} \cdots Y_{1,n-1} \\ Y_{21} & \cdots \cdots Y_{2,n-1} \\ & \vdots \\ Y_{n-1,1} & \cdots \cdots Y_{n-1,n-1} \end{bmatrix} \begin{bmatrix} E_1 \\ E_2 \\ \vdots \\ E_{n-1} \end{bmatrix}$$

For computing the angle and frequency trajectory the swing equation discussed in Sect. 1 Eq. (16) is solved and the results are shown in Sect. 3.

5 Problem Formulation and System Studies

The IEEE 9 bus system is used to perform the transient analysis. The network equation similar to load flow is used to formulate performance equations. The N-R method is used to solve the performance equation, where the elements of admittance matrix represent the connection between the bushes including load. The single line diagram if IEEE 9 bus system is shown in Fig. 1 where the system consists of three generating units G1, G2 and G3 connected at buses 1, 2 and 3 respectively. The bus 1 is considered as slack bus whereas 2 and 3 are PV buses. The loads are connected at buses 5, 6 and 8 are considering as load buses. The total generation is 330 MVA whereas the total load is 315 MW.

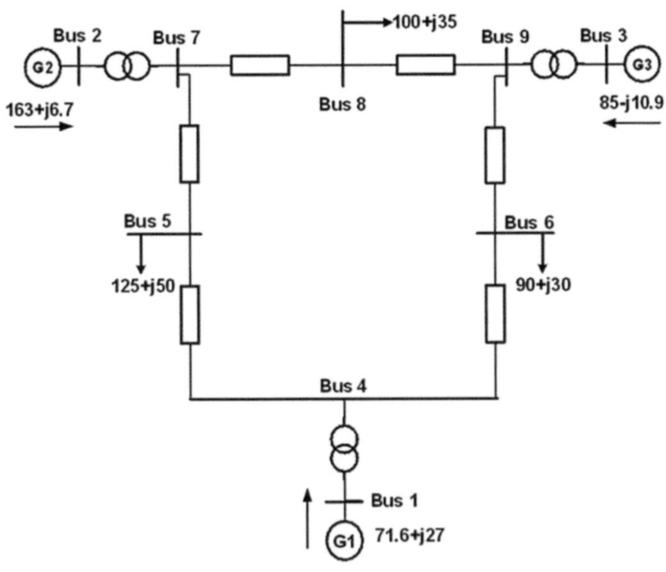

Fig. 1 One line diagram of IEEE 9 bus system

The problem associated with this system is transmission of power during disturbances from one machine group to another. During disturbances, the machines of same group swings together and also machines of other group are greatly deviated with their relative positions. The balanced three phase faults are considered at bus 3, 2 and 1 one by one and the critical clearing time, relative stability and excitation control to improve system stability during each case is observed by using Power world software.

Case I: Three phase fault at bus 3, the critical clearing time for this fault is $t_{clr} =$ 550 ms in Fig. 2. The frequency deviation over different buses with maximum at $f_{bus2} > f_{bus1} > f_{bus3}$ shown in Fig. 3.
Case II: 3-phase fault at bus 2, the fault should be cleared before $t_{clr} = 310$ ms. With the similar deviation in rotor angle, frequency (+1.43 Hz) and excitation as in case I. The shift in rotor angle is plotted for fault clearing time $t < t_{clr}$ when all machines are swing together and remain in synchronism.

For time $t > t_{clr}$, the result are plotted, shows that the machine is out of synchronism and not swinging together. Frequency variation is uncontrolled and the excitation control hits its boundary value at bus 2 and is not able to maintain the voltages within limits.

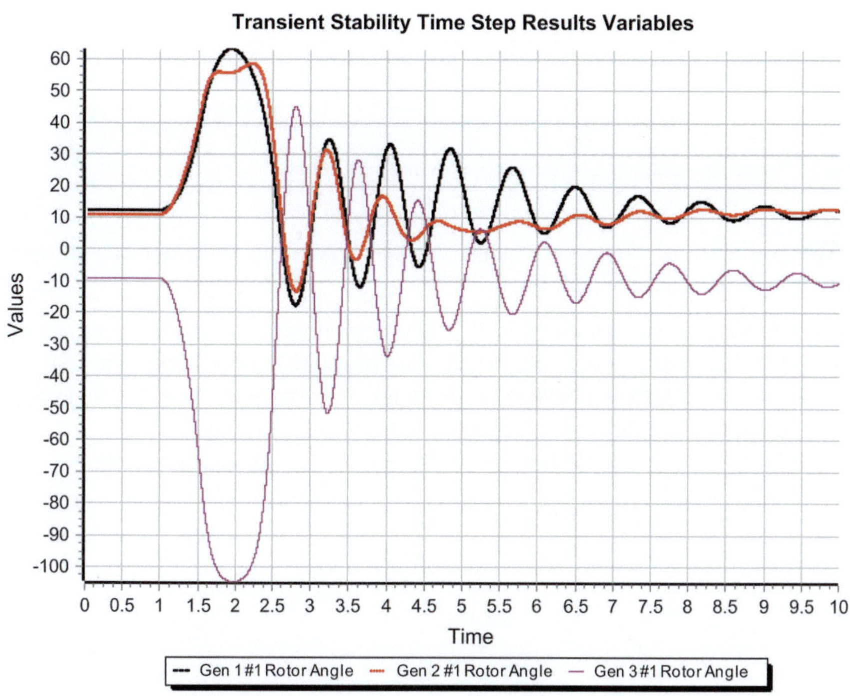

Fig. 2 Rotor angle variations for fault clearing time t < 550 ms

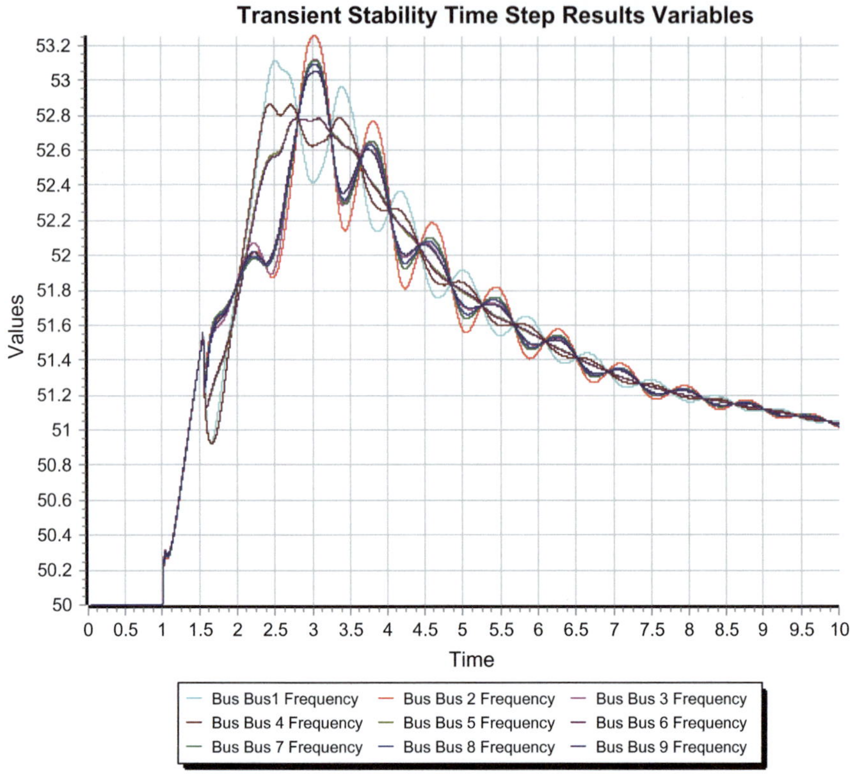

Fig. 3 Frequency variations at different buses during fault at bus 3

The variation of excitation voltage due to variation in generator bus voltages during fault bus 2 is shown in Fig.6.

Case III: Three phase fault at bus 1, it is observed that it would be most critical 3-phase fault on the given system which requires least fault clearing time of $t_{clr} =$ 290 ms or less. The three faults persist for more than 16 cycles will leads to system to be unstable. The rotor angle, frequency and excitation voltage variation for fault clearing time $t < t_{clr}$ are similar as in case I and II. For time $t > t_{clr}$ is show in the figures respectively.

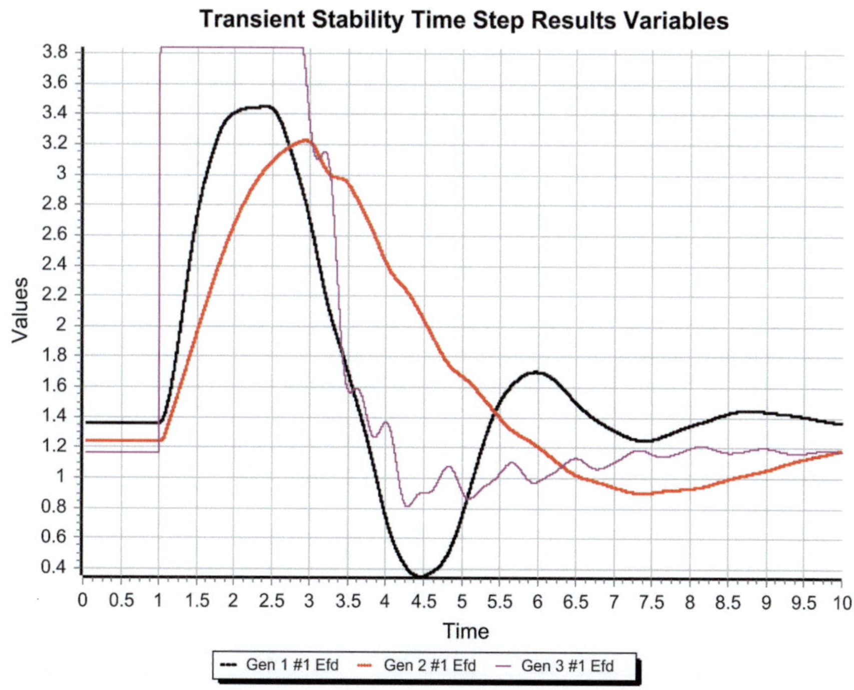

Fig. 4 Variation of excitation control of G1, G2, and G3 before fault, during fault and after clearing fault for $t < t_{clr}$

6 Discussion

Several factors affect critical clearing time. We have considered fault location. Three fault locations were considered for the same damping and inertia constants. A three-phase fault is done on all the generator buses i.e. on bus 1, bus 2 and bus 3 of the IEEE 9-bus system. Observations are obtained on Figs. 2, 3, 4, 5 and 6.

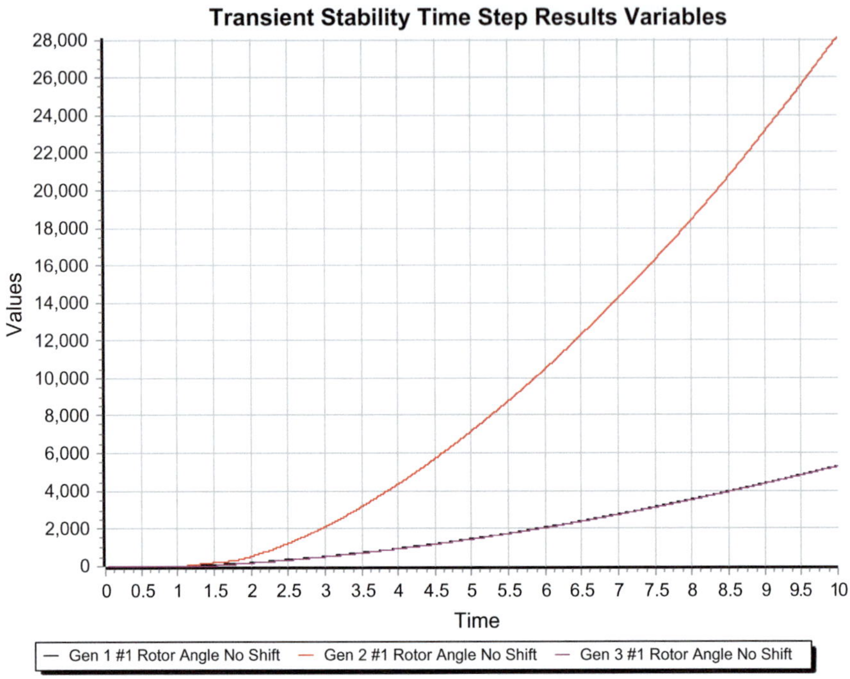

Fig. 5 Rotor angle shift

In this paper, the stability of the IEEE 9-bus system has been studied by plotting the curves of rotor angle variations, frequency variations, rotor angle shift and variations in excitation. These variations are obtained for the 3-phase fault on generator buses. It is recommended to that power system engineers must do proper analysis w.r.t. transient stability.

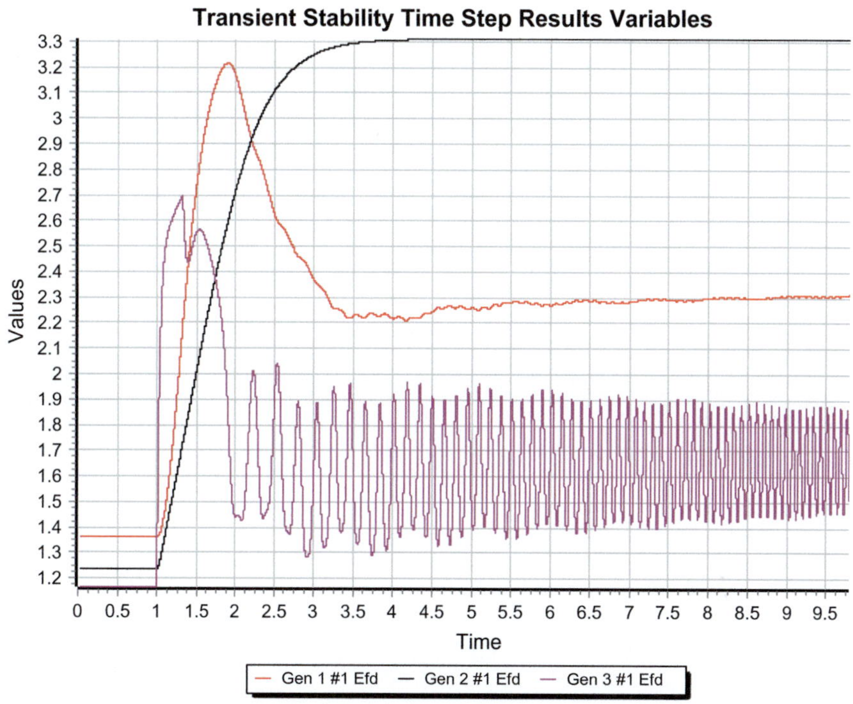

Fig. 6 Variation of excitation E_{fd} for fault clearing time at 320 ms > t_{clr}

References

1. Nallagalva, S.K., Kirar, M.K., Agnihotri, G.: Transient stability analysis of the ieee 9-bus electric power system. Int. J. Sci. Eng. Technol. **1**(3), 161–166 (2012)
2. Iyambo, P.K., Tzoneva, R.: Transient stability analysis of the IEEE 14-bus electric power system. In: AFRICON 2007, pp. 1–9. IEEE, New York (2007)
3. Price, W.W., Wirgau, K.A., Murdoch, A., Mitsche, J.V., Vaahedi, E., El-Kady, M.: Load modeling for power flow and transient stability computer studies. IEEE Trans. Power Syst. **3**(1), 180–187 (1988)
4. Vittal, V., Martin, D., Chu, R., Fish, J., Giri, J.C., Tang, C.K., Eugenio Villaseca, F., Farmer, R.G.: Transient stability test systems for direct stability methods. IEEE Trans. Power Syst. **7**(1), 37 (1992)

Discrete Time Minimum Variance Control of Satellite System

Deepali Y. Dube and Hiren G. Patel

Abstract This paper is concerned with the types of stochastic disturbances affecting the potential of the aerial system. The satellite system for continuous and discrete time domain is discussed. A phase lead compensator completes the orientation successfully. Astrom's single-input single-output (SISO) model is implemented with using the minimum variance control strategy. The separation principle then provides the optimal control law which curtails the cost function to a value as small as possible. The satellite system is positioned for one quarter revolution with the co-ordination of generalized minimum variance controller (GMVC) and standard generalized dual controller (GDC) based on certainty equivalence assumption. The revolutions in radians are tracked as output of the system for the input specified in degrees to the system. The controller proved useful in reducing the overshoot and atmospheric disturbances which allows a stable motion even for larger time delays.

Keywords Discrete time domain · Single-input single-output · Optimal control law

Nomenclatures

e	Independent vector
k'	Gain factor
$u(t)$	Control signal
x	Radial perturbation
$y(t)$	System output
ω_d	Damping frequency
ξ	Damping ratio

D. Y. Dube (✉)
National Institute of Technology, Surat, GJ 395007, India
e-mail: deepalidube404@gmail.com

H. G. Patel
Department of Electrical Engineering, National Institute of Technology, Surat, GJ 395007, India
e-mail: hpatel@svnit.eed.in

S. Manna et al. (eds.), *Mathematical Modelling and Scientific Computing with Applications*, Springer Proceedings in Mathematics & Statistics 308,
https://doi.org/10.1007/978-981-15-1338-1_25

ξ_t Gaussian white noise
G_{LA} Longitudinal disturbance
G_P Pressure disturbance
G_T Temperature disturbance
I_q Covariance matrix
T_s Time delay
V Loss function
W_x Weiner process
Y_r Desired signal

1 Introduction

The existence of stochastic disturbances in the real practical problems has encouraged the prerequisite to control the system optimally. The resolution to the said problem is required in the areas involving economy, reliability, stability and security of the whole system. These also depend on the performance of the regulation. This has emerged as the main reason of huge progress observed in the area of stochastic control engineering. Theoretical results are made in use with proper hardware and software conditions and a powerful control algorithms is also made to utilization. Astrom introduced the minimum variance control in order to design optimal control loop. This method has some nicer properties as:

(1) *The control signal deviation around the required value is reduced,*
(2) *It can be applied to the cases where the parameters are known and unknown of a given system,*
(3) *Being simple; it does not consume much time,*
(4) *Disturbances occurring in practical situations are also considered,*
(5) *No prior knowledge about the process is required,*
(6) *The respective variation of the system parameters with the desired value is followed.*

A framework introduced by Hunek and Latawiec [1] of implementing minimum variance control of both discrete and continuous time linear time-invariant multi-input multi-output systems became popular. It concluded that the continuous processes can be synthesized without the theory of output predictor emulation.

Whereas, Soroka and Shaked [2] the problems dealing with the finite-time, reduced-order, minimum-variance, full-state estimation of linear and continuous time invariant systems were of main interest. This was done for the cases where the output measurement is free of white noise components. An expression of transfer function provided good insight into the mechanism of the optimal estimation.

Yamamoto et al. [3] discussed the problem of camera vibration in aerial photography. It involved the RC helicopter system where the image is affected. Hence the position of the camera is first controlled by PD tuning method. The estimation done by least square method involving the later computations by the generalized minimum variance control.

A non-linear operator approach to estimation was described and discussed by Grimble and Naz [4]. The assumed measurements being corrupted by colored noise signal, a nonlinear filter proposed in paper reduces the generality of the problem and leads to a very practical non-linear estimation of algorithm. Silveria and Coelho [5] suggested that while dealing with systems having longtime delays the design procedures require only a small amount of work. The goal was to propose a state-space design technique for the generalized minimum variance control. The results were verified by linear algebra. Two examples were provided to prove the effectiveness of proposed technique.

The problem of minimum variance event triggered output feedback control of LTI processes was given main concern. It involved the presence of white gaussian noise too. Also, the optimal event generated was separate from the controller configuration and was determined by an optimal stopping algorithm. Hence, the paper was able to minimize the asymptotic average variance of the regulated output of LTI plant.

Analytical results were obtained by Yokoyama and Masuda [6] for the iterative data-driven PID gain tuning based on generalized minimum variance regulatory control. It achieved better PID gains due to generalized minimum variance evaluation. A numerical simulation also proved this technique. The need of controlling various industrial processes with an extension of the principle of the minimum variance control have been in abundance supported by above nicer properties of stochastic control methodology. Also the merit of the method is not to require any advance statistics about the plant and disturbance action.

In this paper, the problems occurring due to the stochastic disturbance are simplified in Sect. 2. Mathematical representation of the system model in Sect. 3. Design of controller is discussed in Sect. 4 and simulation results in Sect. 5, whereas the conclusion is quoted in Sect. 6.

2 Problem Formulation

It was found that there were meteoric swings on earth's atmospheric density with distractions. This has to be taken in while modelling the rigid body dynamics. This is developed from Newtonian mechanics of securing the satellite in circular orbit [8, 9].

$$\ddot{x} + b(1 + a\xi_t)\dot{x} + (1 + a\xi_t)\sin x - c\sin 2x = 0 \tag{1}$$

here x is radial perturbation about the given orbit, ξ_t is Gaussian white noise and a, b, c are constants. The above equation is modeled as 2-dimensional Ito stochastic differential equation

$$d\begin{pmatrix} X_t^1 \\ X_t^2 \end{pmatrix} = \begin{pmatrix} X_t^2 \\ -bX_t^2 - \sin X_t^1 + c\sin 2X_t^1 \end{pmatrix}dt + \begin{pmatrix} 0 \\ -abX_t^2 - b\sin X_t^1 \end{pmatrix}dW_t \tag{2}$$

The problem due to the atmospheric distractions or swings were modeled in (3–5). Above equations were evolved virtually in MATLAB with a 2-dimensional Ito-stochastic differential equation. The weiner process *W(x)* being analysed with randomness in the vector (Brownian motion). This concluded a different output each time the system is manipulated. Atmospheric chaos are necessary for design of both inlet and engine flight controls and for studying link between the propulsion and the vehicle structural dynamics for supersonic vehicles. In the referred paper, a more accurate model was developed by George to represent fractional order of atmospheric disturbances. Atmospheric model was accomplished by first ascending the kolmogorov spectral to convert them into finite energy von-karman forms. Hence, the objective lies for given parameters and the atmospheric conditions with all prior information, (the poles and zeros describing disturbances for respective acoustic velocity, temperature, pressure and density) the appropriate time domain simulations are evaluated. These disturbances also contribute to the problems cited in this paper. Their respective transfer functions as in [7]:

$$G_{LA}(s) = 70\varepsilon^{2/9}\frac{(s/9.2+1)(s/55+1)(s/335.5+1)}{(s/1.46+1)(s/30.1+1)(s/85.7+1)(s/1593.1+1)} \tag{3}$$

$$G_T(s) = \frac{(s/33+1)(s/45.6+1)(s/602.4+1)}{(s/1.1+1)(s/25.1+1)(s/109.8+1)(s/816.3+1)} \tag{4}$$

$$G_P(s) = \frac{(s/33+1)(s/45.6+1)(s/602.4+1)}{(s/1.1+1)(s/25.1+1)(s/109.8+1)(s/816.3+1)} \tag{5}$$

here G_{LA}, G_T and G_P are the atmospheric disturbance transfer function for longitudinal, temperature and pressure. These exhibit behavioral pattern when in the system and may cause large instability in satellite control system operations. As shown in Fig. (1), the response is completely unwanted. It reflects that the disturbance is very inclined in the right half plane. A specific scaling will allow the satellite to hover at an altitude for executing spinning maneuver and also communicate the required messages through sensing.

3 Satellite System

In order to position a satellite communication system at a specific angular location in such a way that the communication link to the earth station is maintained. This requires following considerations

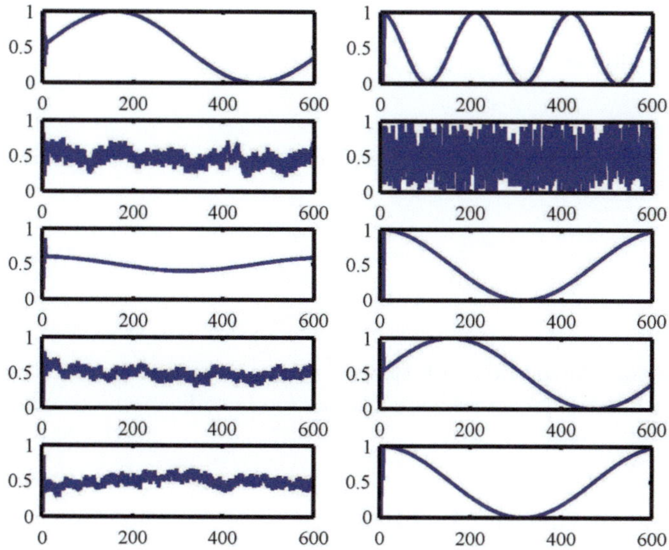

Fig. 1 The affect of stochastic disturbance at the output (Right hand side) via input (left hand side) of the satellite system

- A pair of bi-directional rockets. These are equipped on the outside wall of the satellite. It allows the corrections required to be accommodated in the presence of disturbances offset of the angular position,
- The satellite is a rotating control system object placed in a frictionless space,
- The electrical input voltage controls the torque deployed by a pair of the rockets,
- The perfect position of the satellite system is predicted by the position information available in the form of electrical signal.

Here, in the paper a digital controller *D(z)* of first order phase lead type is utilized. This makes the satellite position more stable. The sample rate $T_s = 3$–6 s respectively. To build the model in a discrete time domain some certain design criteria's must be followed: $\xi > 0.6$, since $\omega_d < 0.1\omega_s$, $\omega_d T_s < 0.2\pi = 360$, $z = e^{-0.2} = 0.8187$. The transfer functions utilized in forward and feedback path in continuous time domain are

$$G(s) = \frac{1 - e^{-sT_s}}{s} \times \frac{0.02}{s^2} \tag{6}$$

Using the residue method above plant is simplified in discrete time domain as

$$G(z) = \frac{0.0001(z + 1)}{(z - 1)^2}, \tag{7}$$

A phase lead compensator is involved for the design of satellite position orientation of a general form as

$$D(z) = k\frac{z - z_0}{p - p_0} \tag{8}$$

$$= \frac{k'}{0.036}\frac{z - z_0}{z - z_p} \tag{9}$$

$$= 11.11\frac{z - 0.8}{z + 0.2} \tag{10}$$

here gain is considered as $k' = 0.4$. Hence the response for a 90° angle as input signal is obtained. It is a step response of the closed loop satellite position control system.

4 Minimum Variance Control

The Astrom model presented a SISO system described by:

$$y(t) = \frac{B(z^{-1})}{A(z^{-1})}u(t - T_s) + \lambda\frac{C(z^{-1})}{A(z^{-1})}e(t) \tag{11}$$

where $B(z^{-1})/A(z^{-1})$ is the process transfer function, $\lambda C(z^{-1})/A(z^{-1})$ being the transfer function of the disturbance, *u(t)* is the control signal, *y(t)* being the output of noisy system, *e(t)* is the sequence of normal independent variables (having zero mean value of variance one) and T_s is the time delay of the process. Also

$$A(z^{-1}) = 1 + a_1 z^{-1} + \cdots + a_n z^{-n}$$
$$B(z^{-1}) = b_0 + b_1 z^{-1} + \cdots + b_m z^{-m}$$
$$C(z^{-1}) = 1 + c_1 z^{-1} + \cdots + c_k z^{-k}$$

here $t = 0, 1, 2, \ldots, q$. The main intension of minimum variance control is to decide the control signal *u(t)* in a way that the loss function described as

$$V = E\{[y(t + d) - y_r]^2\} \tag{12}$$

is as small as possible with yr as the desired value of controlled signal. When considering the multi-variable system an extension of Astrom model is used

$$A(z^{-1})y(t) = B(z^{-1})u(t - d) + C(z^{-1})\Delta e(t), \tag{13}$$

where y is the q-dimensional output vector, u is the q dimensional control vector, A, B and C are polynomial matrices of z^{-1}:

$$A(z^{-1}) = I_q + \sum_{j=1}^{n} z^{-j} A_j,$$
$$B(z^{-1}) = \sum_{j=0}^{m} z^{-j} B_j,$$
$$C(z^{-1}) = I_q + \sum_{j=0}^{k} z^{-j} C_j$$

here e is a sequence of normally distributed independent vector variables with zero mean value and covariance matrix I_q. Here I_q is a $q \times q$ unit matrix. Also, the loss function which is to be minimized

$$V = [y(t+d) - y_r]^T [y(t+d) - y_r], \tag{14}$$

Using the separation principle, the optimal control law obtained:

$$u(t) = B_0^{-1}\left[C(z^{-1})v_r - G(z^{-1})v(t) - \widetilde{B}(z^{-1})u(t)\right], \tag{15}$$

A generalized dual control (GDC) in a non-linear form is obtained after minimizing the cost function

$$u_e(k) = \frac{\hat{f}_0(k)w(k) - [\hat{f}_0(k)\hat{p}_0^T(k) + (1-\lambda)p_{f0p0}^T(k)]m_o(k)}{\hat{f}_0^2(k) + (1-\lambda)p_{f0}(k)} \tag{16}$$

5 Simulation Results

The simulations are carried out to position the satellite at 900 (one quarter revolution). The input provided to the system is in degrees with a response explaining the revolutions in radians for a zero at 0.8 and a pole at 0.2 as the lead compensator into the system. From Fig. 1, it is observed that the presence of atmospheric disturbances create an abnormal curve on the behavior of input-output estimates. Hence, there lies an urgent need of suppression of such stochastic nonlinearities from the system. The standard adaptive generalized minimum variance (GMV) controller based on the certainty equivalence (CE) assumption is displayed in Fig. 2. The results of the simulation of the designed adaptive dual version of the GMV controller are shown in Fig. 3. The response for the generalized dual controller (GDC) is also studied. As can be seen in Fig. 3, the adaptive dual controller provides better control performance and a smaller overshoot. The control of the same plant with a larger time delay can be considered in future.

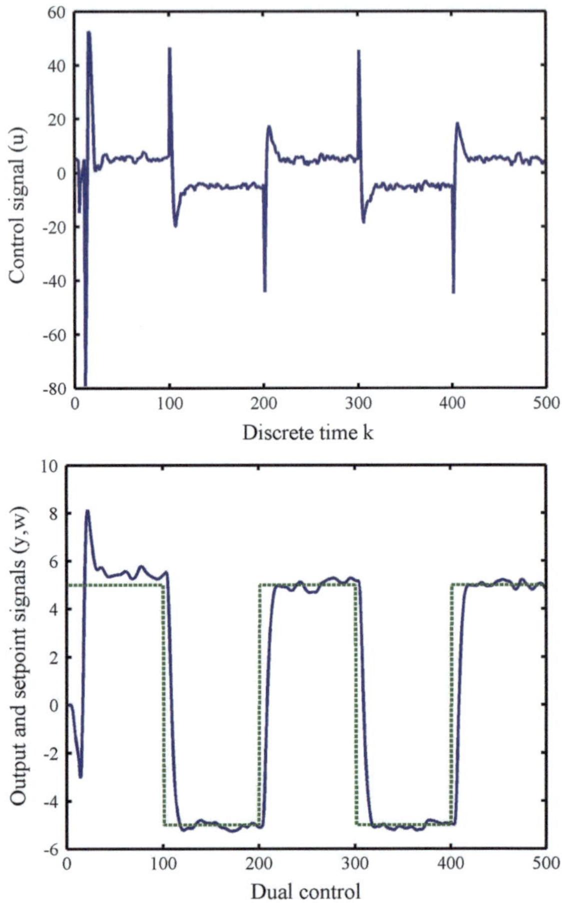

Fig. 2 Simulation results for the standard adaptive GMV controller based on the CE assumption ($T_s = 4$) for w (reference signal, green) and y (actual estimated signal, blue)

6 Conclusion

In this paper, the satellite system has been described and applied for the synthesis of the dual version of the generalized minimum variance controller (GMV). The reward of the depicted controller and the collation with the dual approach have been emphasized. It's properties are applied to position the satellite considering angle as an input. The enhanced potential of the adaptive dual controller has been evinced. The second order discrete system known to time-varying have been cast-off to experiment the renewed controller. The advanced controller have been validated and collate using minimum phase plant. Also, simulation results predict better response for the said problem for atmospheric disturbances of stochastic nature. The depicted controllers indicate competent capability for the said problem.

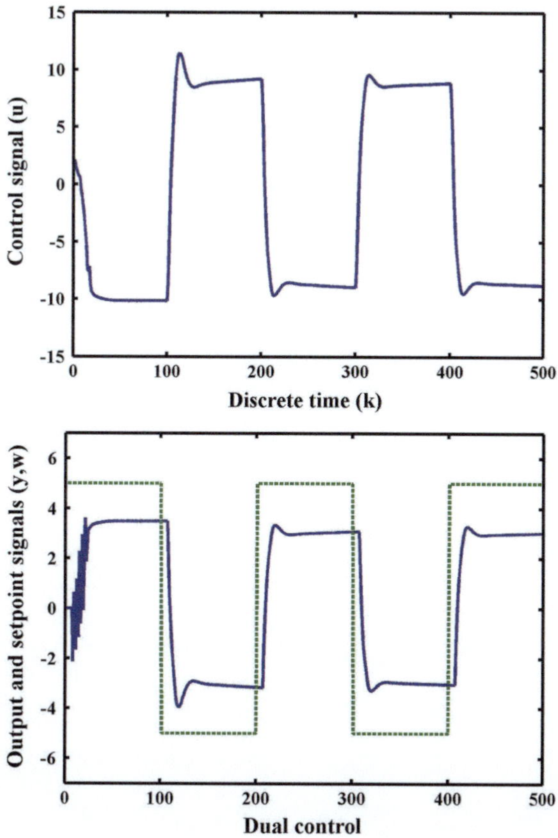

Fig. 3 Simulation results for the standard GMV controller based on the CE assumption with a time delay of $T_s = 6$ for w (reference signal, green) and y (actual estimated signal, blue)

References

1. Hunek, W.P., Latawiec, K.J.: Minimum variance control of discrete time and continuous-time LTI MIMO systems a new unified framework. J. Control Cybern. **38**(3), 609624 (2009)
2. Soroka, E., Shaked, U.: The properties of reduced-order minimum-variance filters for systems with partially perfect measurements. IEEE Trans. Autom. Control **330**(11), 1022–1034 (1988)
3. Shiino, T., Kawada, K., Yamamoto, T., Komichi, M., Nishioka, T.: Gimbals control with the camera for aerial photography in RC helicopter. In: International Conference on Control, Automation and Systems, vol. 120, no. 5, pp. 1135–1139, 14–17 Oct 2008
4. Grimble, M.J., Naz, S.A.: Nonlinear minimum variance estimation for discrete-time multi-channel systems. IEEE Trans. Signal Process. **57**(7), 2437–2444 (2009)
5. Silveira, A.S., Coelho, A.A.R.: Generalised minimum variance control state-space design. IET Control Theor. Appl. **5**(15), 1709–1715 (2016)
6. Yokoyama, R., Masuda, S.: Convergence property for iterative data driven PID gain tuning based on generalized minimum variance regulatory control. In: 6th International Symposium on Advanced Control of Industrial Processes (AdCONIP), Taipei, Taiwan, pp. 511–516, 28–31 May 2017
7. Kopasakis, G.: Atmospheric turbulence modeling for aero vehicles: fractional order fits. NASA Center for Aerospace Information, Glenn Research Center Cleveland, Ohio, pp. 1–12, Dec 2010

8. Sagirow, P.: Stochastic Methods in the Dynamics of Satellite. Int. Centre for Mechanical Sciences, Springer, Berlin (1970)
9. Kloeden, P., Platen, E.: Numerical Solution of Stochastic Differential Equation. Springer, Berlin (2013)

Hand Talk System for Deaf and Dumb Person

Vikash Kumar, Sanjeev Kumar Raghuwanshi and Ankit Kumar

Abstract This project is a prototype of data glove which has ability to convert the movement of figure into visual display and audible sound of predefined language. I have seen many people who find it difficult to communicate with other due to language problem or due to problem in vocal chord. In a diverse country like INDIA where various types of mother tongues is spoken. People form one region of country when travelled to other region find it difficult to convey their message, which make it tedious to survive them in other region. So this data glove can convert the speaker language into the listener language which make it possible to easy communication between different language speaker. There is approximately 70 million people in world who are deaf and dumb. These people use sign language to communicate with each other. But the person with no disability in vocal chord does not learn sign language, this means that not everyone can understand sign language. So it will be a tedious task for deaf or dumb person to communicate with other person having no knowledge of sign language. This data glove can be used to replace code language which is used during war. A particular command can be converted into a gesture and the commanding officer have to make particular gesture using his hand and the command is displayed on the command receiving personnel. Data glove consists of flex sensor. In this prototype I have used 3 flex sensor which produce 8 output result. You can increase no of flex sensor up to 5 on one glove which can produce 32 output. The output produce by combination of flex sensor on data glove is in analog form which is feeded to ADC channel of Arduino microcontroller. Analog to digital converter of microcontroller convert the analog input into digital form. This digital pattern is compared with stored data and then according to comparing result a

V. Kumar (✉)
GLA University, Mathura, India
e-mail: vikas.en@gla.ac.in

S. K. Raghuwanshi
IIT(ISM) Dhanbad, Dhanbad, India
e-mail: sanjeevrus77@gmail.com

A. Kumar
GLA University, Mathura, India
e-mail: ankit.kumar_en13@gla.ac.in

S. Manna et al. (eds.), *Mathematical Modelling and Scientific Computing with Applications*, Springer Proceedings in Mathematics & Statistics 308,
https://doi.org/10.1007/978-981-15-1338-1_26

particular sentence or word is displayed on a 16 * 2 LCD and same word or sentence is played on an 8 O speaker. Speaker and LCD output is for dumb people communication while only LCD is only sufficient for deaf person. This model consists of SD card which make it possible for storing as much audio as required means there will be no limitation of memory.

Keywords Arduino microcontroller · Flex sensor · Sign language · Ultrasonic sensor

1 Introduction

Dumb people use sign language to communicate with each other. Now it comes in mind that what is sign language? In all over the world different sign languages in these sign languages some different spoken languages is also available, these spoken Sign language, any means of communication through bodily movements, especially of the movement of finger, hands and arms [1–3], these sign languages is useful for the deaf and dumb person those has not able to speak and heard the sound, with using Sign language we can coarsely expressed as mere shrugs, pointing or grimaces, or it also delicately nuanced Combination of coded manual signals reinforced by facial expression and perhaps augmented by words spelled out in a manual alphabet. Wherever vocal communication is impossible, as between speakers of mutually unintelligible languages or when one or more would-be communicators is deaf, sign language can be used to bridge the gap. But this bridge has of no use if both the speaker who are communicating does not know sign language. Thus there is a need to develop some device which will convert sign language into voice and display. So this data glove compensate this need. In this project we have used a Arduino Uno board, five op-amp circuits as buffer, LCD, speaker [4–6].

2 Details of Components

A. *Flex sensors*:

Flex sensors are similar to analog potentiometer which can be implemented as a voltage divider circuit. Flex sensor has carbon resistive element inside them in the form of thin substrate. More carbon means less resistance. When the substrate is bent the sensor produces a resistance output relative to the bend radius. Thus flex sensor measures the amount of bending in terms of change in its resistance. The flex which I have used in my project is 4.5 inch long (Fig. 1).

This flex sensor have resistance of about 9.3 kΩ at straight position and when it bend above 60° [7]. It will change its resistance to 33.3 kΩ. One terminal of flex sensor is connected to 5 V pin of microcontroller and other pin is connected to op amp

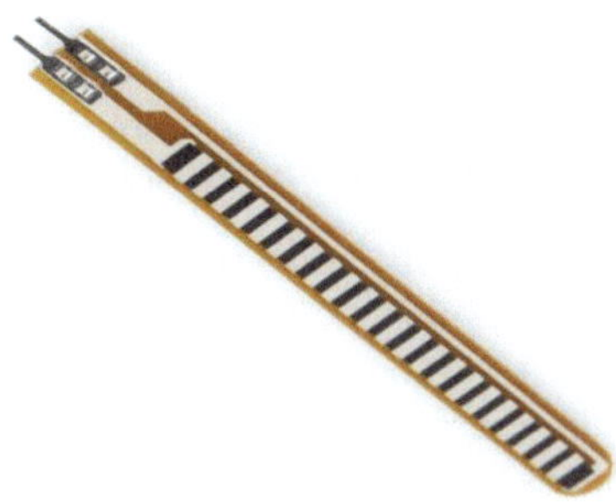

Fig. 1 Flex sensor used for gesture analysis of design hardware set-up

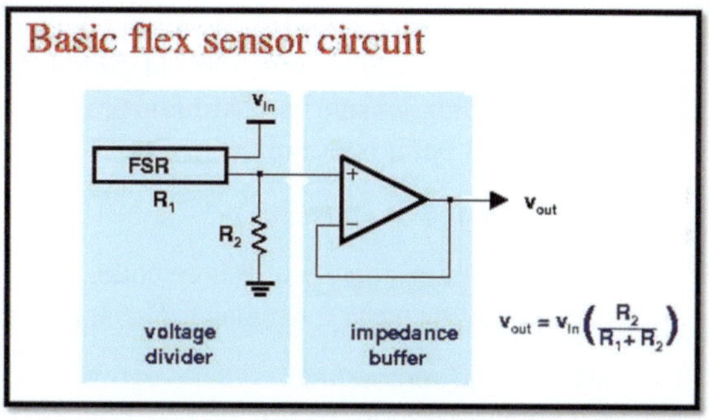

Fig. 2 Flex sensor connected with impedance buffer circuit

IC. Cost of one flex sensor is Rs. 850. This flex sensor is unidirectional 2 dimension sensor. In market bidirectional flex sensor is also available (Fig. 2).

B. *Impedance buffer*:

Buffer amplifier is simply a circuit which provides impedance transformation from input to output with the aim of the source signal being unaffected due to change of load. There are two types of buffer: the voltage buffer and the current buffer. Here in this project what we are using is a voltage buffer because we need voltage values to be fed to Arduino to take further actions with these values. A voltage buffer amplifier is used to transfer a voltage from a first circuit, having a high output impedance level, to a second circuit with a low input impedance level. The impedance buffer amplifier prevents the second circuit from loading the first circuit unacceptably and interfering with its desired operation. In the ideal voltage buffer in the diagram, the input resistance is infinite, the output resistance zero (impedance of an ideal voltage source is zero).

Voltage Relation of input and output signal of impedance amplifier.

$\text{Vout} = \{R2/(R1 + R2)\} * \text{VIN}$
Vout = Output voltage
V_{in} = input voltage

$R_2 = 10\ K\Omega$
$R_1 =$ flex sensor.

C. *Arduino Uno atmega328*:

Microcontroller is the brain of this project. Microcontroller provides platform for the interaction of SD card module, speaker, LCD, flex sensor. The analog voltage of each flex is input to the ADC channel of microcontroller. In my project I have used 3 flex sensor. Each flex sensor is connected to the impedance buffer circuit. And the output of each buffer is connected to analog pin say A0, A1, A2. Now each analog input is mapped to three different range. I have mapped A0 between 0 and 255, A1 between 255 and 512, A1 between 512 and 768. This mapping is necessary to avoid any mismatching of different digital pattern. Now the input of each sensor is set to high or low by comparing input signal with predefined value. Now with three sensor at high or low state we can have 8 different digital pattern (Table 1 and Fig. 3).

Connections of flex sensors with Arduino are shown in Fig. 4. Although the buffer is skipped in Fig. 4 but it is use in the practical circuit.

D. *Flow chart of source code*

Now it's turn for discussion about source code.

Source code follow basic c language. The source code will have following step:

Write code for LCD interfacing
Write code for SD card interfacing
Write code for Speaker interfacing
Write code for Analog voltage input.

Table 1 Configuration of Arduino Uno microcontroller

Pin no.	Function	Connected to Arduino pin no.
1	Ground (0 V)	Arduino gnd
2	Supply voltage; 5 V	Arduino +5
3	Contrast adjustment	10 k potentiometer
4	Low to write to the register; High to read from the register	8 no digital pin of Arduino
5	R/W adjustment	Ground of Arduino
6	Sends data to data pins when a high to low pulse is given	10 no digital pin of Arduino
D4		5 digital pin
D5		7 digital pin
D6		3 digital pin
D7		2 digital pin
15		5 V
16		Ground

Fig. 3 Arduino Uno controller

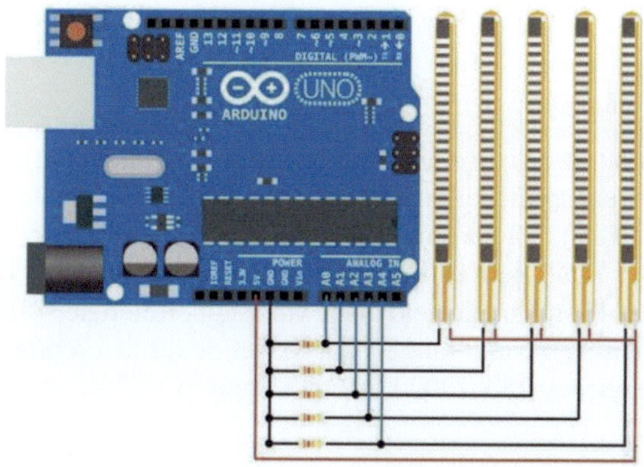

Fig. 4 Arduino Uno controller connect with flex sensor

You have to connect SD card first then connect LCD with Arduino Uno controller. As SD card pin have predefined pin connection but you can connect LCD according to available pin Fig. 5.

E. *LCD*:

LCD (Liquid Crystal Display) screen is an electronic display module which is easily programmable and economical. I have used 16 * 2 LCD it indicate the 16 character for every line for available 2 lines indicated character across LCD IS 5×7 pixel matrix. Usable LCD in our project work has two register his name is command register and data register. Use of command register in LCD has to store the command instruction and also more work like initializing LCD, setting the position of cursor, clearing screen of LCD, control the display. Second register is the data register is uses the tore the display data as in the form of ASCII code.

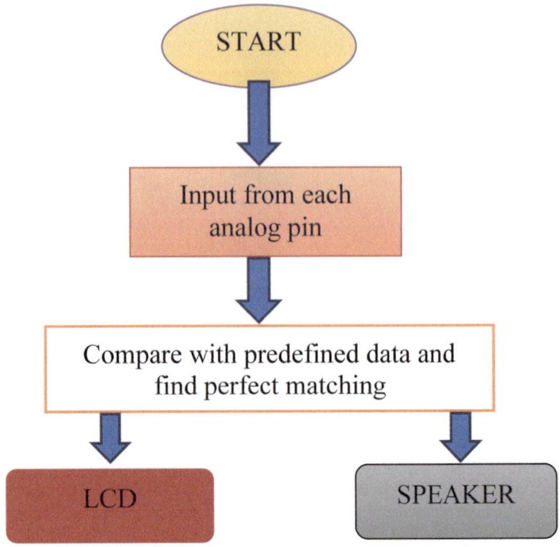

Fig. 5 Flow chart of source code

F. *Speaker*:

In our project we used the common dynamic loudspeaker, his loudspeaker had following specification a lightweight diaphragm, flexible suspension, commonly called a spider, that constrains a voice coil to move axially through a cylindrical magnetic gap. Loudspeaker is worked if input signal is in the form of electrical current is created the magnetic field in the voice coil. Reference [8] in our hardware research work we uses the mid range loudspeaker driver that reproduces the sound in the range of frequency 250–2000 Hz (Figs. 6 and 7).

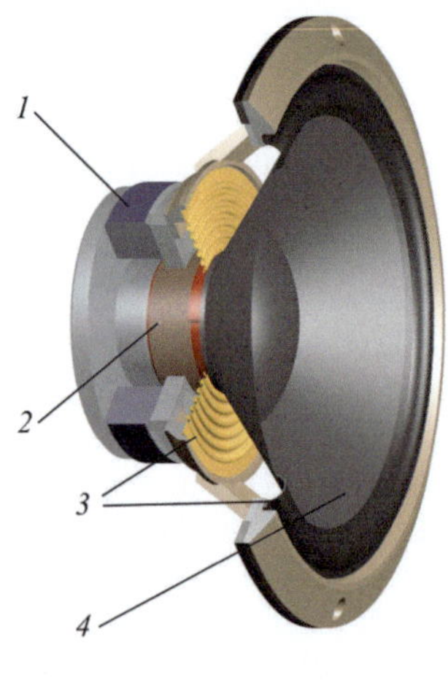

Fig. 6 Speaker is used to create the sound of fingure gesture logic

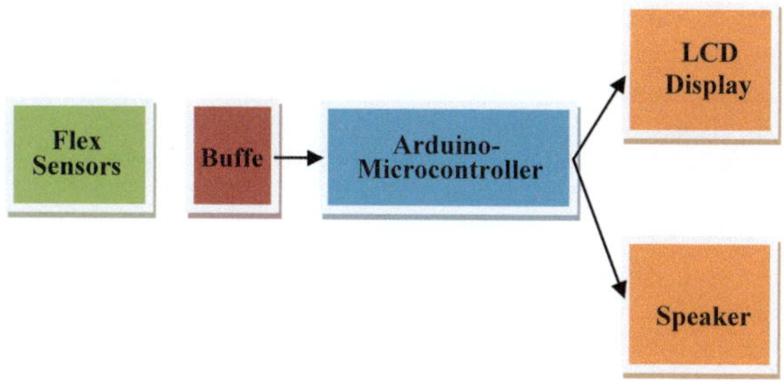

Fig. 7 Block diagram of Hardware setup for analysis of fingure gesture

3 Block Diagram for Purposed Hardware Set-up

The Flex sensors used on the hand gloves system. This hand gloves are uses by the person. Here the flex sensors can detect the fingure movement is done by the person. For movement of the fingure resistance across the flex sensor changes due to changes of resistance across flex sensor it generate electrical signal. In this process basically fingure movements convert the signal into the voltage. Conversion of voltage and resistance range after movement of fingure for each movement is recorded and it is programmed in the Arduino Uno at mega 328 as a reference voltage. The output coming from the flex sensor through buffer output is given to the Arduino Uno controller for the comparison with the reference voltage. After comparison, the measured voltage would lie in the particular range as programmed in the Arduino Uno controller. Indication of the alpha numeric number would be displayed on the LCD with using terminal software. Complete system is worked with 5 V DC supply.

4 Result and Discussion for Gesture Analysis for Hardware Set-Up

The given gesture analysis follow the various possible gestures that can be used in the project but we have used only two levels for a flex sensor. First level is when it is in straight condition and second level is when it is bent from straight position. So in this manner we can get two possible gestures with one flex sensor. Therefore total possible number of gestures with five flex sensors = 2^5 = 32 and corresponding to these gestures the recordings for various essential needs of a person (whether he needs water or food etc.) can be saved to the sd card (Fig. 8).

Although we have used only 32 gestures but the accuracy is high in our project and the reason for high accuracy can be understood as follows: as we go on increasing more levels to each flex sensor then a slight movement, bent or vibration could change the gesture and the Arduino would think of it as a change in gesture thus the output will not remain stable for slight movements and accidental vibrations in the hand.

Use hardware architecture for 3 flex sensor:

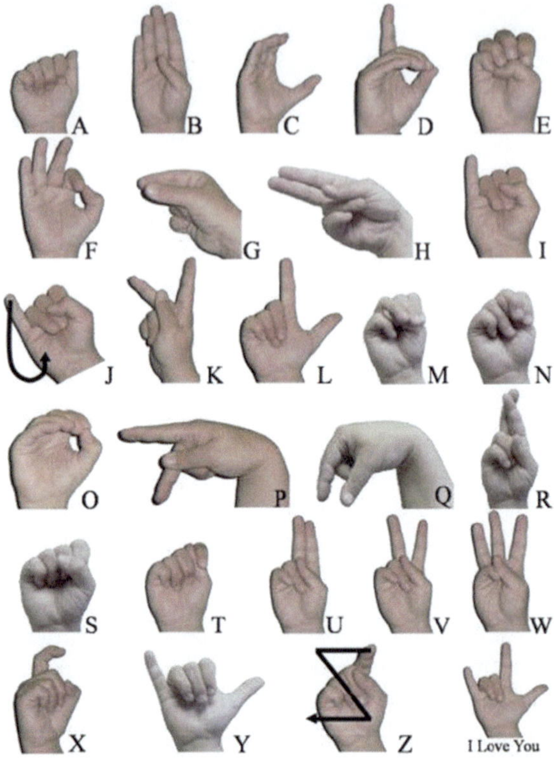

Fig. 8 Movement of fingure gesture to create the alphanumeric number that has display on LCD and create the sound through speaker

The Arduino converts the given input voltage (which is less than or equal to 5 V) into corresponding discrete values between 0 and 1023. Table 2 shows the values attained by these flex sensors corresponding to our 8 gestures.

Here N represents normal state of a flex sensor or the state when it is in straight position and B represents the bent position of a flex sensor. So for example a gesture

Table 2 Flex sensor out put for different figure gesture configuration

Gesture	Sensor1 (Ω)	Sensor2 (Ω)	Sensor3 (Ω)
NNN	106,108	350,351	606,608
NNB	100,101	346,348	605,606
NBN	116,117	354,355	611,612
NBB	148,149	349,350	606,607
BNN	112,113	348,349	605,606
BNB	118,119	351,352	608,609
BBN	123,124	352,353	609,610
BBB	124,128	352,355	609,611

written as NNB would mean that it is a gesture where first and second flex sensors are in straight position and the third flex sensor is in bend position.

Result for flex sensor output:

Deaf people face difficulty in communication with other people therefore for this purpose they may require interpreters. However interpreters cannot be with always with them because of their high costs and due to many problems faced while finding a qualified interpreter. This project "figure gesture recognizing glove" can improve and make their life significantly easy. The Aim of this project always remained to be a real world device that translates the motion and movements of the fingure in recognizable signals and messages. The proposed method can also be extended to recognize more number of gestures via using accelerometer. And 1024 gestures can be used by one glove for each hand and with the help of accelerometer the number of gesture may be raised beyond.

5 Gesture Analysis Use to Control the PC/Laptop

Hand gesture System can be used to control the laptop and PC by using two Ultrasonic sensors is connected with our purposed hardware system. Here sensor power supply is giving through Voltage regulator of Arduino. Our purposed hardware system of the Arduino microcontroller connected to the Laptop or PC for given the power and also for Serial communication. Configured connection will be done on PC as shown below. We used a double side tape to stick it on my PC but you can use your own creativity [9] (Fig. 9).

Purposed hardware system has using the python program in laptop and computer. Now you can play any movie on your computer using the VLC media player and use your hand gesture to control the movie and sound of the play video is given in Fig. 10 [10].

Fig. 9 Ultrasonic sensor connection with Microcontroller

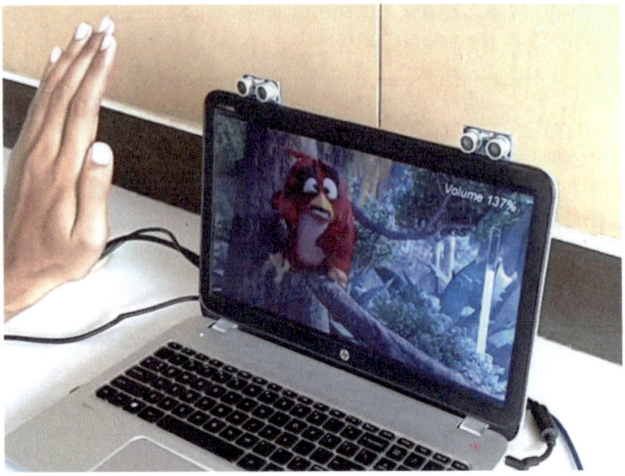

Fig. 10 Analysis of gesture use to control the volume of laptop and open the video clip

6 Conclusion

Dumb people cannot able to speak, our purposed hardware system can solve the problem by using on sign language it interpreters for communication. However, interpreters can solve the problem of Deaf and Dumb but in everyday life mainly due to high costs and difficulty in finding and scheduling qualified interpreters. This system will help them in improving their quality of life significantly. The goal of this project is to design a useful and fully functional real-world product that efficiently translates the movement of the finger with stable output. Our motivation is to help deaf and dumb people communicate more easily. The proposed method can also be extended to recognize more number of gestures via using accelerometer. And 1024 gestures can be used by one glove for each hand and with the help of accelerometer the number of gesture may be raised beyond. Our purposed hardware can be apply on to control the Laptop/Pc by using ultrasonic sensor.

References

1. Otiniano, R., Amara, C.: Finger spelling recognition from RGB-D information using kernel descriptor. IEEE Trans. Neural Syst. Rehabil. Eng. **28**(8), 124–184 (2006)
2. Zhengmao, Z., Prashan, R., Monaragala, N., Malin, P.: Dynamic hand gesture recognition system using moment invariants. IEEE Trans. Neural Netw. Comput. **21**(1), 1034–1320 (2010)
3. Preetham, C., Ramakrishnan, G., Kumar, S., Tamse, A., Krishnapura, N.: Hand talk-implementation of a gesture recognizing glove
4. Anetha, K., Rejina P.J.: Hand talk-a sign language recognition based on accelerometer and SEMG data
5. Ong, S.C.W., Ranganath, S.: Automatic sign language analysis: a survey and the future beyond lexical meaning. IEEE Trans. Pattern Anal. Mach. Intell. **27**(6), 873–891 (2005)
6. Mayura, S.K., Prasad R.K.: Gesture controlled robotic hand using flex sensor. IJSRD—Int. J. Sci. Res. Dev. **3**(7), 2321–0613 (2015) ISSN (online)

7. Saggio, G., Riillo, F., Sbernini, L., Quitadamo, L.R.: Resistive flex sensors: a survey. Smart Mat. Struct. **25**(1), 013001 (2015)
8. Kekre, M., Vaishali, K.: Speaker identification by using vector quantization. IEEE Trans. Mechatronics **15**(1), 1034–1620 (2010)
9. Elleuch, H., Wali, A., Samet, A. Alimi, A.M., 2015, December. A static hand gesture recognition system for real time mobile device monitoring. In: 2015 15th International Conference on Intelligent Systems Design and Applications (ISDA), IEEE, pp. 195–200 (December 2015)
10. Lopes, N.V., Pinto, F., Furtado, P. Silva, J.: IoT architecture proposal for disabled people. In 2014 IEEE 10th International Conference on Wireless and Mobile Computing, Networking and Communications (WiMob), IEEE, pp. 152–158 (October 2014)

Vikash Kumar received the M. tech degree in electrical engineering with specialisation in instrumentation from Indian Institute of Technology Kharagpur, India, in July 27, 2013 currently he has pursuing Ph.D. in Electronics Engineering from Indian Institute of Technology Dhanbad, India in Aug. 8 2103, he is joined an Assistant Professor in Electrical Engineering, GLA University Mathura where he is involved in several projects related to Biomedical instrumentation, MEMS and Control. His research interest includes embedded systems design and instrumentation.

Sanjeev Kumar Raghuwanshi received the Bachelor's degree in electronic and instrumentation engineering from S.G.S. I.T.S. Indore, Madhya Pradesh, India and the Master's degree in Solid State Technology from Indian Institute of Technology, Kharagpur, in Aug. 1999 and Jan. 2002, respectively. Since July 2009, he has been obtained PhD degree in the field of optics from the Department of Electrical Communication Engineering of Indian Institute of Science, Bangalore India. He is an Assistant Professor in Electronics Engineering Department of Indian Institute of Technology (Indian School of Mines) Dhanbad, India. He was the Post-doctoral Research Fellow during 2014–2015 at Instrumentation and Sensor Division, School of Engineering and Mathematical Sciences, City University London, Northampton Square, London. He receipts the Erasmus Mundus Scholarship for his Post Doc study. He is a Fellow of the Optical Society of India (OSI), Life member of IETE, member of IEEE (USA) and a Life Member of the International Academy of Physics Sciences.

PSI (ψ) Invariant Features for Face Recognition

Ajaykumar S. Cholin, A. Vinay, Aditya D. Bhat, Arnav Ajay Deshpande, K. N. B. Murthy and S. Natarajan

Abstract Over last few decades, mathematics has played a crucial role in developing efficient algorithms for Face Recognition (FR) used in biometric systems. FR using Machine Learning (ML) techniques has impacted FR systems tremendously, towards efficient and accurate models for FR. Existing FR systems used in biometrics use ML techniques to learn patterns in the images by extracting various features from them and often require pre-processed face image data for the learning process. In this paper, we have used various pre-processing techniques and compared them in the deployed FR framework. It was observed that the Steerable Pyramid (SP) filter was the most efficient pre-processing technique among all techniques used for pre-processing in this work. Though existing feature extraction methods such as SIFT (Scale-Invariant Feature Transform), SURF (Speeded-Up Robust Features), ORB (Oriented FAST and Rotated BRIEF) have been used in the past, they have not been accurate enough in various vision based biometric systems. Hence, a novel PSI (Pose Scale and Illumination) invariant SURF-RootSIFT technique is proposed by extending the well known SIFT-RootSIFT feature extraction technique which is achieved by calculating the Bhattacharya Coefficient between the feature vectors. A framework which uses the proposed novel feature extraction technique is deployed in this work. This paper demonstrates that the novel SURF-RootSIFT based framework is proven to perform more accurately and efficiently than the other techniques, with 99.65, 99.74 and 97.93% accuracy on the Grimace, Faces95 and Faces96 databases respectively.

Keywords Face recognition · Image pre-processing · SIFT · SURF · RootSIFT · Bhattacharya coefficient · VLAD

A. S. Cholin (✉) · A. Vinay · A. D. Bhat · A. A. Deshpande · K. N. B. Murthy · S. Natarajan
Centre for Pattern Recognition and Machine Intelligence, PES University, 100 Feet Ring Road, Banashankari Stage III, Bengaluru, Karnataka 560085, India
e-mail: cholinajay@gmail.com

S. Manna et al. (eds.), *Mathematical Modelling and Scientific Computing with Applications*, Springer Proceedings in Mathematics & Statistics 308,
https://doi.org/10.1007/978-981-15-1338-1_27

1 Introduction

Face recognition, the problem of identifying the face of a person by digital means has been a trending area of research for quite some time now. Face recognition is applied extensively in lots of applications in our day to day lives, for example, a secured lock system for mobile phones, authenticating bank transactions, face tracking in public places for surveillance purposes, etc. The techniques proposed to this day have been shown to be quite successful in solving problems like identifying a person based on his/her face. Over the years, quite a lot of research has been done on using different machine learning techniques for face recognition using some classification method for learning the faces of different people in a given data. Present-day techniques still pose a plethora of problems, some of them being low accuracy of classification due to noise in the image and varying illumination levels. So, a lot of work is being done in the field of face recognition and image processing to tackle such problems still posed by the present-day techniques and frameworks. Therefore, in this paper we propose Steerable Pyramid filter (SP) as a pre-processing technique for input faces along with the novel SURF-RootSIFT as a feature extraction technique which together have proven to be PSI (Pose Scale and Illumination) invariant and can be used in a real-time FR systems.

2 Background and Related Work

FR was first attempted in the late 90s. Back then, not much importance was given to building efficient FR techniques. FR was thought to be as simple as detecting nose, mouth, eyes etc. Later on, due to the revolutionary improvement in the field of mathematics, people started applying mathematics to improve the FR methods. The concept of interest points (key points), various metric measures such as Euclidean distance, Manhattan distance, Minkowski distance, etc. were used for FR [1]. From the past one or two decades, the concept of pattern recognition and machine intelligence is ruling the world. Due to the availability of humongous image data, the concept of big data, machine learning, deep learning are often employed in FR techniques to build efficient FR models.

Traditionally, ML based FR has been defined as a process made up of three stages: (1) Image pre-processing, (2) Feature extraction and (3) Training a model on the image features followed by classification. Image pre-processing is one of the most important stages in FR to get the fine grained image which is robust to changes in illumination and rotation. A lot of pre-processing techniques have been used in FR like applying dimensionality reduction such as PCA [2] and LDA [3, 4] to get the Eigenfaces and Fisherfaces respectively. Also, various filters are applied to face images such as Gaussian filter [5], Laplacian filter [6] etc. Though these pre-processing techniques seem to be working fine, they still cannot produce efficient results. So, they cannot be used in real time applications where accuracy and time is of at most concern. Therefore, in this paper, we propose a pipeline for face recognition which makes use of the Steerable Pyramid filter [7] at the pre-processing stage which is proven in the results section to perform better both in terms of speed and accuracy, producing accurate results. We also did a comparative study considering other filters such as Gabor Filter [8] and Cellular Neural Networks [9, 10] at the pre-processing stage.

Feature Extraction from the pre-processed images decides the parameters of learning for ML models and also gives the complete description of the image with its descriptors. SIFT [11–13], SURF [14, 15] and ORB [16] are the well known feature extraction techniques used in FR. SIFT and SURF are both computationally inefficient in terms of time and accuracy. Though ORB is quick enough to satisfy the time constraints thereby performing better than SIFT and SURF, it still does not produce accurate results. In this paper we propose a novel technique SURF-RootSIFT by extending the well known SIFT-RootSIFT [17] feature extraction technique. It has been shown in results section that SURF-RootSIFT is the most accurate feature extraction technique.

The Steerable Pyramid [7] is a linear multi-scale, multi-orientation image decomposition that provides a useful front-end for image-processing and computer vision applications. This is suitable for pre-processing images in an image classification system.

The existing work on face recognition done by Abhishree et al. [18] employs Gabor Filters for the feature extraction process. Gabor Filters with varying filter parameters were used for extracting features out of an image to tackle the problem of distortions due to varying illumination, pose and expression. Sharif et al. [19] proposed a method of using Gabor Filters for face recognition where they filtered each image using a set of Gabor Filters and used points with maximum intensity for face matching. In our work, we use the Gabor Filter for pre-processing the input image and detecting the edges in the image.

VLAD (Vector of Locally Aggregated Descriptors) [20, 21] is an extension of the BOW (Bag of Words) [21] representation, where the residuals of descriptors with respect to the centroid of it's closest cluster is accumulated. This adds more information in the feature vector and improves the discrimination ability of the classifier.

Random Forest [22] is an ensemble learning method which uses multiple decision trees for classification and regression. Multiple decision trees are constructed by taking subsets of the data. During the testing phase, the classification of a query instance is obtained based on the classifications predicted by the individual trees. This work uses Random Forest for classification and selects the classification that had the majority vote among the trees as the classification of the query instance.

3 Proposed Framework

The block diagram in Fig. 1 describes the proposed pipeline for face recognition. This section elaborates the methodologies and approaches adopted at every stage.

Pre-processing is done on the input image using Gabor Filter/Cellular Neural Networks/Steerable Pyramids. Feature extraction is then carried out on the pre-processed image using the proposed SURF-RootSIFT technique. The features extracted are clustered and then aggregated using VLAD, which is then used to train a Random Forest Classifier. Later, the match is classified as TP (true-positive), TN (true-negative), FP (false-positive) or FN (false-negative).

3.1 Pre-processing

3.1.1 Gabor Filter

Gabor filters [8] are a class of bandpass filters used to perform texture analyses, edge detection, feature extraction etc. The filters make use of the Gabor transform which is a kind of Fourier transform. For this work, we have used a two-dimensional Gabor filter which is basically a product of a complex sinusoid and a Gaussian function. It can be said that a Gabor filter is a Gaussian function modulated with a complex sinusoid.

The response function of the Gabor Filter can be written as:

$$g(x, y, \lambda, \theta, \psi, \sigma, \gamma) = \mathrm{e}^{\frac{-1}{2\sigma^2}(x'^2+\gamma^2 y'^2)}\mathrm{e}^{i\left(\frac{2\pi x'}{\lambda}+\psi\right)} \tag{1}$$

where,

$$x' = x\cos(\theta) + y\sin(\theta) \tag{2}$$

$$y' = -x\sin(\theta) + y\cos(\theta) \tag{3}$$

λ is the wavelength of the sinusoid,
θ is the rotation of the filter,
ψ is the phase shift,

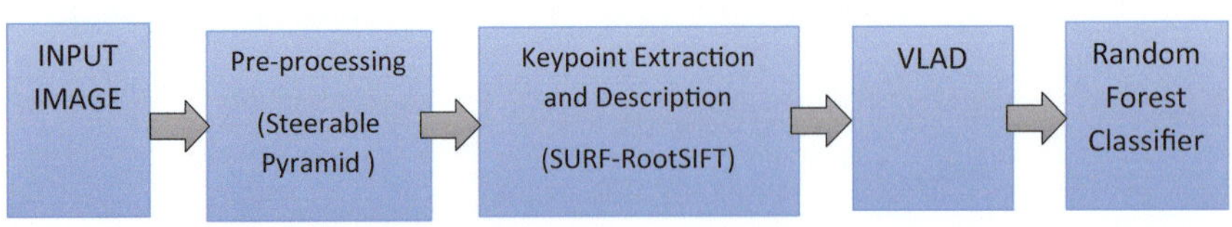

Fig. 1 Block diagram of proposed framework

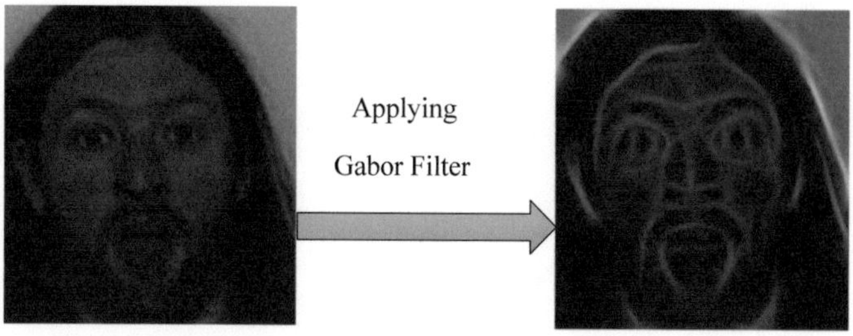

Fig. 2 Image after applying Gabor Filter

σ is the standard deviation of the Gaussian function and
γ is a parameter that defines the shape of the filter, namely the spatial aspect ratio.

The input images were pre-processed with the help of sixteen different Gabor Filters with θ varying from 0 to $\frac{15\pi}{16}$ in steps of $\frac{\pi}{16}$. The other parameters used ($\lambda = 10$, $\psi = 0$, $\sigma = 4$, $\gamma = 0.5$) were kept the same for all the sixteen filters. The output image was obtained by taking a combination of the convolution of the sixteen filters with the input image. Figure 2 shows the result of applying this pre-processing step to an input image.

3.1.2 Cellular Neural Networks

Cellular Neural Networks (CNN) [9, 10] are a class of neural networks in which communication is allowed between neighbouring units only. Chua and Yang introduced them in their two-part 1988 article [9, 10].

A cellular neural network of size M × N is a network with MN units in M rows and N columns. Each unit is connected to only its neighbouring units. These units are called cells. The r-neighbourhood of a cell is defined as:

$$N_r = \{C(k, l);\ \max(|k - i|, |l - j|) \leq r \forall k \in \{1 \ldots M\} \forall l \in \{1 \ldots N\}\} \quad (4)$$

In this paper, we use the CNN model for pre-processing the image introduced by Crounse et al. [23]. The notations in the following equations are borrowed from the field of pattern recognition. The behaviour of the CNN is modeled using the following equation:

$$\frac{dx_{ij}(t)}{dt} = -x_{ij}(t) + \sum_{C(k,l) \in N_r(i,j)} A(i, j; k, l) y_{kl}(t) + \sum_{C(k,l) \in N_r(i,j)} B(i, j; k, l) u_{kl}(t) + I_{ij} \quad (5)$$

and the output is modeled as,

$$y_{ij} = f(x_{ij}) = \frac{1}{2}(|x_{ij} + 1| - |x_{ij} - 1|) \tag{6}$$

where,

y_{kl} is the set of outputs from the cells in N_r,
u_{kl} is the set of inputs from the cells in N_r,
x_{ij} is the state value of the cell $C(i, j)$,
$A(i, j; k, l)$ are the template parameters for the feedback of the outputs from neighbouring units,
$B(i, j; k, l)$ are the template parameters for the inputs from the neighbouring units and
I_{ij} is the bias of the cell.
Initially, $x_{ij} = 0$ and $u_{ij} = 0$ for all integers i and j, if not defined for the given data.
For a linear system $|x_{ij}| < 1$, and so, $y_{ij} = x_{ij}$

The above equation for the dynamics of the CNN can be now written as:

$$\frac{dx_{ij}(t)}{dt} = -x_{ij}(t) + \sum_{C(k,l) \in N_r(i,j)} A(i, j; k, l) x_{kl}(t) + \sum_{C(k,l) \in N_r(i,j)} B(i, j; k, l) u_{kl}(t) + I_{ij} \tag{7}$$

Figure 3 shows the working of the CNN system. The input to the function f(x) is the solution of the state equation, where A in Eq. (7) corresponds to the feedback template and B in Eq. (7) corresponds to the control template.

x_{ij} is calculated and then fed as input to the block labeled $f(x)$ which calculates y_{ij}, the output value.

The parameters, and together form a template. By changing the template parameters, we can carry out different kinds of operations on the image. In this work,

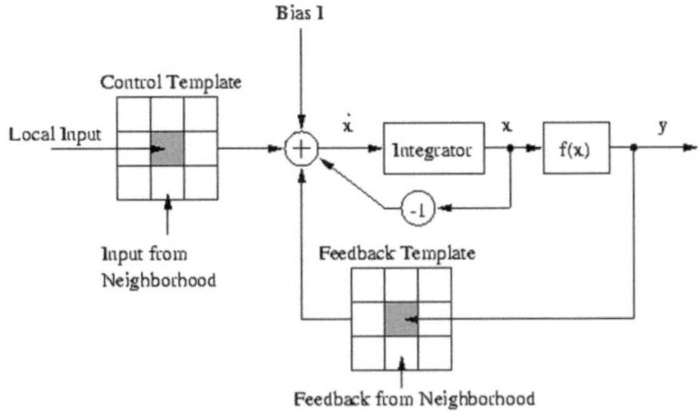

Fig. 3 Working of a cellular neural network

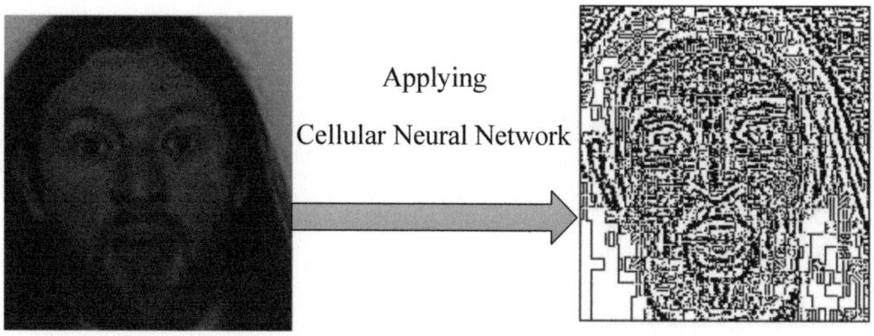

Fig. 4 Image after applying cellular neural network

edge detection on the input image is carried out as the pre-processing step with the template parameters for the cellular neural network defined as follows.

$$A = \begin{pmatrix} 0 & 0 & 0 \\ 0 & 1 & 0 \\ 0 & 0 & 0 \end{pmatrix}, \quad B = \begin{pmatrix} -1 & -1 & -1 \\ -1 & 8 & -1 \\ -1 & -1 & -1 \end{pmatrix}, \quad z = -1 \tag{8}$$

Figure 4 shows the usage of the above template parameter on an input image.

3.1.3 Steerable Pyramids

Steerable Pyramids (SP) [7] is a type of band-pass filter that provides multi-scale, multi-orientation decomposition of an image and thus can be used effectively in pre-processing images for computer vision applications. Here, a bank of steerable pyramid filters is used at each level of the pyramid.

The Kth order directional derivative in $K+1$ orientations forms the basis function of the steerable pyramid and thus can be used to form any number of orientation bands. It is over complete but has the advantage of being a tight transform and also being able to eliminate aliasing [7].

The polar form of the Fourier magnitude of the ith bandpass filter is given by,

$$B_i(\vec{\omega}) = A(\theta - \theta_i)B(\omega) \tag{9}$$

where, $\theta = \tan^{-1}(\omega_y/\omega_x)$, $\theta_i = \frac{2\pi}{k}$ and $\omega = |\vec{\omega}|$.

The desired derivative order is used to determine the angular portion of decomposition $A(\theta)$. The directional derivative in the spatial domain, can be thought of as multiplication by a ramp function in the Fourier domain, and is given in polar co-ordinates as,

$$-j\omega_x = -j\omega\cos(\theta) \tag{10}$$

For the Nth order directional derivative, the angular portion of the filter is $\cos(\theta)^N$.

The Radial function $B(\omega)$ is constrained by the need to prevent aliasing as well as the need to build the decomposition recursively.

The constraints on the filter are as follows:

$$\text{Bandlimiting: } L_1(\omega) = 0 \; for \; |\omega| > \pi/2 \tag{11}$$

$$\text{Flat system response: } |H_0(\omega)|^2 + |L_0(\omega)|^2 \lfloor |L_1(\omega)|^2 + |B(\omega)|^2 \rfloor = 1 \tag{12}$$

$$\text{Recursion: } |L_1(\omega/2)|^2 = |L_1(\omega/2)|^2 \lfloor |L_1(\omega)|^2 + |B(\omega)|^2 \rfloor \tag{13}$$

Figure 5a shows the block diagram for decomposition. Initially, L_0 *and* H_0 filters are used to separate the image into low and high pass sub-bands respectively. Band-pass images obtained by pre-processing an image by using Steerable Pyramids are shown in Fig. 5b.

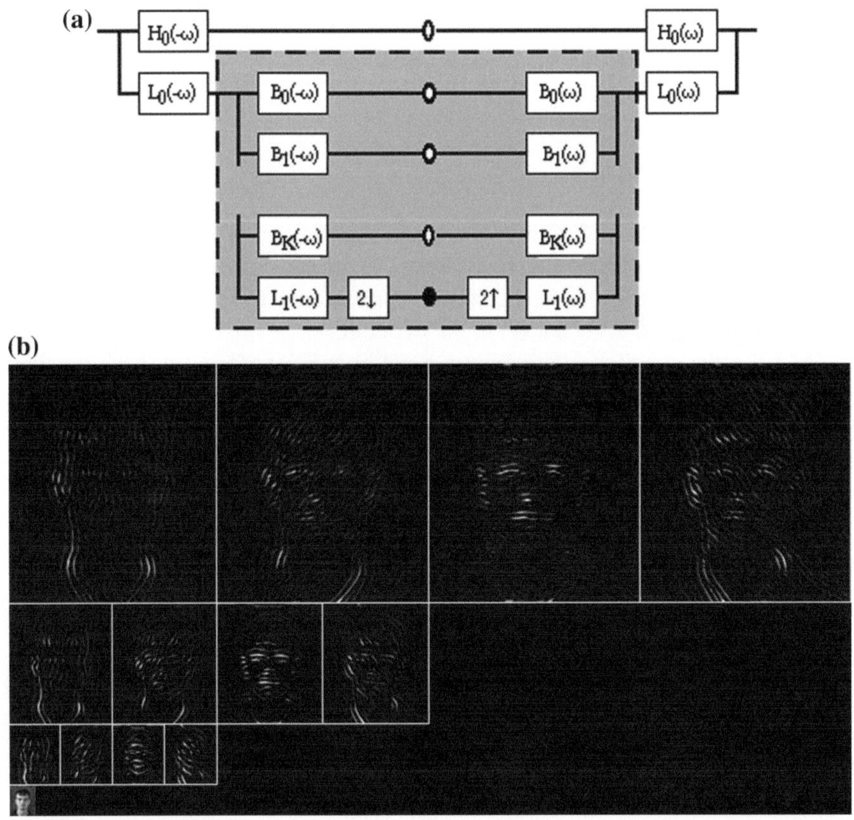

Fig. 5 **a** Block diagram for decomposition. **b** Bandpass images obtained by applying steerable pyramid to an image

3.2 *Key-Point Extraction and Description*

3.2.1 Root-SIFT

It is well known that in the design of SIFT [11–13], Euclidean distance is used as the metric measure. In a domain such as image categorization or texture classification, the performance of Euclidean distance to compare histograms has been unsatisfactory when compared to other distance measures like chi square or Hellinger. The fact that SIFT is a histogram [13], drives us to examine whether it is beneficial to try out the other histogram distance measures [17]. It is indeed observed that Hellinger Kernel is more efficient when compared with Euclidean distance. The standard dependencies between the metrics (distances) and kernels are found to be helpful as seen below.

Let us consider x and y as n-vectors each having unit Euclidean norm ($\|x\|_2 = 1$). The Euclidean distance and the similarity (kernel) between them is represented as $d_E(x, y)$ and $S_E(x, y)$ respectively. Then, the relation between $d_E(x, y)$ and $S_E(x, y)$ is given by,

$$d_E(x, y)^2 = \|x - y\|_2^2 = \|x\|_2^2 + \|y\|_2^2 - 2x^T y = 2 - 2S_E(x, y) \tag{14}$$

where $S_E(x, y) = x^T y$, and $\|x\|_2^2 = \|y\|_2^2 = 1$. Our interest is to make use of the Hellinger kernel instead of Euclidean kernel. To achieve this, the $L1$ normalization of the histograms (SIFT descriptors) is computed and then the Hellinger kernel (also known as Bhattacharyya's coefficient) for these $L1$ normalized histograms x and y $\left(i.e. \sum_i^n x_i = 1\, and\, x_i \geq 0\right)$, is defined as:

$$H(x, y) = \sum_{i=1}^{n} \sqrt{x_i y_i} \tag{15}$$

By employing a simple two step algebraic manipulation, the computed SIFT vectors (descriptors) can be compared with Hellinger Kernel. The first step involves L1 normalization of the SIFT vector and the second involves taking the square root of each element resulting in,

$$S_E\left(\sqrt{x}, \sqrt{y}\right) = \sqrt{x}^T \sqrt{y} = H(x, y) \tag{16}$$

It is seen that the resulting vectors are L2 normalized since $S_E\left(\sqrt{x}, \sqrt{x}\right) = \sum_i^n x_i = 1$. Hence, we ended up taking the element-wise square root of the L1 normalized SIFT vectors, thereby defining a new descriptor termed RootSIFT. The crucial significance here is that using the Euclidean kernel (distance) for comparing RootSIFT descriptors is analogous to comparing original SIFT vectors by making use of Hellinger Kernel. This is illustrated by the following equation,

$$d_E(\sqrt{x}\sqrt{y})^2 = 2 - 2H(x, y) \tag{17}$$

RootSIFT is favorable to be employed in any FR framework in the place of SIFT to improve it's accuracy. In fact, some of the well-known methods such as k-means which are based on Euclidean distance work effortlessly well with Root-SIFT descriptors for constructing visual vocabulary (since RootSIFT is also based on Euclidean distance). In addition to this, some of the other techniques that pose Euclidean distance as the only requirement on SIFT such as approximate nearest neighbor methods (in systems that require larger vocabularies), query expansion, soft assignment of descriptors to visual words [24, 25] and its derivatives [26–29] can still work very well with RootSIFT. The considerable improvements in efficiency and performance of RootSIFT has been shown in Sect. 6. These enhancements come with no additional storage or cost as the conversion from SIFT to RootSIFT seemed to be an easy task [17].

3.2.2 SURF-Root-SIFT

SURF (Speeded-Up Robust Features) [14, 15] has proven to be a computationally less expensive substitute to SIFT. Considering this, we make use of SURF to compute the keypoints in the image. These keypoints can now be used with Root-SIFT as explained above in Sect. 3.2.1 to generate descriptors corresponding to the keypoints computed by SURF. From the results shown in Sect. 6, we can see that this method gives better result than other keypoint detection and description methods.

3.3 VLAD

Techniques like Bag of words (BOW) [21] is based on a Natural Language Processing concept of representing a document as a collection of words. This concept is extended to image classification where a vocabulary is created to simplify images by representing it as a collection of features that can best describe it. A clustering technique such as k-means is used to create a codebook of k-centroids. The BOW representation is then a histogram of a number of descriptors assigned to each centroid. This representation has an the advantage of generating a fixed length vector irrespective of the number of detections and also benefits from powerful local descriptors.

VLAD (Vector of Locally Aggregated Descriptors) [20, 21] is a first-order extension of the BOW model. As in BOW, we first create a codebook of k-centroids using k-means given as:

$C = \{c_1, \ldots, c_k\}$, where k is the number of centroids.

The local descriptor of the image, x is assigned to its nearest neighbor. Feature quantization is now carried out by accumulating the residuals $x - c_i$, for each x, where $c_i = NN(x)$.

The VLAD vector v of the image is now given by:

$$v_{i,j} = \sum_{x\,such\,that\,NN(x)=c_i} x_j - c_{i,j} \tag{20}$$

where $i = 1, 2, 3, \ldots, k\ and\ j = 1, 2, 3, \ldots, d$

Here, d is the dimension of the local descriptor and k is the number of clusters. The dimension of our representation is given by $D = k \times \mathrm{d}$.

L_2 norm is applied on the vector v,

$$v = \frac{v}{\|v\|_2} \tag{21}$$

3.4 Random Forest

Random forest (RF) is one of the machine learning algorithms used for supervised learning. It falls under the category of ensemble learning methods and can be used for both classification and regression. For classification, the random forest generates multiple classification trees. These trees are used to classify an input query instance into a class. Each classification tree predicts a class for the instance and votes for that class. The class with the majority vote is assigned to the query instance and is the prediction of the random forest for that query instance.

This work involves use of random forests for image classification. It has been observed in the results section that Random Forest method has higher accuracy compared to other classification techniques namely SVM (Support Vector Machines) and Gaussian Naïve Bayes (GNB) on the considered benchmark datasets.

4 Algorithm

Training phase:
1. For each image in the dataset:
Apply pre-processing on the image using Gabor Filter / Cellular NN / Steerable Pyramids.
2. For each class in dataset (after pre-processing):
Extract and aggregate Root-SIFT/SURF-RootSIFT descriptors of the image.
3. Construct a vocabulary of k-clusters from the aggregated descriptors by using K-means algorithm.
4. For each image:
Initialize a VLAD Matrix(K x D)
For each feature in the image:
Find the centroid closest to the feature vector by using the L2 norm.
Calculate the difference between the feature vector and the centroid.
Add the difference to the row of the VLAD matrix.
Use L2 norm to normalize the VLAD matrix.
Reshape the matrix to a 1×KD row matrix.
5. The VLAD vectors and the labels corresponding to them are used to train a Random Forest Classifier.
Testing phase:
1. For each image : apply pre-processing using Gabor Filter / Cellular NN / Steerable Pyramids.
2. For each image :
Generate Root-SIFT/ SURF-RootSIFT descriptors.
Calculate the VLAD vector as given above.
Use the Random Forest Classifier to predict the label corresponding the the obtained vector.

5 Databases

To evaluate the framework we used three benchmark databases namely Faces96 [30], Grimace [31] and Faces95 [32]. Each database consists of images of equal size. The Faces96 database consists of images of varying inclination, illumination and size of face area. Grimace consists of sequence of images similar to Faces96 where the subject adjusts his/her head and makes grimaces. The Faces95 database consists on images of varying illumination, background, head-scale, lighting and expression. Figure 6 illustrates a sample of images (Fig.7).

6 Results and Inferences

In this section, we report the experimental results obtained on benchmark databases making use of the proposed technique. Comparison of accuracy obtained using the proposed techniques have been tabulated in Table 1. It can be seen that making use of the Steerable Pyramid filter along with the proposed SURF-RootSIFT technique gives a better accuracy. Gabor Filter too, gives good results when compared to results obtained without using any pre-processing. Although Cellular Neural Networks were able to detect highest amount of keypoints as shown in Fig. 8, it gave low accuracy on the benchmark datasets (Fig. 9).

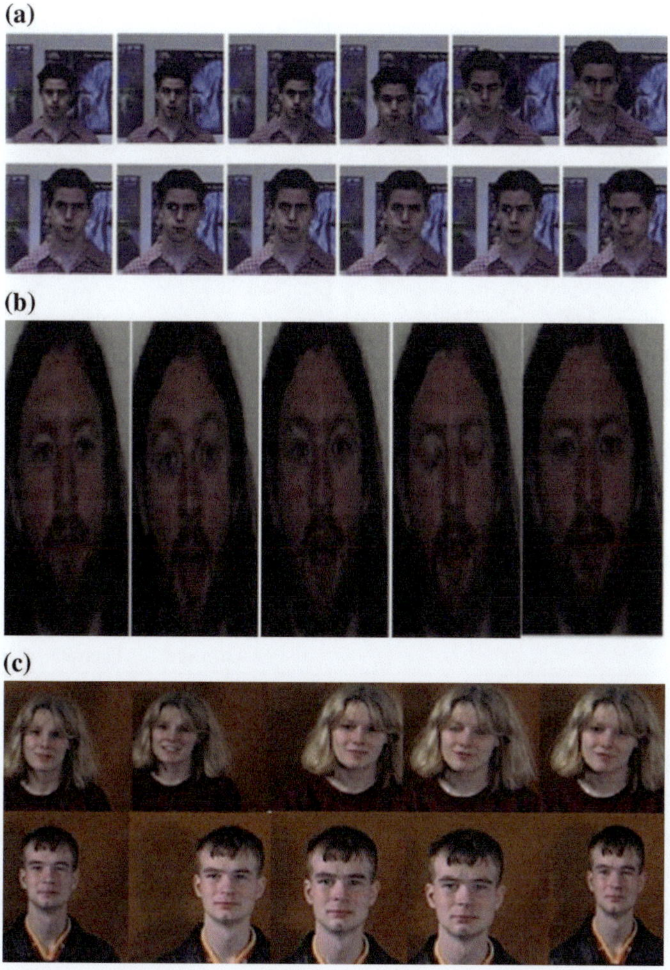

Fig. 6 a Sample of images from Faces96 database; **b** Sample of images from grimace database; **c** Sample of images from Faces95

From Table 2, it can be seen that the proposed technique performs better than other methods. SURF based techniques are seen to perform better than SIFT based techniques and the SURF-RootSIFT technique gives the highest accuracy. VLAD being an extension of BOW gives better results than the latter, and is seen to improve accuracy by about 5%. The proposed technique SP-SURF-RootSIFT-VLAD-RandomForest has the highest accuracy.

From Fig. 7a and b we can see that the number of keypoints detected by using SURF is clearly higher than those detected using SIFT on a normal image. SURF-RootSIFT therefore outperforms SIFT-RootSIFT as it computes descriptors over a

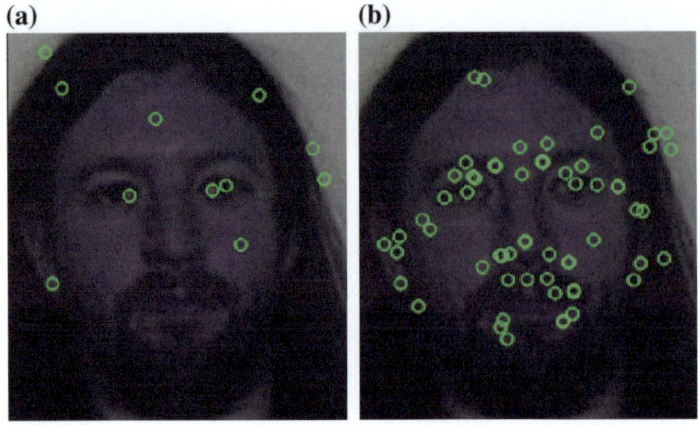

Fig. 7 **a** SIFT keypoints detected on a normal image. **b** SURF keypoints detected on a normal image

Table 1 Comparison of results obtained on benchmark databases

Method	10-fold cross validation accuracy (%)		
	Grimace	Faces95	Faces96
RootSIFT-VLAD-RandomForest	95.27	94.42	89.23
Gabor-RootSIFT-VLAD-RandomForest	97.22	96.49	92.69
Cellular NN-RootSIFT-VLAD-RandomForest	51.88	50.37	45.45
SP-RootSIFT-VLAD-RandomForest	97.46	98.49	95.62
SURF-RootSIFT-VLAD-RandomForest	97.89	98.27	94.61
Gabor-SURF-RootSIFT-VLAD-RandomForest	99.66	99.49	96.53
Cellular NN-SURF-RootSIFT-VLAD-RandomForest	60.38	58.78	54.91
SP-SURF-RootSIFT-VLAD-RandomForest	**99.65**	**99.74**	**97.93**

fairly higher number of keypoints. The results tabulated in Table 1 show that SURF-RootSIFT has a better accuracy. From the figures shown in Fig. 10, we can see that using a suitable pre-processing techniques such as Gabor Filter or Steerable Filter can improve the keypoints detected in the image which can improve the overall accuracy of the classification system.

Figure 8 shows a comparison of average number of keypoints detected in various pre-processing techniques used. We can see that using Cellular Neural Networks increases the number of keypoints detected drastically. Gabor Filter too, increases the number of keypoints when compared to that of a normal image, whereas using Steerable Pyramid gives approximately the same number of keypoints as in a normal image.

Figure 9 shows a comparison of time taken for feature extraction across the various pre-processing techniques used. We can see that using Steerable Pyramids or Gabor

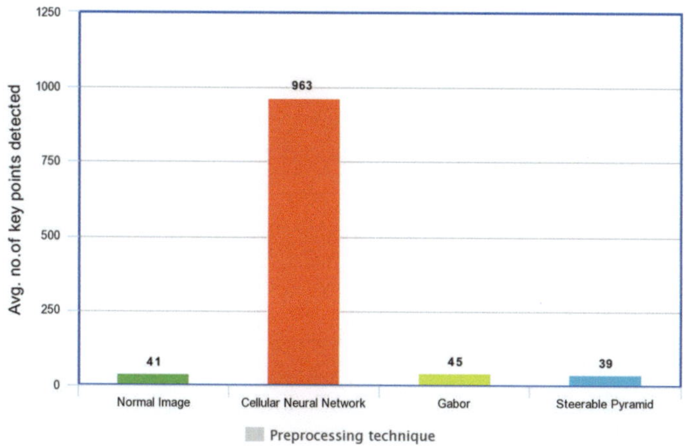

Fig. 8 Comparison of number of keypoints detected for different pre-processing techniques

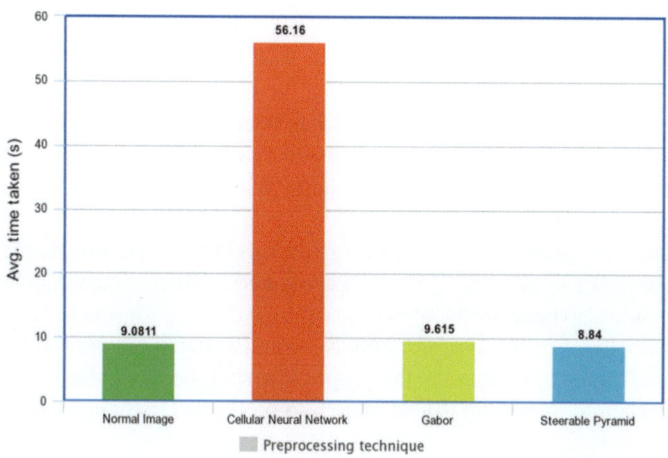

Fig. 9 Comparison of time taken for feature extraction across different pre-processing techniques

Filter decreases the time taken when compared to that of a normal image. Cellular neural networks are seen to take a lot more time and may not be suitable if the application is real-time.

7 Conclusion and Future Work

It is crucial to train classification systems for face recognition on datasets containing images varying in background, lighting, illumination, rotations and translations.

Table 2 Comparison of accuracy of proposed method with some standard techniques

	Accuracy (%)	
Method	10-fold	20-fold
SIFT-BOW-SVM	74.37	79.7
SIFT-BOW-GNB	82.77	88.05
SIFT-BOW-RandomForest	76.38	76.38
SURF-BOW-GNB	92.22	93.05
SURF-BOW-SVM	90.5	90.8
SURF-BOW-RandomForest	84.6	86.38
SIFT-VLAD-RandomForest	87.22	88.8
SIFT-VLAD-SVM	90.56	90.82
SIFT-VLAD-GNB	85.3	88.08
SURF-VLAD-SVM	94.3	94.41
SURF-VLAD-GNB	93.46	93.89
SURF-VLAD-RandomForest	93.05	93.47
SP-SURF-RootSIFT-VLAD-RandomForest	**99.74**	**99.28**

Therefore, our proposed method was tested on benchmark databases (mentioned in Sect. 5). Pre-processing is an important stage in a classification system for FR and affects the overall performance of the system. While it is important to have some pre-processing on the images, in a real-time scenario, care needs to be taken that it doesn't affect the efficiency of classification in terms of time. The proposed technique improved accuracy of classification when compared to other standard techniques by about 5% while also being efficient in terms of time taken (as observed in the results). In a real-time classification system where we would need not only real-time response but also need them to be accurate, the proposed technique would be a suitable solution. The proposed technique is also proven to be PSI invariant accounting for real time applications.

Future work is focused on making the framework more efficient in terms of speed by making use of various optimization techniques. Also, feature fusion could be incorporated into the framework.

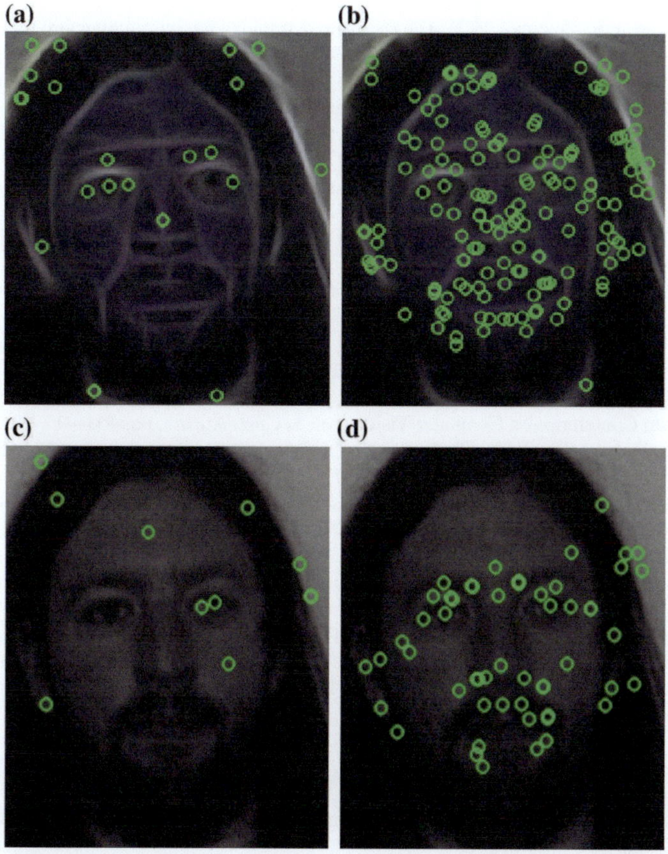

Fig. 10 **a** SIFT keypoint (using Gabor filter). **b** SURF keypoints (using Gabor filter). **c** SIFT keypoints (using steerable pyramid). **d** SURF Keypoints (using steerable pyramid)

References

1. Perlibakas, V.: Distance measures for PCA-based face recognition. Pattern Recog. Lett. **25**(6), 711–724 (2004). ISSN 0167-8655
2. Mudrová, M., Procházka, A.: Principal component analysis in image processing (2018)
3. Chen, L.F., Liao, H.Y., Ko, M.T., Lin, J.C., Yu, G.J.: New LDA-based face recognition system which can solve the small sample size problem. Pattern Recogn. **33**, 1713–1726 (2000). https://doi.org/10.1016/S0031-3203(99)00139-9
4. Yang, J., Zhang, D., Yong, X., Yang, J.Y.: Two-dimensional discriminant transform for face recognition. Pattern Recog. **38**(7), 1125–1129 (2005). ISSN 0031-3203
5. Tan, X., Triggs, B.: Enhanced local texture feature sets for face recognition under difficult lighting conditions. In: Zhou, S.K., Zhao, W., Tang, X., Gong, S. (eds.) Analysis and Modeling of Faces and Gestures. AMFG 2007. Lecture Notes in Computer Science, vol. 4778. Springer, Berlin, Heidelberg (2007)
6. He, X., Yan, S., Hu, Y., Niyogi, P., Zhang, H.J.: Face recognition using laplacian faces. IEEE Trans. Pattern Anal. Mach. Intell. **27**, 328–340 (2005). https://doi.org/10.1109/TPAMI.2005.55

7. Simoncelli, E.P., Freeman, W.T.: The steerable pyramid: a flexible architecture for multi-scale derivative computation. In ICIP, p. 3444 (1995)
8. Adak, C.: Gabor filter and rough clustering based edge detection. In: 2013 International Conference on Human Computer Interactions (ICHCI), Chennai, pp. 1–5 (2013)
9. Chua, L., Yang, L.: Cellular neural networks: theory. Circuits Syst. IEEE Trans. **35**, 1257–1272 (1988). https://doi.org/10.1109/31.7600
10. Chua, L.O., Yang, L.: Cellular neural networks: applications. IEEE Trans. Circuits Syst. **35**(10), 1273–1290 (1988)
11. Lowe, D.G.: Distinctive image features from scale-invariant keypoints. Int. J. Comput. Vision **60**(2), 91–110 (2004)
12. Jurie, F., Schmid, C.: Scale-invariant shape features for recognition of object categories. In: CVPR, vol. 2, pp. 90–96 (2004)
13. Geng, C., Jiang, X.: Face recognition using sift features. In Image Processing (ICIP), 2009 16th IEEE International Conference on, IEEE, pp. 3313–3316 (2009)
14. Bay, H., Tuytelaars, T., Van Gool, L.: Surf: Speeded up robust features. In the proceedings of European Conference on Computer Vision, vol. 1, Graz, Austria, pp. 404–417 (May 2006)
15. Bay, H., Ess, A., Tuytelaars, T., Van Gool, L.: Speeded-up robust features (SURF). Comput. Vis. Image Underst. **110**(3), 346–359 (2008)
16. Rublee, E., Rabaud, V., Konolige, K., Bradski, G.: ORB: an efficient alternative to SIFT or SURF. In: Computer Vision (ICCV), 2011 IEEE International Conference on, IEEE, pp. 2564–2571 (2011)
17. Arandjelovic, R., Zisserman, A.: Three things everyone should know to improve object retrieval (2012)
18. Abhishree, T.M., Latha, J., Manikantan, K., Ramachandran, S.: Face recognition using Gabor filter based feature extraction with anisotropic diffusion as a pre-processing technique. Procedia Computer Sci. **45**, 312–321 (2015). ISSN 1877-0509, https://doi.org/10.1016/j.procs.2015.03.149
19. Sharif, M., Khalid, A., Raza, M., Mohsin, S.: Face recognition using Gabor filters. J. Appl. Comput. Sci. Math. **11** (2011)
20. Arandjelovic, R., Zisserman, A.: All about VLAD. In: Proceedings of the IEEE Computer Society Conference on Computer Vision and Pattern Recognition, pp. 1578–1585 (2013). https://doi.org/10.1109/cvpr.2013.207
21. Jegou, H., Perronnin, F., Douze, M., Sanchez, J., Perez, P., Schmid, C.: Aggregating local image descriptors into compact codes. IEEE Trans. Pattern Anal. Mach. Intell. **34**, 1704 (2012)
22. Vishwakarma, V., Srivastava, A.: Performance improvements in face classification using random forest. Int. J. Eng. Res. Appl. **2**(3), 2384–2388 (2012)
23. Crounse, K.R., Chua, L.: Methods for image processing and pattern formation in cellular neural networks: a tutorial. Circuits Syst. I: Fundam. Theory Appl., IEEE Trans. **42**, 583–601 (1995). https://doi.org/10.1109/81.473566
24. J´egou, H., Douze, M., Schmid, C.: Improving bag-of-features for large scale image search. IJCV **87**(3), 316–336 (2010)
25. Philbin, J., Chum, O., Isard, M., Sivic, J., Zisserman, A.: Lost in quantization: improving particular object retrieval in large scale image databases. In Proceedings CVPR (2008)
26. J´egou, H., Douze, M., Schmid, C.: Hamming embedding and weak geometric consistency for large scale image search. In: Proceedings ECCV (2008)
27. J´egou, H., Douze, M., Schmid, C., P´erez, P.: Aggregating local descriptors into a compact image representation. In: Proceedings CVPR (2010)
28. Mikulik, A., Perdoch, M., Chum, O., Matas, J.: Learning a fine vocabulary. In: Proceedings ECCV (2010)
29. Philbin, J., Isard, M., Sivic, J., Zisserman, A.: Descriptor learning for efficient retrieval. In: Proceedings ECCV (2010)
30. http://cswww.essex.ac.uk/mv/allfaces/faces96.html
31. http://cswww.essex.ac.uk/mv/allfaces/grimace.html
32. http://cswww.essex.ac.uk/mv/allfaces/faces95.html

Performance Analysis of a Modified Savonius Hydrokinetic Turbine

Sourish Singha and R. P. Saini

Abstract Extraction of hydrokinetic energy from in stream flow can be an effective, sustain able and environmental friendly replacement of the conventional energy resources. Savonius hydrokinetic turbine is a vertical axis turbine, which has a poor efficiency and its operating range under various loading conditions is very short. The coefficient of performance value curves reported under different investigations on Savonius hydrokinetic turbine are highly fluctuated in respect of tip speed ratios (TSR). In order to enhance the efficiency of the turbine and to get less fluctuations in coefficient of power output curve for smooth operation under higher range of tip speed ratios, this paper aims to model a modified design of Savonius turbine with twisted blades and analyze its performance on various operating conditions. For CFD analysis commercial unsteady Reynolds-Averaged Navier-Stokes (URANS) solver in conjunction with SST k-ω turbulence model have used. Due to symmetrical cross sections on different geometrical positions 3D transient simulations are conducted to find out the average torque. Coefficient of torque and coefficient of power are analyzed and discussed. The result of this study concluded as very less deviation in power coefficient curve up to tip speed ratio 1.4, which indicates the increment of operating range and beyond that TSR value it falls down rapidly. The maximum power coefficient obtained 0.30 corresponding to a TSR value 1.4 for input water velocity of 2 m/s.

Keywords Numerical modelling · Twisted Bach type Savonius hydrokinetic turbine · Computational fluid dynamics (CFD) · Coefficient of torque · Coefficient of power

---

S. Singha (✉) · R. P. Saini
Indian Institute of Technology Roorkee, Roorkee, Uttarakhand, India
e-mail: alpsoupr.pah2016@iitr.ac.in

R. P. Saini
e-mail: rajsafah@iitr.ac.in

S. Manna et al. (eds.), *Mathematical Modelling and Scientific Computing with Applications*, Springer Proceedings in Mathematics & Statistics 308,
https://doi.org/10.1007/978-981-15-1338-1_28

1 Introduction

In the field of conventional small hydro the power output is directly related to the discharge and the net head available at the site. But it is not easy to get proper head and discharge every time for power generation. So the ample and easily available resources of hydro energy can't be fully utilized. Therefore harnessing hydrokinetic energy from locally available canal or high energy flows could be a sustainable and reliable way to electrify the rural and remote areas. Hydrokinetic technologies have their geneses in wind energy extraction technologies. Hydrokinetic turbines can also be set up for a wide range of applications including industrial outflows, irrigation canals, rivers and tidal streams. The villages which are located in close proximity to rivers hydrokinetic technology is best suitable for those areas than traditional hydropower generation [1]. The interest and development of hydrokinetic energy has grown up from last couple of decades [2].

Savonius hydrokinetic turbine is a vertical axis turbine which was developed by Finnish engineer S. Savonius in 1931 for wind applications [3]. Savonius hydrokinetic turbine, drag-type rotor starts rotating at very low fluid velocity as compared to the conventional hydraulic turbines. It has an ability to accept fluid from any direction with good starting characteristics at very low speed. Despite such advantages, Savonius hydrokinetic turbines face low efficiency and large static torque variation. According to betz limit the theoretical maximum extraction of kinetic energy is 0.59 [4].

2 Literature Reviews and Problem Formulation

The literatures of Savonius hydrokinetic turbine has been reviewed for getting the design idea of modified turbine blade structure. Zhao Z et al. reported in his numerical study that for a two bladed helical (180°) Savonius rotor the optimum C_P value can be obtained as 0.20 for 0.3 overlap ratio [5]. For an experimental investigation conducted by A. J. Alexander et al. concluded an optimum C_P value of 0.243 while using end plate and flat shield [6]. M. A. Kamoji reported C_P values ranging from 0.11 to 0.19 at wind velocities of 4 m/s up to 14 m/s respectively for a helical Savonius turbine with 90° twist angle [7]. Another experimental result obtained by M.A. Kamoji confirm the C_p values spanning between 0.171 and 0.205 for Reynolds number 8 × 104 up to 15 × 104 respectively for a modified Savonius rotor with out shaft [8]. A. Kianifar reported a C_P value 0.254 at Reynold's number 1.5 × 105 for a 2 bladed Savonius turbine in his numerical and experimental study [9]. By 2D unsteady numerical study S. Roy and U. K. Saha concluded a static C_P of 0.224 at wind velocity of 10.44 m/s [10]. C. Jian et al. concluded C_P values to be 0.173 and 0.145 for a two bladed two stages Savonius turbine [11]. O. Mojola in his experimental study under field condition resulted a C_P value of 0.267 for a TSR value of 1.0 and aspect ratio 1.52 [12]. R. E. Sheldahl reported a C_P value of 0.24 at

Reynold's number 8.64×10^5 [13]. Another experimental investigation on 2 bladed Savonius rotor conducted by M. A. Kamoji reported that C_P value of 0.157 can be obtained for Reynold's number 80,000 and aspect ratio of 1.0 [14]. U. K. Saha and M. J. Rajkumar's experimental investigation on 3 bladed Savonius turbine with aspect ratio of 1.83 concludes maximum C_p is 13.99 for twist angle 15° [15]. S. V. Ghatage and J. B. Joshi's numerical and experimental research concludes a C_P value of 0.216 for a 2 bladed rotor with twist angle of 30° [16]. J. V. Akwa et al. reported that for Reynolds number ranging from 216,750 to 867,000 calculated C_T values are found in between 0.09 and 0.22 [17].

Reviewing all those researches, which are related to Savonius turbine both in wind energy extraction field and hydrokinetic energy harnessing field it can be concluded that optimum C_P values have been obtained in the medium range of tested TSR domain. So, the turbines are optimum in medium loading conditions. Thus it is necessary to increase its optimum performance up to certain value of TSR for different loading conditions. The following research will aim for a CFD analysis of a modified bach type Savonius turbine which will be efficient, will have less fluctuations in C_P value and will be applicable for a wider range of TSR.

3 Geometry Modification and Selection of Parameters

As Savonius turbines have very poor efficiencies, researchers have modified its geometrical structure since its invention. A conventional semi-circular blade was designed by the inventor, which has received many changes during investigations by the researchers. Summing up all the changes a new turbine blade has designed for the study. The Fig. 1 shows the gradual modifications of the modern Savonius turbine. The lower most design has used for this study, which is in complex structure with respect to the conventional designs.

There are various system parameters that are used to design a Savonius blade. Twist angle of a Savonius rotor is an angle that is between the horizontal surface to the twist blade surface. Blade arc angle (ψ) is the angle between blade profiles. The number of stage indicates the number of same or different Savonius models that are attached in parallel.

Aspect ratio (α) is the ratio of height of the rotor (H) divided by its diameter (D). Blade shape factor (p/q) is defined as the length of straight blade part (p) to the blade arc radius (q) for a bach type Savonius blade.

The design of modified bach type Savonius [18] is chosen for the study because of its improving efficiency capability if some amount of twist angle is provided into it. As per as the literature [19] the optimum twist angle test was done on the conventional Savonius geometry and it is found to be 12.5°. The system parameters taken for this modelling are discussed in Table 1.

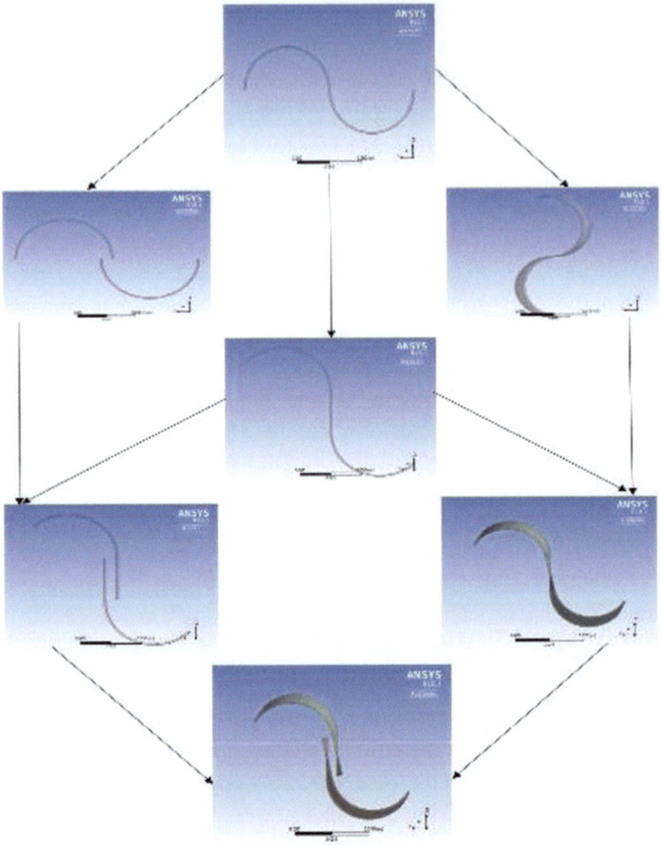

Fig. 1 Evolution of blade design of Savonius turbine used for the study

Table 1 System parameters

Aspect ratio (H/D)	1.59
Rotor diameter (D)	159 mm
Rotor height (H)	253 mm
End plate diameter (D_0)	175 mm
Enclosure diameter (D_r)	239 mm
Channel cross section (A)	550 mm × 650 mm
Number of blades	2
Number of stages	1
Blade arc angle	135°
Blade shape factor (p/q)	1
Twist angle (α)	12.5°

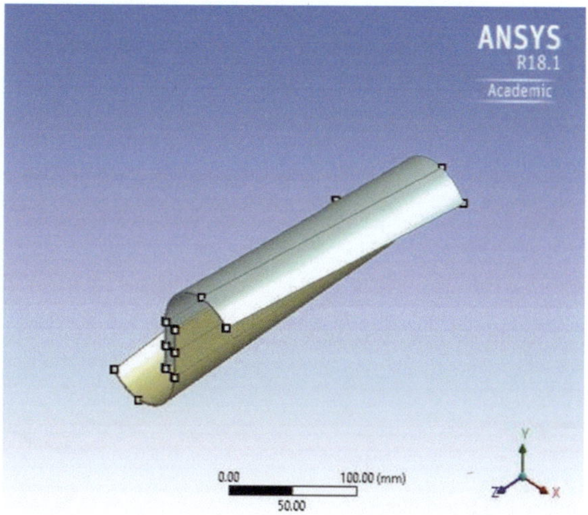

Fig. 2 Isometric view of modified twisted Savonius blade

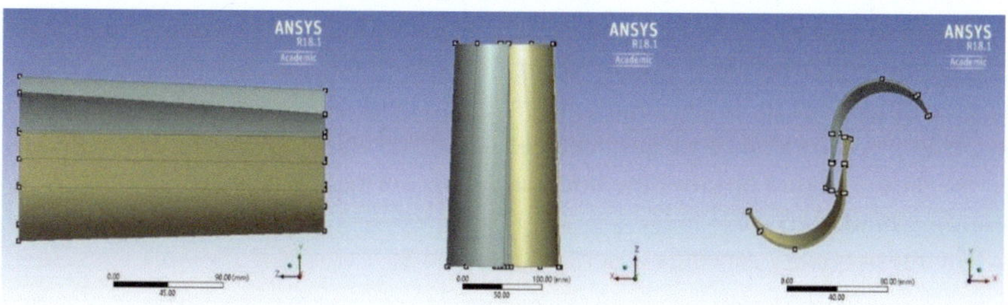

Fig. 3 The blade is shown from the X, Y and Z directions respectively

4 Modeling

The diameter of the blade (D) is kept as 159 mm for the turbine. optimum twist angle 12.5° is provided into the blade. The blade is shown in Figs. 2 and 3.

End plate diameter is taken as 1.1D and rotating zone diameter as 1.5D. The channel inlet and outlet height is 650 mm and width is 550 mm and channel length is 3000 mm. The base geometry of the model and the end plates can be seen in Fig. 4.

5 Grid Generation

The entire fluid domain is exported to ANSYS MESH (V 18.1) for non conformal unstructured grid generation. The tetrahedral elements are formed due to meshing

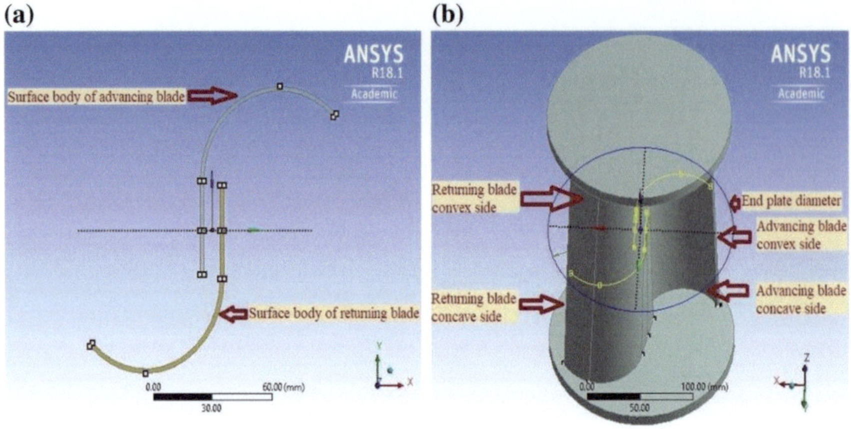

Fig. 4 **a** Base geometry and **b** end plate

of each computational subdomain. Figure 5a shows the rotating zone and stationary zone and Fig. 5b shows the dense mesh near the blade surface.

Boundary layer has been created near the blade surface for improving quality and to precisely describe the boundary layer flow characteristics. y+ value less than 1 was recommended to fix the height of first prism layer above the blade surface [20]. Accordingly, the value of y+ has been considered as less than one under the present investigation. Figure 6 shows the boundary layer formation near the advancing and returning blade surface.

The mesh quality aspect is described in Table 2:

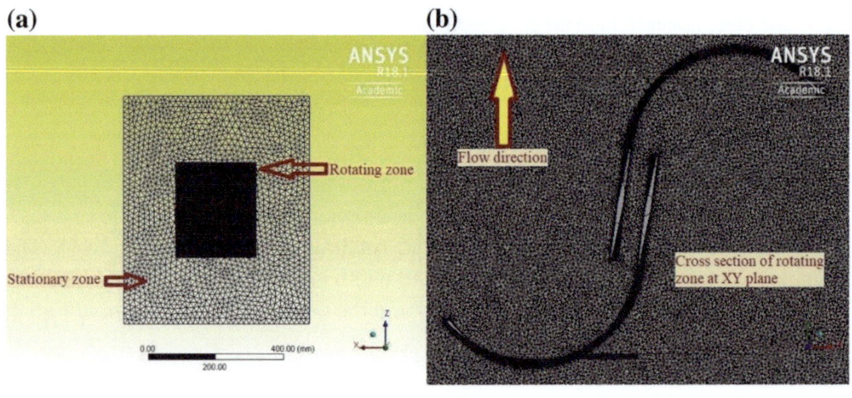

Fig. 5 **a** Rotating and stationary zone, **b** dense mesh near blade surface

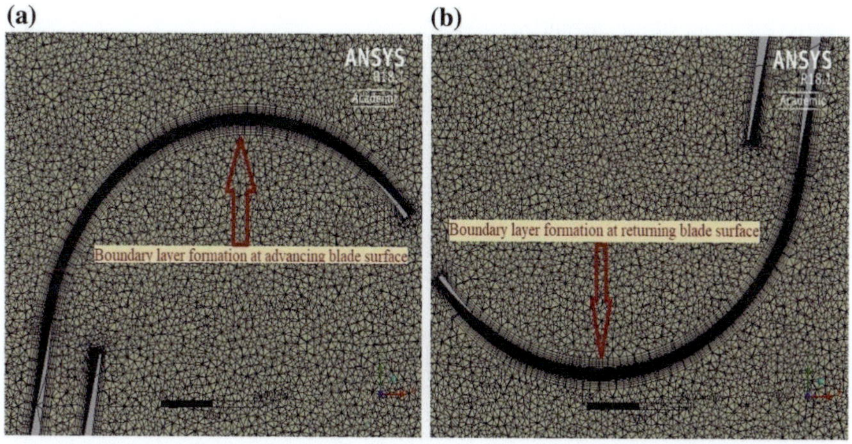

Fig. 6 Boundary layer near, **a** advancing blade and **b** returning blade

Table 2 Quality aspect

Quality aspect	Max value
Skewness	0.84
Aspect ratio	15
Orthogonal quality	0.88

6 Simulation Procedure

ANSYS CFX solver (V18.1) is used to solve the unsteady incompressible Navier-Stokes equations. The generated mesh is imported into the solver. In order to obtain a numerical solution, the case with necessary boundary conditions is defined as Table 3.

The TSR range taken for this analysis is from 0.6 up to 1.6. The TSR is calculated as,

$$\mathrm{TSR}(\lambda) = \frac{\text{Circumferential velocity}}{\text{fluid velocity}} = \frac{\omega \mathrm{D}}{2\mathrm{V}} \quad (1)$$

where,

ω is angular velocity,

Table 3 Boundary conditions

Name	Boundary type	Boundary condition
Inlet	Inlet velocity	1, 1.5 & 2 m/s
Outlet	Pressure flow	Pressure flow
Channel walls	Free slip wall	Stationary
Rotating zone	No slip wall	Rotates at desired RPM

D is the diameter of the turbine,
V is the velocity of water.

Depending upon the geometry and Reynolds number various simulation processes are available. To calculate the power, torque is considered as the output of the simulation. Shear stress transport k-ω model is used for this simulation.

The kinematic eddy viscosity is calculated as,

$$\nu_T = \frac{a_1 k}{\max(a_1\omega, SF_2)} \tag{2}$$

Turbulence kinetic energy is calculated as,

$$\frac{\partial k}{\partial t} + U_j\frac{\partial k}{\partial x_j} = P_k - \beta^* k\omega + \frac{\partial}{\partial x_j}\left[(\nu + \sigma_k\nu_T)\frac{\partial k}{\partial x_j}\right] \tag{3}$$

Specific dissipation rate is calculated as,

$$\frac{\partial \omega}{\partial t} + U_j\frac{\partial \omega}{\partial x_j} = \alpha S^2 - \beta\omega^2 + \frac{\partial}{\partial x_j}\left[(\nu + \sigma_\omega\nu_T)\frac{\partial \omega}{\partial x_j}\right] + 2(1 - F_1)\sigma_{\omega 2}\frac{1}{\omega}\frac{\partial k}{\partial x_i}\frac{\partial \omega}{\partial x_i} \tag{4}$$

Closure coefficients and auxiliary relations can be governed by the following equations,

$$F_2 = \tanh\left[\left[\max\left(\frac{2\sqrt{k}}{\beta^*\omega y}, \frac{500\nu}{y^2\omega}\right)\right]^2\right] \tag{5}$$

$$P_k = \min\left(\tau_{ij}\frac{\partial U_i}{\partial x_j}, 10\beta^* k\omega\right) \tag{6}$$

$$F_1 = \tanh\left\{\left\{\min\left[\max\left(\frac{\sqrt{k}}{\beta^*\omega y}, \frac{500\nu}{y^2\omega}\right), \frac{4\sigma_{\omega 2}k}{CD_{k\omega}y^2}\right]\right\}^4\right\} \tag{7}$$

$$CD_{k\omega} = \max\left(2\rho\sigma_{\omega 2}\frac{1}{\omega}\frac{\partial k}{\partial x_i}\frac{\partial \omega}{\partial x_i}, 10^{-10}\right) \tag{8}$$

$$\phi = \phi_1 F_1 + \phi_2(1 - F_1) \tag{9}$$

where,

$\alpha_1 = \frac{5}{9}, \alpha_2 = 0.44$
$\beta_1 = \frac{3}{40}, \beta_2 = 0.0828$
$\beta^* = \frac{9}{100}$

$\sigma_{k1} = 0.85, \sigma_{k2} = 1$
$\sigma_{\omega 1} = 0.5, \sigma_{\omega 2} = 0.856$
ω is the specific dissipation rate,
k is the turbulent kinetic energy,
F1 is the first blending function,
F2 is the second blending function,
S is the invariant measure of the strain rate,
α, β^* is the empirical constants of the SST mode, y is the distance to the nearest wall,
ν_t is the turbulence eddy viscosity etc.

7 Results and Discussion

For the water velocity 1 m/s the maximum C_T obtained is 0.78 and minimum C_T value obtained is −0.44. The average C_T values for tip speed ratios of 0.6 to 1.6 are ranging between 0.39 and 0.17. The average torque values are spanning between 0.626 and 0.269 N-m for increasing values of TSR from 0.6 to 1.6. The input power is constant which is 20.053 W. The angular velocities are calculated for the rotation of the fluid domain are staring from 72.070 RPM to 192.187 RPM. RPM values are taken up to 5 decimal places. The time step size taken for each 15° of rotation are calculated for the different TSR values spanning between 0.0347 and 0.0130 s.

The C_T or coefficient of torque is calculated from Eq. (10).

$$C_T = \frac{T}{0.5\rho A V^2 R} \tag{10}$$

where,

C_T is coefficient of torque,
T is the output torque,
ρ is the density of water, V is the velocity of water,
R is the radius of the turbine,
A is the swept area of the turbine.

For the flow velocity of 1 m/sec the corresponding Reynold's number is found out to be 17.79×10^4. Variation of C_T with respect to rotational angle is shown in Fig. 7.

The next analysis is done over the flow velocity of 1.5 m/s under different TSR values. RPM values are changed from 108.105 to 288.281. The input power is same throughout this case which is 67.679 W. The time step values taken are spanning between 0.0231 and 0.0087 depending upon the TSR ranging from 0.6 to 1.6. The average torque is found out to be spanning between 1.43 and 0.612 N-m for the same TSR values. The maximum C_P value for this case is approximately 0.298 for TSR 1.4.

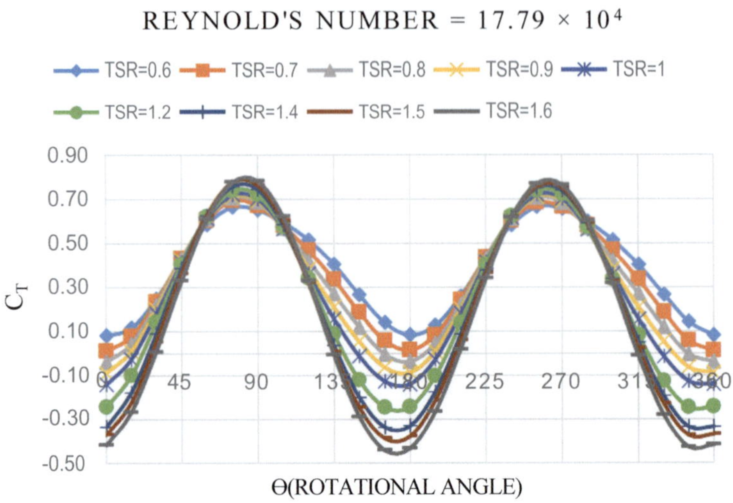

Fig. 7 Variation of C_T with respect to rotational angle for 1 m/s water velocity

The maximum C_T found out for this flow analysis of 1.5 m/s is 0.78 and minimum C_T value is −0.43. The Reynold's number for this flow is calculated and it's value is 26.73×10^4. In Fig. 8 the variation of C_T with respect to rotational angle is shown.

The next simulation results are obtained for 2 m/s water velocity. This case gives the maximum coefficient of power output, which is 0.30. The angular velocity values are ranging between 15.094 rad/s up and 40.252 rad/s depending upon the increasing TSR values. The average torque obtained from this study is found out to be ranging between 2.55 and 1.09 N-m for various TSR values which are ranging between 0.6

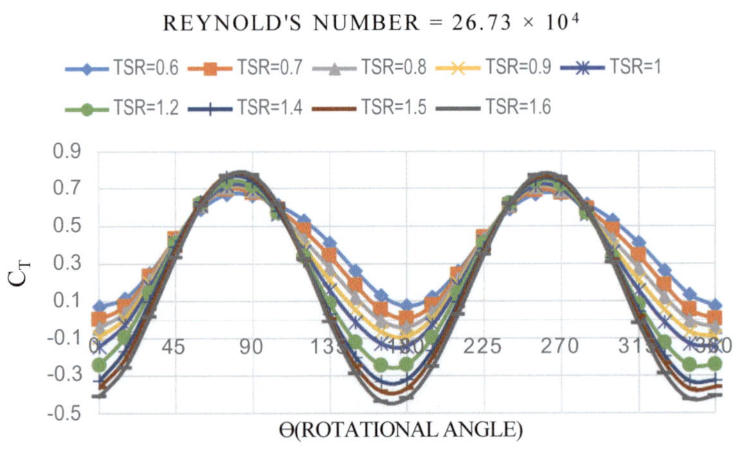

Fig. 8 C_T versus θ for water velocity of 1.5 m/s

and 1.6. The time step values are taken for 15° of rotation are between 0.0065 and 0.0173 depending upon the different RPM values.

The Reynold's number found out for this flow of 2 m/s is 35.68×10^4. Maximum C_T is 0.77 and minimum is −0.43. Figure 9 shows the variation graph between C_T for different angle of rotation.

In Fig. 10 the results of C_T versus rotational angle θ is shown which was conducted by Roy and Saha in order to extract the wind energy [18].

The coefficient of power (C_P) is calculated as,

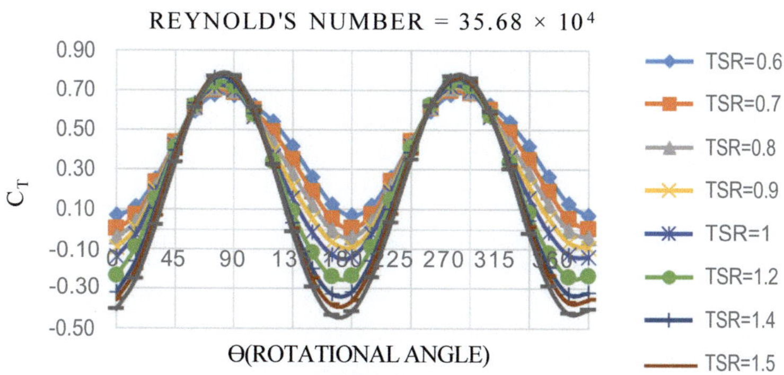

Fig. 9 C_T versus θ for various TSR for 2 m/s flow velocity

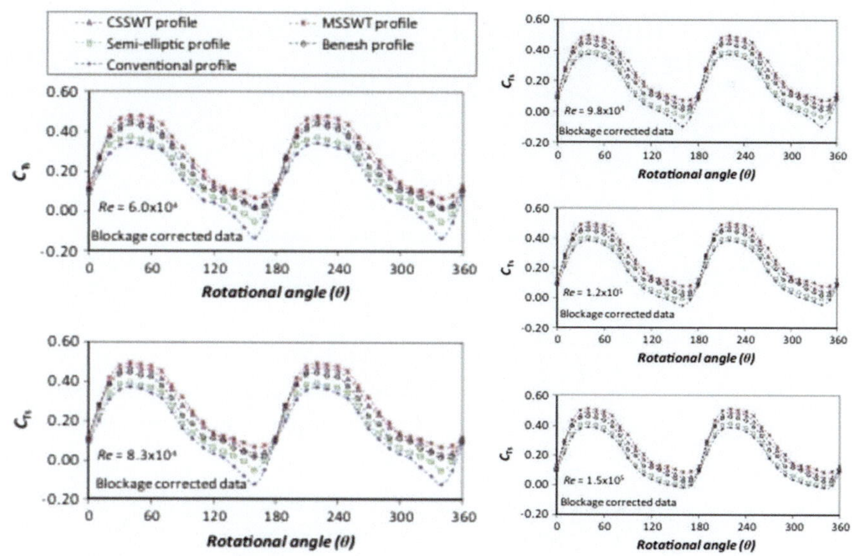

Fig. 10 C_T versus rotational angle θ [18]

Table 4 C_T and C_P values for different input conditions

TSR	Velocity	C_P	C_T	Velocity	C_P	C_T	Velocity	C_P	C_T
0.6	1	0.236	0.393	1.5	0.239	0.399	2	0.240	0.400
0.7	1	0.254	0.363	1.5	0.258	0.368	2	0.258	0.369
0.8	1	0.269	0.337	1.5	0.272	0.340	2	0.272	0.340
0.9	1	0.279	0.310	1.5	0.279	0.309	2	0.279	0.310
1	1	0.284	0.284	1.5	0.284	0.284	2	0.285	0.285
1.2	1	0.291	0.242	1.5	0.293	0.244	2	0.295	0.246
1.4	1	0.293	0.210	1.5	0.298	0.213	2	0.300	0.213
1.5	1	0.287	0.191	1.5	0.290	0.193	2	0.292	0.194
1.6	1	0.270	0.169	1.5	0.273	0.171	2	0.273	0.171

$$C_P = \frac{P_{\text{turbine}}}{P_{\text{available}}} = \frac{T\omega}{0.5\rho A V^3} = \frac{T}{0.5\rho A V^2 R}\frac{R\omega}{V} = \mathrm{C_T} \times \mathrm{TSR} \qquad (11)$$

In Table 4 the average C_P and C_T values are plotted under different TSR and flow velocities.

The results of this study can be compared with previous investigations. Figure 11 shows C_T versus TSR experimental result by Roy and Saha [18] and Fig. 12 shows the average C_T variation for different TSR values by this study. The C_T variation which was experimentally conducted by Roy and Saha has shown in Fig. 13 [18]. Figure 14 indicates the C_P variation under different TSR ranges on a study conducted by Kailash et al. [21]. The C_P variation with respect to TSR on various velocities has been plotted in Fig. 15 as a result of this study.

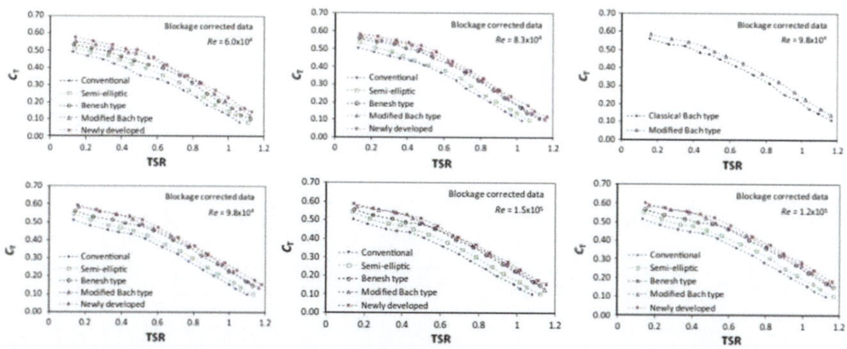

Fig. 11 C_T versus TSR from the study by Roy and Saha [18]

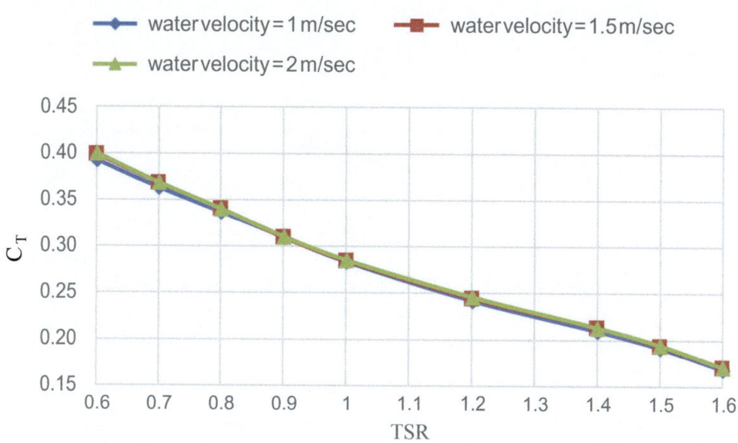

Fig. 12 C_T versus TSR on various water velocity input from this study

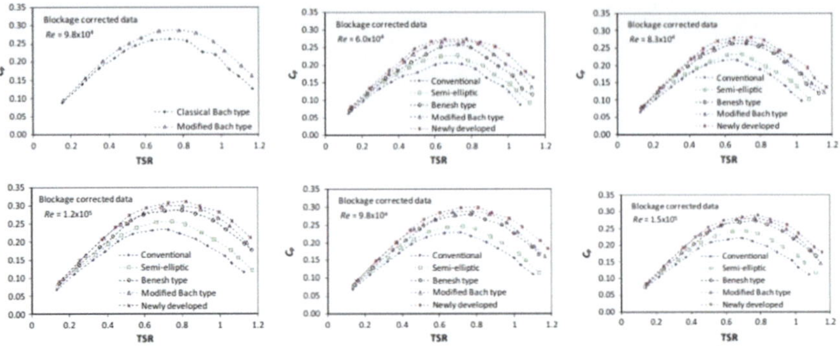

Fig. 13 C_P versus TSR for various Reynold's number conducted by Roy and Saha for Savonis wind turbine [18]

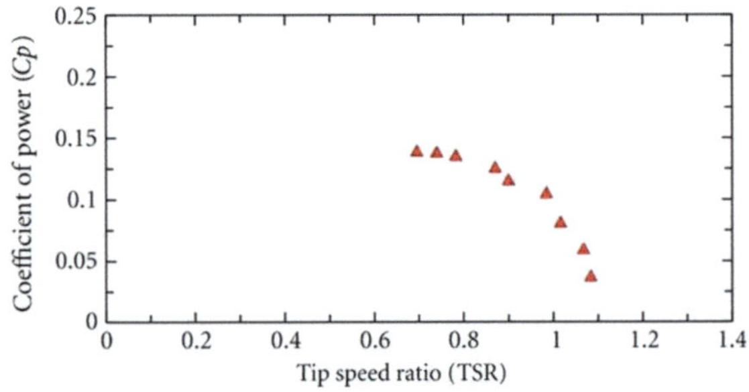

Fig. 14 C_P versus TSR graph obtained by Kailash et al. [21]

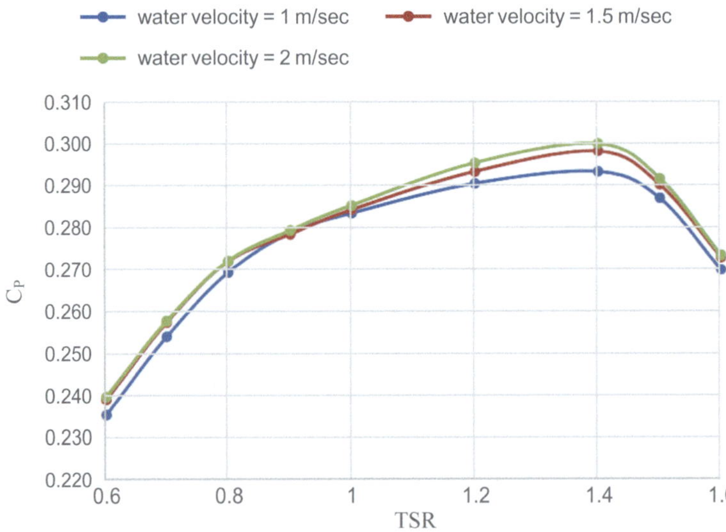

Fig. 15 C_P versus TSR on various water velocities from this study

8 Conclusions

From the results obtained by this numerical approaches followings can be concluded.

(1) The C_P values are significantly higher than the previous investigations for most of the TSR values.
(2) Fluctuation of C_P is very less compared to other researches done on Savonius in hydro kinetic energy extraction field.
(3) The maximum efficiency point for most of the researches lies almost at the middle of the tested TSR domain, but for this case with increase of TSR values C_P values are less deviated and maximum C_P of 0.30 was achieved up to a value of 1.4 TSR for 2 m/s flow velocity in the channel.
(4) This maximum efficiency point for a wider range of TSR values dominantly signifies the increment of operating zone under different loading conditions for smooth and feasible operation of Savonius hydrokinetic turbine.

References

1. Kusakana, K., Vermaak, H.J.: Hydrokinetic power generation for rural electricity supply: case of South Africa. Renew. Energy **55**, 467–473 (2013)
2. Kumar, D., Sarkar, S.: A review on the technology, performance, design optimization, reliability, techno-economics and environ mental impacts of hydrokinetic energy conversion systems. Renew. Sustain Energy Rev. **58**, 796–813 (2016)
3. Savonius, S.: The S-rotor and its applications. Mech. Eng. **53**(5), 333–338 (1931)
4. Riglin, J., Schleicher, W.C., Liu, I.H., Oztekin, A.: Characterization of a micro-hydrokinetic turbine in close proximity to the free surface. Ocean Eng. **110**, 270–280 (2015)
5. Zhao, Z., Zheng, Y., Xu, X., Liu, W., Hu, G.: Research on the improvement of the performance of Savonius rotor based on numerical study. In: Proceedings of International Conference on Sustainable Power Generation and Supply (SUPERGEN), pp. 1–6 (2009)
6. Alexander, A.J., Holownia, B.P.: Wind tunnel test on a Savonius rotor. J. Wind Eng. Ind. Aerodyn. **3**(4), 343–351 (1978)
7. Kamoji, M.A., Kedare, S.B., Prabhu, S.V.: Performance tests on helical Savonius rotors. Renew. Energy **34**(3), 521–529 (2009)
8. Kamoji, M.A., Kedare, S.B., Prabhu, S.V.: Experimental investigations on single stage modified Savonius rotor. Appl. Energy **86**((7–8)), 1064–1073 (2009)
9. Kianifar, A., Anbarsooz, M.: Blade curve influences on the performance of Savonius rotors: experimental and numerical. Proc. Inst. Mech. Eng. Part A: J. Power Energy **225**, 343 (2011)
10. Roy, S., Saha, U.K.: Computational study to assess the influence of overlap ratio on static torque characteristics of a vertical axis wind turbine. Proc. Eng. **51**, 694–702 (2013)
11. Jian, C., Kumbernuss, J., Linhua, Z., Lin, L., Hongxing, Y.: Influence of phase-shift and overlap ratio on Savonius wind turbine's performance. J. Sol Energy Eng. **134**, 1–9 (2012)
12. Mojola, O.: On the aerodynamic design of the Savonius wind mill rotor. J. Wind Eng. Ind. Aerodyn. **21**, 223–231 (1985)
13. Sheldahl, R.E., Feltz, L.V., Blackwell, B.F.: Wind tunnel performance data for two- and three-bucket Savonius rotors. AIAA J. Energy **2**(3), 160–164 (1978)
14. Kamoji, M.A., Kedare, S.B., Prabhu, S.V.: Experimental investigations on single stage, two stage and three stage conventional Savonius rotor. Int. J. Energy Res. **32**, 877–895 (2008)

15. Saha, U.K., Rajkumar, M.J.: On the performance analysis of Savonius rotor with twisted blades. Renew. Energy **31**(11), 1776–1788 (2006)
16. Ghatage, S.V., Joshi, J.B.: Optimisation of vertical axis wind turbine: CFD simulations and experimental measurements. Can. J. Chem. Eng. **90**(5), 1186–1201 (2012)
17. Akwa, J.V., Vielmo, H.A., Petry, A.P.: A review on the performance of Savonius wind turbines. Renew. Sustain Energy Rev. **16**(5), 3054–3064 (2012)
18. Roy, S., Saha, U.K.: Wind tunnel experiments of a newly developed two-bladed Savonius-style wind turbine. Appl. Energy **137**, 117–125 (2015)
19. Kumar, A., Saini, R.P.: Performance analysis of a Savonius hydrokinetic turbine having twisted blades. Renew. Energy **108**, 502–522 (2017)
20. Zhou, T., Rempfer, D.: Numerical study of detailed flow field and performance of Savonius wind turbines. Renew. Energy **51**, 373–381 (2013)
21. Kailash, G., Eldho, T.I., Prabhu, S.V.: Performance study of modified Savonius water turbine with two deflector plates. Int. J. Rotat. Mach. **2012**, 12 (2012)

Facial Analysis Using Jacobians and Gradient Boosting

A. Vinay, Abhijay Gupta, Vinayaka R. Kamath, Aprameya Bharadwaj, Arvind Srinivas, K. N. B. Murthy and S. Natarajan

Abstract Security and identity have become one of the primary concerns of the people in this digital world. Person authentication and identification is transforming the way these services are provided. Earlier it was mainly achieved through passwords and patterns but with significant advancements in face recognition technologies, it has been exploited in providing authentication in smart phones and computers. Face Recognition (FR) extends its use in applications like face tagging in social media, surveillance system at theaters, airports and so on. The proposed mathematical model employs linear algebra and mathematical simulations for the task of person identification. Kernel singular value decomposition is used to denoise the facial image which is then passed to a feature detector and descriptor based on nonlinear diffusion filtering. The obtained descriptors are quantized into a vector using an encoding model called VLAD which uses k-means++ for clustering. Further, classification is done using Gradient boosting decision trees. The pipeline proposed aims at reducing the average computational power and also enhancing the efficiency of the system. The proposed system has been tested on the three benchmark datasets namely Face 95, Face 96 and Grimace.

Keywords Linear algebra · Kernel-SVD · Feature quantization · Gradient boosted decision tree

1 Introduction

The need for recognizing a person among the masses is very much important for various applications. With a wide variety of purposes, face recognition is gaining prominence in almost all the fields where there is a need for interaction between humans and machines. Face recognition involves recognizing a facial biometric blueprint among

A. Vinay · A. Gupta (✉) · V. R. Kamath · A. Bharadwaj · A. Srinivas · K. N. B. Murthy · S. Natarajan
Centre for Pattern Recognition and Machine Intelligence, PES University, Bengaluru, India
e-mail: abhijayvgupta@gmail.com

S. Manna et al. (eds.), *Mathematical Modelling and Scientific Computing with Applications*, Springer Proceedings in Mathematics & Statistics 308,
https://doi.org/10.1007/978-981-15-1338-1_29

an already existing database of individuals. This mechanism can be achieved by several means. The images of these subjects may be subjected to several mathematical operations so as to contain these blueprints in the desired format. This technique which is maneuvered on images of the individuals determines the extent of the correctness of application which utilizes the design. The system imposed also decides the computational demand the pipeline expects in making the decision of recognizing the subjects. Hence choosing the right mathematical models plays a prominent role in directly influencing the end results of the process. These steps involve choosing right preprocessing techniques, optimal feature extraction algorithm as well as representing the extracted features in an appropriate form. This apparatus experiences a lot of problems when used in an application. Numerous factors affect the image and impart variations in the end result. Variations in illumination, pose of the subject constitute a major hurdle to overcome. Recognizing these variations and mapping them to the same class is another challenge to tackle. The procedure may give appropriate results for frontal views of the person while running poorly when the vision of the subject is progressed away from the source that is capturing the image. Facial hair such as mustache and beard can cause loss of features in the lower half of the face, contributing in delivering imperfect output. The distortions in the image may be due to the subject itself or variations in the background. These need to be handled to avoid variations in the decisions. Even though modern techniques claim to overcome these challenges, the scope for making this process better is perennial. Making the system computationally superficial is also a characteristic one has to keep in mind before designing a model. Numerous face recognition algorithms assume that a large number of samples per person are available for training. This is another setback which they oversee and is very distant to the real world scenario. There is also a need for making face recognition resistant to aging. The wrinkles developed over time can bring significant changes in the outlook of the person. This as a whole can be broken down into three important key aspects: detecting the region of interest, extracting the prominent key features and illustrating these features for classification. Most of the obstacles mentioned hinder the perseverance of the descriptors.This brings out the need for strategies which can conquer the problems listed. It may be difficult to design a model which can overcome all the hitches. But contraptions which can excel at specific conditions can be drafted to fit certain use cases closer to the real world.

2 Related Work

D. Suter and K. Schindler in their recent work have used incremental kernel SVD [1] to achieve face recognition with image sets. They have put forth a popular linear subspace updating algorithm to the nonlinear case by using the kernel methodology and apply a reduced set construction method to generate sparse expressions [2] for the derived subspace basis in order to maintain constant processing speed and memory usage.

In [3], KPCA was used to extract feature descriptors from numerous images for utilization in mobile robot navigation and localization. RS expansions are constructed to compress the KPCA-derived bases to reduce computational load during KPC utilization. In [4], AKAZE was used in remote sensing [5] image matching. Distortions caused by the orientation change of camera were first modeled by various tilts images; then the key points were localized by improved Accelerated-KAZE (AKAZE) algorithm. The feature points are detected in a nonlinear space constructed by Fast Explicit Diffusion (FED) with the help of variable conductance function, and the resulting feature points are described by improved SIFT [6] descriptor. In the end euclidean distance metric was used to determine the correspondences and Random Sample Consensus (RANSAC) [7] algorithm was used in eliminating the false matches as well.

Given a collection of local features taken from an image, VLAD is generated by quantizing local features with a visual vocabulary, recollecting the residual statistics of quantized features for each of the generated centroids and by summing up the aggregated residual vectors from each centroid. The search accuracy can be optimized by increasing the size of vocabulary can prove to be costly both in terms of memory and sheer computational power. Demonstrating a remarkable accuracy-efficiency trade-off, VLAD has gained prominence from the community and large number of extensions have been proposed. In [8] an attempt to make an in-depth analysis of the framework which aimed at increasing the thorough understanding of its various processing steps and inflating its overall performance was made. It involved the evaluation of various existing and novel extensions along with the study of the consequences of several unexplored parameters. It focused on exploring more productive local features, making the aggregated representation better and tuning the indexing scheme to get better results. The authors successfully managed to produce various insights into extensions that contributed, and multiple others that do not.

In [9], discriminative feature descriptors were constructed as an application of Vector of Locally Aggregated Descriptors. A hierarchical multi-VLAD was introduced to interpret the trade off between descriptor discriminability and computation complexity. A tree-structured hierarchical quantization (TSHQ) was constructed to speed up the VLAD computation with a large vocabulary. If at all quantization error propagates from root to leaf node with TSHQ, multi-VLAD can be used; by constructing a VLAD descriptor for each level of the vocabulary tree to cope up with the quantization error [10] at that level. Intense analysis with regards to various benchmark datasets has proven that the proposed approach was far better the state-of-the-art methods in terms of retrieval accuracy, fast extraction, as well as light memory cost. A mechanism was proposed for pose based human action recognition using Extreme Gradient Boosting [11] by Vina Ayumi. A clear insight on comparisons for gesture recognitions using SVM [12], Naive Bayes [13] and XGBoost was also delivered. Clustering is basically a strategy of grouping a set of objects in such a way that objects which share similar properties are likely to be more closer to each other in comparison to those data point or objects in different assessments or clusters. A comparative study of partition-predicated clustering techniques, such as K-Means [14], K-Means++ [15] and object predicated Fuzzy C-Means clustering algorithm was presented by Kapoor and Singhal. A methodology for obtaining better results by application of sorted and unsorted data into the algorithms. Gradient Boosting Decision Tree (GBDT) has a lot of effective implementations such as XGBoost and pGBRT. The implementations are not satisfactory when there is a lot of data to process. To overcome this hurdle, two techniques were designed by researchers from Microsoft: Gradient-based One-Side Sampling (GOSS) and Exclusive Feature Bundling (EFB). LightGBM [16] was used to speed up the training process of conventional GBDT by over 20 times without compromising the accuracy.

3 Proposed Methodology

This sections explains about the procedures implemented to achieve face recognition. The pipeline of the proposed model is represented in Fig. 1.

3.1 *Image De-Noising*

In order to evaluate a bunch of images which are numerous and are compound in nature the requirement of subspaces which are nonlinear in nature are required for analysis. To achieve this, the input data is mapped in a non linear fashion to a space which is of higher dimension by making use of the function $\varphi : R^m \Rightarrow$ J and then singular value decomposition which is in J. The kernel function K introduces a mapping φ which takes the inner product amongst the input data which is mapped and contained in the feature space. A = [$\varphi(x_1)$ ... $\varphi(x_n)$] is obtained by the alteration of input data by utilizing φ. Let M = A^TA , by performing decomposition using eigen values we can deduce that M = QδQ^T where M is a matrix which is obtained by taking the inner product of the corresponding columns in matrix A and is evaluated by using the kernel regression function. By performing singular factorization of A the rank r of the matrix is given by the equation:

$$A^r = [AQ^r(\delta^r)^{-1/2}][(\delta^r)^{1/2}][(Q^r)^T] = U^r\sigma^r(V^r)^T. \quad (1)$$

where $Q^r = Q[:, 1:r]\, and \delta^r = \delta[1:r, 1:r]$

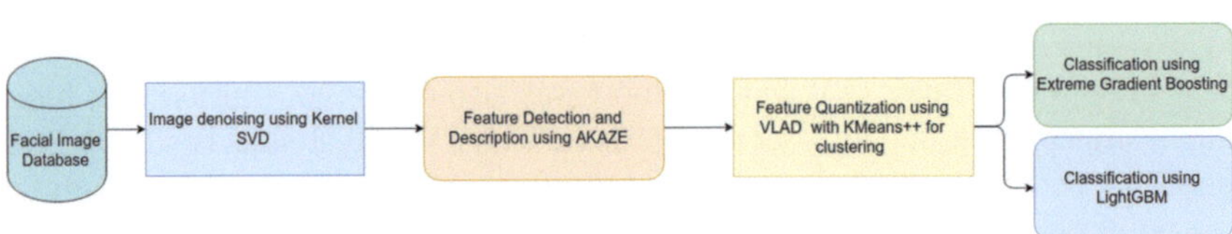

Fig. 1 Schematic of the proposed pipeline

The matrix M is positive in nature and is semi-definite which is obtained from the utilization of the Mercer kernel. Basis U^r is given by the equation:

$$U^r = AQ^r(\delta^r)^{-1/2} := A\alpha \tag{2}$$

U^r which is the basis is obtained by the linear combination of B which is the input data mapped. One more basis X_r = Bμ which is acquired from B by applying KSVD to it. The resulting output D can be computed and is given by the equation:

$$D = (U^r)^T X^r = \alpha^T A^T B\mu \tag{3}$$

Kernel function particularly that of ATB is used in the above equation. On applying SVD on D the equation becomes:

$$Y^T DZ = \theta \tag{4}$$

Here $diag(\theta) = \{\theta_1, \Delta\Delta\Delta, \theta_r\}$ are the primary angles formed by span (U^r) and span (X^r). Those functions which are a function of θ are often used to measure the distance amongst subspaces. Figure 2 shows denoising of images.

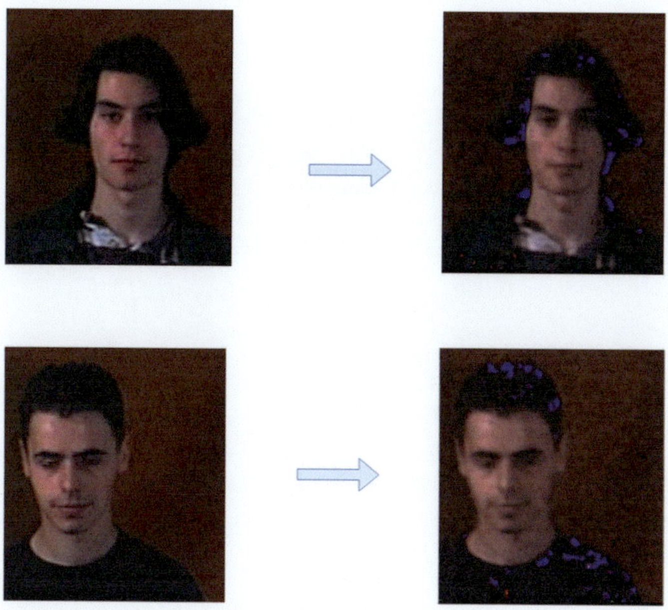

Fig. 2 Image de-noising using Kernel-SVD

3.2 Feature Detection and Description

In this method, as a preliminary step the input image is gray scaled. This helps in removing unwanted features that arise due to variation in color. To detect facial features in the image at different scale levels, the determinant of the Hessian matrix is computed for each image.

$$L^{i}_{Hessian} = \sigma^{2}_{i,norm}(L^{i}_{xx}L^{i}_{yy} - L^{i}_{xy}L^{i}_{xy}) \tag{5}$$

Here L represents a filtered image in a non linear scale space. The Fast Explicit Diffusion technique (FED) to enhance the non linear scale space computation. The features extracted from the FED technique is used describe the different characteristics of the image. Here, robustness, rotation-invariance is achieved by doing binary tests between the mean of areas and an estimation of orientation of facial interest point in KAZE and the rotation of the grid of local binary descriptor respectively. The information about gradients and intensity are very important for fast detection and description of features in the image. The descriptor obtained is now used for feature aggregation and then for classification. Key point detection using AKAZE can be seen in Fig. 3.

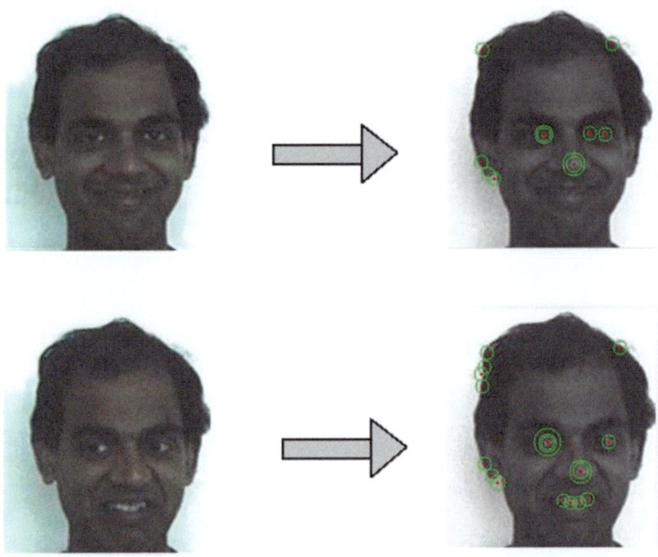

Fig. 3 Keypoint detection using AKAZE

3.3 VLAD

When the all the vectors of features are obtained after the descriptors are applied on the images, they now have to be aggregated. These vectors are of large lengths and contain a lot information about the keypoints being described such as color, location, intensity, information about neighboring pixels and a lot more. However, all this information might not be needed for classification. If these features are not ignored classification becomes harder and the time complexity also increases. Hence, in order to prevent this we quantize the features using vector of locally aggregated features VLAD in a non probabilistic Fisher kernel which uses a codebook, computed using k-means++ algorithm. Each descriptor x_t is affiliated with its closest visual word in the codebook. Let $\mu_1, \mu_2 ..., \mu_k$ represent codewords. The difference $x_t - \mu_i$ to a vector l_i is cumulated. The algorithm for feature quantization is described below: L2 normalization is applied to V. In the above algorithm the 2nd FOR loop represents the cumulation of descriptors and the 3rd FOR loop is used for power normalization. The obtained d dimensional vector is quantized version of our facial descriptors and is apt for the classification job.

Algorithm 1 Computing the descriptor V from a set of descriptors $x_1, x_2, ..., x_t$. Given codewords $\mu_1, \mu_2 ..., \mu_k$ computed using k-means++ algorithm

for $i = 1, ..., K$ **do**
  $l_i := 0_d$
end for
for $t = 1, ..., T$ **do**
  $i = argmin_j ||x_t - \mu_j||$
  $l_i := l_i + x_i - \mu_i$
end for
$V = [l_1^T ... l_K^T]$
for $u = 1, ...Kd$ **do**
  $V_u := sign(V_u)|V_u|^\alpha$
end for

3.4 Classification Using Gradient Boosted Trees

After the features are quantized into a vector, they are passed to tree boosting algorithms for person classification. Two implementations based on gradient boosted decision tree have been used in this paper. They provide a comparison between the speed and accuracy of the model.

The first one is a scalable end-to-end tree boosting algorithm called XGBoost. Boosting combines a set of relatively weak learners to form a complex predictor which tends to have a low error rate as they learn from the mistakes of the previous learner. The previous learners weights are also accounted for and at each iteration they are updated with respect to the residual weights. Multiple decision trees are constructed with a specific number of terminal nodes in the decision tree, six in our case. This allows intercommunication of node values within tree resulting in better feature under-standing. Gradient descent is used to minimize the error. This algorithm was designed as a sparsity aware algorithm providing robust and inexpensive computation.

The second implementation of the gradient boosted decision tree is LightGBM which uses Gradient-based One-Side Sampling (GOSS) and Exclusive Feature Bundling (EFB). In order to compute information gain at a faster rate, GOSS samples data instances with only large gradients as they contribute more to the final classification. EFB is a greedy technique which bundles exclusive features, that is, it reduces features to provide robustness at the cost of accuracy.

These two implementations of gradient boosted decision tree provides a robust and reliable method for classification of facial features.

4 Datasets Used

4.1 Faces95

Faces95 as shown in Fig. 4 is a collection of facial images of 72 individuals with large head scale variations. The dataset provides images of resolution 180 × 200 pixels in portrait mode. The collection contains images of both male and female subjects, thus delivering a challenge in the upper half of the region of interest as well. No agitation is offered in terms of background disfigurements, but slight variations are observed in the red background because of shadows and changes in illuminations. The person is not stationary and is subjected to slight movements. The reverberation of this is observed in brightness changes in the region of interest. The artificial lighting system adds up to the intensified changes in the glare. The dataset offers meager variations in the expressions of the subjects which is not a significant obstacle. The same hairstyle is maintained on all the sample images for a single test subject.

4.2 Faces96

Using a still camera 152 subjects were photographed with 20 images per person. The database possesses images of size 192 × 192 pixels of both male and female candidates. However one of the paramount hurdles offered by faces96 is the variations in the background which does not lie in the region of interest. Major challenge constituted by the dataset is the variations in the trivial parts of the images. The individuals proffer slightest changes in the expressions while tremendous changes in the head scale are exhibited. The person also locomotes towards the camera after

Fig. 4 Sample record from the Faces95 dataset

every test image delivering changes in the lighting conditions as well. They dataset demonstrates variations in the head tilt, turn, and slant but is not significant enough. The collection was designed and maintained by Dr. Libor Spacek under Computer Vision Science Research Projects.

4.3 Grimace

The individual moves his/her head after every picture and makes grimaces on their faces which gets drastic towards the end of every sequence. A set of 18 individuals are put through this process to form a collection of faces. The images are of size 180 × 200 pixels which have a considerate amount of information for further processing. No variations in terms of background and head scale are presented. Although a considerable amount of discrepancy is unveiled head tilt and turns. Very little variability is presented in terms of illumination, which is slightly better in case of the positioning of the head in the image. The database as shown in Fig. 5 offers excessive fluctuations in the expressions of the figurines which delivers a major challenge for all mathematical models.

Fig. 5 Subject exhibiting variations in facial expressions

5 Results and Conclusion

We executed our proposed model over every group of images present in the three benchmark databases, namely, FACES95, FACES96 and GRIMACE. The results obtained using our technique on the three datasets are tabulated in Table 1. On FACES95 and FACES96 our model predicts on an average 91% of the time correctly. Even though the above mentioned datasets contains a high number of classes (72 and 152), our model is still able to predict each individual with a good confidence level. GRIMACE contains faces with extreme variations in expression, illumination and translation, our model is able to correctly identify the person with an accuracy of 93.33%.

Kernel-SVD plays an important role in denoising the image, reducing a few features and thereby helping to lower computation cost in further steps.

Changing the vector size of the feature aggregator impacts the accuracy a lot. On increasing the number of features in the quantization step, the accuracy of the model increases till a certain extent which can be seen in Table 2.

Table 3 shows a direct comparison of XGBoost with LightGBM for feature classification. Both of these algorithms ran for 500 epochs with maximum depth of each tree created set to 7 and learning rate set to 0.5. From Table 3 it is understood that LightGBM outperforms XGBoost on all the three datasets by an average margin of 2%.

Table 1 Performance grid of XGBoost

Dataset	Accuracy	Recall	Precision	Prediction time (100 images)
Faces95	0.9214	0.9214	0.9312	0.42 ms
Faces96	0.90	0.9010	0.9134	0.94 ms
Grimace	0.9333	0.9333	0.9467	0.13 ms

Table 2 Feature aggregation with vector of size 100 versus 200

Dataset	200	100
Faces95	0.9214	0.8927
Faces96	0.90	0.8533
Grimace	0.9333	0.9167

Table 3 Comparison of XGBoost with LightGBM

Dataset	LightGBM	XGBoost
Faces95	0.94	0.9214
Faces96	0.9367	0.90
Grimace	0.9633	0.9333

The framework proposed is a mathematical model employing Single valued decomposition for the purpose of image denoising. Further an hessian matrix is computed to extract features from a facial image. The use of gradient boosting algorithm for classification by minimizing a loss function. The aim of the paper is completed using mathematical models and hence do justice to the theme of the conference.

A direct comparison of our proposed method with the state of the art models cannot be made because of the following reasons. Implementation and testing of the modules are performed on different hardware which might result in different efficiency results. Datasets used and pre-processing steps followed in the state of the art models and our proposed methods are different which leads to variation in results. [17] developed a method for deep hypersphere embedding for face recognition to achieve a remarkable accuracy of 95% on the YTF dataset. Chain code based local descriptors are proposed in [18] for the task of face recognition. Karczmarek et al. [18] was tested on CAS-PEAL, ColorFERET and FG-NET resulting in an average accuracy of 98%. A deep learning approach for face recognition was developed in [19] where a trunk branch ensemble convolutional neural network was designed to solve the problem of pose variation and occlusions resulting in an average accuracy of 95% on PaSC, COX Face and YouTube faces datasets. A different approach to face recognition was proposed in [20] using multi-resolution wavelet combining discrete cosine transform and Walsh transform which resulted in an accuracy of 99.24% on FACES94 dataset.

References

1. Chin, T.-J., Schindler, K., Suter, D.: Incremental Kernel SVD for face recognition with image sets. In: 7th International Conference on Automatic Face and Gesture Recognition (FGR06), Southampton, pp. 461–466 (2006)
2. Wang, Z.W., Huang, M.W., Ying, Z.L.: The performance study of facial expression recognition via sparse representation. In: 2010 International Conference on Machine Learning and Cybernetics, Qingdao, pp. 824–827 (2010)
3. Meltzer, J., Yang, M.-H., Gupta, R., Soatto, S.: Multiple view feature descriptors from image sequences via kernel pca. In: ECCV, pp. 215–227 (2004)
4. Liu, Y., Lan, C., Yao, F., Li, L., Li, C.: Oblique remote sensing image matching based on improved AKAZE algorithm. In: 2016 Sixth International Conference on Information Science and Technology (ICIST), Dalian, pp. 448–454 (2016)
5. Sakarya, U., et al.: A short survey of hyperspectral remote sensing and hyperspectral remote sensing research at tbtak Uzay. In: 2015 7th International Conference on Recent Advances in Space Technologies (RAST), Istanbul, pp. 187–192 (2015)
6. Liu, J., Wu, Q., Li, X.: Research on image matching algorithm based on local invariant features. In: 2013 Ninth International Conference on Intelligent Information Hiding and Multimedia Signal Processing, Beijing, pp. 113–116 (2013)
7. Hu, M., Chen, J., Shi, C.: Three-dimensional mapping based on SIFT and RANSAC for mobile robot. In: 2015 IEEE International Conference on Cyber Technology in Automation, Control, and Intelligent Systems (CYBER), Shenyang, pp. 139–144 (2015)
8. Spyromitros-Xioufis, E., Papadopoulos, S., Kompatsiaris, I.Y., Tsoumakas, G., Vlahavas, I.: A comprehensive study over VLAD and product quantization in large-scale image retrieval. IEEE Trans. Multimedia **16**(6), 1713–1728 (2014)

9. Wang, Y., Duan, L.Y., Lin, J., Wang, Z., Huang, T.: Hierarchical multi-VLAD for image retrieval. In: 2015 IEEE International Conference on Image Processing (ICIP), pp. 4629–4633. QC, Quebec City (2015)
10. Lu, Z., Xu, F., Tian, Q.: Research on quantization errors of stability for model-based networked control system. In: 2012 Proceedings of International Conference on Modelling, pp. 867–872. Wuhan, Hubei, China, Identification and Control (2012)
11. Ayumi, V.: Pose-based human action recognition with extreme gradient boosting. In: 2016 IEEE Student Conference on Research and Development (SCOReD), Kuala Lumpur, pp. 1–5 (2016)
12. Shubo, Q., Shuai, G., Tongxing, Z.: Research on paper defects recognition based on SVM. In: 2010 WASE International Conference on Information Engineering, pp. 177–180. Hebei, Beidaihe (2010)
13. Qiang, G.: Research and improvement for feature selection on Naive Bayes text classifier. In: 2010 2nd International Conference on Future Computer and Communication, Wuhan, pp. V2-156–V2-159 (2010)
14. Na, S., Xumin, L., Yong, G.: Research on k-means clustering algorithm: an improved k-means clustering algorithm. In: 2010 Third International Symposium on Intelligent Information Technology and Security Informatics. Jinggangshan, pp. 63–67 (2010)
15. Kapoor, A., Singhal, A.: A comparative study of K-Means, K-Means++ and Fuzzy C-Means clustering algorithms. In: 2017 3rd International Conference on Computational Intelligence & Communication Technology (CICT), Ghaziabad, pp. 1–6 (2017)
16. Ke, G., Meng, Q., Finley, T., Wang, T., Chen, W., Ma, W., Ye, Q., Liu, T.-Y.: LightGBM: a highly efficient gradient boosting decision tree. In: Conference on Neural Information Processing Systems (2007)
17. Liu, W., Wen, Y., Yu, Z., Li, M., Raj, B., Song, L.: Sphereface: deep hypersphere embedding for face recognition. In: The IEEE Conference on Computer Vision and Pattern Recognition (CVPR), vol. 1. (2017)
18. Karczmarek, P., Kiersztyn, A., Pedrycz, W., Dolecki, M.: An application of chain code-based local descriptor and its extension to face recognition. Pattern Recogn. **65**, 26–34 (2017)
19. Ding, C., Tao, D.: Trunk-branch ensemble convolutional neural networks for video-based face recognition. IEEE Trans. Pattern Anal. Mach, Intell (2017)
20. Choudhary, A., Vig, R.: Face recognition using multiresolution wavelet combining discrete cosine transform and Walsh transform. In: Proceedings of the 2017 International Conference on Biometrics Engineering and Application, pp. 33–38. ACM (2017)

Facial Image Classification Using Rotation, Illumination, Scale and Expression Invariant Dense Features and ENN

A. Vinay, Ankur Singh, Nikhil Anand, Mayank Raj, Aniket Bharati, K. N. B. Murthy and S. Natarajan

Abstract Face Recognition is immensely proliferating as a research area in the paradigm of Computer Vision as it provides an extensive choice of applications in surveillance and commercial domains. This paper throws light upon the comparison of various dense feature descriptors (Dense SURF, Dense SIFT, Dense ORB) with each other and also with their classical counterparts (SURF, SIFT, ORB) using a novel technique for recognition. This proposed technique uses Laplacian of Gaussian filter for enhancement of the image. It applies various dense and classical feature descriptors on the enhanced image and outputs a feature vector. In order to achieve high performance, this feature vector is given to Fisher vector since Fisher Vector is a feature patch-aggregation method. Finally, extended nearest neighbor Classifier is used for classification over the orthodox k-nearest classifier. Experiments were carried out on three diverse datasets—ORL, Faces94, and Grimace. On scrutinizing the results, Dense SIFT and Dense ORB were found to be preeminent as measured by various performance metrics. 98.44 on Grimace, 98.15 on Faces94.

Keywords Scale invariant feature transformation · Speed up robust feature · Oriented FAST and rotated BRIEF · Extended nearest neighbor · Laplacian of Gaussian

Nomenclature

SIFT Scale Invariant Feature Transformation
SURF Speed Up Robust Feature

A. Vinay (✉) · A. Singh · N. Anand · M. Raj · A. Bharati · K. N. B. Murthy · S. Natarajan
Centre for Pattern Recognition and Machine Intelligence, PES University,
100 Feet Ring Road, Banashankari Stage III, Bengaluru, Karnataka 560085, India
e-mail: a.vinay@pes.edu

M. Raj
e-mail: rajmike.mike183@gmail.com

S. Manna et al. (eds.), *Mathematical Modelling and Scientific Computing with Applications*, Springer Proceedings in Mathematics & Statistics 308,
https://doi.org/10.1007/978-981-15-1338-1_30

ORB Oriented FAST and Rotated BRIEF
ENN Extended Nearest Neighbor
LoG Laplacian of Gaussian

1 Introduction and Related Work

In the current digital era, protecting sensitive information has become a cumbersome task. Research shows that biometrics are more prominent than the traditional passwords for authentication and authorization. Face recognition is a class of biometrics that maps a person's facial features mathematically and stores the information as a faceprint. Face Recognition even surpasses other biometric modalities because it is non-intrusive and can identify a distant subject. Face recognition unlike other physiological modalities does not require any special hardware component. Any modern-day camera can be used for face recognition. Extensive research in the domain of face recognition has led to various classical techniques like FisherFace, Elastic Graph Matching, EigenFace etc.

Feature detection and description are one of the most crucial steps for an image processing task. Over the last decade, Scale Invariant Feature Transform(SIFT) which was suggested by Lowe [1], Speed-Up Robust Features(SURF) which was originally proposed by Herbert Bay [2] and Oriented FAST and Rotated BRIEF(ORB) [3] have been widely used for face recognition. Some of the popular works include—adaptation of SIFT Features for Face Recognition under Varying Illumination [4], SURF-Face [5] and ORB-PCA based feature extraction technique for Face Recognition [6]. The algorithms are subjective to the type of problem that has to be handled. SIFT is a robust classical algorithm which intents to produce scale and orientation invariant features [1] with descriptors which will perform well in matching the state of the image processing pipeline [7]. Analogously, SURF is computationally less exorbitant and mathematically less complicated [2, 8]. It is preeminent because of its standout facets like scale and rotation invariance, repeatability, distinctiveness, and robustness [2]. Similarly, ORB is more efficient than SURF because it uses binary descriptor for feature detection [3, 8]. But for the scale and rotation invariance, it is not as much robust as SURF [3, 7].

However, all these feature descriptors need the facial images to be properly aligned and have a decent contrast. Otherwise, very limited number of key points are detected in the image which produces poor results. Recently an alternative to the traditional SIFT descriptor called the Dense SIFT (DSIFT) descriptor was proposed by Wang [9]. The DSIFT descriptor increases the number of keypoints in an image [9, 10] which in turn enhances the performance of the Face Recognition system. Thus, we propose to exploit DSIFT [9], Dense SURF (DSURF) [5] and Dense ORB(DORB) feature descriptors with a novel pipeline constituting of Laplacian of Gaussian (LoG) filter [11, 12] for enhancing an input image, Fisher vector (FV) for image feature patch aggregation and extended nearest neighbor (ENN) classifier [13] for classification, in this paper. To evaluate the performance of the proposed descriptors comparisons were made with each other and also with the traditional descriptors (SIFT [1], SURF [2] and ORB [3] descriptors). To the best of our knowledge, the application of Dense SURF (DSURF) [5] and DORB on Face Recognition and their comparison with the classical techniques has not been explored yet.

The paper is laid out in the following manner: Sect. 2 describes our proposed system. It contains a detailed explanation of various steps involved along with their usage in our pipeline. Section 3 describes the experimental design. It discusses the various datasets used. Section 4 contains experimental results and their graphical visualization. Section 5 contains various conclusions and inferences that were drawn from the paper. We have also discussed the future enhancements.

2 Proposed System

This section discusses the different steps involved in this proposed method including necessary theoretical and mathematical background of each step. The various steps involved in this approach are depicted below. In the suggested approach, LoG filter is applied to enhance an input image [11] i.e. improve contrast and brightness of the image (Fig. 1).

This is depicted in the image below.

Then the enhanced image is passed to various dense feature descriptors. These descriptors return a feature vector for each of the keypoints in the image. These obtained feature vectors are passed to the Fisher vector which in turn enhances these feature vectors and returns the enhanced feature vectors which are more suitable for classification. Finally, Extended Nearest Neighbour classifier [13] is used to classify the image. The enhanced feature vectors of all the images in the training dataset and their corresponding labels are used to fit the classifier model. The resultant model can then be used to classify any query image.

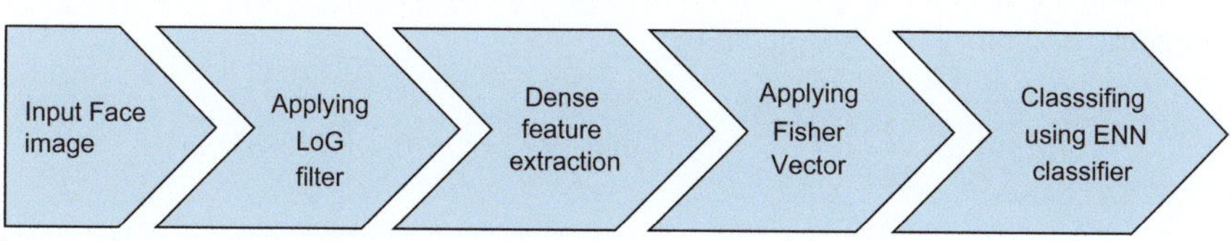

Fig. 1 Pipeline for the applied methodology

The implementation involves several tactical changes in the existing SIFT [1], SURF [2] and ORB [3] feature descriptors to produce three novel descriptors: DSIFT [9, 10], DSURF [5] and DORB. The pipeline also includes ENN classifier which is an improved version of the popular K-Nearest Neighbour (KNN) classifier [13]. The results obtained are compared with the classical techniques to state the proficiency of the proposed system.

2.1 *Laplacian of Gaussian*

2.1.1 Background

Laplacian filter is a second order differential mask [11] which is generally used to find edges in an image [12]. Laplacian operator is isotropic in nature I.e. it is impartial and applies uniformly in all directions in an image. It measures the amount of change in image intensity per change in image position [11].

The Laplacian operator is defined as the dot product of two gradient vector operators [11]

$$L(x, y) = \left[\frac{\partial}{\partial x}\frac{\partial}{\partial y}\right] \cdot \left[\frac{\partial}{\partial x}\frac{\partial}{\partial y}\right]^T = \frac{\partial^2}{\partial x^2} + \frac{\partial^2}{\partial y^2}$$

The Laplacian operator L(x, y) when applied on an image with intensity values I(x, y), is defined as

$$\mathrm{L(x, y)} = \frac{\partial^2 I}{\partial x^2} + \frac{\partial^2 I}{\partial y^2}$$

A convolution filter can be used to approximate the Laplacian operator. For doing so, a discrete kernel is required that can approximate the second order derivatives used by the Laplacian operator. But, these kernels are highly susceptible to noise [11]. To overcome this, noise within an image needs to be reduced. Smoothing filters reduce the noise in an image and generate a less pixelated image [11].

Generally, the Gaussian smoothing filter is used to reduce the sensitivity of an image to noise. The Gaussian operator is a two-dimensional convolution operator [14] that blurs an image and removes some details and noise in the process. It uses a kernel which has a bell-shaped representation. The Gaussian operator is a circularly symmetric operator [14]. It is given by

$$\mathrm{G(x, y)} = \frac{1}{2\psi^2}\mathrm{e}^{-\frac{x^2+y^2}{2\sigma^2}} \quad (\sigma\, represents\ standard\ deviation)$$

The distribution is represented as

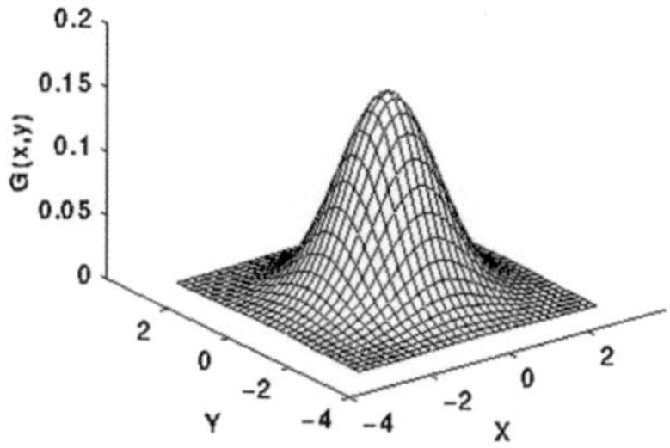

(*mean is* (0,0) *and standard deviation is* σ)

The Gaussian operator blurs out any point-like object (in this case a pixel) to a three-dimensional image with certain minimal size and shape. Since the image is represented using discrete pixel values so, before performing convolution a discrete approximation of the Gaussian function must be found. Theoretically, the Gaussian function is always greater than zero, which implies to an infinitely large convolution kernel. But, practically the Gaussian distribution becomes negligible (approximately 0) beyond 3 standard deviations from the mean. So, the convolution kernel can be terminated at this point.

Once an appropriate kernel is obtained, standard convolution techniques can be used to perform Gaussian smoothing. By decomposing the Gaussian kernel into x and y components [14], we can speed up the convolution step. Thus, we can perform the two-dimensional convolution by first convolving in the x-direction using the one-dimensional x component and then convolving in the y-direction using the one-dimensional y component. The Gaussian operator is the only operator which can be divided in such a way [14].

Since convolution is associative in nature, the Gaussian smoothing filter can be convolved with the Laplacian filter [11] and then this LoG filter can be convolved with the image to produce the desired results. LoG function is defined as

$$LoG(x, y) = -\frac{1}{\pi\sigma^4}\left[1 - \frac{x^2 + y^2}{2\sigma^2}\right]e^{-\frac{x^2+y^2}{2\sigma^2}} \quad (mean\,is\,(0, 0)\,and\,standard\,deviation\,is\,\sigma)$$

The LoG function is represented as

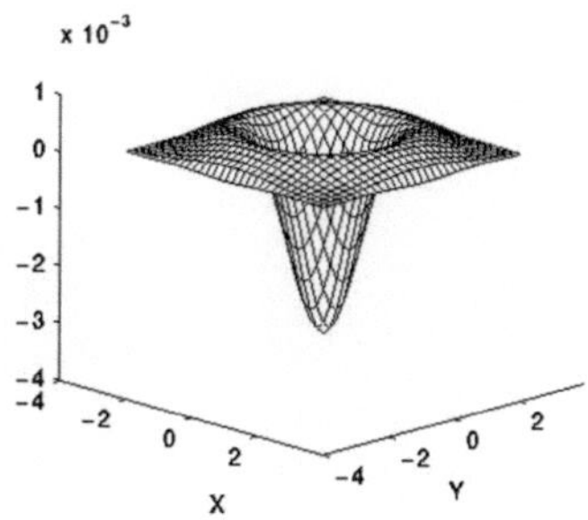

(*The x and y axes are marked in standard deviations* (σ))

LoG filter has many advantages like: (1) Generally Laplacian and Gaussian kernels are much smaller than the image, so LoG filter requires fewer arithmetic operations. (2) The LoG kernel can be precomputed so, that it can be directly convolved with the image at run-time. Thus, only one convolution is performed per image at run-time.

2.1.2 Usage

Pre-processing images is an integral part of Face Recognition systems. Input images were enhanced by improving the contrast and brightness, in order to optimize the performance of the proposed Face Recognition system.

Knowing the advantages of the LoG filter over the traditional Laplacian and Gaussian filter, LoG filter was chosen for pre-processing the images. LoG filter measures the amount of change of image intensity per change in image position [11]. So, the response of the LoG filter will be zero for all the image patches having a constant pixel intensity. On the other hand, whenever the intensity changes the LoG filter will return a positive response on the darker side and negative response on the lighter side [11]. This is depicted in the image below.

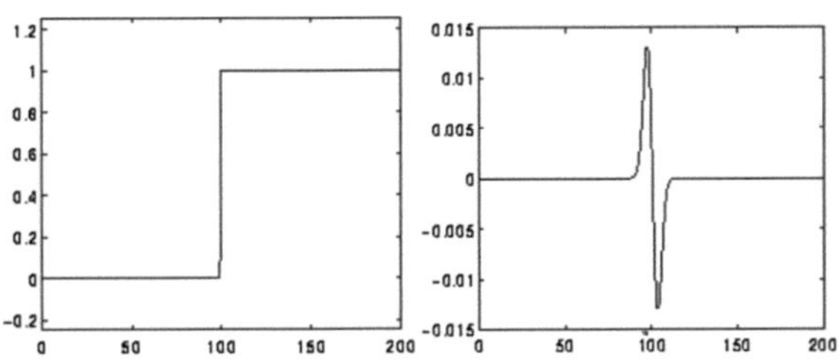

So, basically LoG filter is used to highlight all the edges present in an image (since intensity changes across an edge). This is depicted in the image in Fig. 2. Gaussian filter removes the additional details and noise from the input image and then the Laplacian filter predicts the edges in the image. Now, when the filtered image is subtracted from the original image then, the edges in the resulting image are much sharper and have higher contrast [11]. So, this enhances the image. This is depicted in the image in Fig. 2.

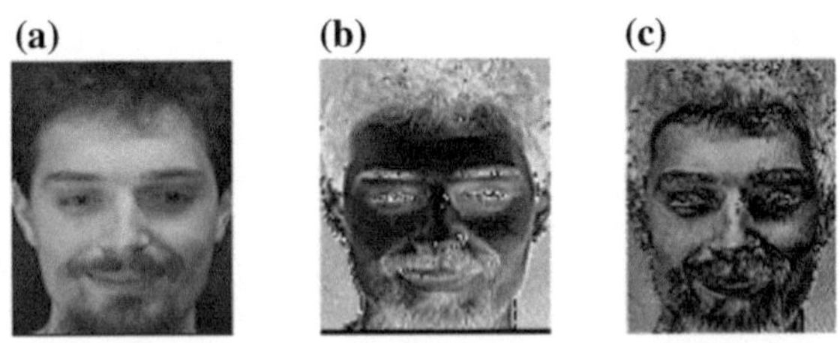

Fig. 2 **a** Original image, **b** image obtained after applying LoG filter and **c** image obtained after subtracting the filtered image from the original image

2.2 *Feature Descriptors*

2.2.1 Dense Sift

SIFT is a feature extraction algorithm which helps in detecting stable feature points in an image. The sole purpose of SIFT algorithm is to obtain the feature descriptors that overcome several computer vision challenges such as rotation invariance, scale invariance and robust to variations in geometric transformations [7]. SIFT extracts features from a given image by detecting interest points in the image [7]. SIFT detector is implemented by the Difference-of-Gaussian function. DoG finds possible interest points that are invariant to scale and rotation [7].

$$G(x, y, \sigma) = \frac{1}{2\pi\sigma^{2c}}\left(x^2 + y^2\right)/2\sigma^2$$

DoG is accomplished by the convolving the Gaussian Filter on the image at different scales [7]. DoG image is described as below:

$$L(x, y, \sigma) = G(x, y, \sigma) * l(x, y)$$

Where the term L(x, y,) represents the convolved image. Eventually, the difference between successive Gaussian-blurred images is calculated [7]. The operation of the DoG function is shown below:

$$D(x, y, \sigma) = (G(x, y, k\sigma) - G(x, y, \sigma)) * l(x, y) = L(x, y, k\sigma) - L(x, y, \sigma)$$

These key-points reveals detailed information about the location, orientation, and scale. It then computes the local descriptor for the local region around the keypoint. The combination of all these computed descriptors gives the entire feature descriptor for an image [7].

But, SIFT has many limitations. SIFT detector can't detect enough number of keypoints if an image is ill-illuminated [9]. The classical SIFT detector is generally used on large images to make sure that enough number of interest points are detected [9]. Dense SIFT overcomes these problems by making use of dense pixel grid representation of images and considering the regular image grid points as keypoints [9]. Thus, DSIFT is able to detect a sufficient number of keypoints irrespective of the illumination and size of the image. DSIFT descriptor computes feature descriptors for each of these keypoints producing a dense representation of facial features. These descriptors are finally concatenated to form the feature vector for the face [9].

2.2.2 Dense SURF

SURF was proposed to speed up the computation required by feature detection and extraction [2, 15]. It is made up of a scale and in-plane rotation invariant feature detector and descriptor [2]. The feature detector does the job of detecting keypoints in an image and the is used to describe the features of these detected keypoints by constructing feature vectors.

SURF feature detector uses the determinant of the approximate Hessian matrix as the underlying principle [2]. It calculates the determinant at all the points in the image and detects droplet-like structures wherever the determinant is at maximum [2]. But, these calculations are quite expensive. So, SURF uses integral images to reduce the computation time. For any point x = (x, y) in an image at scale, the Hessian matrix H(x,) is calculated as:

$$H(\mathrm{x},\sigma) = \begin{bmatrix} L_{xx}(\mathrm{x},\sigma) & L_{xy}(\mathrm{x},\sigma) \\ L_{xy}(\mathrm{x},\sigma) & L_{yy}(\mathrm{x},\sigma) \end{bmatrix}$$

where $L_{xx}(x,\sigma)$, $L_{xy}(x,\sigma)$, $L_{yy}(y,\sigma)$ are defined as convolutions of Gaussian second order partial derivatives on point x in image I. In order to reduce the computation cost a set of box filters is used by SURF to approximate the Gaussian and represent the lowest scale for computing the droplets (blobs) [2]. These are denoted by $D_{xx}(x,\sigma), D_{xy}(x,\sigma)\, and\, D_{yy}(x,\sigma)$. The result produced is:

$$\det\left(H_{appox}\right) = D_{xx}D_{yy} - \left(\omega D_{xy}\right)^2$$

where ω is the weight used for conserving energy between Gaussian kernels and approximated Gaussian kernels. The value of ω can be calculated as:

$$\omega = \frac{\left|L_{xy}(1.2)\right|_F \left|D_{yy}(9)\right|_F}{\left|L_{yy}(1.2)\right|_F \left|D_{xy}(9)\right|_F} = 0.912 \approx 0.9$$

Here, |XF| is Frobenius Norm.

For incorporating scale invariance, like SIFT, SURF also generates a pyramid scale space. But it does this in a unique way. Since SURF makes use of box filters and integral images so it generates the scale space by directly varying the scale of box filters [2].

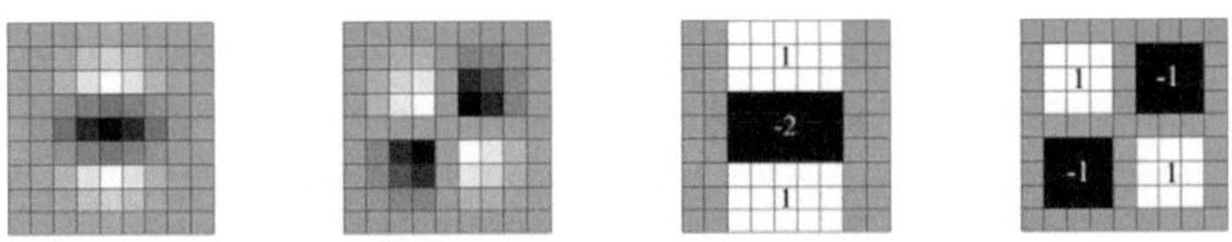

SURF feature descriptor is based on the local Haar wavelet responses [2]. It calculates the sum of Haar wavelet responses and uses it to describe the feature of a keypoint. To compute the descriptor a square region centered at the key point is constructed and oriented along the direction given by the orientation selection method [2]. Now the square region is divided into smaller 4 × 4 square sub-regions. Now each sub-region is further split into 5 × 5 squares and Haar wavelet response is calculated for each of these squares. Haar wavelet response in x-direction and y-direction are denoted by dx and dy respectively. To increase robustness towards errors, the responses d_x and d_y are weighted with a Gaussian centered at the keypoint [2].

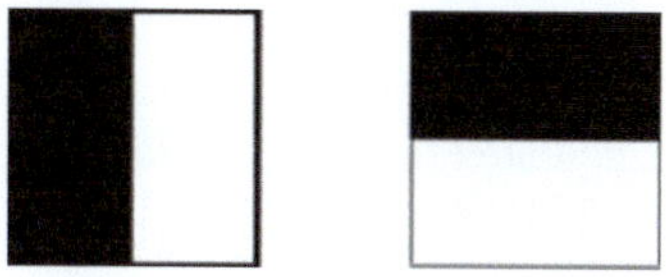

Then the sum of wavelet responses d_x and d_y is computed over all the sub-regions. These form the first entries of the feature descriptor of the keypoint [2]. Other entries are also made in order to capture various types of information.

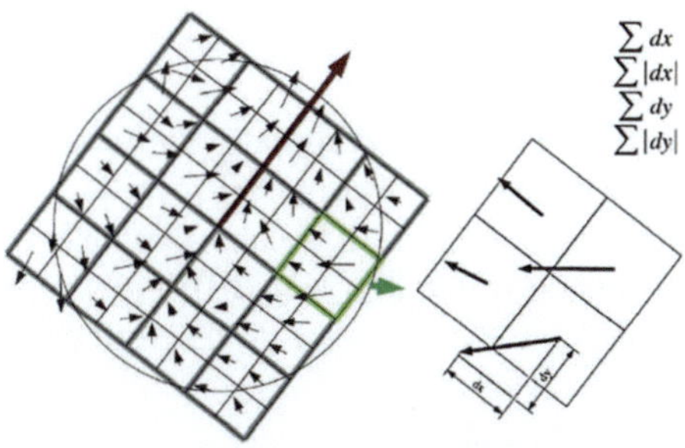

But, SURF faces problems when an image is small, does not have a proper orientation or is ill-illuminated. DSURF is an enhanced version of SURF. The main problem with the classical SURF detector is that the number of false positives is high [5]. SURF extracts image features by detecting keypoints in the image. But, if the image is not properly oriented or illuminated then very few keypoints are detected in the image leading to very few descriptors [5]. So, DSURF overcomes this limitation by using a dense pixel grid representation for images [5]. It considers the regular image grid points as keypoints and generates descriptors for them. So, DSURF is able to generate a good number of descriptors for every image irrespective of the conditions under which it is captured. Experimental results show that this modified version of SURF is better as it makes keypoint detection invariant to illumination and orientation.

2.2.3 Dense ORB

ORB makes use FAST feature detector and BRIEF descriptor [3]. It adds an orientation component to the well known FAST descriptor by using the Intensity Centroid approach [6] and creates a variant of the classical BRIEF descriptor which is rotation invariant [6].

The Intensity Centroid approach uses a robust measure of corner orientation. The centroid is calculated using the moments of an image patch [6]. The $(p+q)$th order moment whose intensity function is $I(x, y)$, can be calculated as:

$$m_{pq} = \sum_{x,y} x^p y^q I(x, y)$$

Once the moments are calculated then the centroid is given by:

$$C = \left(\frac{m_{10}}{m_{00}}, \frac{m_{01}}{m_{00}}\right)$$

Now, a vector joining the center and centroid is constructed and the orientation of the patch is calculated by:

$$\theta = a\tan2(m_{01}, m_{10})$$

where $a$tan2 is arctan. This approach incorporates illumination invariance as angle measures are independent of the type of corner [6].

Secondly, ORB includes a rotation invariant component called r-BRIEF [6] which is an improved version of the classical BRIEF descriptor. To achieve rotation invariance, ORB steers the BRIEF in the direction of orientation of key-points [6].

This is achieved in the following way:

Suppose that for any binary feature set constituting of n tests at a point (x_i, y_i) results in a matrix represented as:

$$S = \begin{pmatrix} x_1, \ldots, x_n \\ y_1, \ldots, y_n \end{pmatrix}$$

Now, by utilizing the patch orientation (Θ) and corresponding rotation matrix (R_Θ) a steered version of the original S can be obtained.

$$S_\Theta = R_\theta S$$

Subsequently, the steered BRIEF operator is defined as:

$$g_n(p, \Theta) := f_n(P) | (x_i y_i) \in S_\Theta$$

But, ORB faces problems when an input image is not properly illuminated. ORB uses FAST detector with some modifications to make it invariant to orientation but, it does not handle illumination invariance. So, if an image is ill-illuminated or has low contrast then FAST detects only a few keypoints and is not able to describe the image features properly. DORB overcomes this limitation by using a dense pixel grid representation for images. It increases the number of keypoints in an image by considering regular image grid points as keypoints. So, the number of keypoints detected by DORB is independent of the conditions under which the image is captured. So, now the r-BRIEF descriptor is able to describe every image properly irrespective of its illumination and contrast.

2.3 *Fisher Vector*

2.3.1 Background

Fisher Vector

Patch-aggregation techniques have proved to be effective in recent past, revealing high performance for a variety of computer vision tasks. Fisher Vector (FV) is another patch-aggregation technique which uses Fisher Kerne (FK)l as its underlying principle [16]. FK framework derives a kernel by characterizing an image based on the deviation from a generative data model [17]. The FV is represented vectorially, which is obtained by the calculating the slope of the log-likelihood to the model parameters [13, 17, 18]. FV is a high-dimensional vector formed by aggregating vast set of feature vectors extracted by various feature descriptors (e.g. DSIFT, DSURF, DORB).

Fisher Kernel

FK is used because of its potential of being used in learning a model when the training objects have a different underlying graph structure. It is based on the concept of having similar log-likelihood gradients for similarly structured objects in a generative model [17, 18].

Let $X = \{x_n, t = 1, 2, \ldots, N\}$ where $x_n \in \chi$ is a set of D-dimensional local descriptors, like DSIFT, DSURF or DORB descriptors [17]. By the theory of information geometry, a Riemannian manifold M_A with a local metric is derived by the Fisher Information Matrix(FIM) $F_\lambda \in \mathbb{R}^{M \times M}$

$$F_\lambda = E_{X \sim u_\lambda}\left[G_\lambda^X G_\lambda^{X^T}\right]$$

where u_λ is the probability density function for the elements in χ where $\lambda = \lambda_1, \lambda_2, \ldots, \lambda_M \in \mathbb{R}^M$ which represents a vector with M parameters of u_λ.

FKl for two samples X and Y is defined as:

$$K(X, Y) = G_\lambda^{X^T} F_\lambda^{-1} G_\lambda^Y.$$

By the Cholesky decomposition, equation can be written as a dot product:

$$K_{FK}(X,Y) = \widehat{G}_{\lambda}^{X^T} \widehat{G}_{\lambda}^{Y}$$

where $\widehat{G}_{\lambda}^{X} = L_{\lambda} G_{\lambda}^{X} = L_{\lambda} ?_{\lambda} \log u_{\lambda}(X)$, $\widehat{G}_{\lambda}^{X}$ is known as the Fisher Vector of X. Let us assume that samples are independent, we can write the equation as below:

$$\widehat{G}_{\lambda}^{X} = \sum_{n=1}^{N} L_{\lambda} \nabla_{\lambda} \log u_{\lambda}(x_n).$$

According to the assumption, FV is a sum of the normalized gradient for each descriptor. The contribution by each x_n can be inferred as an embedding of local descriptors x_n in a high-dimensional space. Gaussian Mixture Model is selected as u_{λ} [13, 17]. We are denoting T-component GMM by $\lambda = \{w_t, u_t, \Sigma_t, t = 1, \ldots, T\}$ where w_t, u_t, Σ_t are mixture weight, mean vector and covariance matrix of Gaussian t.

L_{λ} is calculated by taking square-root of the inverse of FIM. The normalized gradients can be formulated by performing coordinate-wise normalization of the gradient vectors. Initially, the accumulators are initialized as $S_t^0 \leftarrow 0, S_t^1 \leftarrow 0, S_t^2 \leftarrow 0$ for $\forall \{t \in \mathbb{R} \mid 1 \leq t \leq T\}$. For each of the local image descriptors, posterior probability is derived by $\gamma_n(t) = \frac{w_t u_t(x_n)}{\sum_{j=1}^{T} w_j u_j(x_n)}$, then update the accumulators with the S_t^0, S_t^1, S_t^2 with $\gamma_n(t), \gamma_n(t)x_n$ *and* $\gamma_n(t)x_n^2$ respectively [17]. In terms of statistics, these computed normalised gradients can be written in the form of 0th-order, 1st-order and 2nd-order statistics:

$$S_t^0 = \sum_{n=1}^{N} \gamma_n(t)$$

$$S_t^1 = \sum_{n=1}^{N} \gamma_n(t) x_n$$

$$S_t^0 = \sum_{n=1}^{N} \gamma_n(t) x_n^2$$

After the statistics are computed, the signature of the Fisher Vectors for all the t components of the GMM needs to be accounted by the following equations:

$$\widehat{G}_{\alpha_t}^{X} = \left(S_t^0 + N w_t\right) / \sqrt{w_t}$$

$$\widehat{G}_{\mu_t}^{X} = \left(S_t^1 + \mu_t S_t^0\right) / \sqrt{w_t} \sigma_t$$

$$\widehat{G}_{\sigma_t}^{X} = \left(S_t^2 - 2\mu_t S_t^1 + \left(\mu_t^2 - \sigma_t^2\right) S_t^0\right) / \sqrt{2 w_t} \sigma_t^2$$

where α_t is the re-parametrization of the following the definition of soft-max formalism. Using the Eq. (19), the components are calculated separately. All the FV components are concatenated to form a single vector representing FV.

To improve the results with various linear classifiers it is a necessity to use normalization techniques. Different normalization techniques have been proposed in past [13, 17]. Some of them are l2-normalization, power normalization. FV depends on some percentage of the image-specific proportion (ω). Accordingly, this can be inferred from the fact that two images having the same objects but different scales have different signatures. l2-normalization is used to eliminate the dependence on ω.

$$\widehat{\mathrm{G}}_\lambda^X = \widehat{\mathrm{G}}_\lambda^X / \sqrt{\widehat{\mathrm{G}}_\lambda^{X^T} \widehat{\mathrm{G}}_\lambda^{X^X}}$$

Power normalization is applied for all i = 1, ..., T(2D + 1) of the form:

$$\left[\widehat{\mathrm{G}}_\lambda^X\right]_i - \mathrm{sign}\left(\left[\widehat{\mathrm{G}}_\lambda^X\right]_i\right)\sqrt{\left|\widehat{\mathrm{G}}_\lambda^X\right|_i}\,\Bigg]$$

In the experiments performed, power coefficient ρ has been set to 1/2. This adjustment is also referred to "signed square rooting" and has been found advantageous for image representations [13, 16].

2.4 Extended Nearest Neighbor Classifier

As the name suggests, this classifier is an extension of the well known KNN classifier. It approximates the optimal Bayes theorem and enhances the performance of KNN and weighted-KNN classifiers [8].

Classifiers are broadly classified into two types namely parametric classifiers and non-parametric classifiers. ENN classifier comes under non-parametric classifier. In non-parametric classifiers, the classification rules are independent of the underlying distribution of input data [8]. Non-parametric classifiers have been used extensively recently.

Talking about the KNN classifiers, they have numerous advantages such as simple implementation, great performance on the data independent of the underlying data distribution.

But they have a lot of shortcomings, like determining the optimal value of k. A straightforward approach to solve this would be to try out different values of k and choose the one which produces optimal results. The second problem is choosing an appropriate distance measure.

KNNs are influenced heavily by the distribution of predefined classes [19]. The outcome, i.e. the classification of the test data is more likely to be decided by the class with higher density. Suppose there are two classes A and B, and class A has a lower variance which means that data points appear to be more concentrated and class B has a distribution which is more spread out. This clearly leads to misclassification of the test data points since the nearest neighbors from class A will be more dominant.

ENNs works independently of the fact that whether the data points of the class are well spread or they have a concentrated distribution. ENN doesn't only classify the test samples by just finding the nearest neighbors of the predefined classes but also takes into account the test samples as which are their nearest neighbors [8].

Defining the general class wise T_i as the following:

$$T_i = \frac{1}{n_i k} \sum_{x \in S} \sum_{r=1}^{k} I_r(x, A = A_1 \cup A_2)$$

where, A_1 and A_2 denote the samples belonging to the class 1 and class 2. And A is the union of the A_1 and A_2, k is the number of nearest neighbor. I is the indicator function, sees if both the sample x and it's rth nearest neighbor are part of the same class, defined as follows:

$$I_r(x, A) = \begin{cases} 1, if\ x \in AA_i\ and\ NN_r(x, A) \epsilon A_i \\ 0,\ \text{otherwise} \end{cases}$$

where NNr denotes the rth nearest neighbor of x in A.

The intra-class coherence is defined as follows:

$$\theta^j = \sum_{i=1}^{2} T_i^j$$

$$f_{ENN} = \arg\max \sum_{t=1}^{2} T_i^j = \arg\max_{j \in 1,2} \theta^j$$

ENN.V1

$$T_i^j = \frac{1}{n_i' k} \sum_{x \in A_i \cup z} \sum_{r=1}^{k} I_r\left(x, A' = A_1 \cup A_2 \cup \{z\}\right).$$

when i = j we have,

$$T_i^j = \frac{\left(n_i k T_i + \Delta n_i^j + k_i\right)}{(n_i + 1)k}$$

and when,

$$T_i^j = T_i - \Delta n_i^j / n_i k$$

Therefore we have,

$$f_{\mathrm{ENN}} = \arg_{j\in 1,2,\ldots,N} \max \sum_{i=1}^{N} \left(T_i^j - T_i\right)$$

$$f_{\mathrm{ENN}} = \arg_{j\in 1,2,\ldots,N} \max \left\{ \left(T_i^j - T_i\right)_{i=j} + \sum_{i\neq 1}^{N} \left(T_i^j - T_i\right) \right\}$$

The ENN decision rule can be formulated as:

$$f_{ENN} = \arg_{j\in 1,2,\ldots,N} \max\left\{\Delta n_j + k_j - kT_j\right\}$$

3 Experimental Design

3.1 Face Datasets

3.1.1 ORL (Olivetti Research Laboratory) Dataset

The dataset consists of 40 subjects with 10 distinct images per subject, totaling to 400 images. This dataset is created specifically for Face Recognition. This dataset consists of very diverse images, captured under various lighting conditions. The dataset also captures a wide range of facial expressions which makes it a good choice for unconstrained face recognition (pose, expression, and illumination invariant) applications.

3.1.2 Faces94

This dataset consists of 153 subjects with 20 images per subject, totaling to 3060 images. The dataset consists of 133 male and 20 female subjects. The images are taken from a fixed distance by the camera under the same lighting conditions, so there are no scale or illumination variations. The subjects are speaking so, there are considerable expression variations. So, this dataset is generally preferred for expression invariant applications.

3.1.3 Grimace

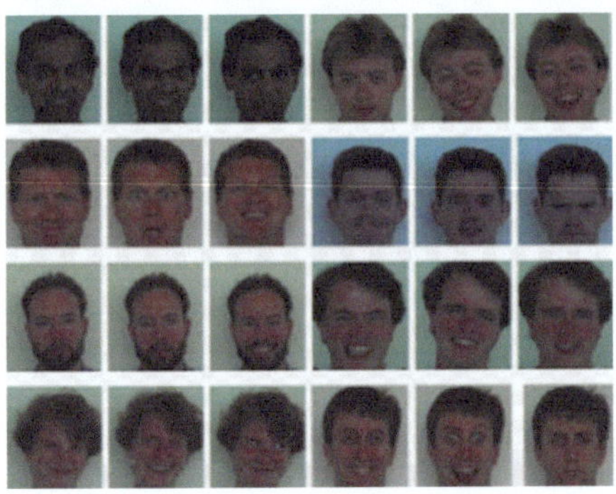

This dataset consists of 18 subjects with 20 images per subject, totaling to 360 images. All the images of a subject are taken in a single session with a 0.5-s interval between two consecutive image captures. During the session, subjects try to make grimaces by varying their poses and facial expressions. So, this dataset is generally preferred for pose and expression invariant applications.

4 Experimental Results and Visualization

Various experiments were carried out in order to evaluate the performance of our proposed system. We used Accuracy, Precision and Recall as the performance metrics. We compared the results obtained from different dense feature descriptors. We also compared the results obtained from the dense descriptors with their traditional counterparts I.e. we compared the results of DSIFT [9] with SIFT [1], DSURF [5] with SURF [2] and DORB with ORB [3] descriptor.

4.1 Dense SIFT

The traditional SIFT descriptor fails to describe an ill-illuminated, ill-oriented image properly [9]. Actually the SIFT detector is not able to generate enough number of keypoints for such an image. DSIFT detector increases the number of keypoints in the image by making use of regular image grid points as interest points and passes these new keypoints to the DSIFT descriptor [9]. This is depicted in the figure below. The DSIFT detector takes a parameter which determines the grid size used to represent the input images. It's value is dependent on the training dataset. We tuned this parameter to achieve the optimal results. For the ORL dataset grids containing squares of size 5 *pixel* × 5 *pixel* gave the best results. For the Faces94 dataset grids containing squares of size 4 *pixel* × 4 *pixel* produced the best results. Whereas for Grimace dataset grids containing squares of 3 *pixel* × 3 *pixel* gave optimal results.

4.2 Dense SURF

The classical SURF descriptor fails to describe an ill-illuminated, ill-oriented image properly [5]. SURF detector is not able to generate enough number of keypoints for such an image. DSURF detector increases the number of keypoints in the image by making use of regular grid points as keypoints and passes these new keypoints to the DSURF descriptor [5]. This is depicted in the figure below. The dense SURF descriptor takes a parameter which determines the grid size used to represent the input images. Its value is dependent on the training dataset. We tuned this parameter to achieve the optimal results. For the ORL dataset grids containing squares of size 11 *pixel* × 11 *pixel* gave the best results. For the Faces94 dataset grids containing squares of size 10 *pixel* × 10 *pixel* produced the best results whereas for Grimace dataset grids containing squares of size 15 *pixel* × 15 *pixel* gave optimal results.

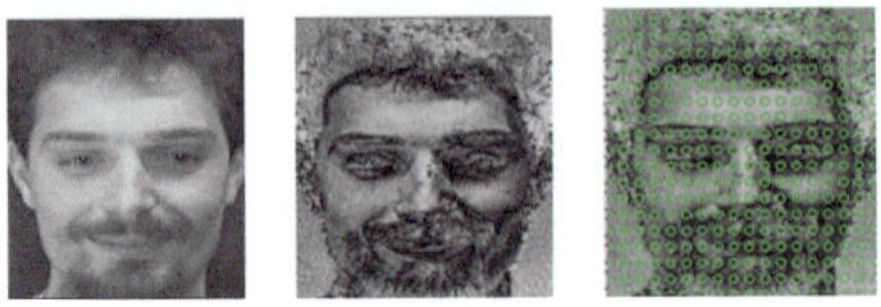

4.3 Dense ORB

ORB employs a FAST detector which is rotation invariant [3]. But, it fails to incorporate illumination invariance. Because of this, ORB fails if the images are ill-illuminated or have a low contrast. DORB is able to counter this by using regular image grid points as keypoints. This way it can detect keypoints even in a poorly lit image. This is depicted in the figure below. The FAST detector present in DORB, takes a parameter which determines the grid size used to represent the input images. For the ORL, Faces94 and Grimace datasets grids containing squares of size 3 *pixel* × 3 *pixel* produced optimal results.

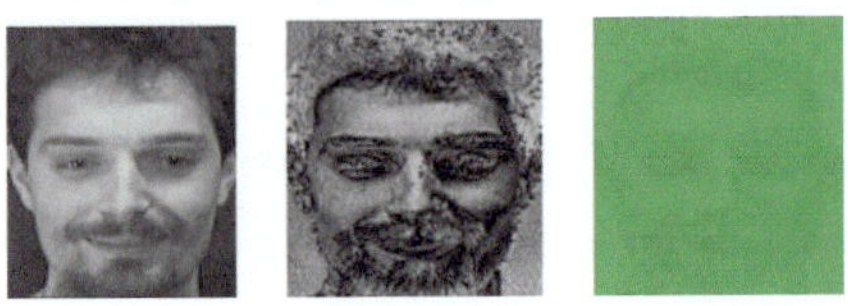

4.4 Performance Evaluation

On the ORL dataset, the proposed DSIFT descriptor and DORB descriptor performed quite well. DSIFT gave better results than DORB. These two descriptors surpassed all other descriptors. DSIFT outperformed DORB by an accuracy margin of 0.54% and DSURF by an accuracy margin of 16.96%. Also, DSIFT outperformed SIFT by an accuracy margin of 5.41%, DSURF outperformed SURF by an accuracy margin of 1.39% and DORB outperformed ORB by an accuracy margin of 3.26%.

On the Faces94 dataset, the proposed DSIFT descriptor and DORB descriptor performed quite well. DORB gave better results than DSIFT. DORB outperformed DSIFT by an accuracy margin of 0.79% and DSURF by an accuracy margin of 13.56%. Also, DSIFT outperformed SIFT by an accuracy margin of 4.02%, DSURF outperformed SURF by an accuracy margin of 0.23% and DORB outperformed ORB by an accuracy margin of 1.48%.

On the Grimace dataset, the proposed DSIFT descriptor and DORB descriptor performed quite well. DSIFT gave better results than DORB. DSIFT outperformed DORB by an accuracy margin of 1.89% and DSURF by an accuracy margin of 17.21%. Also, DSIFT outperformed DSIFT by an accuracy margin of 6.35%, DSURF outperformed SURF by an accuracy margin of 1.01% and DORB outperformed ORB by an accuracy margin of 2.3%.

Feature descriptors	Accuracy (%)	Precision (%)	Recall (%)
Dense SIFT	94.25	92.10	97.25
Dense SURF	77.29	78.76	84.30
Dense ORB	93.71	91.27	95.00
SIFT	88.84	89.73	91.34
SURF	75.90	77.90	81.14
ORB	90.45	90.89	93.67

Feature descriptors	Accuracy (%)	Precision (%)	Recall (%)
Dense SIFT	97.36	97.72	98.11
Dense SURF	84.59	83.20	90.23
Dense ORB	98.15	98.07	99.44
SIFT	93.34	92.50	95.00
SURF	84.36	84.20	88.47
ORB	96.67	96.23	96.16

Feature descriptors	Accuracy (%)	Precision (%)	Recall (%)
Dense SIFT	98.44	96.85	99.32
Dense SURF	81.23	83.48	88.91
Dense ORB	96.55	94.67	97.30
SIFT	92.09	93.43	94.17
SURF	80.22	79.65	84.73
ORB	94.25	94.5	95.23

4.5 Performance Comparison

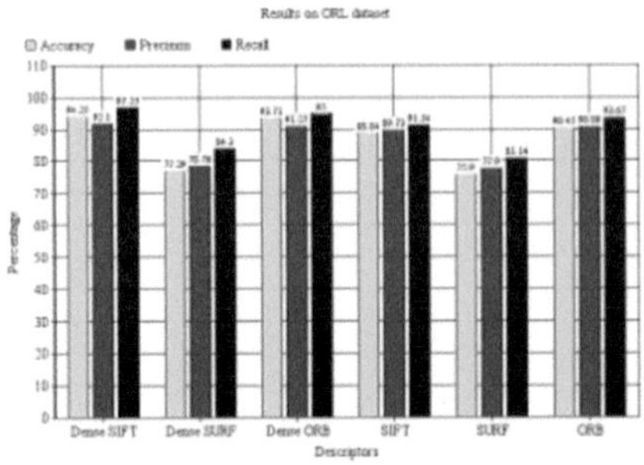

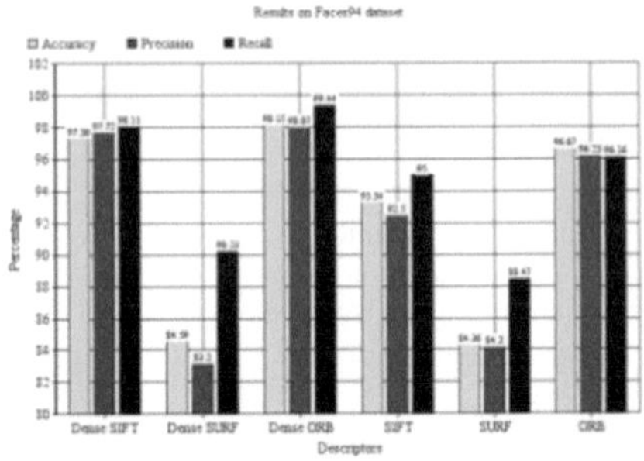

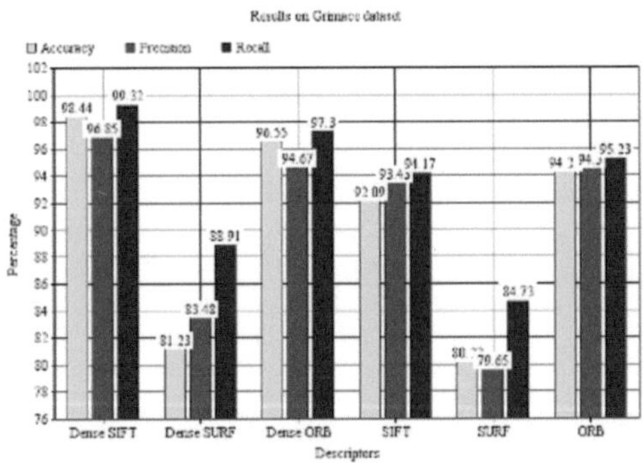

5 Conclusion and Future Work

This paper introduces a novel pipeline for Face Recognition. It employs dense feature descriptors for feature extraction and extended nearest neighbor classifier for the classification task. This paper also provides a detailed comparison of various dense feature descriptors (DSIFT, DSURF, and DORB) with themselves and with their classical counterparts (SIFT, SURF, and ORB). Upon extensive experimentation, we are able to conclude that DSIFT and DSURF surpass other feature descriptors in terms of accuracy, precision, and recall. Therefore, these are better suited for face recognition.

In future, we would focus on making the model more robust and making it work under unconstrained scenarios i.e. invariant to scaling, illumination, occlusion, and age.

References

1. Lowe, D.G.: Distinctive image features from scale invariant keypoints. Int. J. Comput. Vis. **60**, 91–110 (2004)
2. Bay, H., Ess, A., Tuytelaars, T., Van Gool, L.: Surf: speeded up robust features. Comput. Vis. Image Underst. (CVIU) **110**(3), 346–359 (2008)
3. Rublee, E., Rabaud, V., Konolige, K., Bradski, G., ORB: an Efficient Alternative to SIFT or SURF. Willow Garage, Menlo Park, California
4. Križaj, J., Štruc, V., Pavešić, N.: Adaptation of SIFT Features for Face Recognition Under Varying Illumination, May 24–28. MIPRO 2010, Opatija, Croatia (2010)
5. Dreuw, P., Steingrube, P., Hanselmann, H., Ney, H.: SURF-Face: Face Recognition Under Viewpoint Consistency Constraints. Human Language Technology and Pattern Recognition RWTH Aachen University Aachen, Germany (2009)
6. Vinay, A., Kumar, CA., Shenoy, G.R., Murthy, KNB., Natarajan, S.: ORB-PCA based feature extraction technique for face recognition. In: Second International Symposium on Computer Vision and Internet (2015)
7. Vinay, A., Hebbar, D., Shekhar, V.S., Murthy, K.N.B., Natarajan, S.: Two novel detector-descriptor based approaches for face recognition using SIFT and SURF. In: 4th International Conference on Eco-friendly Computing and Communication Systems, ICECCS (2015)
8. Panchal, P.M., Panchal, S.R., Shah, S.K.: A comparison of SIFT and SURF. Int. J. Innov. Res. Comput. Commun. Eng. **1**(2) (2013)
9. Wang, J.G., Li, J., Lee, C.Y., Yau, W.Y.: Dense SIFT and Gabor descriptors-based face representation with applications to gender recognition. In: 2010 11th International Conference on Control Automation Robotics & Vision, pp. 1860–1864. Singapore (2010)
10. Geng, C., Jiang, X.: Face recognition using sift features. In: 16th IEEE International Conference on Image Processing (ICIP), pp. 3313–3316. IEEE (2009)
11. Fisher, R., Perkins, S., Walker, A., Wolfart, E.: Laplacian/Laplacian of Gaussian. Web article published in Image Processing Learning Resources (2004)
12. Darkos, N.: Laplacian of Gaussian (LoG). Computer Based Learning Unit, University of Leeds (1996)
13. Simonyan, K., Parkhi, O.M., Vedaldi, A., Zisserman, A.: Fisher Vector faces in the wild. Web article published by Visual Geometry Group Department of Engineering Science University of Oxford
14. Fisher, R., Perkins, S., Walker, A., Wolfart, E.: Gaussian Smoothing. Web article published in Image Processing Learning Resources (2004)
15. Kokkinos, I., Bronstein, M., Yuille, A.: Dense Scale Invariant Descriptors for Images and Surfaces [Research Report] RR-7914, INRIA. <hal-00682775> (2012)
16. Perronnin, F., Sánchez, J., Mensink, T.: Improving the fisher kernel for large-scale image classification. In: Proceedings of ECCV (2010)
17. Sanchez, J., Perronnin, F., Mensink, T., Verbeek, J.: Image classification with the fisher vector: theory and practice. Int. J. Comput. Vis. **105**(3), 222–245 (2013). Springer
18. Perronnin, F., Sánchez, J., Mensink, T.: Improving the Fisher Kernel for large-scale image classification. In: Daniilidis, K., Maragos, P., Paragios, N. (eds.) ECCV 2010. LNCS, vol. 6314, pp. 143–156. Springer, Heidelberg (2010)
19. Tang, B., He, H.: ENN: extended nearest neighbor method for pattern recognition (Research Frontier). IEEE Comput. Intell. Mag. **10**(3), 52–60 (2015)
20. Du, G., Su, F., Cai, A.: Face recognition using SURF features. In: Multimedia Communication and Pattern Recognition Labs, School of Information and Telecommunication Engineering, Beijing University of Posts and Telecommunications, Beijing 100086, China (2009)
21. Jose, J.P., Poornima, P., Kumar, K.M.: A novel method for color face recognition using KNN classifier. In: 2012 International Conference on Computing, Communication and Applications, pp. 1–3, Dindigul, Tamil Nadu (2012)

Aggregation of LARK Vectors for Facial Image Classification

A. Vinay, Vinayaka R. Kamath, M. Varun, Nidheesh, S. Natarajan and K. N. B. Murthy

Abstract Face recognition is prevailing to be a key aspect wherever there is a need for interaction between humans and machines. This can be achieved by containing a set of sketches for all the possible individuals and then cross-validating at necessary circumstances. We propose a mechanism to fulfil this task which is centred on locally adaptive regression kernels. A comparative study has been presented at encoding stages as well as at the classification stages of the pipeline. The results are cautiously examined and analyzed to deduce the best mechanism out of the proposed methodologies. All the ideologies have been tested for multiple iterations on benchmark datasets like ORL, grimace and faces 95. The vectorized descriptors have been subjected to encoding using slightly refined methods of feature aggregation and clustering to assist classifiers in imputing the test subjects to their respective classes. The encoded vectors are classified using Gaussian Naive Bayes, Stochastic Gradient Descent classifier, linear discriminant analysis and K Nearest Neighbour to accomplish face recognition. An inference on sparse nature of locally adaptive regression kernels was made from the experimentation. A rigorous study regarding the discrepancies of the performance of LARK descriptors is reported.

Keywords Adaptive Kernels · Bayesian · Classifier · Feature aggregation · Sparse features · Image · Classification

1 Introduction

Facial recognition is aimed at computing the similar and dissimilar features of an individual by combining the digital image data with the features extracted beforehand. The input image is compared with a library consisting of a collection of images which might not be similar in all respects to the compared image. This image will be contrasted with all the images of the library and then list out a collection of similar images, which often helps us recognise the input image. It can classify the

A. Vinay · V. R. Kamath (✉) · M. Varun · Nidheesh · S. Natarajan · K. N. B. Murthy
Center for Pattern Recognition and Machine Intelligence, PES University, Bengaluru, India
e-mail: vinayakkamath2010@gmail.com

S. Manna et al. (eds.), *Mathematical Modelling and Scientific Computing with Applications*, Springer Proceedings in Mathematics & Statistics 308,
https://doi.org/10.1007/978-981-15-1338-1_31

dataset that the image is obtained from. A set of unique and recognisable features are extracted from the images of individuals and are fed into the matchers. This system identifies the nodal points of prime importance. These points act as main features which highlight on primary facets such as the distance and breadth of the nose, the depth of the eye sockets and the measurement of the cheekbones. These systems work by collating data of nodal points on the digital image of an individual's face and storing the data for further interpretation. These face prints are used as a scale to contrast with data obtained from numerous other faces present in an image or video. A plethora of applications can be tied up to face recognition. Uses include fraud detection in visas and passports, increased security which maps facial data of the card user against ATM's and banks, tracking of criminals; prevent voters from committing fraud and to maintain a record of attendance.

Conventionally, the security factor is what most facial recognition systems work on. There are several advancements in the field of feature extraction and their description which has spread across multiple domains including face recognition, object detection and automation. These algorithms have played a key role in several applications as well. Consumer digital imaging requires several features to be considered. Putting up with uncontrolled lighting conditions, large pose variations, facial expressions, makeup, changes in facial hair, aging, partial occlusions, loss in pixels and many more parameters can be a tough nut to crack. This paper is an attempt to exhibit a pipeline which is not only computationally inexpensive but highly accurate as well. Locally adaptive regression kernels have proven to be capable descriptors and have shown significant potential to participate in simple and accurate classification models. But selectively aggregating features using a clustering approach is very important to extract feature descriptors for those regions and restrict to regions that is likely to contain specific interest points. Hence the strategy is to find possible clusters in the vector space acquire region descriptors from them and match these vectors based upon their region. The trained classifiers can help us in categorizing these vectors into our interest regions. The proposed pipeline has shown accuracy up to 96% on validation on different benchmark datasets. This paper is an attempt to make a comparative study to explore the properties of LARKs in order to perform facial image classification. A post-processing measure has been implemented on the vectorized LARKs in order to enable us to achieve better results by eliminating overlapping features and obtain more concise results.

Section 2 deals with relevant works that is associated with face recognition and any of the techniques used in the proposed system. Section 3 is an attempt to explain the mechanisms involved in the pipeline to the core. The results are presented and briefly critiqued in the Sect. 4. The outcome of the pipeline is analysed and conclusions are made on the basis of the observations in the Sect. 5.

2 Related Work

Locally adaptive regression kernels [1] have been widely used for non-parametrized training free object detection [2] in a real world application. They have squealed by providing higher accuracy than their counter parts. Several papers used LARK representation of the vectors to achieve face verification within limited amount of computational resources. A popular variant of descriptor called local binary patterns (LBPs) [3] have also shown remarkable outcomes in achieving face recognition. Since LBPs came into light, various version of LBP as three-patch LBP (TPLBP) [4], and four-patch LBP (FPLBP) [5] have been proposed by a different set of individuals. These mapped with the OSS measure helped in growth of "one-shot learning" techniques. Various mechanisms which used learning based descriptors also gained popularity overtime.

The discourse of face verification based on aging effects has been done by Ramanathan and Chelappa with the help of Bayesian and Probabilistic Eigen space [6]. This gave a staggering result with only an average error rate of 8.5%. In [7], Ismail and El-Khoribi with the application of HMT (Hidden Markov Tree) obtained a specification on numerous databases of face images by a of age difference of 5 months which was further divided into 4 junctures. With a range disparity of 20 months, promising results reaching 98% were acquired. An attempt to classify facial images was done by Turk and Pentland using eigen faces [8]. A simple yet affective approach assumed face recognition as two dimensional problems and the frame work designed learned to recognize new faces in an unsupervised manner. Locally adaptive regression kernels have been used for target detection and localization [9] as well. Bag-of-words model, which inspired the design of Bag of lark features model [10], has been used for object recognition by Soon Wei Jun and Safirin Karis. The algorithm learnt new patterns from the code book it creates and learns to classify using those features. Locally-constrained linear coding [11] was used instead of VQ coding in traditional SPM. This performs significantly better than its counter parts on several benchmarks. The time efficiency of LLC helped it to gain popularity in short time. In [12], Zhang and Feng evaluated the performance of naive bayes in text classification applications and gave an improvement over orthodox approach. The new technique exhibited better results that the conventional method.

Chen and Wang focused on achieving multi-face detection system in real time with a bit of hardware acceleration using FPGA [13]. The method used naive bayes for classification and focused on achieving the task using low-memory and in real-time. A slightly different approach was proposed in [14], which combined local features and selected them for naive Bayesian classification. K Nearest neighbour was optimised to objective function based sparse representation to generate locally linear k nearest neighbours (LLK) [15]. The mechanism used two classifiers, an LLK-based classifier and a locally linear nearest mean-based classifier. Novel theoretical analysis was also presented which included the nonnegative constraint, group regularization and also threw light on the computational efficiency of the LLK method. In [16], various classification techniques were benchmarked which helped us rule out the classifiers and limit our study to the one being used in this pipeline. A case study was performed using SVM, KNN, LDA and KNN with PCA along with a thorough analysis of the results to deduce a conclusion of the superiority of the classifiers.

3 Proposed Methodology

3.1 Overview

The region of interest is cropped out from the raw images of an individual to prevent any distortions in the background. As a further measure LARKs are extracted from the images and are converted into vectors by removing overlapping from the raw LARKs which gives a visual impression of the generated LARKs. These vectors act as descriptors of keypoints for the raw image. Eventually they are used for classification and image recognition after aggregation if these features. HOGSVD is then applied to these vectors for the purpose of dimensionality reduction and making them computationally inexpensive. Bag of Lark Features (BOLF) is a clustering algorithm which is used to cluster and compile elements having similar features. Locally constrained linear coding (LLC) is used as a replacement of SPM approach which is on similar grounds as the BOLF. Fischer vectors [17] is an extension of the bag of visual words feature based on visual vocabulary built in low level feature space. This concept is extended to vectors that describe the keypoints. These algorithms make the vectors ready to be fed into classifiers. LDA [18] finds a linear combination of features which selects and classifies objects or lists from analyzed objects. Naive Bayes classifier is based on the conditional probability of classification which uses the previous knowledge obtained. Stochastic Gradient Descent (SGD) is a classification algorithm used to measure analytically the degree of relation two given amongst values or images. Using K nearest neighbors [19] an object is classified based on the majority of votes obtained it's from its neighbours. These classifiers help to come to a judgement with regards to recognition (Fig. 1).

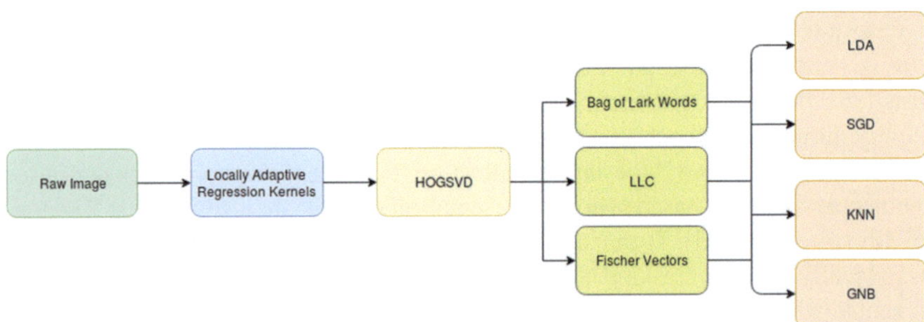

Fig. 1 Diagrammatic illustration of the pipeline

3.2 Locally Adaptive Regression Kernel (LARK)

Kernel Regression is a non-parametric technique to provide a general idea to the user about the conditional expectation of a completely random variable. The LARK [1] does not require any prior training to identify the image. In normal regression kernels, we usually use spatial differences to weigh the input values. However, in the case of locally adaptive regression kernels, we make use of not only spatial differences but also the difference in data (pixel gradients). Locally adaptive regression kernels denote and represent the local structure of an image taken into consideration. It helps give us a measure of local pixel similarities.

In order to recreate a low quality image on a high resolution image there is a need for a classic regression kernel, for denoising and deblurring the low quality image. The kernel regression framework used in the LARK features is explained as follows:

$$y_i = z(x_i + \varepsilon_i x_i \in \omega, \quad i = l \ldots S) \tag{1}$$

y_i is a denoised sample measured at $x_i = [x_{1i}, x_{2i}]^T$ where Z(x) is the required regression function, ε_i is an independently and identically distributed zero mean noise. P is the total number of samples in an arbitrary "window" ω around a position of interest X.

$$z(x_i) \approx z(x) + (\nabla z(x))^T (x_i - x) + 1/2!(x_i - x)^T Hz(x)(x_i - x) + \cdots \tag{2}$$

$$\approx \beta_0 + \beta_1^T (x_i - x)\beta_2^T vech\big((x_i - x)(x_i - x)^T\big) + \cdots \tag{3}$$

where ∇ and H are gradient and Hessian operators, while vech is the half-vectorization operator that lexicographically orders the lower triangular portion of the symmetric matrix into a column stacked vector. β_1 and β_2 can be mathematically defined as:

$$\beta_1 = \left[\frac{\partial z(x)}{\partial x_1}, \frac{\partial z(x)}{\partial x_2}\right]^T \tag{4}$$

$$\beta_2 = 1/2\left[\frac{\partial^2 z(x)}{\partial x_1^2}, \frac{\partial^2 z(x)}{\partial x_2^2}\right]^T \tag{5}$$

The vech operation can be illustrated as below,

$$\text{vech}\left(\begin{bmatrix} a & b \\ b & d \end{bmatrix}\right) = \begin{bmatrix} a & b & d \end{bmatrix}^T$$

$$\text{vech}\left(\begin{bmatrix} a & b & c \\ b & e & f \\ c & f & i \end{bmatrix}\right) = \begin{bmatrix} a & b & c & f & e & i \end{bmatrix}^T$$

Locally adaptive regression kernel can be formulated as follows:

$$K(C_i, x_i, x) = \exp\{-(x_i - x)C_l(x_i - x)\} \tag{6}$$

where,

$$C_i = \sum_{K \in \omega} \begin{bmatrix} z_{x_1}^2(x_k) & z_{x_1}x(k)z_{x_2}x(k) \\ z_{x_1}x(k)z_{x_2}x(k) & z_{x_2}^2(x_k) \end{bmatrix} \tag{7}$$

Vectorized LARKs act as key points and descriptors for the image. Overlapping patches are removed from the vectorized LARKs, hence giving a visual impression of the generated LARKs. This in turn can be used to plot an image. Different set of key points can be obtained from the pre-processed images by varying smoothness, window size and sensitivity, each time resulting in a slightly different set of vectorized version of LARKs. A unique set of LARKs is obtained every time the parameters are tweaked. Significant variations are observed in the visual LARKs based on the input image. These vectors are later exposed to some dimensionality reduction technique to attain uniformity in processing (Fig. 2).

3.3 Higher-Order Generalized Singular Value Decomposition

The post-processing of these vectors of interest is used to conduct steps that will reduce the complexity and increase the accuracy of the applied algorithm. We cannot write a unique algorithm for each of the condition in which an image is taken, thus, when we acquire an image, we tend to convert it into a form that would allow a general algorithm to solve it. The acquired image is also noisy (inherent in a signal) and thus de-noising it is also a crucial step. Most pre-processing steps that are implemented are either to reduce the noise, to reconstruct an image, to perform morphological operations and to convert the image to binary/greyscale so that operations can be easily implemented on the image. Here HOGSVD will help in reducing the computational intensity and help in escalating the process of feature aggregation (Fig. 3).

This algorithm provides a generalization of the matrix obtained by singular value decomposition for matrices of order N > 2. It is represented as $D_i \epsilon R^{m \times n}$ each having a full column rank. Every matrix can be split into components $D_i = U_i \sum_i V^T$ where V similar in all its components is acquired from the eigensystem $SV = V\Lambda$ by acquiring the arithmetic mean S of all pairwise quotients of the matrices $A_i A_i^T$, where i ≠ j. It's eigen values satisfy the inequality $\lambda_k \geq 1$. This equality is valid only if its corresponding eigen vector V_k is a right basis vector of identical significance in all the matrices D_i and D_j where $\sigma_{i,k}/\sigma_{j,k} = 1$ for all i and j, and its respective left basis vector $U_{i,k}$ is orthogonal to all other vectors in U_i for all i.

HOGSVD of these N matrices are:

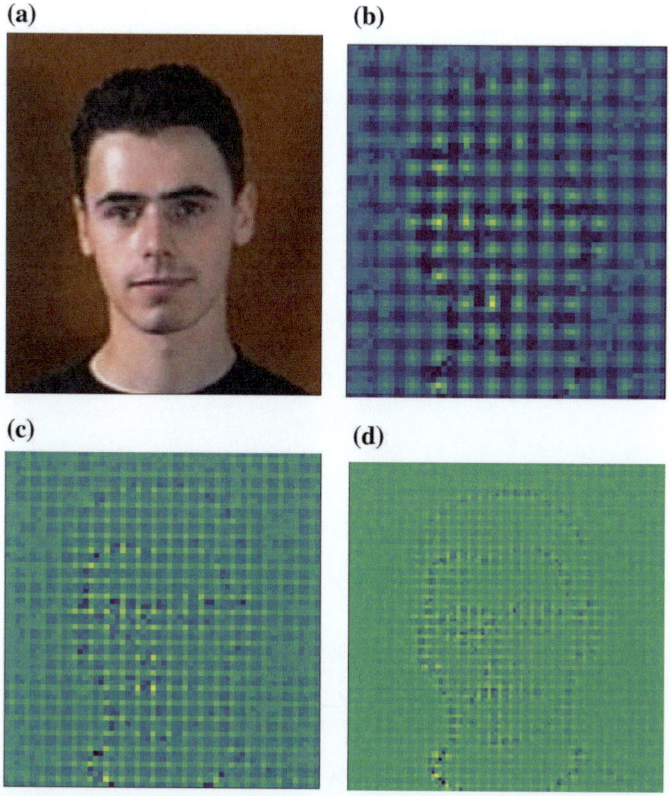

Fig. 2 **a** Sample image from database. **b** Visual LARK with low window size. **c** Visual LARK with high sensitivity. **d** Visual LARK with low sensitivity

$$\mathrm{D_1} = \mathrm{U_1} \sum_{1} \mathrm{V^T} \tag{8}$$

$$\mathrm{D_2} = \mathrm{U_2} \sum_{2} \mathrm{V^T} \tag{9}$$

$$\mathrm{D_N} = \mathrm{U_N} \sum_{\mathrm{N}} \mathrm{V^T} \tag{10}$$

$A_i = D_i^T * D_i$ equivalently for all S_{ij},

$$S_{ij} = \frac{1}{2} * \left(A_i * A_j^{-1} + A_j * A_i^{-1}\right) \quad \text{where } i! = j \tag{11}$$

$$\mathrm{S} = \frac{1}{N(N-1)} \sum_{i=1}^{N} \sum_{j>i}^{N} \left(\mathrm{A_i A_j^{-1} + A_j A_i^{-1}}\right)$$

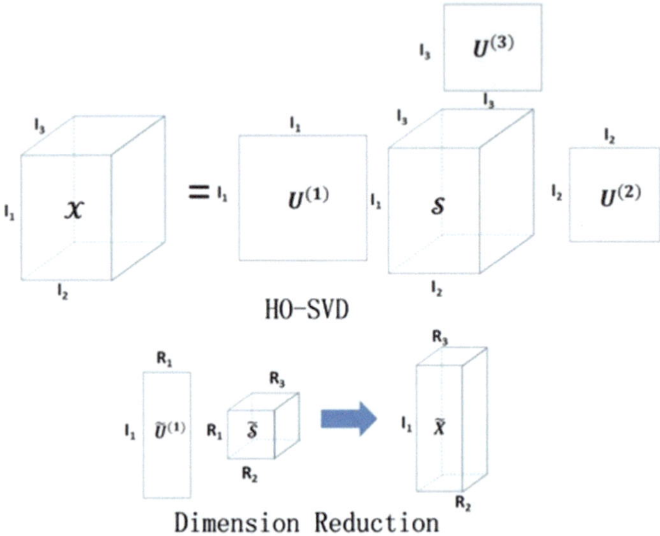

Fig. 3 Dimensionality reduction using HOGSVD

$$= \frac{2}{N(N-1)} \sum_{i=1}^{N} \sum_{j>i}^{N} \mathrm{S}_{\mathrm{ij}} \tag{12}$$

$SV = V\Lambda$ where $S = (v_1 \ldots v_n)$ and $\Lambda = diag(lambda_k)$.
Characteristics and applications of HOGSVD include:

1. HOGSVD is used to the extract the key information from multi-way arrays. Data analysis, recognition and synthesis problems are multilinear tensor problems based on the fact that most data that is observed are results of several causal factors of data formation, and are well suited for multi-modal data tensor analysis.
2. Currently it is being used in signal processing and big data which includes genomic signal processing.
3. Collation of HOSVD and SVD has been applied to detect real time events which are obtained from complex data streams.
4. HOGSVD was considered one of the best to be applied to multi-view data analysis and was successfully applied to discover silico drug from gene expression (Fig. 4).

Once the vectorized LARKs are prepared for feature aggregation, the set of vectors are processed using Bag of LARK features, LLC and Fischer vectors.

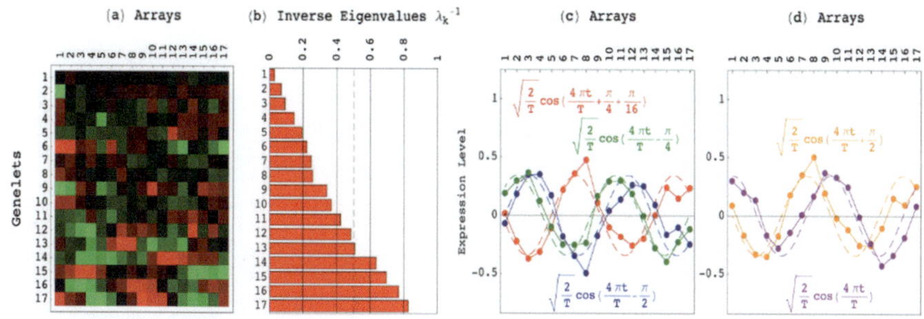

Fig. 4 Overview of the process

3.4 *Feature Aggregation*

3.4.1 Bag of LARK Features

The bag of LARK features is a logical way of representing data while modelling the dataset with various processing algorithms. It is a clustering algorithm which compiles elements of similar features. However, these clustering algorithms cannot work with the raw image which we consider as an input. It must be represented as a kernel after multiplying the Kernel RBF—PCS with the locally adaptive regression Kernel (LARK).

The bag of LARK features is a way of extracting a particular feature from a dataset of input features present in a codebook. It is called as a "bag" of LARK features as any information about the origin of the vector in the plain space is disregarded. It is only concerned whether the particular variable occurs in the cluster present in the code book.

In the Bag of LARK Features model which we have considered, the set of local variables from the vectorized version of the locally adaptive regression kernel into the final set of images is done in a succession of two steps: Clustering and Pooling.

1. Clustering: The clustering part in the original Bag of LARK Features model is the formation of clusters consisting of similar vectors when the vectorized model of the locally adaptive regression kernel is plotted. Since this low-level combination has a large impact on performance, the results are reported to be over 40% similarity for images with pre-processing done and up to 90% similarity for images which have not undergone any pre-processing.
2. Pooling: Once the clustering is finished and we have obtained a new locally adaptive regression kernel with every vector replaced with the special vector from the cluster, all images from the data set are plotted once more and the same processes is repeated to obtain images with a higher level of accuracy. The algorithm involving a combination of these two features is the Bag of LARK features algorithm (Fig. 5).

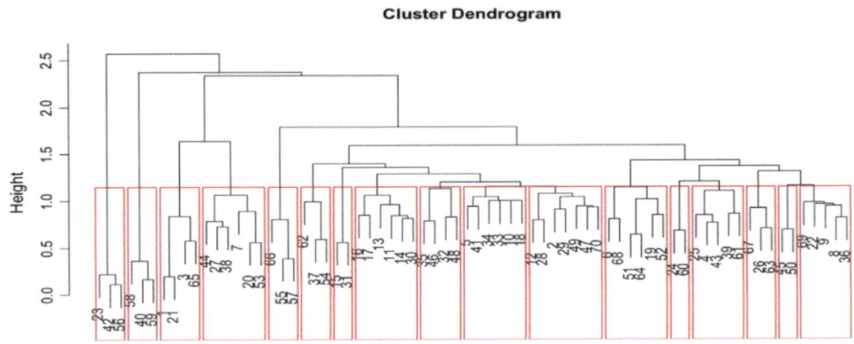

Fig. 5 Cluster Dendrogram from bag of LARK features

Using this algorithm, the user can extract images similar to the input image fed from the given dataset. The Bag of LARK features is a novel scheme of image classification using mid-level parameters such as codebooks and normalization. The codebooks are the most significant parameters as they allow to group images with a richer density to obtain more accurate results.

3.4.2 Locally Constrained Linear Coding (LLC)

The locally constrained linear coding is a clustering technique which is simple but extremely effective. It can be used as a suitable replacement to the SPM approach based on the bag-of-features (BoF) approach which requires non-linear classifiers to achieve a good image classification performance.

The locally constrained linear coding algorithm constraints to project each of the image descriptor which is the vectorized form of the locally adaptive regression kernel (LARK) for our case into its local database system. The projected co-ordinate vectors are then max-pooled to generate the final representation of the image.

Let X be a set of D Dimensional local descriptors extracted from an image. i.e. $X = [x_1, x_2, \ldots, x_n] \in \mathbb{R}^{D\times N}$. Given a codebook with M entries from the image vectors of locally adaptive regression kernel considered, $B = [b_1, b_2, \ldots, b_m] \in \mathbb{R}^{D\times M}$. Different coding schemes convert each image descriptor into M-Dimensional code to generate the final image representation.

The locally constrained linear coding incorporates the locality constraint instead of the sparsity constraint which leads to several favourable properties. Specifically, the LLC uses the following criteria:

$$\min \sum_{i=0}^{n} \| x_i - B_{ci} x^2 + \lambda \| d_i \odot C_i \| \tag{13}$$

where $\circledcirc$ denotes element wise multiplication and $d_i \in IR^M$ is locality adaptor that gives freedom for each basis vector.

The locally constrained linear coding algorithm is preferred as it provides a superior image classification performance compared to other clustering or classification techniques.

Once we process the images using the Bag of LARK Features algorithm and the Locally Constrained Linear Coding algorithms, we then teach the computer the datasets which we have considered in the paper. The classification and identification of the data input image is done using the proposed techniques.

3.4.3 Fischer Vectors

The Fisher Vector (FV) representation of images can be seen as an extension of the popular bag-of-visual word (BOV). Both of them are based on an intermediate representation, the visual vocabulary built in the low-level feature space. If a probability density function (in our case a Gaussian Mixture Model) is used to model the visual vocabulary, we can compute the gradient of the log likelihood with respect to the parameters of the model to represent an image. The Fisher Vector is the concatenation of these partial derivatives and describes in which direction the parameters of the model should be modified to best fit the data. This representation has the advantage to give similar or even better classification performance than BOV obtained with supervised visual vocabularies, being at the same time class independent.

We model the visual vocabulary with a Gaussian mixture model (GMM) where each Gaussian corresponds to a visual word. Let $\lambda = \{\omega_i, \mu_i, \sum i,\ i = 1 \ldots N\}$ be the set of parameters of p where ω_i, μ_i, $\sum i$ denote the weight, mean vector and covariance matrix in the LARK.

$$p(x|\lambda) = \sum_{i=1}^{N} \omega_i p_i(x|\lambda) = \sum_{i=1}^{N} \omega_i N\left(x|\mu_i \sum i\right) \tag{14}$$

Let $\{x_t, x_t, \in \mathbb{R}^D, t = 1 \ldots T\}$ be the set of local descriptors of the image, then by using Baye's formula we have.

$$\gamma_i(x_t) = \frac{\omega_i p_i(x|\lambda)}{\sum_{j=1}^{N} \omega_j p_j(x|\lambda)} \tag{15}$$

In the BOV representation, the low-level descriptor is hence transformed into a high level N-dimensional descriptor.

$$\gamma_t = \left[\gamma_1 x_t, \gamma_2 x_t, \ldots, \gamma_n x_t,\right] \tag{16}$$

where $\sum_{n=1}^{N} \gamma_n x_t = 1$ is an accumulation of these probabilities over low level descriptors.

3.5 *Classifiers*

3.5.1 Stochastic Gradient Descent

The Stochastic Gradient Descent (SGD) is basically an optimization algorithm which is used to calculate analytically the degree of relation between two given data values or images for the particular case that we have considered. It has a simple goal to best estimate a target function (f) that maps input data (x) onto output variable (y). It also describes the basic algorithm of all classification and regression problems. It provides a process of optimization to find the set of coefficients that result in the best estimate of the target file. Gradient descent is a slow technique which cannot often be run on very large datasets because of the time constraints. In these cases, we use the Stochastic Gradient Descent. Usually, every descent of the gradient algorithm has a prediction for each instance in the dataset. This is not recommended as there maybe millions of instances present. However, in case of the Stochastic Gradient Descent, update to the coefficients is performed for each training instance rather than at the end of the batch of the data instances. It utilises a single new sample data in each iteration and processes the end data in a stream-like fashion. SGD optimization is linearly scalable in time and the computational time can be sped up to two or three times in magnitude.

Consider a supervised learning model, where we are given a set of samples $(a, b) \in A \times B$ taken from the probability distribution $P(a, b)$. The conditional probability represents the relation between the input variable a and the output variable b. The difference between the estimated variable $\hat{b}$ and the true variable b is represented by a loss function $l(b, b)$. Using SGD algorithm, we try to estimate the function f that minimises this expected risk.

$$E(f) = \int l(f(a), b) dP(a, b) = \mathbb{E}[l(f\{a), b)] \tag{17}$$

Due to its incremental behaviour, SGD has features that support online adaptation of classification functions and a classification model that is available at any given point of time. This enables us to give complete solutions in cases where the time constraints do not allow us to give a retraining of the classification model (Fig. 6).

3.5.2 Gaussian Naive Bayes

A Naive Bayes is a classifier with an inbuilt powerful algorithm for the classification of millions of data files and records with only a limited number of attributes. The Bayes theorem is an integral part of the Naive Bayes classification system. It is based on conditional probability which is calculating the probability of an even occurring based on prior knowledge. The formulae for calculating conditional probability can be explained as follows:

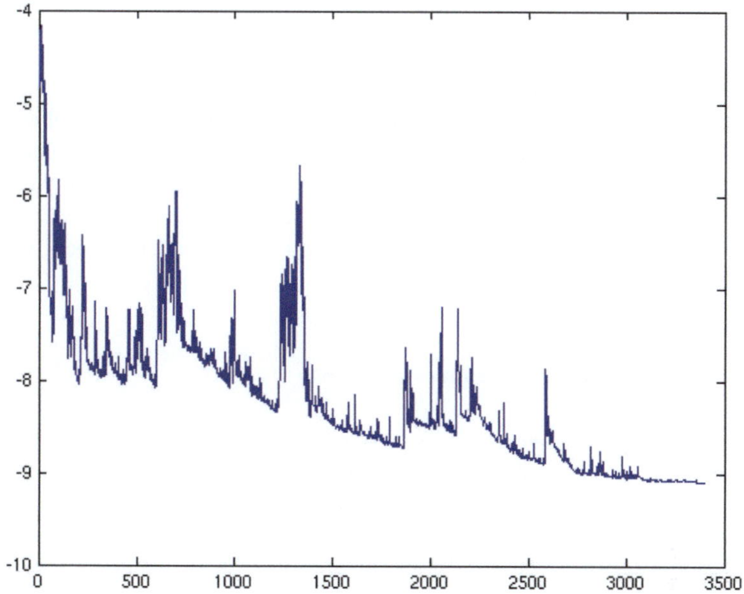

Fig. 6 Tuning alpha value in stochastic gradient descent using a plot

$$P(H|E) = \frac{P(E|H) * P(H)}{P(E)} \tag{18}$$

where,

$P(H)$ is the probability of the hypothesis H being true
$P(E)$ is the probability of evidence.
$P(E|H)$ is probability of evidence given that the hypothesis is true.
$P(H|E)$ is probability of hypothesis given evidence is present.

The Naive Bayes classifier predicts the probabilities for each of the class that is considered such as the probability of the given data or record belongs to a particular class. Hence the class with the highest probability is considered as the most likely class. When an image is processed and the features are extracted, the Naive Bayes naturally assumes that all features are unrelated to each other. One feature being present or absent does not affect any other feature in any way.

Gaussian Naive Bayes is a particular type of Naive Bayes algorithm that considers all the attribute values to be continuous and an assumption is made that all the values associated with each other are grouped into a Normal Distribution. The basic theorem upon which the Gaussian Naive Bayes algorithm works is given as:

$$P(x_i|y) = \frac{1}{\sqrt{2\pi\sigma_y^2}} \exp\left(-\frac{(x_i - \mu_y)^2}{2\sigma_y^2}\right) \tag{19}$$

The Gaussian Naive Bayes algorithm [17] is a fast and reliable algorithm that can be used for Binary and Multiclass classification. It can also be easily trained on a small dataset but is not recommended for a larger dataset due to the time constraints. It can also deal with missing attributes in a given data file. Consider a set of image files that we have taken as an input, here each image is represented by an individual vector representation called as the Bag of LARK Features representation. We then fit the Gaussian Naive Bayes algorithm by initially teaching a dataset to the computer memory and then the target elements are classified by comparing them with the elements implemented into the database.

3.5.3 Fischer Linear Discriminant Analysis

Fischer Linear Discriminant Analysis or commonly known as Linear Discriminant Analysis [20] is a classification technique that is used in statistics, pattern recognition and machine learning to find a linear combination of features which selects or classifies objects or lists from the analysed objects. The transformation is based on maximizing mean square error between original data vectors and data vectors that can be estimated from the reduced dimensionality vectors (Fig. 7).

The left plot shows the samples from two classes along with the histograms resulting from the projection onto the line joining the class means. The right plot shows the corresponding projection based on the Fischer linear discriminant, showing the greatly improved class separation.

Assume we have a set of D-dimensional samples $X = \{x^{(1)}, x^{(2)}, \ldots x^{(m)}\}$, N_1 of which belong to class C_1 and N_2 of which belong to class C_2. We also assume the mean vector of the two classes in X-space.

$$u_k = \frac{1}{N_k} \sum_{i \in C_k} x^{(i)} \quad where\ k = 1, 2, \ldots. \tag{20}$$

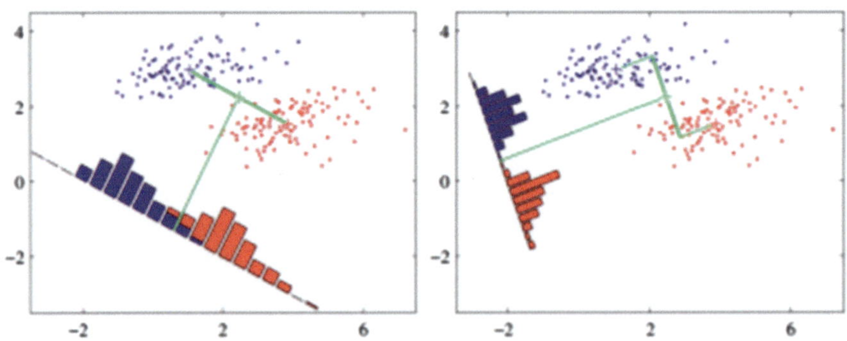

Fig. 7 LDA on a custom dataset

And in y-space:

$$\hat{u}_k = \frac{1}{N_k}\sum_{i\in C_k}\theta^T x^{(i)} = \theta^T u_k \quad where\ k = 1, 2, \ldots \tag{21}$$

One way to define a measure of separation between two classes is to choose the distance between the projected means, which is in y-space, so the between class variance is:

$$\hat{u}_2 - \hat{u}_1 = \theta^T(u_2 - u_1) \tag{22}$$

Also, we define the within–class variance for each class C_k is:

$$\hat{s}_k^2 = \sum_{i\epsilon C_k}\left(y^{(i)} - \hat{u}_k\right)^2 \quad where\ k = 1, 2, \ldots \tag{23}$$

Then, we get the between-class variance and within-class variance; we can define our objective function $J(\theta)$ as:

$$J(\theta) = \frac{\left(\hat{u}_2 - \hat{u}_1\right)^2}{\hat{s}_1^2 + \hat{s}_2^2} \tag{24}$$

If maximising the objective function J, we are looking for a projection where examples from the class are projected very close to each other and at the same time, the projected means are as farther apart as possible.

For the Multi-Classes Problems we see that the fisher's LDA generalizes gracefully. Assuming we still have a set of D-dimensional samples $X = \left\{x^{(1)}, x^{(2)}, \ldots x^{(m)}\right\}$ and there are totally C classes. Instead of one projection y as mentioned we will seek $(C - 1)$ projections where:

$$y_i = \theta_i^T X \Rightarrow y = \Theta^{\mathrm{T}}\mathrm{X} \tag{25}$$

We will use the scatters in space-x as follows:

Within-class scatter matrix:

$$S_W = \sum_{i=1}^{C} S_i \quad where\ S_i = \sum_{i\epsilon C_i}\left(x^{(i)-u_i}\right)\left(x^i - u_i\right)^T,$$
$$u_i = \frac{1}{N_i}\sum_{i\epsilon C_i} x^{(i)} \tag{26}$$

Between-Class scatter matrix:

$$S_B = \sum_{i=1}^{C} N_i(u_i - u)(u_i - u)^T \quad where\, u = \frac{1}{m}\sum_{i=1}^{m} x^{(i)} = \frac{1}{m}\sum_{i=1}^{C} N_i u_i \tag{27}$$

Total scatter matrix:

$$S_T = S_B + S_W \tag{28}$$

3.5.4 K Nearest Neighbours

This is a non-parametric methodology to perform classification and regression. Input comprises of k-nearest examples which are to be trained in feature space. Output is either the one obtained by classification or regression. It works on the principle of representative based learning where approximation on the function is discharged provincially and the resulting calculation is often delayed in unit organisation. The neighbours are obtained from a collection of objects where the category of the class, the property and value of the object is familiar in nature.

The tuples (A, B),(A_1, B_1), …, (A_n, B_n) where the values are in $R^{d}*\{1, 2\}$, B being the class identifier for A the equation is given by:

$$\frac{A}{B} = r \sim P_r \quad where\, r = 1, 2, \ldots \tag{29}$$

Here P_r denotes a probability distribution.

By interchanging the tuples ($A_{(1)}B_{(1)}$), …, ($A_{(n)}B_{(n)}$) in a way that $||A_{\{1\}}$ − all $\Leftarrow \cdots \Leftarrow ||A_{\{n\}}$ − all.

The similarity of KNN is obtained by measuring the distance between two points using distance metrics between data points. The Euclidean distance is given by the equation:

$$d(x, x') = ((x_1 - x_1')^2 + (x_2 - x_2')^2 + \cdots + (x_n - x_n')^2)^{\frac{1}{2}} \tag{30}$$

From the above graph the boundary between the red and blue interface becomes smoother for increasing values of k. As the value of k tends to infinity it becomes either blue or red in colour depending completely upon the larger proportion (Fig. 8).

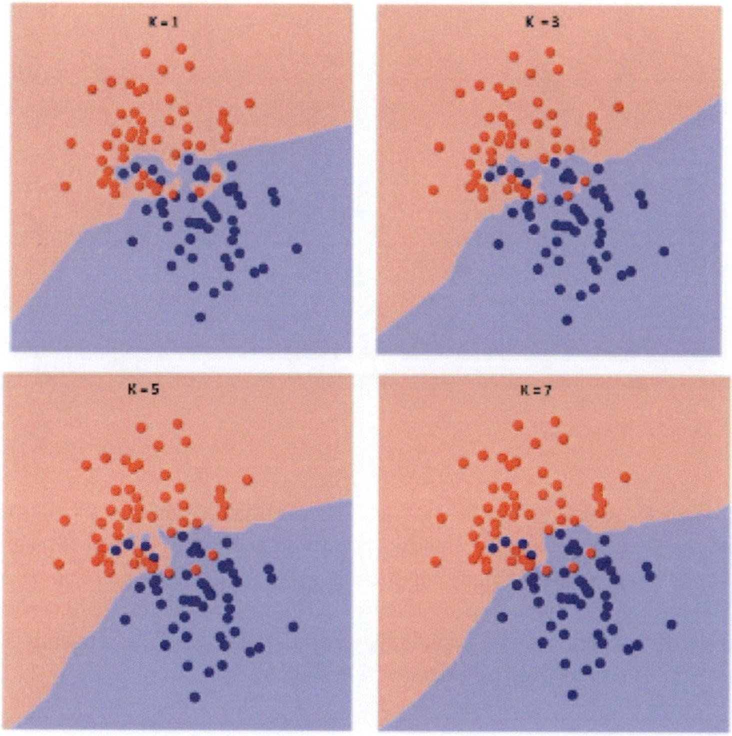

Fig. 8 KNN output for different values of k on a custom dataset

3.6 Datasets and Experimentation

To test the righteousness of the methodology for variations, corresponding datasets were used. They helped in the testing and the analysis of the methodology that was used in the paper.

3.6.1 ORL Faces

Formerly known as 'The ORL Dataset of Faces', the dataset holds images from the early 1990s captured at the Cambridge University Computer Laboratory. It contains ten unique images of 40 different individuals, subjected to various variations such as the time of capture of the image, lighting of the images, facial expressions of the individuals and other accessories worn by the individuals. Each of the images has a standard 92 × 112 pixels with a set 256 grey levels per unit. The dataset was a unique dataset as lots of different image variations were considered while taking the images.

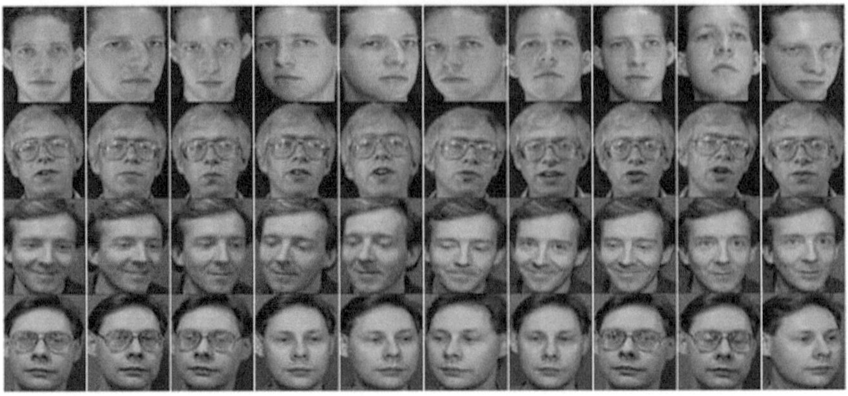

Fig. 9 A sample ORL dataset

The image is quantized to 256 grey levels and stored as unsigned 8-bit integers; the loader will convert these to floating point values on the interval [0, 1], which are easier to work with for many algorithms. The "target" for this database is an integer from 0 to 39 indicating the identity of the person pictured; however, with only 10 examples per class, this relatively small dataset is more interesting from an unsupervised or semi-supervised perspective (Fig. 9).

3.6.2 Grimace

This unique dataset is an assembly of 18 different individuals designed and maintained by Dr. Libor Spacek. Grimace [20] has a main objective to focus on variations between male and female candidates. The dataset contains 20 portraits of each of the candidate considered at a resolution of 180 × 200 pixels. The background is kept same throughout all the images with small uniform head scale variations. The lighting changes are minimal and little to no variations in hairstyle of the considered candidates (Fig. 10).

3.6.3 Faces95

Once again, a Brain Child [21] of Dr. Libor Spacek, this particular dataset contains portraits of 72 different and distinct subjects. Sequences of 20 images were captured while the subject was asked to step towards the camera after every snap that was taken. This kind of a special dataset offers a huge head scale variation and minor variations due to the difference in the depth of the shadows that is varied each time the subject takes a step towards the camera. This results in a discrepancy in red background. Noticeable changes in lighting occur due to the artificial lighting systems used.

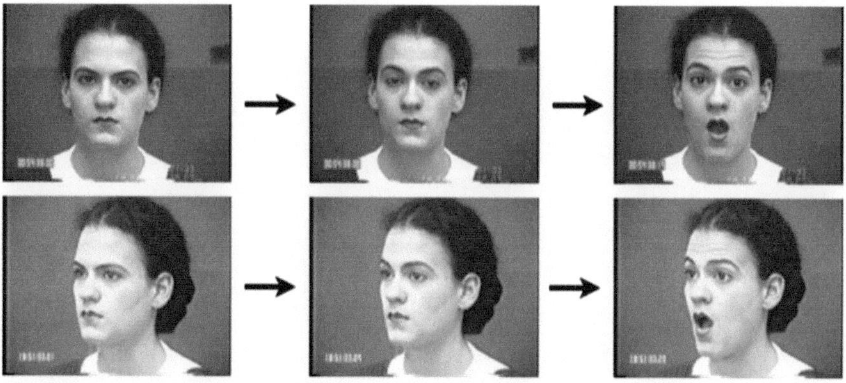

Fig. 10 A sample Grimace faces dataset

4 Results and Inference

The results obtained are presented below in a concise manner in the table that is shown in Table 1.

Accuracy assessment is an important path of any classification project. It compares an input image with another image that is present in the dataset which is classified and gives an accurate report of the matching between the two images that have been considered. Recall is also known as sensitivity in image processing and classification. It is the fraction of relevant instances that have been taken or considered in the total instances that are present in the dataset. It is basically a measure of relevance. Following observations were made from the obtained values during experimentation.

1. LARK needs a lot of variance in the data. Any preprocessing done to the input image is likely to affect the result of the classification.
2. Some data sets like the FACE95 datasets have low precision/recall because they do not have a lot of variations or variables when compared to the other datasets
3. The challenges that are offered by the datasets need to be handled better and all the variables need to be considered. Only then we will be able to achieve the precision and accuracy that we require.

Out of three feature aggregation techniques LLC always yielded poor results compared to other two techniques. We can deduce that LLC is not a suitable pair for clustering of LARK vectors. Bag of LARK features showed remarkable results on several iterations for all the datasets with an aggregate of over 15% better accuracy than other techniques. It was clearly observed that HOGSVD as a post processing step did not help much in increasing the accuracy. On an average all the combinations of aggregation mechanism and classifiers yielded 10% higher accuracy without HOGSVD. This proves that the dimensionality reduction does lead to loss of features and LARKs give high covariance values in all the dimensions. The precision as well as recall was exceptionally high for grimace dataset which proves that the

Table 1 Results obtained from ORL dataset

	HOGSVD		Precision				Recall				F1-score			
			GNB	SGD	LDA	KNN	GNB	SGD	LDA	KNN	GNB	SGD	LDA	KNN
ORL	No	BOLF	0.85	0.86	0.94	0.85	0.75	0.8	0.88	0.78	0.75	0.81	0.88	0.77
		LLC	0.67	0.68	0.74	0.67	0.57	0.71	0.71	0.62	0.6	0.66	0.69	0.6
		FV	0.57	0.96	0.95	0.81	0.56	9.95	0.91	0.81	0.56	0.96	0.91	0.79
	Yes	BOLF	0.64	0.71	0.77	0.73	0.69	0.7	0.74	0.57	0.67	0.68	0.73	0.59
		LLC	0.65	0.54	0.69	0.56	0.6	0.54	0.65	0.6	0.61	0.54	0.64	0.57
		FV	0.53	0.98	0.92	0.75	0.48	0.97	0.85	0.74	0.47	0.97	0.86	0.72
Grimace	No	BOLF	0.97	0.96	0.95	0.94	0.96	0.96	0.96	0.95	0.96	0.95	0.95	0.94
		LLC	0.92	0.97	0.93	0.97	0.9	0.95	0.96	0.95	0.92	0.96	0.94	0.95
		FV	0.96	0.98	0.95	0.96	0.91	0.97	0.94	0.92	0.94	0.97	0.94	0.95
	Yes	BOLF	0.88	0.86	0.91	0.85	0.87	0.81	0.89	0.81	0.87	0.78	0.89	0.81
		LLC	0.81	0.8	0.82	0.84	0.78	0.77	0.87	0.82	0.79	0.78	0.83	0.82
		FV	0.91	0.82	0.9	0.84	0.86	0.83	0.85	0.83	0.87	0.82	0.87	0.84
Faces95	No	BOLF	0.98	0.91	0.97	0.89	0.96	0.88	0.95	0.86	0.96	0.87	0.96	0.86
		LLC	0.91	0.94	0.96	0.91	0.83	0.92	0.95	0.88	0.84	0.92	0.95	0.87
		FV	0.96	0.97	0.95	0.96	0.96	0.95	0.97	0.94	0.96	0.96	0.95	0.94
	Yes	BOLF	0.85	0.71	0.91	0.71	0.8	0.66	0.88	0.64	0.8	0.62	0.88	0.64
		LLC	0.72	0.75	0.8	0.78	0.75	0.76	0.78	0.76	0.72	0.75	0.77	0.75
		FV	0.76	0.85	0.84	0.68	0.76	0.82	0.79	0.64	0.72	0.82	0.79	0.6

pipeline can handle changes in facial expressions very well. The results for ORL dataset was just above average for any combination. This reflects that the pipeline does not perform well when the region of interest shift is small or the distortions in the background are very high. There is a scope making the pipeline completely scale and orientation invariant as ORL does offer head scale and orientation invariance to certain extent. Out of all the classifiers LDA performed outstanding with every combination the pipeline has to offer.

5 Conclusion and Future Work

Locally Adaptive Regression Kernels have proven to be capable descriptors. It can be concluded that LARK descriptors are sparse in nature and are suffice themselves. Any form of processing to achieve dimensionality reduction will lead to degeneration of features and degrade the performance of the system. They show little to no variance on repeated trails and give high accuracy when paired with different classifiers. From the experimentation, it is deduced that any form of pre-processing or post-processing in the form of dimensionality reduction or denoising does lead to loss of features and declined accuracy. The methodology requires feature rich images and loss of data in any form is not tolerated by the mechanism. Feature aggregation helped us in clustering the vectors into groups, similar to our interests. Since the regional descriptors were procured from the constructed clusters vectors using these aggregated vectors for classification yielded better results.

A comparative study performed has shown that Bag of LARK features and stochastic gradient descent comprise the best combination in the pipeline used. It was observed that LLC delivered lesser accuracy when paired with Higher-Order generalized singular value decomposition. Grimace dataset posed least challenges to the system and consistently provided very good results on several iterations of the mechanism. The other classifiers performed up to the mark on faces95, while ORL dataset delivered harder set of challenges.

The need for a better algorithm with respect to pre-processing the images is of high prominence. A suitable algorithm which denoises the image without the loss of significant features is essential. The model looks promising and can deliver better results if worked on. There is scope for better classifiers as well. Future work includes pairing these post-processed LARKs with artificial neural networks for better classification, designing an algorithm for pre-processing the images before using locally adaptive kernels on them. The encoding system which aggregates the features can elevate the rightness if worked on.

References

1. Seo, H.J., Milanfar, P.: Face verification using the LARK representation. IEEE Trans. Inf. Forensics Secur. **6**(4), 1275–1286 (2011)
2. Meena, K., Suruliandi, A.: Local binary patterns and its variants for face recognition. In: 2011 International Conference on Recent Trends in Information Technology (ICRTIT), Chennai, Tamil Nadu, pp. 782–786 (2011)
3. Hadid, A.: The local binary pattern approach and its applications to face analysis. In: 2008 First Workshops on Image Processing Theory, Tools and Applications, Sousse, pp. 1–9 (2008)
4. Ahonen, T., Hadid, A., Pietikainen, M.: Face description with local binary patterns: application to face recognition. IEEE Trans. Pattern Anal. Mach. Intell. **28**(12), 2037–2041 (2006)
5. Ramanathan, N., Chellappa, R.: Face verification across age progression. IEEE Trans. Image Process. **15**(11), 3349–3361 (2006)
6. Osman, A.A.E., El-Khoribi, R.A., Shoman, M.E., Shalaby, M.A.W.: Trajectory learning using posterior hidden Markov model state distribution. Egypt. Inform. J. (2017)
7. Turk, M.A., Pentland, A.P.: Face recognition using eigenfaces. In: Proceedings of IEEE Conference Computer Vision and Computer Vision (CVPR), pp. 586–591 (1991)
8. He, K., Zhou, D., Nie, R., Jin, X., Wang, Q.: Image specific target detection and localization based on locally adaptive regression kernels algorithm. In: 2016 8th IEEE International Conference on Communication Software and Networks (ICCSN), Beijing, pp. 647–651 (2016)
9. Ali, N.M., Jun, S.W., Karis, M.S., Ghazaly, M.M., Aras, M.S.M.: Object classification and recognition using Bag-of-Words (BoW) model. In: 2016 IEEE 12th International Colloquium on Signal Processing & Its Applications (CSPA), Malacca City, pp. 216–220 (2016)
10. Wang, J., Yang, J., Yu, K., Lv, F., Huang, T., Gong, Y.: Locality-constrained Linear Coding for image classification. In: 2010 IEEE Computer Society Conference on Computer Vision and Pattern Recognition, San Francisco, CA, pp. 3360–3367 (2010)
11. Zhang, W., Gao, F.: Performance analysis and improvement of naïve Bayes in text classification application. In: IEEE Conference Anthology, China, pp. 1–4 (2013)
12. Chen, Y.P., Liu, C.H., Chou, K.Y., Wang, S.Y.: Real-time and low-memory multi-face detection system design based on naive Bayes classifier using FPGA. In: 2016 International Automatic Control Conference (CACS), Taichung, pp. 7–12 (2016)
13. Ouarda, W., Trichili, H., Alimi, A.M., Solaiman, B.: Combined local features selection for face recognition based on Naïve Bayesian classification. In: 13th International Conference on Hybrid Intelligent Systems (HIS 2013), Gammarth, pp. 240–245 (2013)
14. Liu, Q., Liu, C.: A novel locally linear KNN method with applications to visual recognition. IEEE Trans. Neural Netw. Learn. Syst. **28**(9), 2010–2021 (2017)
15. Parveen, P., Thuraisingham, B.: Face recognition using multiple classifiers. In: 2006 18th IEEE International Conference on Tools with Artificial Intelligence (ICTAI'06), Arlington, VA, pp. 179–186 (2006)
16. Uchida, Y., Sakazawa, S.: Image retrieval with fisher vectors of binary features. In: 2013 2nd IAPR Asian Conference on Pattern Recognition, Naha, pp. 23–28 (2013)
17. Chelali, F.Z., Djeradi, A., Djeradi, R.: Linear discriminant analysis for face recognition. In: 2009 International Conference on Multimedia Computing and Systems, Ouarzazate, pp. 1–10 (2009)
18. Taneja, S., Gupta, C., Goyal, K., Gureja, D.: An enhanced K-nearest neighbor algorithm using information gain and clustering. In: 2014 Fourth International Conference on Advanced Computing & Communication Technologies, Rohtak, pp. 325–329 (2014)
19. Putranto, E.B., Situmorang, P.A., Girsang, A.S.: Face recognition using eigenface with Naive Bayes. In: 2016 11th International Conference on Knowledge, Information and Creativity Support Systems (KICSS), Yogyakarta, pp. 1–4 (2016)
20. Dr Libor SpaceK.: Collection of Facial Images: Grimace (Online) (2007). Retrieved from: http://cswww.essex.ac.uk/mv/allfaces/grimace.html
21. Dr Libor SpaceK.: Collection of Facial Images: Faces95 (Online) (2007). Retrieved from:http://cswww.essex.ac.uk/mv/allfaces/faces95.html

MIX
Papier aus verantwortungsvollen Quellen
Paper from responsible sources
FSC® C105338

If you have any concerns about our products,
you can contact us on
ProductSafety@springernature.com

In case Publisher is established outside the EU,
the EU authorized representative is:
Springer Nature Customer Service Center GmbH
Europaplatz 3, 69115 Heidelberg, Germany

Printed by Libri Plureos GmbH
in Hamburg, Germany